1.7 PROPERTIES OF REAL NUMBERS

If a, b, and c are real numbers, then

Closure properties:

$a + b$ is a real number

$a - b$ is a real number

ab is a real number

$\dfrac{a}{b}$ is a real number $(b \neq 0)$

Commutative properties:

$a + b = b + a$

$ab = ba$

Associative properties:

$(a + b) + c = a + (b + c)$

$(ab)c = a(bc)$

Distributive property:

$a(b + c) = ab + ac$

Identity elements:

0 is the additive identity.

1 is the multiplicative identity.

Additive and multiplicative inverses:

a and $-a$ are additive inverses.

a and $\dfrac{1}{a}$ $(a \neq 0)$ are multiplicative inverses.

2.1 INTRODUCTION TO EQUATIONS

Let a, b, and c be real numbers. Then:

If $a = b$, then $a + c = b + c$.

If $a = b$, then $a - c = b - c$.

If $a = b$, then $\dfrac{a}{c} = \dfrac{b}{c}$ $(c \neq 0)$.

If $a = b$, then $ca = cb$.

If r is the rate, b is the base, and p is the percentage, then

$p = r \cdot b$

2.2 SOLVING MORE EQUATIONS

Retail price = cost + markup

Markup = percent of markup · cost

2.3 SIMPLIFYING EXPRESSIONS TO SOLVE EQUATIONS

To combine like terms, add their numerical coefficients and keep the same variables and exponents.

2.7 SOLVING INEQUALITIES

Let a, b, and c be real numbers. Then:

If $a < b$, then $a + c < b + c$.

If $a < b$, then $a - c < b - c$.

If $a < b$, and $c > 0$, then $ac < bc$.

If $a < b$, and $c < 0$, then $ac > bc$.

If $a < b$, and $c > 0$, then $\dfrac{a}{c} < \dfrac{b}{c}$.

If $a < b$, and $c < 0$, then $\dfrac{a}{c} > \dfrac{b}{c}$.

3.1 POLYNOMIALS

Properties of exponents: If a is a natural number, then

$$x^n = \overbrace{x \cdot x \cdot x \cdot \cdots \cdot x}^{n \text{ factors of } x}$$

If m and n are natural numbers and there are no divisions by 0, then

$$x^m x^n = x^{m+n} \qquad (x^m)^n = x^{mn}$$

$$(xy)^n = x^n y^n \qquad \left(\frac{x}{y}\right)^n = \frac{x^n}{y^n}$$

$$\frac{x^m}{x^n} = x^{m-n} \quad \text{provided } m > n$$

3.2 ZERO AND NEGATIVE INTEGRAL EXPONENTS

If x is any nonzero real number and n is a natural number, then

$$x^0 = 1, \quad x^{-n} = \frac{1}{x^n}, \quad \frac{x^m}{x^n} = x^{m-n}$$

3.3 SCIENTIFIC NOTATION

A number is written in scientific notation if it is written as a product of a number between 1 (including 1) and 10 and an integer power of 10.

Beginning Algebra

Books in the Gustafson/Frisk Series

4

Beginning Algebra

R. David Gustafson
Peter D. Frisk

Rock Valley College

Brooks/Cole Publishing Company

I(T)P™ An International Thomson Publishing Company

Pacific Grove • Albany • Bonn • Boston • Cincinnati • Detroit • London • Madrid • Melbourne
Mexico City • New York • Paris • San Francisco • Singapore • Tokyo • Toronto • Washington

GWO A Gary W. Ostedt Book

Sponsoring Editor: *Gary W. Ostedt*
Editorial Associate: *Carol Ann Benedict*
Production Editor: *Ellen Brownstein*
Production Service: *Hoyt Publishing Services*
Manuscript Editor: *David Hoyt*
Permissions Editor: *May Clark*
Interior and Cover Design: *E. Kelly Shoemaker*
Marketing Team: *Patrick Farrant and Jean Vevers Thompson*

Interior Illustration: *Lori Heckelman*
Cover Photo: *Koji Kitagawa/Super Stock, Inc.*
Photo Editor: *Kathleen Olson*
Typesetting: *The Clarinda Company*
Cover Printing: *Color Dot*
Printing and Binding: *R. R. Donnelley & Sons, Crawfordsville*

For more information, contact:

BROOKS/COLE PUBLISHING COMPANY
511 Forest Lodge Road
Pacific Grove, CA 93950
USA

International Thomson Publishing
Berkshire House 168-173
High Holborn
London WC1V 7AA
England

Thomas Nelson Australia
102 Dodds Street
South Melbourne, 3205
Victoria, Australia

Nelson Canada
1120 Birchmount Road
Scarborough, Ontario
Canada M1K 5G4

International Thomson Editores
Campos Eliseos 385, Piso 7
Col. Polanco
11560 México D.F. México

International Thomson Publishing Gmbh
Königwinterer Strasse 418
53227 Bonn
Germany

International Thomson Publishing Asia
221 Henderson Road #05-10
Henderson Building
Singapore 0315

International Thomson Publishing—Japan
Hirakawacho-cho Kyowa Building, 3F
2-2-1 Hirakawacho-cho
Chiyoda-ku, Tokyo 102
Japan

Printed in the United States of America.

10   9   8   7   6   5   4   3   2

Library of Congress Cataloging-in-Publication Data

Gustafson, R. David (Roy David), [date]
    Beginning algebra / R. David Gustafson, Peter D. Frisk.—4th ed.
        p.   cm.
    Includes index.
    ISBN 0-534-24618-4
    1. Algebra.   I. Frisk, Peter D., [date].   II. Title.
QA152.2.G85   1994
512.9—dc20   94-29956
                                                                            CIP

Photo credits: P. 4, The British Museum; **p. 81,** Ken Eward, Science Source; **p. 151,** Martin Bond, Science Photo Library; **p. 371,** Courtesy of Texas Instruments; **p. 415,** Courtesy of International Business Machines Corporation; and **p. 509,** Courtesy of Princeton University.

To
Caitlin Mallory Barth
Nicholas Connor Barth
Prescott Alexander Heighton
Laurel Marie Heighton
Daniel Mark Voeltner
and Tyler, too

■ ■ ■ ■ ■ ■ ■ PREFACE FOR THE INSTRUCTOR

Beginning Algebra, Fourth Edition, is written for students studying algebra for the first time and for those who need a review of basic algebra. It presents all of the topics associated with a first course in algebra, providing students with a thorough foundation in the basic skills of algebraic manipulation and equation solving.

Our goal was to write a book that

1. is enjoyable to read,

2. is easy to understand,

3. is relevant, and

4. will develop the necessary skills for success in future academic courses or on the job.

The Fourth Edition retains the basic philosophy and organization of the highly successful previous edition. However, we have made several improvements in line with the NCTM standards and the current trends in mathematics reform. For example, much more emphasis has been placed on problem solving.

Changes in the Fourth Edition

To make the book more enjoyable to read, we have:

- used a new and more open four-color design.

- enlarged section heads to make them easier to find. Sections are now divided into subsections.

- redrawn all art, most with added color. Art that accompanies application problems is now much more representational.

- placed application exercises in a two-column format. Each application now has a title.

- added a Perspective to each chapter. These Perspectives give brief, interesting stories that pertain to the material in the chapter.

- included pictures of famous mathematicians to provide a flavor of mathematics history.

To make the book easier to understand, we have:

- revised the explanations and simplified the language.

- added many more Author's Notes to explain more steps in the problem-solving process.

- added ⬦ **Warning!** notices to warn students of common errors and misconceptions.

- added Getting Ready exercises at the beginning of each section. These exercises review skills that will be necessary in the section.

- added Oral Exercises before each exercise set. These problems enable the instructor to check student understanding before assigning homework.

To make the book more relevant, we have:

- opened each chapter with an application problem that can be solved by using the techniques developed in the chapter. The problem is solved in detail at the end of the chapter.

- strengthened the problem-solving emphasis in the book by adding many more application problems and distributing them throughout the book. Percent problems are now introduced early, providing meaningful applications, such as discount, percent change, markup and markdown, etc. An index of applications has been included.

- emphasized geometry throughout the text, particularly the concepts of perimeter, area, and volume.

To develop the necessary skills for success in future academic courses or on the job, we have:

- added Writing Exercises to each exercise set. These exercises will help students clarify ideas.

- added Something to Think About exercises to each exercise set. These exercises require extra thought and insight.

- added material using graphing calculators. Although this material is integrated throughout the book, it can be omitted with no loss in continuity.

- added Cumulative Review Exercises after every third chapter. These are in addition to the Review Exercises in each exercise set, the Chapter Review Exercises, the Sample Chapter Tests, and the Sample Final Examination.

- added a Project to each chapter, to be used for extended assignments or for cooperative learning.

Features of This Edition

3.6 MULTIPLYING POLYNOMIALS   **181**

3.6 Multiplying Polynomials

■ The FOIL Method ■ Multiplying Binomials to Solve Equations

GETTING READY   *Simplify.*

1. $(2x)(3)$ **2.** $(3xxx)(x)$ **3.** $5x^2 \cdot x$ **4.** $8x^2x^3$

Use the distributive property to remove parentheses.

5. $3(x + 5)$ **6.** $x(x + 5)$ **7.** $4(y - 3)$ **8.** $2y(y - 3)$

In Section 3.1, we multiplied certain monomials by other monomials. To multiply $4x^2$ by $-2x^3$, for example, we use the commutative and associative properties of multiplication to group the numerical factors and the variable factors together and multiply.

$$4x^2(-2x^3) = 4(-2)x^2x^3$$
$$= -8x^5$$

EXAMPLE 1   **a.** $3x^5(2x^5) = 3(2)x^5x^5$ **b.** $-2a^2b^3(5ab^2) = -2(5)a^2ab^3b^2$
$= 6x^{10}$ $= -10a^3b^5$

c. $-4y^5z^2(2y^3z^3)(3yz) = -4(2)(3)y^5y^3yz^2z^3z$
$= -24y^9z^6$ ■

The previous examples suggest the following rule.

Multiplying Two Monomials   To multiply two monomials, first multiply the numerical factors and then multiply the variable factors.

To find the product of a monomi...
we use the distributive property. To m...
as follows:

$$3x(x + 4) = 3x \cdot x + 3x \cdot 4$$
$$= 3x^2 + 12x$$

MATHEMATICS IN ECOLOGY   Many types of bacteria cannot survive in air. In one of the steps in waste treatment, sewage is exposed to the air by placing it in large, shallow, circular aeration pools.

One sewage processing plant has two such pools, with diameters of 40 and 42 meters. To meet new clean water standards, the plant must double its capacity, which includes building another aeration pool. How large a pool should the design engineers specify to double the capacity of this phase of sewage treatment?

■ ■ ■ ■ ■ ■ ■ ■   After reading this chapter, you will be able to answer this question.

◀ Section heads are easier to find, and sections are divided into subsections.

◀ Getting Ready exercises help prepare the student for the material in the section.

◀ Examples worked on videotape are marked with an icon.

Each chapter begins with an application ▶ to motivate the material in the section.

4.3 FACTORING THE DIFFERENCE OF TWO SQUARES   **221**

Writing Exercises ■ *Write a paragraph using your own words.*

1. Explain why $a - b$ and $b - a$ are negatives of each other.

2. Explain how you would factor $x(a - b) + y(b - a)$.

Something to Think About ■ 1. Factor $ax + ay + bx + by$ by grouping the first two terms and the last two terms. Then rearrange the terms as $ax + bx + ay + by$, and factor again by grouping the first two and the last two. Do the results agree?

2. Factor $2xy + 2xz - 3y - 3z$ by grouping in two different ways.

Review Exercises ■ *Simplify each expression. Write all results without using negative exponents.*

1. $u^3 u^2 u^4$

2. $\dfrac{y^6}{y^8}$

3. $\dfrac{a^3 b^4}{a^2 b^5}$

4. $(3x^5)^0$

4.3 **Factoring the Difference of Two Squares**
■ Solving Equations

GETTING READY *Multiply the binomials.*

1. $(a + b)(a - b)$

2. $(2r + s)(2r - s)$

3. $(3x + 2y)(3x - 2y)$

4. $(4x^2 + 3)(4x^2 - 3)$

Whenever we multiply a binomial of the form $x + y$ by a binomial of the form $x - y$, we obtain another binomial:

$$(x + y)(x - y) = x^2 - y^2$$

◄ Writing Exercises, Something to Think About exercises, and Review Exercises follow each exercise set.

Application problems ►
are in a two-column format.

All applications have a title. ►

All art has been redrawn ►
and made more colorful. ►

5.9 RATIO AND PROPORTION   **323**

Recommended	Gasoline	Oil
50 to 1	6 gal	16 oz

Are these instructions correct? (*Hint:* There are 128 ounces in 1 gallon.)

74. **Height of a tree** A tree casts a shadow of 26 feet at the same time as a 6-foot man casts a shadow of 4 feet. (See Illustration 2.) Find the height of the tree.

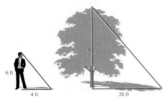

ILLUSTRATION 2

75. **Height of a flagpole** A man places a mirror on the ground and sees the reflection of the top of a flagpole, as in Illustration 3. The two triangles in the illustration are similar. Find the height, h, of the flagpole.

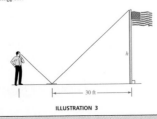

ILLUSTRATION 3

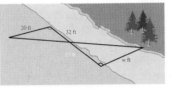

ILLUSTRATION 4

77. **Flight path** An airplane ascends 100 feet as it flies a horizontal distance of 1000 feet. How much altitude will it gain as it flies a horizontal distance of 1 mile? (See Illustration 5.) (*Hint:* 5280 feet = 1 mile.)

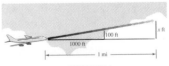

ILLUSTRATION 5

78. **Flight path** An airplane descends 1350 feet as it flies a horizontal distance of 1 mile. How much altitude is lost as it flies a horizontal distance of 5 miles?

79. **Ski runs** A $\frac{1}{2}$-mile long ski course falls 100 feet in every 300 feet of horizontal run. Find the height of the hill.

80. **Mountain travel** A mountain road ascends 375 feet in every 2500 feet of travel. By how much will the road rise in a trip of 10 miles?

81. **Recommended dosage** The recommended child's dose of the sedative hydroxine is 0.006 gram per kilogram of body mass. Find the dosage for a 30-kg child.

P R O J E C T ■ **Pictures of Polynomial Products**

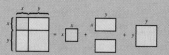

Figure 4-3

Because the length of each side of the largest square in Figure 4-3 is $x + y$, its area is $(x + y)^2$. This area is also the sum of four smaller areas, which illustrates the factorization

◀ Projects are included for cooperative learning.

■ **P E R S P E C T I V E**

Much of the mathematics we have inherited from the ancients is the result of teamwork. In a battle early in the 12th century, control of the Spanish city of Toledo was taken from the Mohammedans, who had ruled for four centuries. Libraries in this great city contained many books written in Arabic, full of knowledge that was unknown in Europe.

The Archbishop of Toledo wanted to share this knowledge with the rest of the world. He knew that these books should be translated into Latin, the universal language of scholarship. But

what European scholar could read Arabic? The archbishop wasn't concerned. The citizens of Toledo could read both Arabic and Spanish, and most scholars of Europe could understand Spanish.

Teamwork saved the day. A citizen of Toledo read an Arabic text aloud, in Spanish. The scholar listened to the Spanish version and wrote in Latin. One of these scholars was an Englishman, Robert of Chester. It was he who translated al-Khowarazmi's book, *Ihm al-jabr wa'l muqabalah*, the beginning of the subject we now know as *algebra*.

◀ Perspectives are found in each chapter.

The fraction bar in the symbol $\frac{a}{b}$ indicates that a is to be divided by b. The number above the fraction bar is called the **numerator,** and the number below is called the **denominator.**

Warning symbols ▶ appear throughout.

 WARNING!   Remember that the denominator of a fraction can never be 0.

There are three signs associated with every fraction: the sign of the fraction, the sign of the numerator, and the sign of the denominator.

EXAMPLE 2   Graph the equation $y = -|x| + 2$.

Solution   We make a table of values by substituting numbers for   and finding the corresponding values of y. For example, if we substitute -2 for x, we get

$$y = -|x| + 2$$
$$y = -|-2| + 2$$
$$y = -(2) + 2$$
$$y = 0$$

After plotting the points listed in the table shown in Figure 6-24, we obtain the graph of the equation.

All mathematics art ▶ has been made more colorful.

$y = -|x| + 2$

x	y
-3	-1
-2	0
-1	1
0	2
1	1
2	0
3	-1

FIGURE 6-24   ■

■ **Graphing Calculators**

So far, we have graphed equations by making a table of values and plotting points. This method is usually tedious and time-consuming. Fortunately, the task of graphing is made much easier when we use a graphing calculator.

Several brands of graphing calculators are available. Although we will use calculators to graph equations, we will not show the keystrokes of any specific brand. For these details, please consult your owner's manual.

All graphing calculators have a **viewing window,** used to display graphs (see Figure 6-25). To see the proper picture of a graph, we must often set the minimum and maximum values for the x- and y-coordinates. The standard RANGE settings of

Graphing calculators ▶ are included.

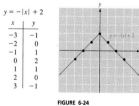

Xmin = -10 Xmax = 10 Ymin = -10 Ymax = 10

indicate that -10 is the minimum x-coordinate and the minimum y-coordinate that will be used in the graph, and that 10 is the maximum x- and y-coordinate that will

FIGURE 6-25

Organization

The book is designed primarily for the chapters to be taught in order. However, for instructors who would like to teach graphing early, the chapters can be taught in the following sequence without loss of continuity:

1, 2, 6, 7, 3, 4, 5, 8, 9

If graphing is taught early, a small number of Review Exercises will have to be omitted. For easy reference, these Review Exercises are noted in the margin of the Annotated Instructor's Edition.

Calculators

The use of calculators is assumed throughout the book. We believe that students should learn calculator skills in the mathematics classroom. They then will be prepared to use calculators in science and business classes, and for nonacademic purposes.

We also have included material on the graphing calculator. We recommend that instructors try it. However, it can be skipped without interrupting the flow of ideas.

The directions within each exercise set indicate which exercises require calculators. There are two symbols, for scientific calculators and for graphing calculators.

Accuracy

To guarantee the accuracy of the answers, each exercise has been worked by both authors and two other problem checkers. The manuscript has been read by many reviewers, and the printed pages have been read by many proofreaders.

Student Support

We have included many features that make the book very accessible to students.

Worked Examples   The text contains more than 400 worked examples. Extensive use of explanatory notes makes the examples easy to follow.

Chapter Summaries   Each chapter concludes with a summary of key words, key ideas, and formulas used in the chapter.

Author's Notes   Author's notes explain the steps in the solutions of most examples.

Functional Use of Color   For easy reference, definitions and theorems are boxed in blue. Strategy boxes are green. In addition, the book uses color to highlight terms and expressions that you would point to in a classroom discussion.

End Papers   Key formulas and ideas are listed inside the front and back covers for easy reference.

Exercises The book includes more than 4500 carefully graded exercises, with answers to the odd-numbered exercises provided in an appendix.

Systematic Review Each exercise set is followed by Review Exercises. In addition, there are Chapter Review Exercises, Chapter Tests, Cumulative Review Exercises, and a Sample Final Examination.

Videotapes Many examples in the book are taught on videotape. These examples are marked with a symbol [⊙⊙] in the book.

Computer Software Students can get additional practice with BCX software. BCX drills students on problems similar to those in the book. BCX will give hints and show complete solutions when necessary.

Ancillaries for the Instructor

Annotated Instructor's Edition The Annotated Instructor's Edition has the answer to every exercise printed next to the exercise. Annotations in the margin give alternate methods of presenting material, as well as teaching hints and strategies.

Test Manual
Teresa Bittner The *Test Manual* contains four ready-to-use forms of every chapter test. Two of the tests are free response and two are multiple choice.

Computer Testing Software
Teresa Bittner Available with the book are two extensive electronic question banks, one free-response and one multiple-choice. Each bank contains approximately 1700 questions and is available for either IBM-compatible or Macintosh computers. The testing program gives you all of the features of state-of-the-art word processors and more, including the ability to see all technical symbols, fonts, and formatting on the screen just the way they will appear when printed. The question banks can be edited.

EXPTEST™ runs on IBM and compatible computers.
ExamBuilder™ runs on Macintosh computers.

Transparencies Color transparencies of key graphics from the book are available to assist the instructor in the classroom.

Videotapes A set of 18 book-specific videotapes is available without charge for adoptions of 100 books or more. The videos include the solutions of all examples in the book marked with a [⊙⊙] symbol. The instructors appearing on the videotapes are David Gustafson, Peter Frisk (the authors), and Diane Koenig.

Computer-Aided Instruction
Teresa Bittner BCX is book-specific tutorial software that drills students on problems similar to those found in the book. There is a set of questions for each section in the book. BCX provides hints to students and, if students cannot answer a question correctly, the complete solution will be displayed. BCX monitors student progress and includes a reporting system.

Ancillaries for the Student

Study Guide
George Grisham
and Robert Eicken

The *Study Guide* provides more explanation, worked examples, practice problems, and practice tests. Available for sale at your college bookstore.

Student Solutions Manual
Diane Koenig

The *Student Solutions Manual* gives complete solutions for the odd-numbered exercises in the book. Available for sale at your college bookstore.

■ ■ ■ ■ ■ ■ ■ PREFACE FOR THE STUDENT

Congratulations. You now own a state-of-the-art textbook that has been written especially for you. To use the book properly, read it carefully, do the exercises, and check your progress with the Review Exercises and the Chapter Tests. Be sure to read and use the following hints on studying algebra.

A *Student Solutions Manual* is available, for sale at your college bookstore, that contains solutions to the odd-numbered exercises. A *Study Guide* that contains additional explanations, worked examples, and practice problems is also available for sale.

When you finish this course, consider keeping your book. It is the single reference source that will keep the information that you have learned at your fingertips. You may need this reference material in future mathematics, science, or business courses.

We wish you well.

Hints on Studying Algebra

The phrase "Practice makes perfect" is not quite true. It is *perfect* practice that makes perfect. For this reason, it is important that you learn how to study algebra to get the most out of this course.

Although we all learn differently, there are some hints on how to study algebra that most students find useful. Here is a list of some things you should consider.

Plan a Strategy for Success   To get where you want to be, you need a goal and a plan. Set a goal of passing this course with a grade of A or B. To meet this goal, you must have a good plan. Your plan should include several points:

- getting ready for class,
- attending class,

- doing homework,
- arranging for special help when you need it, and
- having a strategy for taking tests.

Getting Ready for Class

To get the most out of every class period, you will need to prepare. One of the best things that you can do is to read the material in the book before your instructor discusses it. You may not understand all of what you read, but you will be better able to understand it when your instructor presents the material in class.

Be sure to do your work every day. If you get behind and attend class without understanding prior material, you will be lost and your classroom time will be wasted. Even worse, you will become frustrated and discouraged. Promise yourself that you will always prepare for class and keep your promise.

Attending Class

The classroom experience is your opportunity to learn from your instructor. Make the most of it by attending every class. Sit near the front of the room where you can see and hear well and where you won't be distracted. It is your responsibility to follow the instructor's discussion, even though that might be hard work.

Pay attention to your instructor and jot down the important things that he or she says. However, do not spend so much time taking notes that you fail to concentrate on what your instructor is explaining. It is much better to listen and understand the *big picture* than it is merely to copy solutions to problems.

Don't be afraid to ask questions. If something is unclear to you, it is probably unclear to other students as well. They will appreciate your willingness to ask. Besides, asking questions will make you an active participant in class. This will help you pay attention and keep you alert and involved.

Doing Homework

Everyone knows that it requires practice to excel at tennis, master a musical instrument, or learn a foreign language. It also requires practice to learn mathematics. Since *practice* in mathematics is the homework, homework is your opportunity to practice skills and experiment with ideas.

It is very important to pick a definite time to study and do homework. Set a formal schedule and stick to it. Try to study in a place that is comfortable and quiet. If you can, do some homework shortly after class or at least before you foget what was discussed in class. This quick follow-up will help you remember the skills and concepts your instructor taught that day.

Study Sessions

Each formal study session should include three parts:

1. Begin every study session with a review period. Look over previous chapters and see if you can do a few problems from previous sections. Keeping old skills alive will greatly reduce the amount of time that you will need to cram for tests.

2. After reviewing, read the assigned material. Resist the temptation of diving into the problems without reading and understanding the examples. Instead, work the examples with pencil and paper. Only after you completely understand the underlying principles behind them should you try to work the problems.

Once you begin to work the problems, check your answers with the printed answers in the back of the book. If one of your answers differs from the printed answer, see if you can reconcile the two. Sometimes answers can have more than one form. If you still believe that your answer is incorrect, compare your work to the example in the book that most closely resembles the problem and try to find your mistake. If you cannot find an error, consult the *Student Solutions Manual*. If nothing works, mark the problem and ask about it during your next class meeting.

3. After you complete the written assignment, read the next section. That preview will be helpful when you hear that material discussed during the next class period.

You probably know that the rule of thumb for doing homework is two hours of homework for every hour spent in class. If mathematics is hard for you, plan on spending even more time on homework.

To make homework more enjoyable, study with one or more friends. The interaction will clarify ideas and help you remember them. If you must study alone, try talking to yourself. A good study technique is to explain the material to yourself.

Arranging for Special Help   Take advantage of any special help available from your instructor. Often, your instructor can clear up difficulties in a very short period of time.

Find out if your college has a free tutoring program. Peer tutors also can be of great help. Be sure to use the videotapes and BCX software.

Taking Tests   Students often get nervous before taking a test because they are afraid that they will not do well. The most common reason for this fear is that students are not confident that they know the material.

To build confidence in your ability to work tests, rework many of the problems in the exercise sets, work the Review Exercises at the end of each chapter, and work the Chapter Tests. Check all your answers with the answers printed at the back of the book.

Then guess what the instructor will ask and make up your own tests and work them. Once you know your instructor, you will be surprised at how good you can get at picking test questions. With this preparation, you will have some idea of what will be on the test and will have more confidence in your ability to do well.

When you take a test, work slowly and deliberately. Scan the test and first work the easy problems that you know you can do. This will build confidence. Tackle the hardest problems last.

■ ■ ■ ■ ■ ■ ■ ACKNOWLEDGMENTS

We are grateful to the following people who have reviewed the book at various stages of its development:

Helen Banes
Kirkwood Community College

Theresa Barrie
Texas Southern University

Robert Billups
Citrus Community College

Elaine D. Bouldin
Middle Tennessee State University

David Byrd
Enterprise State Junior College

Baruch Cahlon
Oakland University

Don Cohen
SUNY-Cobleskill

Patricia Cooper
St. Louis Community College-Park Forest

Sally Copeland
Johnson County Community College

Elwin Cutler
Ferris State University

Russell M. Day
Illinois Central College

Elias Deeba
University of Houston-Downtown

Edward Doran
Front Range Community College

Arthur Dull
Diablo Valley College

Robert Eicken
Illinois Central College

Marc Glucksman
El Camino College

George Grisham
Bradley University

Robert G. Hammond
Utah State University

Mitzy Johnson
Northeast Mississippi Community College

Robert Keicher
Delta College

Katherine McLain
Consumnes River College

Laurie McManus
St. Louis Community College-Meramac

Wayne Milloy
Crafton Hills College

Myrna Mitchell
Pima Community College

John Monroe
University of Akron

Carol M. Nessmith
Georgia Southern University

Kent Neuerburg
Consumnes River College

Paul Peck
Glenville State College

Thea Prettyman
Essex Community College

Michael Rosenthal
Florida International University

Jack W. Rotman
Lansing Community College

Irwin Schochetman
Oakland University

Erik A. Schreiner
Western Michigan University

Kenneth Seydel
Skyline College

David Sicks
Olympia College

Willie Taylor
Texas Southern University

Douglas Tharp
University of Houston-Downtown

Lynn E. Tooley
Bellevue Community College

Gary VanVelsir
Anne Arundel Community College

Gerry C. Vidrine
Louisiana State University

Rosalyn Wells
Georgia State University

Clifton T. Whyburn
University of Houston

Hette Williams
Broward Community College

Sonya Woodard
Paradise Valley Community College

We wish to thank Diane Koenig and Robert Hessel, who read the entire manuscript and worked every problem. We also wish to thank Bill Hinrichs, Jerry Frang, George Mader, Michael Welden, Jennifer Dollar, and Rob Clark for their helpful suggestions. We give special thanks to Gary Ostedt, Ellen Brownstein, David Hoyt, Kelly Shoemaker, Lori Heckelman, Kathleen Olson, Audra Silverie, and Carol Benedict for their assistance in the production process.

R. David Gustafson
Peter D. Frisk

■ ■ ■ ■ ■ ■ ■ ■ CONTENTS

Beginning
Algebra

1

Real Numbers and Their Basic Properties

When three mathematics professors attending a convention in Las Vegas registered at the hotel, they were told that the room rate was $120. Each of the three professors paid his $40 share.

Later the desk clerk realized that the cost of the room should have been $115. To fix the mistake, she sent a bellhop to the room to refund the $5 overcharge. Realizing that $5 could not be evenly divided among three professors, and not wanting to start a quarrel, the bellhop refunded only $3 and kept the other $2.

Since each professor received a $1 refund, each paid $39 for the room, and the bellhop kept $2. This gives $39 + $39 + $39 + $2, or $119. What happened to the other $1?

After you have read this chapter, you will be able to answer this question.

1.1 Real Numbers and Their Graphs

■ Equality, Inequality Symbols, and Variables ■ The Number Line
■ Graphing Subsets of the Real Numbers ■ Absolute Value of a Number

GETTING READY

1. Give an example of a number used for counting.

2. Give an example of a number used when dividing a pizza among friends.

3. Give an example of a number used for measuring very cold temperatures.

4. Give an example of a number used to indicate the balance of an overdrawn checking account.

A **set** is a collection of objects. For example, the set

$$\{1, 2, 3, 4, 5\}$$

contains the numbers 1, 2, 3, 4, and 5. The members, or **elements,** in a set are listed within braces {   }.

One basic set of numbers is the set of **natural numbers,** often called the **positive integers.**

Natural Numbers   The **natural numbers** (or the **positive integers**) are the numbers

1, 2, 3, 4, 5, 6, 7, 8, 9, 10, . . .

The three dots in the previous definition, called an **ellipsis,** indicate that the list of natural numbers (positive integers) continues on forever.

We can use the natural numbers (positive integers) to describe many real-life situations. For example, some cars get 30 miles per gallon of gas, and some students might have paid $1750 in tuition.

The set of natural numbers together with 0 form the set of **whole numbers.**

Whole Numbers   The **whole numbers** are the numbers

0, 1, 2, 3, 4, 5, 6, 7, 8, 9, 10, . . .

Numbers that show a loss or a downward direction are called **negative integers,** denoted as -1, -2, -3, and so on. For example, a debt of $1500 can be denoted as $-$1500$, and a temperature of 20° below zero can be denoted as $-20°$.

The set of negative integers together with the set of whole numbers form the set of **integers.**

Integers   The **integers** are the numbers

. . . -5, -4, -3, -2, -1, 0, 1, 2, 3, 4, 5, . . .

Because the set of natural numbers and the set of whole numbers are included within the set of integers, we say that these sets are **subsets** of the set of integers.

Integers cannot describe every real-life situation. For example, a student might study $3\frac{1}{2}$ hours, or a television set might cost $217.37. To describe these situations, we need fractions, more formally called **rational numbers.**

Rational Numbers   A **rational number** is any number that can be written as a fraction with an integer in its numerator and a nonzero integer in its denominator.

Some examples of rational numbers are

$$\frac{3}{2}, \quad \frac{17}{12}, \quad -\frac{43}{8}, \quad 0.25, \quad \text{and} \quad -0.66666 \ldots$$

The decimals 0.25 and $-0.66666\ldots$ are rational numbers because they can be written as fractions: 0.25 is the fraction $\frac{1}{4}$ and $-0.66666\ldots$ is the fraction $-\frac{2}{3}$.

Since every integer can be written as a fraction with a denominator of 1, every integer is also a rational number. Since every integer is a rational number, the set of integers is a subset of the rational numbers.

 WARNING!   Because division by 0 is undefined, expressions such as $\frac{6}{0}$ and $\frac{0}{0}$ do not represent any number.

Some numbers, such as $\sqrt{2}$ and π, are not rational numbers, because they cannot be written as fractions with an integer numerator and an integer denominator. Such numbers are called **irrational numbers.** We can find decimal approximations for the values of irrational numbers with a calculator. For example,

$$\sqrt{2} = 1.414213562\ldots \qquad \text{Press } \boxed{\sqrt{}}\,.$$

$$\pi = 3.14592654\ldots \qquad \text{Press } \boxed{\pi}\,.$$

If we combine the sets of rational numbers and irrational numbers, we have the set of **real numbers.**

PERSPECTIVE

Algebra is an extension of arithmetic. In algebra, the operations of addition, subtraction, multiplication, and division are performed on both numbers and letters, with the understanding that the letters represent numbers.

The origins of algebra are found in a papyrus written before 1600 B.C. by an Egyptian priest named Ahmes. This papyrus contains 84 algebra problems and their solutions. Because the Egyptians did not have a suitable system of notation, however, they were unable to develop algebra completely.

Further development of algebra occurred in the ninth century in the Middle East. In A.D. 830, an Arabian mathematician named al-Khowarazmi wrote a book called *Ihm al-jabr wa'l muqabalah.* This title was shortened to *al-Jabr.* We now know the subject as *algebra.* The French mathematician

The Ahmes Papyrus
(British Museum)

François Vieta (1540–1603) later simplified algebra by developing the symbolic notation that we use today.

Real Numbers   A **real number** is any number that is either a rational number or an irrational number.

Figure 1-1 shows how the various sets of numbers are interrelated.

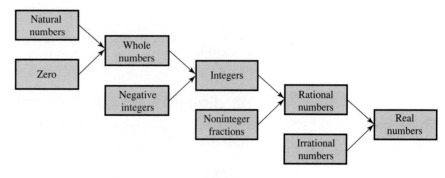

FIGURE 1-1

EXAMPLE 1   List the numbers in the set $\left\{-3, 0, \frac{1}{2}, 1.25, \sqrt{3}, 5\right\}$ that are

a. natural numbers   **b.** whole numbers   **c.** negative integers

d. rational numbers   **e.** irrational numbers   **f.** real numbers

Solution   **a.** The only natural number is 5.   **b.** The whole numbers are 0 and 5.

c. The only negative integer is -3.   **d.** The rational numbers are -3, 0, $\frac{1}{2}$, 1.25, and 5. $\left(1.25 \text{ is rational because } 1.25 \text{ can be written in the form } \frac{5}{4}.\right)$

e. The only irrational number is $\sqrt{3}$.   **f.** All of the numbers are real numbers.   ■

Other subsets of the real numbers also have special names. For example, natural numbers greater than 1 that can only be divided evenly by 1 and themselves are called **prime numbers**. Natural numbers greater than 1 that are not prime are called **composite numbers**.

Prime numbers: {2, 3, 5, 7, 11, 13, 17, 19, 23, 29, . . .}

Composite numbers: {4, 6, 8, 9, 10, 12, 14, 15, 16, 18, 20, 21, 22, . . .}

Integers that can be divided evenly by 2 are called **even integers**. Integers that cannot be divided evenly by 2 are called **odd integers**.

Even integers: {. . . , -10, -8, -6, -4, -2, 0, 2, 4, 6, 8, 10, . . .}

Odd integers: {. . . -9, -7, -5, -3, -1, 1, 3, 5, 7, 9, . . .}

EXAMPLE 2 List the numbers in the set $\{-3, -2, 0, 1, 2, 3, 4, 5, 9\}$ that are

    a. prime numbers   **b.** composite numbers

    c. even integers   **d.** odd integers

Solution   **a.** The prime numbers are 2, 3, and 5.

    b. The composite numbers are 4 and 9.

    c. The even integers are -2, 0, 2, and 4.

    d. The odd integers are -3, 1, 3, 5, and 9. ∎

■ Equality, Inequality Symbols, and Variables

To show that two expressions represent the same number, we use the **equal sign**, written as $=$. Because $4 + 5$ and 9 represent the same number, we can write

    $4 + 5 = 9$ Read as "The sum of 4 and 5 is 9," or "4 plus 5 equals 9."

Likewise, we can write

    $5 - 3 = 2$ Read as "The difference between 5 and 3 is 2," or "5 minus 3 equals 2."

    $4 \cdot 5 = 20$ Read as "The product of 4 and 5 is 20," or "4 times 5 equals 20."

and

    $30 \div 6 = 5$ Read as "The quotient obtained when 30 is divided by 6 is 5," or "30 divided by 6 equals 5."

We can use **inequality symbols** to show that expressions are not equal.

Symbol	Read as
$\neq$	"is not equal to"
$<$	"is less than"
$>$	"is greater than"
$\leq$	"is less than or equal to"
$\geq$	"is greater than or equal to"

EXAMPLE 3   **a.** $6 \neq 9$ Read as "6 is not equal to 9."

    b. $8 < 10$ Read as "8 is less than 10."

    c. $12 > 1$ Read as "12 is greater than 1."

    d. $5 \leq 5$ Read as "5 is less than or equal to 5." (Because $5 = 5$, this is a true statement.)

    e. $9 \geq 7$ Read as "9 is greater than or equal to 7." (Because $9 > 7$, this is a true statement). ∎

Leonardo Fibonacci
(late 12th and early 13th centuries)
Fibonacci, an Italian mathematician,
is also known as Leonardo da Pisa.
In his work *Liber abaci,* he
advocated the adoption of Arabic
numerals, the numerals that we use
today. He is best known for a
sequence of numbers that bears his
name. Can you find the pattern
in this sequence?
1, 1, 2, 3, 5, 8, 13, . . .

Inequality statements can be written so that the inequality symbol points in the opposite direction. For example, the inequality statements

$$5 < 7 \qquad \text{and} \qquad 7 > 5$$

both indicate that 5 is a smaller number than 7. Likewise,

$$12 \geq 3 \qquad \text{and} \qquad 3 \leq 12$$

both indicate that 12 is greater than or equal to 3.

In algebra, we use letters, called **variables,** to represent real numbers. For example,

- If x represents the number 4, then $x = 4$.
- If y represents any number greater than 3, then $y > 3$.
- If z represents any number less than or equal to -4, then $z \leq -4$.

 WARNING! We usually do not use a times sign, $\times$, to indicate multiplication. It might be mistaken for the variable x.

■ The Number Line

We can use the **number line** shown in Figure 1-2 to represent sets of numbers.

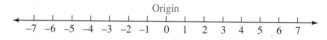

Origin

FIGURE 1-2

The number line continues forever to the left and to the right. Numbers to the left of 0 are negative, and numbers to the right of 0 are positive.

 WARNING! The number 0 is neither positive nor negative.

The number that corresponds to each point on the number line is called the **coordinate** of that point. For example, the coordinate of the **origin** is 0.

Many points on the number line do not have integer coordinates. The point midway between 0 and 1, for example, has the coordinate $\frac{1}{2}$. The point with coordinate $-\frac{5}{2}$ lies midway between -3 and -2, as shown in Figure 1-3.

FIGURE 1-3

Two numbers represented by points that are on opposites sides of the origin and at equal distances from the origin are called **negatives** (or **opposites**) of each other. For example, 5 and −5 are negatives (or opposites). We need parentheses to express the opposite of a negative number. For example, −(−5) represents the opposite of −5, which we know to be 5. Thus,

$$-(-5) = 5$$

In general, we have the following rule.

Double Negative Rule   If x represents a real number, then

$$-(-x) = x$$

On the number line, if one point is to the *right* of a second point, its coordinate is the *greater*. For example, the point with coordinate 5 lies to the right of the point with coordinate −2. Thus, $5 > -2$.

If one point is to the *left* of another, its coordinate is the *smaller*. The point with coordinate −6, for example, is to the left of the point with coordinate 2. Thus, $-6 < 2$.

■ Graphing Subsets of the Real Numbers

Figure 1-4 shows the **graph** of the natural numbers from 2 to 8.

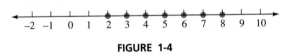

FIGURE 1-4

The point corresponding to the coordinate 4, for example, is called the **graph of 4.**

EXAMPLE 4   Graph the set of integers between −3 and 3 on the number line.

Solution   The integers between −3 and 3 are −2, −1, 0, 1, and 2. The graph is shown in Figure 1-5.

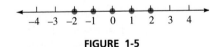

FIGURE 1-5   ■

Graphs of many sets of real numbers are intervals on the number line instead of isolated points. For example, the graph of all numbers x such that $x > -2$ is shown in Figure 1-6.

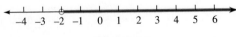

FIGURE 1-6

The open circle at -2 shows that this point is not included in the graph. The arrow pointing to the right shows that all numbers to the right of -2 are included.

Figure 1-7 shows the graph of the set of real numbers between -2 and 4. This is the graph of all numbers x such that $x > -2$ and $x < 4$.

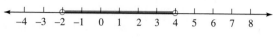

FIGURE 1-7

The open circles at -2 and 4 show that these points are not included in the graph. However, all the numbers between -2 and 4, such as -1, $-\frac{1}{2}$, 0, $\frac{2}{3}$, and 3.99, are included in the graph.

 EXAMPLE 5   Graph all real numbers x such that $x < -3$ or $x > 1$.

Solution   The graph of all real numbers less than -3 includes all points on the number line that are to the left of -3. The graph of all real numbers greater than 1 includes all points that are to the right of 1. The graph is shown in Figure 1-8.

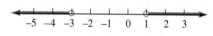

FIGURE 1-8   ∎

 EXAMPLE 6   Graph the set of all real numbers from -5 to -1.

Solution   The set of all real numbers from -5 to -1 includes both -5 and -1 and all the numbers in between. The solid circles at -5 and at -1 show that these points are included in the graph shown in Figure 1-9. This is the graph of all numbers x such that $x \geq -5$ and $x \leq -1$.

FIGURE 1-9   ∎

■ **Absolute Value of a Number**

On a number line, the distance between a number x and 0 is called the **absolute value** of x. For example, the distance between 5 and 0 is 5 units (see Figure 1-10). Thus, the absolute value of 5, denoted as $|5|$, is 5:

$|5| = 5$ Read as "the absolute value of 5 is 5."

The distance between -6 and 0 is 6. Thus,

$|-6| = 6$ Read as "the absolute value of -6 is 6."

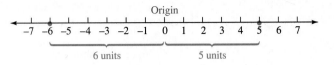

FIGURE 1-10

Because the absolute value of a real number represents the distance on the number line (without regard to direction) that the number is from 0, the absolute value of every real number is either positive or 0. That is, if x is a real number, then

$$|x| \geq 0$$

EXAMPLE 7   Evaluate **a.** $|6|$, **b.** $|-3|$, **c.** $|0|$, and **d.** $|2 + 3|$.

Solution   **a.** $|6| = 6$, because 6 is six units from 0.

b. $|-3| = 3$, because -3 is three units from 0.

c. $|0| = 0$, because 0 is zero units from 0.

d. $|2 + 3| = |5| = 5$ ■

ORALS   *Describe each set of numbers in your own words.*

1. natural numbers   2. whole numbers
3. integers   4. rational numbers
5. real numbers   6. prime numbers
7. composite numbers   8. even integers
9. odd integers   10. irrational numbers

Find each value.

11. $-|15|$ 12. $|-25|$

EXERCISE 1.1

In Exercises 1–12, list the numbers in the set $\left\{-3, -\frac{1}{2}, -1, 0, 1, 2, \frac{5}{3}, \sqrt{7}, 3.25, 6, 9\right\}$ *that are*

1. natural numbers   2. whole numbers   3. positive integers   4. negative integers

5. integers   6. rational numbers   7. real numbers   8. irrational numbers

9. odd integers   10. even integers   11. composite numbers   12. prime numbers

In Exercises 13–20, simplify each expression. Then classify the result as a natural number, an even integer, an odd integer, a prime number, a composite number, and a whole number.

13. $4 + 5$

14. $7 - 2$

15. $15 - 15$

16. $0 + 7$

17. $3 \cdot 8$

18. $8 \cdot 9$

19. $24 \div 8$

20. $3 \div 3$

In Exercises 21–34, place one of the symbols $=$, $<$, or $>$ in the box to make a true statement.

21. $5 \quad\boxed{}\quad 3 + 2$

22. $9 \quad\boxed{}\quad 7$

23. $25 \quad\boxed{}\quad 32$

24. $2 + 3 \quad\boxed{}\quad 17$

25. $5 + 7 \quad\boxed{}\quad 10$

26. $3 + 3 \quad\boxed{}\quad 9 - 3$

27. $3 + 9 \quad\boxed{}\quad 20 - 8$

28. $19 - 3 \quad\boxed{}\quad 8 + 6$

29. $4 \cdot 2 \quad\boxed{}\quad 2 \cdot 4$

30. $7 \cdot 9 \quad\boxed{}\quad 9 \cdot 6$

31. $8 \div 2 \quad\boxed{}\quad 4 + 2$

32. $0 \div 7 \quad\boxed{}\quad 1$

33. $3 + 2 + 5 \quad\boxed{}\quad 5 + 2 + 3$

34. $8 + 5 + 2 \quad\boxed{}\quad 5 + 2 + 8$

In Exercises 35–40, write each statement as a mathematical expression.

35. Seven is greater than three.

36. Five is less than thirty-two.

37. Eight is less than or equal to eight.

38. Twenty-five is not equal to twenty-three.

39. The result of adding three and four is equal to seven.

40. Thirty-seven is greater than or equal to the result of multiplying three and four.

In Exercises 41–52, rewrite each inequality statement as an equivalent inequality in which the inequality symbol points in the opposite direction.

41. $3 \leq 7$

42. $5 > 2$

43. $6 > 0$

44. $34 \leq 40$

45. $3 + 8 > 8$

46. $8 - 3 < 8$

47. $6 - 2 < 10 - 4$

48. $8 \cdot 2 \geq 8 \cdot 1$

49. $2 \cdot 3 < 3 \cdot 4$

50. $8 \div 2 \geq 9 \div 3$

51. $\dfrac{12}{4} < \dfrac{24}{6}$

52. $\dfrac{2}{3} \leq \dfrac{3}{4}$

In Exercises 53–60, graph each pair of numbers on a number line. In each pair, indicate which number is the greater and which number lies farther to the right.

53. $3, 6$

54. $4, 7$

55. $11, 6$

56. $12, 10$

57. $0, 2$

58. $4, 10$

59. $8, 0$

60. $20, 30$

In Exercises 61–72, graph each set of numbers on the number line.

61. The natural numbers between 2 and 8.

62. The prime numbers from 10 to 20.

63. The even integers greater than 10 but less than 20.

64. The even integers that are also prime numbers.

65. The numbers that are whole numbers but not natural numbers.

66. The prime numbers between 5 and 15.

67. The natural numbers between 15 and 25 that are exactly divisible by 6.

68. The odd integers between -5 and 5 that are exactly divisible by 3.

69. The real numbers between 1 and 5.

70. The real numbers greater than or equal to 8.

71. The real numbers greater than 3 or less than or equal to -3.

72. The real numbers greater than -2 and less than 3.

In Exercises 73–80, find each absolute value.

73. $|36|$

74. $|-30|$

75. $|0|$

76. $|120|$

77. $|-230|$

78. $|18 - 12|$

79. $|20 - 12|$

80. $|100 - 100|$

Writing Exercises ■ *Write a paragraph in your own words.*

1. Explain why there is no greatest natural number.

2. Explain why 2 is the only even prime number.

3. Explain how to find the absolute value of a number.

4. Explain why zero is an even integer.

Something to Think About ■ We have studied various sets of numbers: the integers, natural numbers, even and odd numbers, positive and negative numbers, prime and composite numbers, and rational numbers.

1. Find a number that fits in as many of these categories as possible.

2. Find a number that fits in as few of these categories as possible.

Review Exercises ■ *Decide whether the following statements are true.*

1. 6 is an integer. **2.** $\frac{1}{2}$ is a natural number. **3.** 21 is a prime number. **4.** No prime number is even.

5. $8 > -2$ **6.** $-3 < -2$ **7.** $9 \le |-9|$ **8.** $|-11| \ge 10$

1.2 Fractions

■ **Simplifying Fractions** ■ **Multiplying Fractions** ■ **Dividing Fractions**
■ **Adding Fractions** ■ **Subtracting Fractions** ■ **Mixed Numbers** ■ **Decimal Fractions** ■ **Rounding Decimals** ■ **Application**

GETTING READY   *Do the operations.*

1. Add: 132
               45
               73

2. Subtract: 321
                  173

3. Multiply: 437
                  38

4. Divide: $37\overline{)3885}$

In the **fractions**

$$\frac{1}{2}, \quad \frac{3}{5}, \quad \frac{2}{17}, \quad \text{and} \quad \frac{37}{7}$$

the number above the fraction bar is called the **numerator,** and the number below is called the **denominator.**

Fractions are used to indicate parts of a whole. In Figure 1-11(a), a rectangle has been divided into 5 equal parts, and 3 of the parts are shaded. The fraction $\frac{3}{5}$ indicates how much of the figure is shaded. In Figure 1-11(b), $\frac{5}{7}$ of the rectangle is shaded. In either example, the denominator of the fraction shows the total number

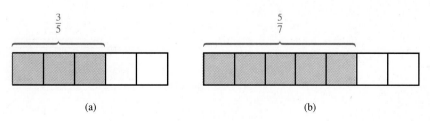

(a)   (b)

FIGURE 1-11

of equal parts into which the whole is divided, and the numerator shows the number of these equal parts that are being considered.

Fractions are also used to indicate division. For example, the fraction $\frac{8}{2}$ indicates that 8 is to be divided by 2:

$$\frac{8}{2} = 8 \div 2$$
$$= 4$$

We note that $\frac{8}{2} = 4$ because $4 \cdot 2 = 8$, and that $\frac{0}{7} = 0$ because $0 \cdot 7 = 0$. However, the fraction $\frac{6}{0}$ is undefined because no number multiplied by 0 gives 6. The fraction $\frac{0}{0}$ is indeterminate because every number multiplied by 0 gives 0.

 WARNING! Remember that the denominator of a fraction cannot be 0.

■ Simplifying Fractions

A fraction is in **lowest terms** when no natural number greater than 1 will divide both its numerator and denominator exactly. The fraction $\frac{6}{11}$ is in lowest terms, because only 1 divides both 6 and 11 exactly. The fraction $\frac{6}{8}$ is not in lowest terms, because 2 divides both 6 and 8 exactly.

We can **simplify** (or **reduce**) a fraction that is not in lowest terms by dividing both its numerator and denominator by the same number. For example, to simplify the fraction $\frac{6}{8}$, we divide both numerator and denominator by 2:

$$\frac{6}{8} = \frac{6 \div 2}{8 \div 2}$$
$$= \frac{3}{4}$$

From Figure 1-12, we see that $\frac{6}{8}$ and $\frac{3}{4}$ are equal fractions, because each represents the same part of the rectangle.

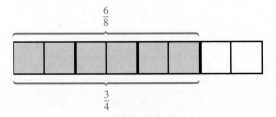

FIGURE 1-12

We can use the fact that composite numbers can be written as the product of other natural numbers to simplify fractions. For example, the composite number 12 can be written as the product of 4 and 3:

$$12 = 4 \cdot 3$$

When a composite number has been written as the product of other natural numbers, we say that it has been **factored.** The numbers 3 and 4 are called **factors** of 12. When a composite number is written as the product of prime numbers, the number is written in **prime-factored form.**

EXAMPLE 1   Write 210 in prime-factored form.

Solution   We can write the number 210 as the product of 21 and 10 and proceed as follows:

$$210 = \mathbf{21} \cdot 10$$
$$210 = \mathbf{3 \cdot 7} \cdot \mathbf{2 \cdot 5} \qquad \text{Factor 21 as } 3 \cdot 7 \text{ and factor 10 as } 2 \cdot 5.$$

Since 210 is now written as the product of prime numbers, its prime-factored form is $210 = 2 \cdot 3 \cdot 5 \cdot 7$.   ∎

To simplify a fraction, we factor its numerator and denominator and then divide out all common factors that appear in both the numerator and denominator. To simplify the fractions $\frac{6}{8}$ and $\frac{15}{18}$, for example, we proceed as follows:

$$\frac{6}{8} = \frac{3 \cdot 2}{4 \cdot 2} = \frac{3 \cdot \overset{1}{\cancel{2}}}{4 \cdot \underset{1}{\cancel{2}}} = \frac{3}{4} \qquad \text{and} \qquad \frac{15}{18} = \frac{5 \cdot 3}{6 \cdot 3} = \frac{5 \cdot \overset{1}{\cancel{3}}}{6 \cdot \underset{1}{\cancel{3}}} = \frac{5}{6}$$

 WARNING!   Remember that a fraction is simplified only when its numerator and denominator have no common factors.

 (a)   **EXAMPLE 2**   **a.** To simplify $\frac{6}{30}$, we factor the numerator and denominator and divide out the common factor of 6.

$$\frac{6}{30} = \frac{6 \cdot 1}{6 \cdot 5} = \frac{\overset{1}{\cancel{6}} \cdot 1}{\underset{1}{\cancel{6}} \cdot 5} = \frac{1}{5}$$

b. To show that $\frac{33}{40}$ is in lowest terms, we must show that the numerator and the denominator share no common factors by writing both the numerator and denominator in prime-factored form.

$$\frac{33}{40} = \frac{3 \cdot 11}{2 \cdot 2 \cdot 2 \cdot 5}$$

Since the numerator and denominator have no common factors, $\frac{33}{40}$ is in lowest terms. ■

The previous examples illustrate the **fundamental property of fractions.**

The Fundamental Property of Fractions	If a represents a real number, and b and x represent nonzero real numbers, then $$\frac{a \cdot x}{b \cdot x} = \frac{a}{b}$$

■ Multiplying Fractions

Multiplying Fractions	To multiply two fractions, we multiply the numerators and multiply the denominators. In symbols, if a, b, c, and d are real numbers, then $$\frac{a}{b} \cdot \frac{c}{d} = \frac{a \cdot c}{b \cdot d} \qquad (b \neq 0 \text{ and } d \neq 0)$$

For example,

$$\frac{4}{7} \cdot \frac{2}{3} = \frac{4 \cdot 2}{7 \cdot 3} \qquad \text{and} \qquad \frac{4}{5} \cdot \frac{13}{9} = \frac{4 \cdot 13}{5 \cdot 9}$$

$$= \frac{8}{21} \qquad\qquad\qquad = \frac{52}{45}$$

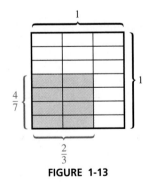

FIGURE 1-13

To justify the rule for multiplying fractions, we consider the square in Figure 1-13. Because the length of each side of the square is one unit and the area is the product of the lengths of two sides, the area is 1 square unit.

If this square is divided into 3 equal parts vertically and 7 equal parts horizontally, it is divided into 21 equal parts, and each part represents $\frac{1}{21}$ of the total area. The area of the shaded rectangle in the square is $\frac{8}{21}$ because it contains 8 of the 21 parts. The width, w, of the shaded rectangle is $\frac{4}{7}$ and its length, l, is $\frac{2}{3}$. Its area, A, is the product of l and w:

$$A = l \cdot w$$

$$\frac{8}{21} = \frac{4}{7} \cdot \frac{2}{3}$$

This suggests that we can find the product of

$$\frac{4}{7} \qquad \text{and} \qquad \frac{2}{3}$$

by multiplying their numerators and multiplying their denominators.

Fractions such as $\frac{8}{21}$ with a numerator that is smaller than the denominator are called **proper fractions.** Fractions such as $\frac{52}{45}$ with a numerator that is larger than the denominator are called **improper fractions.**

EXAMPLE 3   **a.** $\dfrac{3}{7} \cdot \dfrac{13}{5} = \dfrac{3 \cdot 13}{7 \cdot 5}$ Multiply the numerators and multiply the denominators.

$\qquad\qquad\qquad = \dfrac{39}{35}$

b. $5 \cdot \dfrac{3}{15} = \dfrac{5}{1} \cdot \dfrac{3}{15}$ Write 5 as the improper fraction $\frac{5}{1}$.

$\qquad\qquad\quad = \dfrac{5 \cdot 3}{1 \cdot 15}$ Multiply the numerators and multiply the denominators.

$\qquad\qquad\quad = \dfrac{5 \cdot 3}{1 \cdot 5 \cdot 3}$ To attempt to simplify the fraction, factor the denominator.

$\qquad\qquad\quad = \dfrac{\overset{1}{\cancel{5}} \cdot \overset{1}{\cancel{3}}}{1 \cdot \underset{1}{\cancel{5}} \cdot \underset{1}{\cancel{3}}}$ Divide out the common factors of 3 and 5.

$\qquad\qquad\quad = \dfrac{1}{1}$

$\qquad\qquad\quad = 1$ ■

EXAMPLE 4   **European travel**   Out of 36 students in a history class, three-fourths have signed up for a trip to Europe. There are 30 places available on the chartered flight. Will there be room for one more student?

Solution   We first find three-fourths of 36 by multiplying 36 by $\frac{3}{4}$.

$\dfrac{3}{4} \cdot 36 = \dfrac{3}{4} \cdot \dfrac{36}{1}$ Write 36 as $\frac{36}{1}$.

$\qquad\quad = \dfrac{3 \cdot 36}{4 \cdot 1}$ Multiply the numerators and multiply the denominators.

$\qquad\quad = \dfrac{3 \cdot 4 \cdot 9}{4 \cdot 1}$ To simplify the fractions, factor the numerator.

$\qquad\quad = \dfrac{3 \cdot \overset{1}{\cancel{4}} \cdot 9}{\underset{1}{\cancel{4}} \cdot 1}$ Divide out the common factor of 4.

$\qquad\quad = \dfrac{27}{1}$

$\qquad\quad = 27$

Twenty-seven students plan to go on the trip. Because there is room for 30 passengers, there is plenty of room for one more.   ■

■ Dividing Fractions

One number is called the **reciprocal** of another if their product is 1. For example, $\frac{3}{5}$ is the reciprocal of $\frac{5}{3}$ because

$$\frac{3}{5} \cdot \frac{5}{3} = \frac{15}{15} = 1$$

Dividing Fractions	To divide two fractions, we multiply the first fraction by the reciprocal of the second fraction. In symbols, if a, b, c, and d are real numbers, then $$\frac{a}{b} \div \frac{c}{d} = \frac{a}{b} \cdot \frac{d}{c} = \frac{a \cdot d}{b \cdot c} \qquad (b \neq 0, c \neq 0, \text{ and } d \neq 0)$$

EXAMPLE 5   **a.**

$$\frac{3}{5} \div \frac{6}{5} = \frac{3}{5} \cdot \frac{5}{6}$$ Multiply $\frac{3}{5}$ by the reciprocal of $\frac{6}{5}$.

$$= \frac{3 \cdot 5}{5 \cdot 6}$$ Multiply the numerators and multiply the denominators.

$$= \frac{3 \cdot 5}{5 \cdot 2 \cdot 3}$$ Factor the denominator.

$$= \frac{\overset{1}{\cancel{3}} \cdot \overset{1}{\cancel{5}}}{\underset{1}{\cancel{5}} \cdot 2 \cdot \underset{1}{\cancel{3}}}$$ Divide out the common factors of 3 and 5.

$$= \frac{1}{2}$$

b.

$$\frac{15}{7} \div 10 = \frac{15}{7} \div \frac{10}{1}$$ Write 10 as the improper fraction $\frac{10}{1}$.

$$= \frac{15}{7} \cdot \frac{1}{10}$$ Multiply $\frac{15}{7}$ by the reciprocal of $\frac{10}{1}$.

$$= \frac{15 \cdot 1}{7 \cdot 10}$$ Multiply the numerators and multiply the denominators.

$$= \frac{3 \cdot \overset{1}{\cancel{5}}}{7 \cdot 2 \cdot \underset{1}{\cancel{5}}}$$ Factor the numerator and denominator.

$$= \frac{3}{14}$$ Divide out the common factor of 5.   ■

■ Adding Fractions

Adding Fractions with the Same Denominator

To add two fractions with the same denominator, we add the numerators and keep the common denominator. In symbols, if a, b, and d are real numbers, then

$$\frac{a}{d} + \frac{b}{d} = \frac{a+b}{d} \qquad (d \neq 0)$$

For example,

$$\frac{3}{7} + \frac{2}{7} = \frac{3+2}{7} \qquad \text{Add the numerators and keep the common denominator.}$$

$$= \frac{5}{7}$$

Figure 1-14 shows why $\frac{3}{7} + \frac{2}{7} = \frac{5}{7}$.

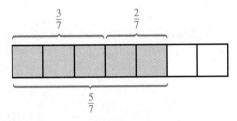

FIGURE 1-14

To add fractions with unlike denominators, we rewrite the fractions so that they have the same denominator. For example, we can multiply both the numerator and denominator of the fraction $\frac{1}{3}$ by 5, to obtain an equal fraction with a denominator of 15:

$$\frac{1}{3} = \frac{1 \cdot 5}{3 \cdot 5}$$

$$= \frac{5}{15}$$

Similarly, we can rewrite the fraction $\frac{1}{5}$ as an equal fraction with a denominator of 15. We multiply both the numerator and the denominator by 3:

$$\frac{1}{5} = \frac{1 \cdot 3}{5 \cdot 3}$$

$$= \frac{3}{15}$$

Because 15 is the smallest number that can be used as a denominator for $\frac{1}{3}$ and $\frac{1}{5}$, it is called the **least** or **lowest common denominator.**

To add the fractions $\frac{1}{3}$ and $\frac{1}{5}$, we rewrite each fraction as an equal fraction having a denominator of 15 and add the results:

$$\frac{1}{3} + \frac{1}{5} = \frac{1 \cdot 5}{3 \cdot 5} + \frac{1 \cdot 3}{5 \cdot 3}$$

$$= \frac{5}{15} + \frac{3}{15}$$

$$= \frac{5 + 3}{15}$$

$$= \frac{8}{15}$$

EXAMPLE 6   Add $\dfrac{3}{10}$ and $\dfrac{5}{28}$.

Solution   To find the least common denominator (LCD), we find the prime factorization of both denominators and use each prime factor the greatest number of times it appears in either factorization:

$$\left. \begin{array}{l} 10 = 2 \cdot 5 \\ 28 = 2 \cdot 2 \cdot 7 \end{array} \right\} \text{LCD} = 2 \cdot 2 \cdot 5 \cdot 7 = 140$$

Since 140 is the smallest number that 10 and 28 divide exactly, we write both fractions as fractions with the least common denominator of 140.

$$\frac{3}{10} + \frac{5}{28} = \frac{3 \cdot \mathbf{14}}{10 \cdot \mathbf{14}} + \frac{5 \cdot \mathbf{5}}{28 \cdot \mathbf{5}} \qquad \text{Write each fraction as a fraction with a denominator of 140.}$$

$$= \frac{42}{140} + \frac{25}{140}$$

$$= \frac{42 + 25}{140} \qquad \text{Add numerators and keep the denominator.}$$

$$= \frac{67}{140}$$

Since 67 is a prime number, it has no common factor with 140. Thus, $\frac{67}{140}$ is in lowest terms and cannot be simplified. ∎

■ Subtracting Fractions

Subtracting Fractions with the Same Denominator
To subtract two fractions with the same denominator, we subtract their numerators and keep their common denominator. In symbols, if a, b, and d are real numbers, then

$$\frac{a}{d} - \frac{b}{d} = \frac{a-b}{d} \qquad (d \neq 0)$$

For example,

$$\frac{7}{9} - \frac{2}{9} = \frac{7-2}{9} \qquad \text{Subtract the numerators and keep the denominator.}$$

$$= \frac{5}{9}$$

To subtract fractions with unlike denominators, we write them as equivalent fractions with a common denominator. For example, to subtract $\frac{2}{5}$ from $\frac{3}{4}$, we write $\frac{3}{4} - \frac{2}{5}$, find the LCD of 20, and proceed as follows:

$$\frac{3}{4} - \frac{2}{5} = \frac{3 \cdot 5}{4 \cdot 5} - \frac{2 \cdot 4}{5 \cdot 4}$$

$$= \frac{15}{20} - \frac{8}{20}$$

$$= \frac{15-8}{20} \qquad \text{Subtract the numerators and keep the denominator.}$$

$$= \frac{7}{20}$$

EXAMPLE 7 Subtract 5 from $\dfrac{23}{3}$.

Solution
$$\frac{23}{3} - 5 = \frac{23}{3} - \frac{5}{1} \qquad \text{Write 5 as the improper fraction } \tfrac{5}{1}.$$

$$= \frac{23}{3} - \frac{5 \cdot 3}{1 \cdot 3} \qquad \text{Write } \tfrac{5}{1} \text{ as a fraction with a denominator of 3.}$$

$$= \frac{23}{3} - \frac{15}{3}$$

$$= \frac{23-15}{3} \qquad \text{Subtract the numerators and keep the denominator.}$$

$$= \frac{8}{3}$$

■

■ Mixed Numbers

The **mixed number** $3\frac{1}{2}$ represents the sum of 3 and $\frac{1}{2}$. We can write $3\frac{1}{2}$ as an improper fraction as follows:

$$3\frac{1}{2} = 3 + \frac{1}{2}$$

$$= \frac{6}{2} + \frac{1}{2} \qquad 3 = \frac{6}{2}.$$

$$= \frac{6 + 1}{2} \qquad \text{Add the numerators and keep the denominator.}$$

$$= \frac{7}{2}$$

To write the fraction $\frac{19}{5}$ as a mixed number, we divide 19 by 5 to get 3, with a remainder of 4. Thus,

$$\frac{19}{5} = 3 + \frac{4}{5}$$

$$= 3\frac{4}{5}$$

EXAMPLE 8 Add $2\frac{1}{4}$ and $1\frac{1}{3}$.

Solution We can change each mixed number to an improper fraction

$$2\frac{1}{4} = 2 + \frac{1}{4} \qquad\qquad 1\frac{1}{3} = 1 + \frac{1}{3}$$

$$= \frac{8}{4} + \frac{1}{4} \qquad\qquad = \frac{3}{3} + \frac{1}{3}$$

$$= \frac{9}{4} \qquad\qquad = \frac{4}{3}$$

and then add the fractions.

$$2\frac{1}{4} + 1\frac{1}{3} = \frac{9}{4} + \frac{4}{3}$$

$$= \frac{9 \cdot 3}{4 \cdot 3} + \frac{4 \cdot 4}{3 \cdot 4}$$

$$= \frac{27}{12} + \frac{16}{12}$$

$$= \frac{43}{12}$$

Finally, we change $\frac{43}{12}$ to a mixed number.

$$\frac{43}{12} = 3 + \frac{7}{12} = 3\frac{7}{12}$$ ■

EXAMPLE 9 **Fencing land** The three sides of a triangular piece of land measure $33\frac{1}{4}$, $57\frac{3}{4}$, and $72\frac{1}{2}$ meters. How much fencing will be needed to enclose the area?

Solution We can find the sum of the lengths by adding the whole number parts and the fractional parts of the dimensions separately:

$$33\frac{1}{4} + 57\frac{3}{4} + 72\frac{1}{2} = 33 + 57 + 72 + \frac{1}{4} + \frac{3}{4} + \frac{1}{2}$$

$$= 162 + \frac{1}{4} + \frac{3}{4} + \frac{2}{4} \qquad \text{Change } \tfrac{1}{2} \text{ to } \tfrac{2}{4} \text{ to obtain a common denominator.}$$

$$= 162 + \frac{6}{4} \qquad \text{Add the fractions by adding the numerators and keeping the denominator.}$$

$$= 162 + \frac{3}{2} \qquad \frac{6}{4} = \frac{2 \cdot 3}{2 \cdot 2} = \frac{2 \cdot 3}{2 \cdot 2} = \frac{3}{2}$$

$$= 162 + 1\frac{1}{2} \qquad \text{Change } \tfrac{3}{2} \text{ to a mixed number.}$$

$$= 163\frac{1}{2}$$

It will require $163\frac{1}{2}$ meters of fencing to enclose the triangular area. ■

■ Decimal Fractions

Rational numbers can always be changed into decimal fractions. For example, to change $\frac{1}{4}$ and $\frac{5}{22}$ to decimal fractions, we use long division:

$$
\begin{array}{r}
0.25 \\
4\overline{)1.00} \\
\underline{8} \\
20 \\
\underline{20} \\
0
\end{array}
\qquad\qquad
\begin{array}{r}
0.22727 \ldots \\
22\overline{)5.00000} \\
\underline{4\,4} \\
60 \\
\underline{44} \\
160 \\
\underline{154} \\
60 \\
\underline{44} \\
16
\end{array}
$$

The decimal fraction 0.25 is called a **terminating decimal,** and the decimal fraction 0.2272727... (often written as $0.2\overline{27}$) is called a **repeating decimal** because it repeats the block of digits 27. Every rational number can be changed into either a terminating or a repeating decimal.

Terminating decimals	*Repeating decimals*
$\dfrac{1}{2} = 0.5$	$\dfrac{1}{3} = 0.33333 \ldots$ or $0.\overline{3}$
$\dfrac{3}{4} = 0.75$	$\dfrac{1}{6} = 0.16666 \ldots$ or $0.1\overline{6}$
$\dfrac{5}{8} = 0.625$	$\dfrac{5}{22} = 0.2272727 \ldots$ or $0.2\overline{27}$

The decimal 0.5 has one **decimal place,** because it has one digit to the right of the decimal point. The decimal 0.75 has two decimal places, and 0.625 has three.

To *add* or *subtract* decimal fractions, we first align their decimal points and then add or subtract.

$$
\begin{array}{r}
25.568 \\
+\ 2.74 \\
\hline
28.308
\end{array}
\qquad
\begin{array}{r}
25.568 \\
-\ 2.74 \\
\hline
22.828
\end{array}
$$

To do the previous two operations with a calculator, we would press these keys:

25.568 $+$ 2.74 $=$ and   25.568 $-$ 2.74 $=$

To *multiply* decimal fractions, we multiply the numbers and then place the decimal point so that the number of decimal places in the answer is equal to the sum of the decimal places in the factors.

$$
\begin{array}{r}
3.453 \\
\times\quad 9.25 \\
\hline
17265 \\
6906 \\
31077 \\
\hline
31.94025
\end{array}
$$

Here there are three decimal places.

Here there are two decimal places.

The product has $3 + 2 = 5$ decimal places.

To do the previous multiplication with a calculator, we would press these keys:

3.453 $\times$ 9.25 $=$

To *divide* decimal fractions, we move the decimal point in the divisor to the right to make the divisor a whole number. We then move the decimal point in the dividend the same number of places to the right.

$$1.23\overline{)30.258}$$

Move the decimal point in both the divisor and the dividend two places to the right.

We align the decimal point in the quotient with the repositioned decimal point in the dividend and then use long division.

$$
\begin{array}{r}
24.6 \\
123\overline{)3025.8} \\
246 \\
\hline
565 \\
492 \\
\hline
73\,8 \\
73\,8 \\
\hline
0
\end{array}
$$

To do the previous division with a calculator, we would press these keys:

30.258 ÷ 1.23 =

■ Rounding Decimals

When decimal fractions are long, we often **round** them to a specific number of decimal places. For example, the decimal fraction 25.36124 rounded to one place (or to the nearest tenth) is 25.4. Rounded to two places (or to the nearest one-hundredth), the decimal is 25.36. To round decimals, we use the following rules.

Rounding Decimals
1. Determine to how many decimal places you wish to round.
2. Look at the first digit to the right of that decimal place.
3. If that digit is 4 or less, drop it and all digits that follow. If it is 5 or greater, add 1 to the digit in the position in which you wish to round, and drop all of the digits that follow.

■ Application

A **percent** is the numerator of a fraction with a denominator of 100. For example, $6\frac{1}{4}$ percent, written $6\frac{1}{4}\%$, is the fraction $\frac{6.25}{100}$, or the decimal 0.0625. In problems involving percent, the word *of* often indicates multiplication. For example, $6\frac{1}{4}\%$ of 8500 is the product 0.0625(8500).

EXAMPLE 10   **Paying tuition**   Juan signs a one-year note to borrow $8500 for tuition. If the rate of interest is $6\frac{1}{4}\%$, how much interest will he pay?

Solution For the privilege of using the bank's money for one year, Juan must pay $6\frac{1}{4}\%$ of $8500. We calculate the interest, i, as follows:

$$i = 6\frac{1}{4}\% \text{ of } 8500$$

$$= 0.0625 \cdot 8500 \qquad \text{The word } of \text{ means } times.$$

$$= 531.25$$

Juan will pay $531.25 interest. ■

ORALS *Simplify each fraction.*

1. $\dfrac{3}{6}$ 2. $\dfrac{5}{10}$ 3. $\dfrac{10}{20}$ 4. $\dfrac{25}{75}$

Do each operation.

5. $\dfrac{5}{6} \cdot \dfrac{1}{2}$ 6. $\dfrac{3}{4} \cdot \dfrac{3}{5}$ 7. $\dfrac{2}{3} \div \dfrac{3}{2}$ 8. $\dfrac{3}{5} \div \dfrac{5}{2}$

9. $\dfrac{4}{9} + \dfrac{7}{9}$ 10. $\dfrac{6}{7} - \dfrac{3}{7}$ 11. $\dfrac{2}{3} - \dfrac{1}{2}$ 12. $\dfrac{3}{4} + \dfrac{1}{2}$

13. $2.5 + 0.36$ 14. $3.45 - 2.21$

15. $0.2 \cdot 2.5$ 16. $0.3 \cdot 13$

Round each decimal to two decimal places.

17. 3.244993 18. 3.24521

E X E R C I S E   1 . 2

In Exercises 1–8, write each fraction in lowest terms. If the fraction is already in lowest terms, so indicate.

1. $\dfrac{6}{12}$ 2. $\dfrac{3}{9}$ 3. $\dfrac{15}{20}$ 4. $\dfrac{22}{77}$

5. $\dfrac{24}{18}$ 6. $\dfrac{35}{14}$ 7. $\dfrac{72}{64}$ 8. $\dfrac{26}{21}$

In Exercises 9–20, do each multiplication. Simplify each result when possible.

9. $\dfrac{1}{2} \cdot \dfrac{3}{5}$ 10. $\dfrac{3}{4} \cdot \dfrac{5}{7}$ 11. $\dfrac{4}{3} \cdot \dfrac{6}{5}$ 12. $\dfrac{7}{8} \cdot \dfrac{6}{15}$

13. $\dfrac{5}{12} \cdot \dfrac{18}{5}$ 14. $\dfrac{5}{4} \cdot \dfrac{12}{10}$ 15. $\dfrac{17}{34} \cdot \dfrac{3}{6}$ 16. $\dfrac{21}{14} \cdot \dfrac{3}{6}$

17. $12 \cdot \dfrac{5}{6}$ 18. $9 \cdot \dfrac{7}{12}$ 19. $\dfrac{10}{21} \cdot 14$ 20. $\dfrac{5}{24} \cdot 16$

In Exercises 21–32, do each division. Simplify each result when possible.

21. $\dfrac{3}{5} \div \dfrac{2}{3}$ **22.** $\dfrac{4}{5} \div \dfrac{3}{7}$ **23.** $\dfrac{3}{4} \div \dfrac{6}{5}$ **24.** $\dfrac{3}{8} \div \dfrac{15}{28}$

25. $\dfrac{2}{13} \div \dfrac{8}{13}$ **26.** $\dfrac{4}{7} \div \dfrac{20}{21}$ **27.** $\dfrac{21}{35} \div \dfrac{3}{14}$ **28.** $\dfrac{23}{25} \div \dfrac{46}{5}$

29. $6 \div \dfrac{3}{14}$ **30.** $23 \div \dfrac{46}{5}$ **31.** $\dfrac{42}{30} \div 7$ **32.** $\dfrac{34}{8} \div 17$

In Exercises 33–60, do each addition or subtraction. Simplify each result when possible.

33. $\dfrac{3}{5} + \dfrac{3}{5}$ **34.** $\dfrac{4}{7} - \dfrac{2}{7}$ **35.** $\dfrac{4}{13} - \dfrac{3}{13}$ **36.** $\dfrac{2}{11} + \dfrac{9}{11}$

37. $\dfrac{1}{6} + \dfrac{1}{24}$ **38.** $\dfrac{17}{25} - \dfrac{2}{5}$ **39.** $\dfrac{3}{5} + \dfrac{2}{3}$ **40.** $\dfrac{4}{3} + \dfrac{7}{2}$

41. $\dfrac{9}{4} - \dfrac{5}{6}$ **42.** $\dfrac{2}{15} + \dfrac{7}{9}$ **43.** $\dfrac{7}{10} - \dfrac{1}{14}$ **44.** $\dfrac{7}{25} + \dfrac{3}{10}$

45. $\dfrac{5}{14} - \dfrac{4}{21}$ **46.** $\dfrac{2}{33} + \dfrac{3}{22}$ **47.** $3 - \dfrac{3}{4}$ **48.** $5 + \dfrac{21}{5}$

49. $\dfrac{17}{3} + 4$ **50.** $\dfrac{13}{9} - 1$ **51.** $\dfrac{3}{15} + \dfrac{6}{10}$ **52.** $\dfrac{7}{5} - \dfrac{2}{15}$

53. $4\dfrac{3}{5} + \dfrac{3}{5}$ **54.** $2\dfrac{1}{8} + \dfrac{3}{8}$ **55.** $3\dfrac{1}{3} - 1\dfrac{2}{3}$ **56.** $5\dfrac{1}{7} - 3\dfrac{2}{7}$

57. $3\dfrac{3}{4} - 2\dfrac{1}{2}$ **58.** $15\dfrac{5}{6} + 11\dfrac{5}{8}$ **59.** $8\dfrac{2}{9} - 7\dfrac{2}{3}$ **60.** $3\dfrac{4}{5} - 3\dfrac{1}{10}$

61. Perimeter of a triangle Each side of a triangle measures $2\dfrac{3}{7}$ centimeters. Find its perimeter, the sum of the lengths of its three sides.

62. Buying fencing Each side of a square field measures $30\dfrac{2}{5}$ meters. How many meters of fencing is needed to enclose the field?

63. Running a race Jim has run $6\dfrac{3}{10}$ kilometers of a 10-kilometer race. How far away is the finish line?

64. Spring plowing A farmer has plowed $12\dfrac{1}{3}$ acres of a $43\dfrac{1}{2}$ acre field. How much more needs to be plowed?

65. Perimeter of a garden The four sides of a garden measure $7\dfrac{2}{3}$ feet, $15\dfrac{1}{4}$ feet, $19\dfrac{1}{2}$ feet, and $10\dfrac{3}{4}$ feet. Find the length of the fence needed to enclose the garden.

66. Making clothes A clothing designer requires $3\dfrac{1}{4}$ yards of material for each dress he makes. How much material will be used to make 14 dresses?

67. Minority population 22% of the 11,431,000 citizens of Illinois are nonwhite. How many are nonwhite?

68. Quality control Reject rates are high in the manufacture of active-matrix color LCD computer displays. If 23% of a production run of 17,500 units are defective, how many units are acceptable?

69. Freeze drying Almost all of the water must be removed when food is preserved by freeze drying. Find the weight of the water removed from 750 pounds of a food that is 36% water.

70. Planning for growth This year, sales at Positronics Corporation totaled $18.7 million. The board of directors predicts 12% annual growth. If the prediction is true, what will be next year's sales?

In Exercises 71–78, do each operation.

71. $23.45 + 135.2$

72. $345.213 - 27.35$

73. $67.235 - 22.45$

74. $12.17 + 3.457$

75. $3.4 \cdot 13.2$

76. $4.21 \cdot 2.73$

77. $0.23)\overline{1.0465}$

78. $4.7)\overline{10.857}$

In Exercises 79–86, use a calculator to do each operation, and round the answer to two decimal places.

79. $323.24 + 27.2543$

80. $843.45213 - 712.765$

81. $55.77443 - 0.568245$

82. $0.62317 + 1.3316$

83. $25.25 \cdot 132.179$

84. $234.874 \cdot 242.46473$

85. $.456)\overline{4.5694323}$

86. $43.225)\overline{32.465758}$

In Exercises 87–100, use a calculator to solve each problem. Round numeric answers to two decimal places.

87. Finding areas Find the area of a square with a side that is 62.17 feet long. (*Hint:* $A = s \cdot s$)

88. Speed skating In tryouts for the Olympics, a speed skater had times of 44.47, 43.24, 42.77, and 42.05 seconds. Find the average time. (*Hint:* Add the numbers and divide by 4.)

89. Cost of gasoline Diego drove his car 15,675.2 miles last year, averaging 25.5 miles per gallon of gasoline. The average cost of gasoline was $1.27 per gallon. Find the fuel cost to drive the car.

90. Paying taxes A businesswoman earns $48,712.32 in taxable income. She must pay 15% tax on the first $23,000 and 28% on the rest. In addition, she must pay a Social Security tax of 15.4% on the total amount. How much tax will she need to pay?

91. Sealing asphalt A rectangular parking lot is 253.5 feet long and 178.5 feet wide. A 55-gallon drum of asphalt sealer covers 4000 square feet and costs $97.50. Find the cost to seal the parking lot. Sealer can only be purchased in full drums.

92. Installing carpet What will it cost to carpet a 23-by-17.5-foot living room and a 17.5-by-14-foot dining room with carpet that costs $29.79 per square yard? One square yard is nine square feet.

93. Inventory costs Each television costs $3.25 per day for warehouse storage. What are the warehousing costs to store 37 television sets for three weeks?

94. Manufacturing profits A manufacturer of computer memory boards has a profit of $37.50 on each standard-capacity memory board and $57.35 for each high-capacity board. The sales department has orders for 2530 standard boards and 1670 high-capacity boards. Which order should production fill first, to receive the greater profit?

95. Dairy production A Holstein cow will produce 7600 pounds of milk each year, with a $3\frac{1}{2}$% butterfat content. Each year, a Guernsey cow will produce about 6500 pounds of milk that is 5% butterfat. Which cow produces more butterfat?

96. Feeding dairy cows Each year, a typical dairy cow will eat 12,000 pounds of food, which is 57% silage. To feed a herd of 30 cows, how much silage will a farmer use in a year?

97. Comparing bids   Two contractors bid on a home remodeling project. The first bids $9350 for the entire job. The second contractor will work for $27.50 per hour, plus $4500 for materials. He estimates that the job will take 150 hours. Which contractor has the lower bid?

98. Choosing a furnace   A high-efficiency home heating system can be installed for $4170, with an average monthly heating bill of $57.50. A regular furnace can be installed for $1730, but monthly heating bills average $107.75. After three years, which system is more expensive?

99. Choosing a furnace   Refer to Exercise 98. Decide which furnace system is the more expensive after five years.

100. Mortgage interest   A mortgage carries an annual interest rate of 9.75%. Each month, the bank's interest charge is one-twelfth of 9.75% of the outstanding balance, currently $72,363. Find the amount of interest that will be paid this month.

Writing Exercises ■ *Write a paragraph using your own words.*

1. Describe how you would find the common denominator of two fractions.

2. Explain how to convert an improper fraction into a mixed number.

3. Explain how to convert a mixed number into an improper fraction.

4. Explain how you would decide which of two decimal fractions is the larger.

Something to Think About ■

1. In what situations would it be better to leave an answer in the form of an improper fraction?

2. When would it be better to change an improper-fraction answer into a mixed number?

3. Can the product of two proper fractions be larger than either of the fractions?

4. How does the product of one proper and one improper fraction compare with the two fractions?

Review Exercises ■ *Place an appropriate symbol in the box to make the statement true.*

1. $3 + 7$ ▨ 10 **2.** $\dfrac{3}{7}$ ▨ $\dfrac{2}{7} = \dfrac{1}{7}$ **3.** $|-2|$ ▨ 2 **4.** $4 + 8$ ▨ 11

1.3   Exponents and Order of Operations

■ Order of Operations ■ Geometry

GETTING READY   *Do the operations.*

1. $2 \cdot 2$ **2.** $3 \cdot 3$ **3.** $3 \cdot 3 \cdot 3$ **4.** $2 \cdot 2 \cdot 2$

5. $\dfrac{1}{2} \cdot \dfrac{1}{2}$ **6.** $\dfrac{1}{3} \cdot \dfrac{1}{3} \cdot \dfrac{1}{3}$ **7.** $\dfrac{2}{5} \cdot \dfrac{2}{5} \cdot \dfrac{2}{5}$ **8.** $\dfrac{3}{10} \cdot \dfrac{3}{10} \cdot \dfrac{3}{10}$

To show how many times a number is to be used as a factor in a product, we use **exponents.** For example, 3 is the exponent in the expression 2^3. The exponent indicates that 2 is to be used as a factor three times:

$$
\overbrace{2^3 = 2 \cdot 2 \cdot 2}^{3 \text{ factors of } 2}
$$
$$
= 8
$$

 WARNING!   Note that $2^3 = 8$. This is not the same as $2 \cdot 3 = 6$.

The exponent 5 in the expression x^5 indicates that x is to be used as a factor five times.

$$
\overbrace{x^5 = x \cdot x \cdot x \cdot x \cdot x}^{5 \text{ factors of } x}
$$

In the expression x^5 (called an **exponential expression** or a **power of x**), 5 is the **exponent** and x is the **base.** In expressions such as x or y, the exponent is understood to be 1:

$$
x = x^1 \qquad \text{and} \qquad y = y^1
$$

In general, we have the following definition.

Natural-Number Exponent   If n is a natural number, then

$$
\overbrace{x^n = x \cdot x \cdot x \cdot \cdots \cdot x}^{n \text{ factors of } x}
$$

 (b,d)   **EXAMPLE 1**   **a.** $4^2 = 4 \cdot 4 = 16$ Read 4^2 as "4 squared" or as "4 to the second power."

b. $5^3 = 5 \cdot 5 \cdot 5 = 125$ Read 5^3 as "5 cubed" or as "5 to the third power."

c. $6^4 = 6 \cdot 6 \cdot 6 \cdot 6 = 1296$ Read 6^4 as "6 to the fourth power."

d. $\left(\dfrac{2}{3}\right)^5 = \dfrac{2}{3} \cdot \dfrac{2}{3} \cdot \dfrac{2}{3} \cdot \dfrac{2}{3} \cdot \dfrac{2}{3} = \dfrac{32}{243}$ Read $\left(\frac{2}{3}\right)^5$ as "$\frac{2}{3}$ to the fifth power." ∎

The base of an exponential expression can be a variable.

[oo] (b,d)   **EXAMPLE 2**

a. $y^6 = y \cdot y \cdot y \cdot y \cdot y \cdot y$ Read y^6 as "y to the sixth power."

b. $x^3 = x \cdot x \cdot x$ Read x^3 as "x cubed" or as "x to the third power."

c. $z^2 = z \cdot z$ Read z^2 as "z squared" or as "z to the second power."

d. $a^1 = a$ Read a^1 as "a to the first power."   ■

■ Order of Operations

The order in which we do arithmetic is important. For example, if we evaluate $2 + 3 \cdot 4$ by multiplying first, we get an answer of 14. However, if we add first, we get an answer of 20.

$$2 + 3 \cdot 4 = 2 + 12 \qquad 2 + 3 \cdot 4 = 5 \cdot 4$$
$$= 14 \qquad\qquad\quad = 20$$

To eliminate the possibility of getting different answers, we will agree to do multiplications before additions. The correct calculation of $2 + 3 \cdot 4$ is

$$2 + 3 \cdot 4 = 2 + 12$$
$$= 14$$

To indicate that additions should be done before multiplications, we use **grouping symbols** such as parentheses ( ), brackets [ ], or braces { }. In the expression $(2 + 3)4$, the parentheses indicate that the addition should be done first:

$$(2 + 3)4 = 5 \cdot 4$$
$$= 20$$

Unless grouping symbols indicate otherwise, exponential expressions are always evaluated before multiplications. The expression $5 + 4 \cdot 3^2$ should be evaluated in the following way:

$$5 + 4 \cdot 3^2 = 5 + 4 \cdot 9 \qquad \text{Evaluate the exponential expression first.}$$
$$= 5 + 36 \qquad\quad \text{Do the multiplication.}$$
$$= 41 \qquad\qquad \text{Do the addition.}$$

To guarantee that calculations will have a single correct result, we use the following set of **priority rules.**

Order of Mathematical Operations   Use the following steps to do all calculations within each pair of grouping symbols, working from the innermost pair to the outermost pair.

1. Find the values of any exponential expressions.

2. Do all multiplications and divisions as they are encountered while working from left to right.

3. Do all additions and subtractions as they are encountered while working from left to right.

When all grouping symbols have been removed, repeat the rules above to finish the calculation.

In a fraction, simplify the numerator and the denominator separately. Then simplify the fraction, whenever possible.

WARNING!   Note that $4(2)^3 \neq (4 \cdot 2)^3$:

$$4(2)^3 = 4 \cdot 2 \cdot 2 \cdot 2 = 4(8) = 32 \quad \text{and} \quad (4 \cdot 2)^3 = 8^3 = 8 \cdot 8 \cdot 8 = 512$$

Likewise, $4x^3 \neq (4x)^3$ because

$$4x^3 = 4xxx \quad \text{and} \quad (4x)^3 = (4x)(4x)(4x) = 64x^3$$

    EXAMPLE 3   Evaluate $5^3 + 2(8 - 3 \cdot 2)$.

Solution   We do the work within the parentheses first and then simplify.

$5^3 + 2(8 - 3 \cdot 2) = 5^3 + 2(8 - 6)$	Do the multiplication within the parentheses.
$= 5^3 + 2(2)$	Do the subtraction within the parentheses.
$= 125 + 2(2)$	Find the value of the exponential expression.
$= 125 + 4$	Do the multiplication.
$= 129$	Do the addition.

EXAMPLE 4   Evaluate $\dfrac{3(3 + 2) + 5}{17 - 3(4)}$.

Solution   We simplify the numerator and the denominator of the fraction separately and then simplify the fraction.

$$\frac{3(3+2)+5}{17-3(4)} = \frac{3(5)+5}{17-3(4)} \qquad \text{Do the addition within the parentheses.}$$

$$= \frac{15+5}{17-12} \qquad \text{Do the multiplications.}$$

$$= \frac{20}{5} \qquad \text{Do the addition and the subtraction.}$$

$$= 4 \qquad \text{Do the division.} \qquad \blacksquare$$

EXAMPLE 5   If $x = 3$ and $y = 4$, evaluate **a.** $3y + x^2$ and **b.** $3(y + x^2)$.

Solution   **a.** $3y + x^2 = 3(4) + 3^2$ Substitute 3 for x and 4 for y.

$$= 3(4) + 9 \qquad \text{Evaluate the exponential expression.}$$

$$= 12 + 9 \qquad \text{Do the multiplication.}$$

$$= 21 \qquad \text{Do the addition.}$$

b. $3(y + x^2) = 3(4 + 3^2)$ Substitute 3 for x and 4 for y.

$$= 3(4 + 9) \qquad \text{Evaluate the exponential expression.}$$

$$= 3(13) \qquad \text{Parentheses indicate that the addition is performed next.}$$

$$= 39 \qquad \text{Do the multiplication.} \qquad \blacksquare$$

EXAMPLE 6   If $x = 4$ and $y = 3$, evaluate $\dfrac{3x^2 - 2y}{2(x + y)}$.

Solution $\dfrac{3x^2 - 2y}{2(x + y)} = \dfrac{3(4^2) - 2(3)}{2(4 + 3)}$ Substitute 4 for x and 3 for y.

$$= \frac{3(16) - 2(3)}{2(7)} \qquad \text{Find the value of } 4^2 \text{ in the numerator and do the addition in the denominator.}$$

$$= \frac{48 - 6}{14} \qquad \text{Do the multiplications.}$$

$$= \frac{42}{14} \qquad \text{Do the subtraction.}$$

$$= 3 \qquad \text{Do the division.} \qquad \blacksquare$$

■ Geometry

Substituting numbers for variables is often required when finding perimeters and areas of geometric figures. The **perimeter** of a geometric figure is the distance around it, and the **area** of a figure is the amount of surface that it encloses. The perimeter of a circle is called its **circumference.**

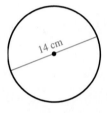 (b)

EXAMPLE 7 **Circles** Find **a.** the circumference and **b.** the area of a circle with a diameter of 14 centimeters.

Solution **a.** The formula for the circumference of a circle is

$$C = \pi D$$

where C is the circumference, π is approximately $\frac{22}{7}$, and D is the diameter—the distance through the center of the circle. We can approximate the circumference of the circle shown in Figure 1-15 by substituting $\frac{22}{7}$ for π and 14 for D in the formula for circumference and simplifying.

$$C = \pi D$$

$$C \approx \frac{22}{7} \cdot \mathbf{14} \qquad \text{Read} \approx \text{as "is approximately equal to."}$$

$$C \approx \frac{22 \cdot \overset{2}{14}}{\underset{1}{7 \cdot 1}} \qquad \text{Multiply the fractions and simplify.}$$

$$C \approx 44$$

The circumference is approximately 44 centimeters. To use a calculator, we would press these keys:

$$\boxed{\pi} \quad \boxed{\times} \quad \boxed{14} \quad \boxed{=}$$

The display will read 43.98229. . . . This answer is not 44 because a calculator uses a better approximation of π than $\frac{22}{7}$.

b. The formula for the area of a circle is

$$A = \pi r^2$$

where A is the area, $\pi \approx \frac{22}{7}$, and r is the **radius** of the circle. (The radius is one-half of the diameter.) We can approximate the area by substituting $\frac{22}{7}$ for π and 7 for r in the formula for area and simplifying.

$$A = \pi r^2$$

$$A \approx \frac{22}{7} \cdot 7^2$$

$$A \approx \frac{22}{7} \cdot \frac{49}{1} \qquad \text{Evaluate the exponential expression.}$$

$$A \approx \frac{22 \cdot \overset{7}{49}}{\underset{1}{7 \cdot 1}} \qquad \text{Multiply the fractions and simplify.}$$

$$A \approx 154$$

14 cm

FIGURE 1-15

The area is approximately 154 square centimeters. To use a calculator, we would press these keys:

$$\boxed{\pi} \quad \boxed{\times} \quad \boxed{7} \quad \boxed{x^2} \quad \boxed{=}$$

The display will read 153.93804 . . . .   ■

Table 1-1 shows the formulas for the perimeter and area of several geometric figures.

Figure	Name	Perimeter	Area
	Square	$P = 4s$	$A = s^2$
	Rectangle	$P = 2l + 2w$	$A = lw$
	Triangle	$P = a + b + c$	$A = \dfrac{1}{2}bh$
	Trapezoid	$P = a + b + c + d$	$A = \dfrac{1}{2}h(b + d)$
	Circle	$C = 2\pi r$	$A = \pi r^2$

TABLE 1-1

The **volume** of a three-dimensional geometric solid is the amount of space it encloses. Table 1-2 shows the formulas for the volume of several solids.

Figure	Name	Volume
	Rectangular solid	$V = lwh$
	Cylinder	$V = Bh$ where B is the area of the base
	Pyramid	$V = \frac{1}{3}Bh$ where B is the area of the base
	Cone	$V = \frac{1}{3}Bh$ where B is the area of the base (If the base is a circle, then $B = \pi r^2$)
	Sphere	$V = \frac{4}{3}\pi r^3$

TABLE 1-2

EXAMPLE 8   **Winter driving**   Find the number of cubic feet of road salt in a conical pile that is 18.75 feet high and covers a circular area 28.60 feet in diameter. Use a calculator and round the answer to two places.

Solution   We can find the area of the circular base of the cone shown in Figure 1-16 by substituting $\frac{1}{2}(28.60)$, or 14.30, for the radius in the formula for the area of a circle.

$$A = \pi r^2$$
$$\approx \pi (14.3)^2$$
$$\approx 642.4242817 \qquad \text{Use a calculator.}$$

We then substitute 642.4242817 for B and 18.75 for h in the formula for the volume of a cone.

$$V = \frac{1}{3}Bh$$

$$\approx \frac{1}{3}(642.4242817)(18.75)$$

$$\approx 4015.151761 \qquad \text{Use a calculator.}$$

To two decimal places, there are 4015.15 cubic feet of salt in the pile.   ■

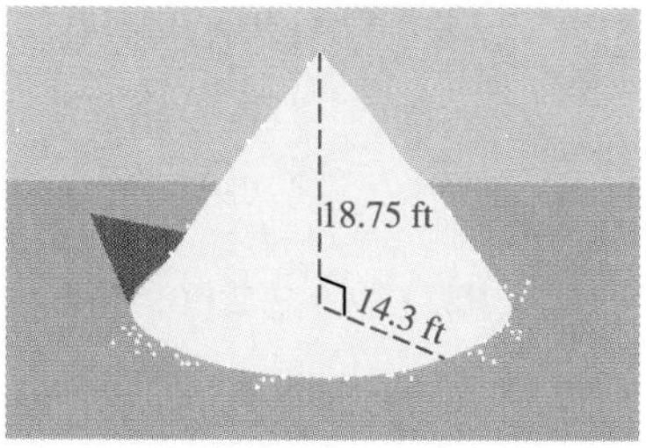

FIGURE 1-16

ORALS   *Find the value of each expression.*

1. 2^5 **2.** 3^4 **3.** 4^3 **4.** 5^3

Simplify each expression.

5. $3(2)^3$ **6.** $(3 \cdot 2)^2$

7. $3 + 2 \cdot 4$ **8.** $10 - 3^2$

9. $4 + 2^2 \cdot 3$ **10.** $2 \cdot 3 + 2 \cdot 3^2$

EXERCISE 1.3

In Exercises 1–6, find the value of each expression.

1. 4^2 **2.** 5^2 **3.** 6^2

4. 7^3 **5.** $\left(\dfrac{1}{10}\right)^4$ **6.** $\left(\dfrac{1}{2}\right)^6$

In Exercises 7–14, write each expression as the product of several factors.

7. x^2 **8.** y^3 **9.** $3z^4$ **10.** $5t^2$

11. $(5t)^2$ **12.** $(3z)^4$ **13.** $5(2x)^3$ **14.** $7(3t)^2$

In Exercises 15–22, find the value of each expression if $x = 3$ and $y = 2$.

15. $4x^2$ **16.** $4y^3$ **17.** $(5y)^3$ **18.** $(2y)^4$

19. $2x^y$ **20.** $3y^x$ **21.** $(3y)^x$ **22.** $(2x)^y$

In Exercises 23–56, simplify each expression by doing the indicated operations.

23. $3 \cdot 5 - 4$

24. $4 \cdot 6 + 5$

25. $3(5 - 4)$

26. $4(6 + 5)$

27. $3 + 5^2$

28. $4^2 - 2^2$

29. $(3 + 5)^2$

30. $(5 - 2)^3$

31. $2 + 3 \cdot 5 - 4$

32. $12 + 2 \cdot 3 + 2$

33. $64 \div (3 + 1)$

34. $16 \div (5 + 3)$

35. $(7 + 9) \div (2 \cdot 4)$

36. $(7 + 9) \div 2 \cdot 4$

37. $(5 + 7) \div 3 \cdot 4$

38. $(5 + 7) \div (3 \cdot 4)$

39. $24 \div 4 \cdot 3 + 3$

40. $36 \div 9 \cdot 4 - 2$

41. $49 \div 7 \cdot 7 + 7$

42. $100 \div 10 \cdot 10 + 10$

43. $100 \div 10 \cdot 10 \div 100$

44. $100 \div (10 \cdot 10) \div 100$

45. $(100 \div 10) \cdot (10 \div 100)$

46. $100 \div [10 \cdot (10 \div 100)]$

47. $3^2 + 2(1 + 4) - 2$

48. $4 \cdot 3 + 2(5 - 2) - 2^3$

49. $5^2 - (7 - 3)^2$

50. $3^3 + (3 - 1)^3$

51. $(2 \cdot 3 - 4)^3$

52. $(3 \cdot 5 - 2 \cdot 6)^2$

53. $\dfrac{3}{5} \cdot \dfrac{10}{3} + \dfrac{1}{2} \cdot 12$

54. $\dfrac{15}{4}\left(1 + \dfrac{3}{5}\right)$

55. $\left[\dfrac{1}{3} - \left(\dfrac{1}{2}\right)^2\right]^2$

56. $\left[\left(\dfrac{2}{3}\right)^2 - \dfrac{1}{3}\right]^2$

In Exercises 57–64, use a calculator to simplify each fraction.

57. $\dfrac{(3 + 5)^2 + 2}{2(8 - 5)}$

58. $\dfrac{25 - (2 \cdot 3 - 1)}{2 \cdot 9 - 8}$

59. $\dfrac{(5 - 3)^2 + 2}{4^2 - (8 + 2)}$

60. $\dfrac{(4^2 - 2) + 7}{5(2 + 4) - 3^2}$

61. $\dfrac{2[4 + 2(3 - 1)]}{3[3(2 \cdot 3 - 4)]}$

62. $\dfrac{3[9 - 2(7 - 3)]}{(8 - 5)(9 - 7)}$

63. $\dfrac{3 \cdot 7 - 5(3 \cdot 4 - 11)}{4(3 + 2) - 3^2 + 5}$

64. $\dfrac{2 \cdot 5^2 - 2^2 + 3}{2(5 - 2)^2 - 11}$

In Exercises 65–92, evaluate each expression given that $x = 3$, $y = 2$, and $z = 4$.

65. $2x - y$

66. $2z + y$

67. $10 - 2x$

68. $15 - 3z$

69. $5z \div 2 + y$

70. $5x \div 3 + y$

71. $4x - 2z$

72. $5y - 3x$

73. $x + yz$

74. $3z + x - 2y$

75. $3(2x + y)$

76. $4(x + 3y)$

77. $(3 + x)y$

78. $(4 + z)y$

79. $(z + 1)(x + y)$

80. $3(z + 1) \div x$

81. $(x + y) \div (z + 1)$

82. $(2x + 2y) \div (3z - 2)$

83. $xyz + z^2 - 4x$

84. $zx + y^2 - 2z$

85. $3x^2 + 2y^2$

86. $3x^2 + (2y)^2$

87. $\dfrac{2x + y^2}{y + 2z}$

88. $\dfrac{2z^2 - y}{2x - y^2}$

89. $\dfrac{2x^3 - (xy - 2)}{2(3y + 5z) - 27}$

90. $\dfrac{2z^2 + xy}{x^3 - y^2(5 - x)}$

91. $\dfrac{x^2[14 - y(x + 2)]}{3[xy - z(5y - 9)]}$

92. $\dfrac{5x + x(y + 1)}{x[yz - y(z - 2)]}$

In Exercises 93–96, insert parentheses in the expression $3 \cdot 8 + 5 \cdot 3$ to make its value equal to the given number.

93. 39

94. 117

95. 87

96. 69

In Exercises 97–100, insert parentheses in the expression $4 + 3 \cdot 5 - 3$ to make its value equal to the given number.

97. 14

98. 10

99. 32

100. 16

In Exercises 101–104, find the perimeter of each figure.

101.
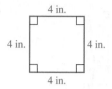
4 in.

4 in.   4 in.

4 in.

102.
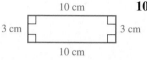
10 cm

3 cm   3 cm

10 cm

103.
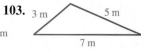
3 m   5 m

7 m

104.
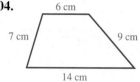
6 cm

7 cm   9 cm

14 cm

In Exercises 105–108, find the area of each figure.

105.

5 m

5m

106.
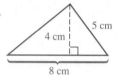
5 cm

4 cm

8 cm

107.

6 ft

10 ft

108.
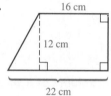
16 cm

12 cm

22 cm

In Exercises 109–110, find the circumference of each circle. Use $\pi \approx \frac{22}{7}$.

109.

14 m

110.

21 cm

In Exercises 111–112, find the area of each circle. Use $\pi \approx \frac{22}{7}$.

111.

42 ft

112.

7 m

In Exercises 113–118, find the volume of each solid. Use $\pi \approx \frac{22}{7}$.

113.
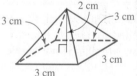
2 cm
3 cm   3 cm

3 cm
3 cm

114.
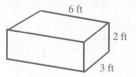
6 ft

2 ft

3 ft

115.

6 m

116.

14 in.

12 in.

117.

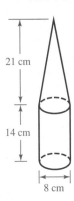

118.

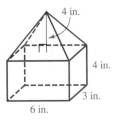

In Exercises 119–124, use a calculator. For π, use the $\boxed{\pi}$ key. Round each answer to two decimal places.

119. Volume of a tank   Find the number of cubic feet of water in a spherical tank with a radius of 21.35 feet.

120. Storing solvents   A hazardous solvent fills a rectangular tank with dimensions of 12 by 9.5 by 7.3 inches. For disposal, it must be transferred to a cylindrical canister 7.5 inches in diameter and 18 inches high. Find how much solvent will be left over.

121. Volume of a classroom   Thirty students are in a classroom with dimensions of 40 feet by 40 feet by 9 feet. How many cubic feet of air are there for each student?

122. Wallpapering   One roll of wallpaper covers about 33 square feet. At $27.50 per roll, how much would it cost to paper two walls 8.5 feet high and 17.3 feet long? (*Hint:* Wallpaper can only be purchased in full rolls.)

123. Focal length   The focal length, f, of a double-convex thin lens is given by the formula

$$f = \frac{rs}{(r + s)(n - 1)}$$

If $r = 8$, $s = 12$, and $n = 1.6$, find f.

124. Resistance   The total resistance, R, of two resistors in parallel is given by the formula

$$R = \frac{rs}{r + s}$$

If $r = 170$ and $s = 255$, find R.

Writing Exercises   ■   *Write a paragraph using your own words.*

1. The symbols $3x$ and x^3 have different meanings. Explain.

2. Students often say that x^n means "x multiplied by itself n times." Explain why this is not correct.

Something to Think About   ■

1. If x were greater than 1, would raising x to higher and higher powers produce bigger numbers or smaller?

2. What would happen in Question 1 if x were a positive number less than 1?

Review Exercises ■

1. On the number line, graph the prime numbers between 10 and 20.

2. Write the inequality $7 \leq 12$ as an inequality using the symbol $\geq$.

3. Classify the number 17 as a prime number or a composite number.

4. Evaluate: $\dfrac{3}{5} - \dfrac{1}{2}$.

1.4 Adding and Subtracting Real Numbers

■ Adding Real Numbers ■ Subtracting Real Numbers

GETTING READY   *Do each operation.*

1. $14.32 + 3.2$ 2. $5.54 - 2.6$ 3. $4.2 - (3.8 - 3)$ 4. $(5.42 - 4.22) - 0.2$

5. $(437 - 198) - 143$ 6. $437 - (198 - 143)$

■ Adding Real Numbers

Since the positive direction on the number line is to the right, positive numbers can be represented by arrows pointing to the right. Negative numbers can be represented by arrows pointing to the left.

To add the integers $+2$ and $+3$, we represent $+2$ with an arrow of length 2, pointing to the right, and we represent $+3$ with an arrow of length 3, also pointing to the right. We then place the arrows end to end as in Figure 1-17. Since the endpoint of the second arrow is the point with coordinate $+5$, we have

$(+2) + (+3) = +5$

The addition problem

$(-2) + (-3)$

can be represented by the arrows shown in Figure 1-18. Since the endpoint of the final arrow is the point -5, we have

$(-2) + (-3) = -5$

Because two real numbers with the same sign can be represented by arrows pointing in the same direction, we have the following rule.

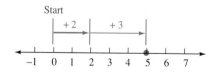

FIGURE 1-17

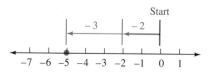

FIGURE 1-18

Adding Real Numbers with Like Signs	To find the sum of two real numbers with the same sign, we add their absolute values and keep their common sign.

EXAMPLE 1

a. $(+4) + (+6) = +(4 + 6)$
$= 10$

b. $(-4) + (-6) = -(4 + 6)$
$= -10$

c. $+5 + (+10) = +(5 + 10)$
$= 15$

d. $-\dfrac{1}{2} + \left(-\dfrac{3}{2}\right) = -\left(\dfrac{1}{2} + \dfrac{3}{2}\right)$
$= -\dfrac{4}{2}$
$= -2$ ∎

Real numbers with unlike signs can be represented by arrows on a number line that point in opposite directions. For example, the addition problem

$$(-6) + (+2)$$

can be represented on a number line as shown in Figure 1-19. Since the endpoint of the final arrow is the point with coordinate -4, we have

$$(-6) + (+2) = -4$$

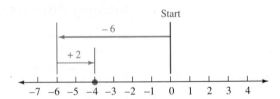

FIGURE 1-19

The addition problem

$$(+7) + (-4)$$

can be represented on a number line as in Figure 1-20. Since the endpoint of the final arrow is the point with coordinate $+3$, we have

$$(+7) + (-4) = +3$$

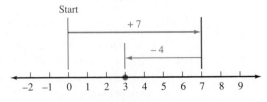

FIGURE 1-20

Because two real numbers with unlike signs can be represented by arrows pointing in opposite directions, we have the following rule.

| **Adding Real Numbers with Unlike Signs** | To find the sum of two real numbers with unlike signs, we subtract their absolute values (the smaller from the larger) and use the sign of the number with the greater absolute value. |

EXAMPLE 2

a. $(+6) + (-5) = +(6 - 5)$
$= 1$

b. $(-2) + (+3) = +(3 - 2)$
$= 1$

c. $+6 + (-9) = -(9 - 6)$
$= -3$

d. $-\dfrac{2}{3} + \left(+\dfrac{1}{2}\right) = -\left(\dfrac{2}{3} - \dfrac{1}{2}\right)$
$= -\left(\dfrac{4}{6} - \dfrac{3}{6}\right)$
$= -\dfrac{1}{6}$ ∎

EXAMPLE 3

a. $[(+3) + (-7)] + (-4) = [-4] + (-4)$ Do the work within the brackets first.
$= -8$

b. $-3 + [(-2) + (-8)] = -3 + [-10]$ Do the work within the brackets first.
$= -13$ ∎

EXAMPLE 4   If $x = -4$, $y = 5$, and $z = -13$, evaluate   **a.** $x + y$ and   **b.** $2y + z$.

Solution   We substitute -4 for x, 5 for y, and -13 for z and simplify.

a. $x + y = (-4) + (5)$
$= 1$

b. $2y + z = 2 \cdot 5 + (-13)$
$= 10 + (-13)$
$= -3$ ∎

Sometimes numbers are added vertically, as shown in the next example.

EXAMPLE 5

a. $\begin{array}{r} +5 \\ +2 \\ \hline +7 \end{array}$ **b.** $\begin{array}{r} +5 \\ -2 \\ \hline +3 \end{array}$ **c.** $\begin{array}{r} -5 \\ +2 \\ \hline -3 \end{array}$ **d.** $\begin{array}{r} -5 \\ -2 \\ \hline -7 \end{array}$ ∎

Words and phrases such as *found, gain, credit, up, increase, forward, rises, in the future,* and *to the right* indicate a positive direction. Words and phrases such as *lost, loss, debit, down, decrease, backward, falls, in the past,* and *to the left* indicate a negative direction.

EXAMPLE 6 **Account balance** The treasurer of a math club opens a checking account by depositing $350 in the bank. The bank debits the account $9 for check printing, and the treasurer writes a check for $22. Find the account balance after these transactions.

Solution The deposit can be represented by $+350$. The debit of $9 can be represented by -9, and the check written for $22 can be represented by -22. The balance in the account after these transactions is the sum of 350, -9, and -22.

$$350 + (-9) + (-22) = 341 + (-22)$$
$$= 319$$

The account balance is $319. ∎

■ Subtracting Real Numbers

In arithmetic, subtraction is a take-away process. For example,

$$7 - 4 = 3$$

can be thought of as taking 4 objects away from 7 objects, leaving 3 objects.
For algebra, a better approach treats the subtraction problem

$$7 - 4$$

as the equivalent addition problem:

$$7 + (-4)$$

In either case, the answer is 3.

$$7 - 4 = 3 \quad \text{and} \quad 7 + (-4) = 3$$

Thus, to subtract 4 from 7, we can add the negative (or opposite) of 4 to 7. In general, we have the following rule.

Subtracting Real Numbers If a and b are two real numbers, then
$$a - b = a + (-b)$$

EXAMPLE 7 Evaluate **a.** $12 - 4$, **b.** $-13 - 5$, and **c.** $-14 - (-6)$.

Solution **a.** $12 - 4 = 12 + (-4)$ **b.** $-13 - 5 = -13 + (-5)$
             $= 8$ $= -18$

     c. $-14 - (-6) = -14 + [-(-6)]$
                     $= -14 + 6$ Use the double negative rule.
                     $= -8$ ∎

(b)

EXAMPLE 8 If $x = -5$ and $y = -3$, evaluate **a.** $\dfrac{y - x}{7 + x}$ and **b.** $\dfrac{6 + x}{y - x} - \dfrac{y - 4}{7 + x}$.

Solution We can substitute -5 for x and -3 for y into each expression and simplify.

a. $\dfrac{y - x}{7 + x} = \dfrac{-3 - (-5)}{7 + (-5)}$

$= \dfrac{-3 + [-(-5)]}{2}$ Use the rule for subtracting two numbers.

$= \dfrac{-3 + 5}{2}$ $-(-5) = +5$

$= \dfrac{2}{2}$

$= 1$

b. $\dfrac{6 + x}{y - x} - \dfrac{y - 4}{7 + x} = \dfrac{6 + (-5)}{-3 - (-5)} - \dfrac{-3 - 4}{7 + (-5)}$

$= \dfrac{1}{-3 + 5} - \dfrac{-3 + (-4)}{2}$ $-(-5) = +5$

$= \dfrac{1}{2} - \dfrac{-7}{2}$

$= \dfrac{1 - (-7)}{2}$

$= \dfrac{1 + [-(-7)]}{2}$ Use the rule for subtracting two numbers.

$= \dfrac{1 + 7}{2}$ $-(-7) = +7$

$= \dfrac{8}{2}$

$= 4$ ■

To use a vertical format for subtracting real numbers, we add the opposite of the number that is to be subtracted by changing the sign of the lower number (called the **subtrahend**) and proceeding as in addition.

EXAMPLE 9 Do each subtraction by doing an equivalent addition.

a. The subtraction $\begin{array}{r} 5 \\ - \underline{-4} \end{array}$ becomes the addition $\begin{array}{r} 5 \\ + \underline{+4} \\ 9 \end{array}$

b. The subtraction $\begin{array}{r} -8 \\ - \underline{+3} \end{array}$ becomes the addition $\begin{array}{r} -8 \\ + \underline{-3} \\ -11 \end{array}$ ■

EXAMPLE 10 Simplify   **a.** $3 - [4 + (-6)]$ and   **b.** $[-5 + (-3)] - [-2 - (+5)]$.

Solution   **a.** $3 - [4 + (-6)] = 3 - (-2)$ Do the addition within the brackets first.

$\qquad\qquad\qquad\quad = 3 + [-(-2)]$ Use the rule for subtracting two numbers.

$\qquad\qquad\qquad\quad = 3 + 2$ $-(-2) = 2$

$\qquad\qquad\qquad\quad = 5$

b. $[-5 + (-3)] - [-2 - (+5)]$

$\qquad = [-5 + (-3)] - [-2 + (-5)]$ Use the rule for subtracting two numbers.

$\qquad = -8 - (-7)$ Do the work within the brackets.

$\qquad = -8 + [-(-7)]$ Use the rule for subtracting two numbers.

$\qquad = -8 + 7$ $-(-7) = 7$

$\qquad = -1$ ∎

EXAMPLE 11 Simplify $[-6 - (-4)] - [-7 + 10]$.

Solution $[-6 - (-4)] - [-7 + 10] = [-6 + 4] - [-7 + 10]$

$\qquad\qquad\qquad\qquad\qquad = -2 - 3$

$\qquad\qquad\qquad\qquad\qquad = -2 + (-3)$

$\qquad\qquad\qquad\qquad\qquad = -5$ ∎

EXAMPLE 12 **Temperature change**   At noon the temperature was 7 degrees above zero. At midnight the temperature was 4 degrees below zero. Find the difference between these two temperatures.

Solution   A temperature of 7 degrees above zero can be represented as $+7$. A temperature of 4 degrees below zero can be represented as -4. To find the difference between these two temperatures, we set up a subtraction problem and simplify.

$\qquad 7 - (-4) = 7 + [-(-4)]$

$\qquad\qquad\quad = 7 + 4$ $-(-4) = 4$

$\qquad\qquad\quad = 11$

The difference between the temperatures is 11 degrees. Figure 1-21 shows this difference.   ∎

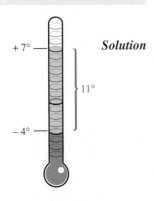

$+7°$

$11°$

$-4°$

FIGURE 1-21

ORALS   *Find each value.*

1. $2 + 3$ **2.** $2 + (-5)$ **3.** $-4 + 7$

4. $-5 + (-6)$ **5.** $6 - 2$ **6.** $-8 - 4$

7. $-5 - (-7)$ **8.** $12 - (-4)$ **9.** $-5 + (3 - 4)$

10. $(-5 + 3) - 4$

EXERCISE 1.4

In Exercises 1–20, find each sum.

1. $4 + 8$

2. $(-4) + (-2)$

3. $(-3) + (-7)$

4. $(+4) + 11$

5. $6 + (-4)$

6. $5 + (-3)$

7. $9 + (-11)$

8. $10 + (-13)$

9. $(-5) + (-7)$

10. $(-6) + (-4)$

11. $(-0.4) + 0.9$

12. $(-1.2) + (-5.3)$

13. $\dfrac{1}{5} + \left(+\dfrac{1}{7}\right)$

14. $\dfrac{2}{3} + \left(-\dfrac{1}{4}\right)$

15. $\left(-\dfrac{3}{4}\right) + \left(+\dfrac{2}{3}\right)$

16. $\dfrac{3}{5} + \left(-\dfrac{2}{3}\right)$

17. $\begin{array}{r} 5 \\ +-4 \\ \hline \end{array}$

18. $\begin{array}{r} -20 \\ +-17 \\ \hline \end{array}$

19. $\begin{array}{r} -1.3 \\ +3.5 \\ \hline \end{array}$

20. $\begin{array}{r} 1.3 \\ +-2.5 \\ \hline \end{array}$

In Exercises 21–34, evaluate each expression.

21. $5 + [4 + (-2)]$

22. $-6 + [(-3) + 8]$

23. $-2 + (-4 + 5)$

24. $5 + [-4 + (-6)]$

25. $[-4 + (-3)] + [2 + (-2)]$

26. $[3 + (-1)] + [-2 + (-3)]$

27. $-4 + (-3 + 2) + (-3)$

28. $5 + [2 + (-5)] + (-2)$

29. $-|-9 + (-3)| + (-6)$

30. $-|8 + (-4)| + 7$

31. $\left|\dfrac{3}{5} + \left(-\dfrac{4}{5}\right)\right|$

32. $\left|\dfrac{1}{6} + \left(-\dfrac{5}{6}\right)\right|$

33. $-5.2 + |-2.5 + (-4)|$

34. $6.8 + |8.6 + (-1.1)|$

In Exercises 35–50, let $x = 2$, $y = -3$, $z = -4$, and $u = 5$. Evaluate each expression.

35. $x + y$

36. $x + z$

37. $x + z + u$

38. $y + z + u$

39. $(x + u) + 3$

40. $(y + 5) + x$

41. $x + (-1 + z)$

42. $-7 + (z + x)$

43. $(x + z) + (u + z)$

44. $(z + u) + (x + y)$

45. $x + [5 + (y + u)]$

46. $y + \{[u + (z + (-6)]\} + y$

47. $|2x + y|$

48. $3|x + y + z|$

49. $|x + z| + |x + y + z|$

50. $|z + z| + |y + y|$

In Exercises 51–70, find each difference.

51. $8 - 4$

52. $-8 - 4$

53. $8 - (-4)$

54. $-9 - (-5)$

55. $-12 - 5$

56. $11 - (+4)$

57. $0 - (-5)$

58. $0 - 75$

59. $\dfrac{5}{3} - \dfrac{7}{6}$

60. $-\dfrac{5}{9} - \dfrac{5}{3}$

61. $-5 - \left(-\dfrac{3}{5}\right)$

62. $\dfrac{7}{8} - (-3)$

63. $-3\dfrac{1}{2} - 5\dfrac{1}{4}$

64. $2\dfrac{1}{2} - \left(-3\dfrac{1}{2}\right)$

65. $-6.7 - (-2.5)$

66. $25.3 - 17.5$

67. $\begin{array}{r} 8 \\ -4 \\ \hline \end{array}$

68. $\begin{array}{r} 8 \\ --3 \\ \hline \end{array}$

69. $\begin{array}{r} -10 \\ --3 \\ \hline \end{array}$

70. $\begin{array}{r} -13 \\ -5 \\ \hline \end{array}$

In Exercises 71–84, evaluate each quantity.

71. $+3 - [(-4) - 3]$ **72.** $-5 - [4 - (-2)]$ **73.** $(5 - 3) + (3 - 5)$ **74.** $(3 - 5) - [5 - (-3)]$

75. $5 - [4 + (-2) - 5]$ **76.** $3 - [-(-2) + 5]$

77. $[5 - (-34)] - [-2 + (-23)]$ **78.** $-5 + \{-3 - [-2 - (+4)]\}$

79. $\left(\dfrac{5}{2} - 3\right) - \left(\dfrac{3}{2} - 5\right)$ **80.** $\left(\dfrac{7}{3} - \dfrac{5}{6}\right) - \left[\dfrac{5}{6} - \left(-\dfrac{7}{3}\right)\right]$

81. $(5.2 - 2.5) - (5.25 - 5)$ **82.** $\left(3\dfrac{1}{2} - 2\dfrac{1}{2}\right) - \left[5\dfrac{1}{3} - \left(-5\dfrac{2}{3}\right)\right]$

83. $-|-9 - (-7)| - (-3)$ **84.** $-|8 - (-4)| - 7$

In Exercises 85–96, let $x = -4$, $y = 5$, $z = -6$. Evaluate each quantity.

85. $y - x$ **86.** $y - z$ **87.** $x - y - z$ **88.** $y + z - x$

89. $x - (y - z)$ **90.** $y + (z - x)$ **91.** $3 - [x + (-3)]$ **92.** $z - (x - y) + 10$

93. $\dfrac{y - x}{3 - z}$ **94.** $\dfrac{y - z}{3y + x}$ **95.** $\dfrac{x - y}{y} - \dfrac{z}{y}$ **96.** $\dfrac{y}{x - z} - \dfrac{x}{8 + z}$

In Exercises 97–104, let $a = 2$, $b = -3$, and $c = -4$. Evaluate each quantity.

97. $a + b - c$ **98.** $a - b + c$ **99.** $b - (c + a)$ **100.** $c + (a - b)$

101. $\dfrac{a + b}{b - c}$ **102.** $\dfrac{c - a}{-(a + b)}$ **103.** $\dfrac{|b + c|}{a - c}$ **104.** $\dfrac{a - b - c}{|a + b|}$

In Exercises 105–108, use a calculator to evaluate each quantity. Let $x = -2.34$, $y = 3.47$, and $z = 0.72$. Round the answers to one decimal place.

105. $x^3 - y + z^2$ **106.** $y - z^2 - x^2$ **107.** $x^2 - y^2 - z^2$ **108.** $z^3 - x^2 + y^3$

In Exercises 109–126, use signed numbers to solve each problem.

109. College tuition   A student owed \$575 in tuition. If she earned a scholarship that would pay \$400 of the bill, what did she still owe?

110. Dieting   Scott weighed 212 pounds, but lost 24 pounds during a diet. What does Scott weigh now?

111. Temperature   The temperature rose 13 degrees in 1 hour, and then dropped 4 degrees in the next hour. What signed number represents the net change in temperature?

112. Mountain climbing   A climbing team climbed 2347 feet one day but then came down 597 feet to a good spot to make camp. What signed number represents their net change in altitude?

113. Temperature   The temperature fell from zero to 14° below one night. By 5:00 P.M. the next day, the temperature had risen 10 degrees. What was the temperature at 5:00 P.M.?

114. History   In 1897 Joseph Thompson discovered the electron. Fifty-four years later, the first fission reactor was built. Nineteen years before the reactor, James Chadwick discovered the neutron. In what year was the neutron discovered?

115. History   The Greek mathematician Euclid was alive in 300 B.C. The English mathematician Sir Isaac Newton was alive in A.D. 1700. How many years apart did they live?

116. Banking   Abdul deposited $212 in a new checking account, wrote a check for $173, and deposited another $312. Find the balance in Abdul's account.

117. Military science   An army retreated 2300 meters. After regrouping, it moved forward 1750 meters. The next day it gained another 1875 meters. What was the army's net gain?

118. Football   A football player gained and lost the following yardage on six consecutive plays: $+5$, $+7$, -5, $+1$, -2, and -6. How many yards were gained or lost?

119. Aviation   A pilot flying at 32,000 feet is instructed to descend to 28,000 feet. How many feet must he descend?

120. Stock market   Tuesday's high and low prices for Transitronics stock were $37\frac{1}{8}$ and $31\frac{5}{8}$. Find the range of prices for this stock.

121. Temperature   Find the difference between a temperature of 32° above zero and a temperature of 27° above zero.

122. Temperature   Find the difference between a temperature of 3° below zero and a temperature of 21° below zero.

123. Stock market   At the opening bell on Monday, the Dow Jones Industrial Average was 3153. At the close, the Dow was down 23 points, but news of a half-point drop in interest rates on Tuesday sent the market up 57 points. What was the Dow average after the market closed on Tuesday?

124. Stock market   On a Monday morning, the Dow Jones Industrial Average opened at 2917. For the week, the Dow rose 29 points on Monday and 12 points on Wednesday. However, it fell 53 points on Tuesday and 27 points on both Thursday and Friday. Where did the Dow close on Friday?

125. Stock splits   A man owned 500 shares of Transitronics Corporation before the company declared a two-for-one stock split. After the split, he sold 300 shares. How many shares does the man now own?

126. Small business   Maria earned $2532 in a part-time business. However, $633 of the earnings went for taxes. Find Maria's net earnings.

In Exercises 127–130, use a calculator.

127. Balancing the books   On January 1, Sally had $437.45 in the bank. During the month, she had deposits of $25.17, $37.93, and $45.26, and she had withdrawals of $17.13, $83.44, and $22.58. How much was in her account at the end of the month?

128. Small business   The owner of a small business has a gross income of $97,345.32. However, he paid $37,675.66 in expenses plus $7537.45 in taxes, $3723.41 in health care premiums, and $5767.99 in pension payments. Find his profit.

129. Closing a real estate transaction   A woman sold her house for $115,000. Her fees at closing were $78 for preparing a deed, $446 for title work, $216 for revenue stamps, and a sales commission of $7612.32. In addition, there was a deduction of $23,445.11 to pay off her old mortage. As part of the deal, the buyer agreed to pay half of the title work. How much money did the woman receive after closing?

130. Winning the lottery   Mike won $500,000 in a state lottery. He will get $\frac{1}{20}$ of the sum each year for the next 20 years. After he receives his first installment, he plans to pay off a car loan of $7645.12 and give his son $10,000 for college. By paying off the car loan, he will receive a rebate of 2% of the loan. If he must pay income tax of 28% on his first installment, how much will he have left to spend?

Writing Exercises ■ *Write a paragraph using your own words.*

1. Explain why the sum of two negative numbers is always negative, and the sum of two positive numbers is always positive.

2. Explain why the sum of a negative number and a positive number could be either negative or positive.

Something to Think About ■

1. Think of two numbers. First, add the absolute values of the two numbers, and write your answer. Second, add the two numbers, take the absolute value of that sum, and write that answer. Do your two answers agree? Can you find two numbers that produce different answers? When do you get answers that agree, and when don't you?

2. "Think of a very small number," requests the teacher. "One one-millionth," answers Charles. "Negative one million," responds Mia. Explain why either answer might be considered correct.

Review Exercises ■ *If $x = 5$, $y = 7$, and $z = 2$, evaluate each expression.*

1. $x + 3(y - z)$ **2.** $(x + 3)(y - z)$ **3.** $x + 3y - z$ **4.** $(x + 3)y - z$

1.5 Multiplying and Dividing Real Numbers

■ Multiplying Real Numbers ■ Dividing Real Numbers

GETTING READY *Find each product or quotient.*

1. 8×7 **2.** 9×6 **3.** 8×9 **4.** 7×9

5. $\dfrac{81}{9}$ **6.** $\dfrac{48}{8}$ **7.** $\dfrac{64}{8}$ **8.** $\dfrac{56}{7}$

Because the times sign, $\times$, looks like the letter x, it is seldom used in algebra. Instead, a dot, parentheses, or no symbol at all is used to denote multiplication. Each of the following expressions indicates the **product** obtained when two real numbers x and y are multiplied.

$$x \cdot y \quad (x)(y) \quad x(y) \quad (x)y \quad xy$$

■ Multiplying Real Numbers

To develop rules for multiplying real numbers, we rely on the definition of multiplication. The expression $5 \cdot 4$ indicates that 4 is to be used as a term in a sum five times. That is,

$$5(4) = 4 + 4 + 4 + 4 + 4$$

$$= 20$$

Read 5(4) as "5 times 4."

Likewise, the expression $5(-4)$ indicates that -4 is to be used as a term in a sum five times. Thus,

$$5(-4) = (-4) + (-4) + (-4) + (-4) + (-4)$$

$$= -20$$

Read $5(-4)$ as "5 times negative 4."

If multiplying by a positive number indicates repeated addition, then it is reasonable that multiplication by a negative number indicates repeated subtraction. The expression $(-5)4$, for example, means that 4 is to be used as a term in a repeated subtraction five times. That is,

$$(-5)4 = -(4) - (4) - (4) - (4) - (4)$$

$$= (-4) + (-4) + (-4) + (-4) + (-4)$$

$$= -20$$

Likewise, the expression $(-5)(-4)$ indicates that -4 is to be used as a term in a repeated subtraction five times. Thus,

$$(-5)(-4) = -(-4) - (-4) - (-4) - (-4) - (-4)$$

$$= -(-4) + [-(-4)] + [-(-4)] + [-(-4)] + [-(-4)]$$

$$= 4 + 4 + 4 + 4 + 4$$

$$= 20$$

The expression $0(-2)$ indicates that -2 is to be used zero times as a term in a repeated addition. Thus,

$$0(-2) = 0$$

Finally, the expression $(-3)(1) = -3$ suggests that the product of any number and 1 is the number itself.

The previous results suggest the following rules:

Rules for Multiplying Signed Numbers

1. The product of two real numbers with like signs is the product of their absolute values.

2. The product of two real numbers with unlike signs is the negative of the product of their absolute values.

3. Any number multiplied by 0 is 0: $a \cdot 0 = 0 \cdot a = 0$.

4. Any number multiplied by 1 is that number itself: $a \cdot 1 = 1 \cdot a = a$

(a, b, c, d)

EXAMPLE 1 Find each product: **a.** $4(-7)$, **b.** $(-5)(-4)$, **c.** $(-7)(6)$, **d.** $8(6)$, **e.** $(-3)(5)(-4)$, and **f.** $(-4)(-2)(-3)$.

Solution
a. $4(-7) = -(4 \cdot 7)$
$\qquad = -28$

b. $(-5)(-4) = +(5 \cdot 4)$
$\qquad = +20$

c. $(-7)(6) = -(7 \cdot 6)$
$\qquad = -42$

d. $8(6) = +(8 \cdot 6)$
$\qquad = +48$

e. $(-3)(5)(-4) = (-15)(-4)$
$\qquad = 60$

f. $(-4)(-2)(-3) = 8(-3)$
$\qquad = -24$ ■

EXAMPLE 2 If $x = -3$; $y = 2$, and $z = 4$, evaluate **a.** $y + xz$ and **b.** $x(y - z)$.

Solution We substitute -3 for x, 2 for y, and 4 for z in each expression and simplify.

a. $y + xz = 2 + (-3)(4)$
$\qquad = 2 + (-12)$
$\qquad = -10$

b. $x(y - z) = -3[2 - 4]$
$\qquad = -3[2 + (-4)]$
$\qquad = -3(-2)$
$\qquad = 6$ ■

EXAMPLE 3 If $x = -2$ and $y = 3$, evaluate **a.** $x^2 - y^2$ and **b.** $-x^2$.

Solution **a.** We substitute -2 for x and 3 for y and simplify.

$x^2 - y^2 = (-2)^2 - 3^2$
$\qquad = 4 - 9$ Simplify the exponential expressions first.
$\qquad = -5$ Do the subtraction.

b. We substitute -2 for x and simplify.

$-x^2 = -(-2)^2$
$\qquad = -4$ Simplify $(-2)^2$.   ■

 (b)

EXAMPLE 4 Find each product: **a.** $\left(-\dfrac{2}{3}\right)\left(-\dfrac{6}{5}\right)$ and **b.** $\left(\dfrac{3}{10}\right)\left(-\dfrac{5}{9}\right)$.

Solution **a.** $\left(-\dfrac{2}{3}\right)\left(-\dfrac{6}{5}\right) = +\left(\dfrac{2}{3} \cdot \dfrac{6}{5}\right)$

$\qquad = +\dfrac{2 \cdot 6}{3 \cdot 5}$

$\qquad = +\dfrac{12}{15}$

$\qquad = +\dfrac{4}{5}$

b. $\left(\dfrac{3}{10}\right)\left(-\dfrac{5}{9}\right) = -\dfrac{3}{10} \cdot \dfrac{5}{9}$

$\qquad = -\dfrac{3 \cdot 5}{10 \cdot 9}$

$\qquad = -\dfrac{15}{90}$

$\qquad = -\dfrac{1}{6}$ ■

EXAMPLE 5 **Temperature change**   If the temperature is dropping 4 degrees each hour, how much warmer was it 3 hours ago?

Solution   A temperature drop of 4 degrees per hour can be represented by -4 degrees per hour. Three hours ago can be represented by -3. The temperature 3 hours ago is the product $(-3)(-4)$.

$$(-3)(-4) = +12$$

The temperature was 12 degrees warmer 3 hours ago.   ■

■ Dividing Real Numbers

We have seen that 8 divided by 4 is 2:

$$\frac{8}{4} = 2 \quad \text{because} \quad 2 \cdot 4 = 8$$

Likewise,

$$\frac{18}{6} = 3 \quad \text{because} \quad 3 \cdot 6 = 18$$

In general, the rule

$$\frac{a}{b} = c \quad \text{if and only if} \quad c \cdot b = a$$

is true for the division of any real number, a, by any nonzero real number, b. For example,

$$\frac{+10}{+2} = +5 \quad \text{because} \quad (+5)(+2) = +10$$

$$\frac{-10}{-2} = +5 \quad \text{because} \quad (+5)(-2) = -10$$

$$\frac{+10}{-2} = -5 \quad \text{because} \quad (-5)(-2) = +10$$

$$\frac{-10}{+2} = -5 \quad \text{because} \quad (-5)(+2) = -10$$

These four examples suggest the rules for dividing real numbers.

Rules for Dividing Signed Numbers

1. The quotient of two real numbers with like signs is the quotient of their absolute values.

2. The quotient of two real numbers with unlike signs is the negative of the quotient of their absolute values.

3. $\dfrac{a}{0}$ is undefined; $\dfrac{0}{0}$ is indeterminate.

4. If $a \neq 0$, then $\dfrac{0}{a} = 0$.

EXAMPLE 6   Find each quotient: **a.** $\dfrac{36}{18}$,   **b.** $\dfrac{-44}{11}$,   **c.** $\dfrac{27}{-9}$,   **d.** $\dfrac{-64}{-8}$,   and   **e.** $\dfrac{6}{0}$.

Solution   **a.** $\dfrac{36}{18} = +\dfrac{36}{18} = 2$ The quotient of real numbers with like signs is the quotient of their absolute values.

b. $\dfrac{-44}{11} = -\dfrac{44}{11} = -4$ The quotient of real numbers with unlike signs is the negative of the quotient of their absolute values.

c. $\dfrac{27}{-9} = -\dfrac{27}{9} = -3$ The quotient of real numbers with unlike signs is the negative of the quotient of their absolute values.

d. $\dfrac{-64}{-8} = +\dfrac{64}{8} = 8$ The quotient of real numbers with like signs is the quotient of their absolute values.

e. $\dfrac{6}{0}$ is undefined.   ■

EXAMPLE 7   If $x = -64$, $y = 16$, and $z = -4$, evaluate   **a.** $\dfrac{yz}{-x}$,   **b.** $\dfrac{z^3 y}{x}$,   and   **c.** $\dfrac{x + y}{-z^2}$.

Solution   We substitute -64 for x, 16 for y, and -4 for z in each expression and simplify.

a. $\dfrac{yz}{-x} = \dfrac{16(-4)}{-(-64)}$ **b.** $\dfrac{z^3 y}{x} = \dfrac{(-4)^3(16)}{-64}$ **c.** $\dfrac{x + y}{-z^2} = \dfrac{-64 + 16}{-(-4)^2}$

$= \dfrac{-64}{+64}$ $= \dfrac{(-64)(16)}{(-64)}$ $= \dfrac{-48}{-16}$

$= -1$ $= 16$ $= 3$ ■

EXAMPLE 8   If $x = -50$, $y = 10$, and $z = -5$, evaluate   **a.** $\dfrac{xyz}{x - 5z}$ and   **b.** $\dfrac{3xy + 2yz}{2(x + y)}$.

Solution We substitute -50 for x, 10 for y, and -5 for z in each expression and simplify.

a. $\dfrac{xyz}{x - 5z} = \dfrac{(-50)(10)(-5)}{-50 - 5(-5)}$ **b.** $\dfrac{3xy + 2yz}{2(x + y)} = \dfrac{3(-50)(10) + 2(10)(-5)}{2(-50 + 10)}$

$= \dfrac{(-500)(-5)}{-50 + 25}$ $= \dfrac{-150(10) + (20)(-5)}{2(-40)}$

$= \dfrac{2500}{-25}$ $= \dfrac{-1500 - 100}{-80}$

$= -100$ $= \dfrac{-1600}{-80}$

$= 20$ ∎

EXAMPLE 9 **Stock reports** In its annual report, a publicly held corporation reports its performance on a per-share basis. When a company with 35 million shares outstanding loses \$2.3 million, what will be the per-share loss?

Solution A loss of \$2.3 million can be represented by $-2{,}300{,}000$. Because there are 35 million shares, the per-share amount lost can be represented by the quotient $\dfrac{-2{,}300{,}000}{35{,}000{,}000}$.

$$\frac{-2{,}300{,}000}{35{,}000{,}000} \approx -0.065714285 \qquad \text{Use a calculator.}$$

The company lost about 6.6 cents per share. ∎

Remember these facts about dividing real numbers.

Division **1.** $\dfrac{a}{0}$ is undefined. **2.** If $a \neq 0$, then $\dfrac{0}{a} = 0$.

3. $\dfrac{a}{1} = a$. **4.** If $a \neq 0$, then $\dfrac{a}{a} = 1$.

ORALS *Find each product or quotient.*

1. $1(-3)$ **2.** $-2(-5)$

3. $-3(-6)$ **4.** $4(-6)$

5. $-2(3)(-4)$ **6.** $-2(-3)(-4)$

7. $\dfrac{-12}{6}$ **8.** $\dfrac{-10}{-5}$ **9.** $\dfrac{3(6)}{-2}$ **10.** $\dfrac{(-2)(-3)}{-6}$

E X E R C I S E   1.5

In Exercises 1–24, find each product.

1. $(+6)(+8)$

2. $(-9)(-7)$

3. $(-8)(-7)$

4. $(9)(-6)$

5. $(+9)(+7)$

6. $(+8)(-5)$

7. $(-7)(7)$

8. $(-11)(-11)$

9. $(+12)(-12)$

10. $(-9)(12)$

11. $\left(\dfrac{1}{2}\right)(-32)$

12. $\left(-\dfrac{3}{4}\right)(12)$

13. $\left(-\dfrac{3}{4}\right)\left(-\dfrac{8}{3}\right)$

14. $\left(-\dfrac{2}{5}\right)\left(\dfrac{15}{2}\right)$

15. $(-3)\left(-\dfrac{1}{3}\right)$

16. $(5)\left(-\dfrac{2}{5}\right)$

17. $(3)(-4)(-6)$

18. $(-1)(-3)(-6)$

19. $(-2)(3)(4)$

20. $(5)(0)(-3)$

21. $(2)(-5)(-6)(-7)$

22. $(-3)(-5)(-5)(-2)$

23. $(-2)(-2)(-2)(-3)(-4)$

24. $(-5)(4)(3)(-2)(-1)$

In Exercises 25–48, let $x = -1$, $y = 2$, and $z = -3$. Evaluate each expression.

25. y^2

26. x^2

27. $-z^2$

28. $-xz$

29. xy

30. yz

31. $y + xz$

32. $z - xy$

33. $(x + y)z$

34. $y(x - z)$

35. $(x - z)(x + z)$

36. $(y + z)(x - z)$

37. $xy + yz$

38. $zx - zy$

39. xyz

40. x^2y

41. y^2z^2

42. z^3y

43. $y(x - y)^2$

44. $z(y - x)^2$

45. $x^2(y - z)$

46. $y^2(x - z)$

47. $(-x)(-y) + z^2$

48. $(-x)(-z) - y^2$

In Exercises 49–60, simplify each expression.

49. $\dfrac{80}{-20}$

50. $\dfrac{-66}{33}$

51. $\dfrac{-110}{-55}$

52. $\dfrac{200}{40}$

53. $\dfrac{-160}{40}$

54. $\dfrac{-250}{-25}$

55. $\dfrac{320}{-16}$

56. $\dfrac{180}{-36}$

57. $\dfrac{8 - 12}{-2}$

58. $\dfrac{16 - 2}{2 - 9}$

59. $\dfrac{20 - 25}{7 - 12}$

60. $\dfrac{2(15)^2 - 2}{-2^3 + 1}$

In Exercises 61–72, evaluate each expression if $x = -2$, $y = 3$, $z = 4$, $t = 5$, and $w = -18$.

61. $\dfrac{yz}{x}$

62. $\dfrac{zt}{x}$

63. $\dfrac{tw}{y}$

64. $\dfrac{w}{xy}$

65. $\dfrac{z + w}{x}$

66. $\dfrac{xyz}{y - 1}$

67. $\dfrac{xtz}{y + 1}$

68. $\dfrac{x + y + z}{t}$

69. $\dfrac{wz - xy}{x + y}$

70. $\dfrac{x^2y^3}{yz}$

71. $\dfrac{yw + xy}{xt}$

72. $\dfrac{tw}{xz - w}$

In Exercises 73–84, evaluate each expression if $x = 4$, $y = -6$, and $z = -3$. Use a calculator.

73. $\dfrac{2x^2 + 2y}{x + y}$

74. $\dfrac{y^2 + z^2}{y + z}$

75. $\dfrac{2x^2 - 2z^2}{x + z}$

76. $\dfrac{8x^3 - 8y^2}{x - z}$

77. $\dfrac{y^3 + 4z^3}{(x + y)^2}$

78. $\dfrac{x^2 - 2xz + z^2}{x - y + z}$

79. $\dfrac{xy^2z + x^2y}{2y - 2z}$

80. $\dfrac{(x^2 - 2y)z^2}{-xz}$

81. $\dfrac{xyz - y^2z}{y(x + z)^4}$

82. $\dfrac{-3x^2 - 2z^2 + x^2}{(x + y + z)^2}$

83. $\dfrac{2(x - y)(y - z)(x - z)}{2x - 3y - 6}$

84. $\dfrac{x^3y - (yz)^2 - 10y - 2}{x^2 + y^2 + z^3}$

In Exercises 85–92, evaluate each expression if $x = \frac{1}{2}$, $y = -\frac{2}{3}$, and $z = -\frac{3}{4}$.

85. $x + y$

86. $y + z$

87. $x + y + z$

88. $y + x - z$

89. $(x + y)(x - y)$

90. $(x - z)(x + z)$

91. $(x + y + z)(xyz)$

92. $xyz(x - y - z)$

In Exercises 93–100, use signed numbers to solve each problem.

93. Temperature change   The temperature is increasing 2 degrees each hour for 3 hours. What product of signed numbers represents the temperature change?

94. Temperature change   If the temperature is decreasing 2 degrees each hour for 3 hours, what product of signed numbers represents the temperature change?

95. Gambling   In Las Vegas, Robert lost $30 per hour playing the slot machines for 15 hours. What product of signed numbers represents the change in his financial condition?

96. Draining a pool   A pool is emptying at the rate of 12 gallons per minute. What product of signed numbers would represent how much more water was in the pool 2 hours ago?

97. Filling a pool   Water from a pipe is filling a pool at the rate of 23 gallons per minute. What product of signed numbers represents the amount of water in the pool 2 hours ago?

98. Mowing lawns   Rafael worked all day mowing lawns and was paid $8 per hour. If he had $94 at the end of an 8-hour day, how much did he have before he started working?

99. Temperature   Suppose that the temperature is dropping at the rate of 3 degrees each hour. If the temperature has dropped 18 degrees, what signed number expresses how many hours the temperature has been falling?

100. Dieting   A man lost 37.5 pounds. If he lost 2.5 pounds each week, how long has he been dieting?

In Exercises 101–104, use a calculator and signed numbers to solve each problem.

101. Stock market   Over a 7-day period, the Dow Jones Industrial Average had gains of 26, 35, and 17 points. In the same period, there were losses of 25, 31, 12, and 24 points. Find the average daily performance over the 7-day period.

102. Astronomy   Light travels at the rate of 186,000 miles per second. How long will it take light to travel from the sun to Venus? (*Hint:* The distance from the sun to Venus is 67,000,000 miles.)

103. Saving for college A student has saved $15,000 to attend graduate school. If she estimates that her expenses will be $613.50 a month while in school, does she have enough to complete an 18-month Master's degree program?

104. Earnings per share Over a five-year period, a corporation reported profits of $18 million, $21 million, and $33 million. In the other two years, it reported losses of $5 million and $71 million. Find the average gain (or loss) per year.

Writing Exercises ■ *Write a paragraph using your own words.*

1. Explain how you would decide whether the product of several numbers is positive or negative.

2. Describe two situations in which negative numbers are useful.

Something to Think About ■

1. If the quotient of two numbers is undefined, what would their product be?

2. If the product of five numbers is negative, how many of the factors could be negative?

3. If x^5 is a negative number, can you decide if x is negative, too?

4. If x^6 is a positive number, can you decide if x is positive, too?

Review Exercises ■ *Solve each problem.*

1. A concrete block weighs $37\frac{1}{2}$ pounds. How much will 30 of these blocks weigh?

2. If one brick weighs 1.3 pounds, how much will a skid of 500 bricks weigh?

3. If $x = -5$, $y = -8$, and $z = 3$, evaluate $x^2 - yz^2$.

4. Put $<$, $=$, or $>$ in the box to make a true statement:
$-2(-3 + 4)$ ▒ $-3[3 - (-4)]$

1.6 Algebraic Expressions

■ **Algebraic Terms**

GETTING READY *If $x = -5$, evaluate each of the following.*

1. $x + 3$ **2.** $-7x$ **3.** $\dfrac{x}{10}$ **4.** $19 - y$

5. $\dfrac{x - 7}{3}$ **6.** $\dfrac{x}{5} - 7$ **7.** $5(x + 2)$ **8.** $5x + 10$

Variables and numbers can be combined with the operations of arithmetic to produce **algebraic expressions**. For example, if x and y are variables, the algebraic expression $x + y$ represents the **sum** of x and y, and the algebraic expression $x - y$ represents their **difference**.

There are many ways to read the sum $x + y$. Some of them are:

- the sum of x and y
- x increased by y
- x plus y
- y more than x
- y added to x

There are many ways to read the difference $x - y$. Some of them are:

- the result obtained when y is subtracted from x
- the result of subtracting y from x
- x less y
- y less than x
- x decreased by y
- x minus y

EXAMPLE 1   Let x represent a certain number. Write an expression that represents   **a.** the number that is 5 more than x and   **b.** the number 12 decreased by x.

Solution   **a.** The number "5 more than x" is the number found by adding 5 to x. It is represented by $x + 5$.

b. The number "12 decreased by x" is the number found by subtracting x from 12. It is represented by $12 - x$. ■

EXAMPLE 2   **Income taxes**   Luciano worked x hours preparing his income taxes. He worked 3 hours less than that preparing his son's return. Write an expression that represents   **a.** the number of hours he spent preparing his son's return   and   **b.** the total number of hours he worked.

Solution   **a.** Because Luciano worked x hours on his return and 3 hours less on his son's return, he worked $(x - 3)$ hours on his son's return.

b. Because he worked x hours on his return and $(x - 3)$ hours on his son's return, the total time he spent on taxes was $[x + (x - 3)]$ hours. ■

There are several ways to indicate the **product** xy in words. Some of them are:

- x multiplied by y
- the product of x and y
- x times y

EXAMPLE 3 Let x represent a certain number. Denote a number that is **a.** twice as large as x, **b.** 5 more than 3 times x, and **c.** 4 less than $\frac{1}{2}$ of x.

Solution **a.** The number "twice as large as x" is found by multiplying x by 2. It is represented by $2x$.

b. The number "5 more than 3 times x" is found by adding 5 to the product of 3 and x. It is represented by $3x + 5$.

c. The number "4 less than $\frac{1}{2}$ of x" is found by subtracting 4 from the product of $\frac{1}{2}$ and x. It is represented by $\frac{1}{2}x - 4$. ■

EXAMPLE 4 **Stock valuation** Jim owns x shares of Transitronic stock, valued at $29 a share, y shares of Positone stock, valued at $32 a share, and 300 shares of Baby Bell, valued at $42 a share. **a.** How many shares of stock does he own? **b.** What is the value of his stock?

Solution **a.** Because there are x shares of Transitronic, y shares of Positone, and 300 shares of Baby Bell, his total number of shares is $x + y + 300$.

b. The value of x shares of Transitronic is $\$29x$, the value of y shares of Positone is $\$32y$, and the value 300 shares of Baby Bell is $\$42(300)$. The total value of the stock is $\$(29x + 32y + 12{,}600)$. ■

If x and y represent two numbers and $y \neq 0$, the **quotient** obtained when x is divided by y is denoted by each of the following expressions:

$$x \div y, \quad x/y, \quad \text{and} \quad \frac{x}{y}$$

EXAMPLE 5 Let x and y represent two numbers. Write an algebraic expression that represents the sum obtained when 3 times the first number is added to the quotient obtained when the second number is divided by 6.

Solution Three times the first number x is denoted as $3x$. The quotient obtained when the second number y is divided by 6 is the fraction $\frac{y}{6}$. Their sum is expressed as $3x + \frac{y}{6}$. ■

EXAMPLE 6 **Cutting a rope** A 5-foot section is cut from the end of a rope that is l feet long. The remaining rope is then divided into three equal parts. Find the length of each of the equal pieces.

Solution   After a 5-foot section is cut from one end of l feet of rope, the rope that remains is $(l - 5)$ feet long. When that remaining rope is cut into 3 equal pieces, each piece will be $\frac{l-5}{3}$ feet long. See Figure 1-22.

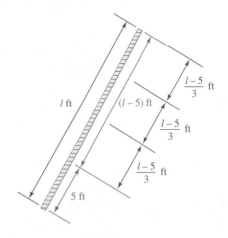

FIGURE 1-22   ■

   Since variables represent numbers, algebraic expressions also represent numbers. We can evaluate an algebraic expression if we know the numbers that its variables represent. For example, if x represents the number 12 and y represents the number 5, we can evaluate the expression $x + y + 2$ as follows:

$$x + y + 2 = 12 + 5 + 2 \qquad \text{Substitute 12 for } x \text{ and 5 for } y.$$
$$= 17 + 2 \qquad \text{Do the addition from left to right.}$$
$$= 19$$

EXAMPLE 7   If x represents the number 8 and y represents the number 10, evaluate the expressions   **a.** $x + y$,   **b.** $y - x$,   **c.** $3xy$ and   **d.** $\dfrac{5x}{y - 5}$.

Solution   We can substitute 8 for x and 10 for y in each expression and simplify.

a. $x + y = 8 + 10$ **b.** $y - x = 10 - 8$
$\qquad = 18$ $\qquad = 2$

c. $3xy = (3)(8)(10)$
$\qquad = (24)(10) \qquad$ Do the multiplications from left to right.
$\qquad = 240$

d. $\dfrac{5x}{y - 5} = \dfrac{5 \cdot 8}{10 - 5}$

$\qquad = \dfrac{40}{5} \qquad$ Simplify the numerator and denominator separately.

$\qquad = 8 \qquad$ Simplify the fraction.

WARNING!   After numbers are substituted for the variables in a product, a dot or parentheses must often be used to show the multiplication. Otherwise, (3)(8)(10), for example, might be mistaken for 3810, and $5 \cdot 8$ mistaken for 58. ■

■ Algebraic Terms

Numbers without variables, such as 7, 21, and 23, are called **constants.** Expressions such as 37, xyz, or $32t$, which are constants, variables, or products of constants and variables, are called **algebraic terms.** Because the expression $3x + 5y$ denotes the sum of the two algebraic terms $3x$ and $5y$, the expression $3x + 5y$ contains two terms. Likewise, the expression $xy - 7$ contains two terms. The expression $3 + x + 2y$ contains three terms. Its first term is 3, its second term is x, and its third term is $2y$.

Numbers that are part of an indicated product are called **factors** of that product. For example, the product $7x$ has two factors, 7 and x. Either factor in this product is called the **coefficient** of the other factor. When we speak of the coefficient in a term such as $7x$, however, we generally mean the **numerical coefficient,** which in this case is 7. For example, the numerical coefficient of the term $12xyz$ is 12. The coefficient of such terms as x, ab, or rst is understood to be 1. Thus,

$$x = 1x, \qquad ab = 1ab, \qquad \text{and} \qquad rst = 1rst$$

EXAMPLE 8   **a.** The expression $5x + y$ has two terms. The numerical coefficient of its first term is 5. The numerical coefficient of its second term is 1.

b. The expression $-17wxyz$ has one term, which contains the five factors -17, w, x, y, and z. Its numerical coefficient is -17.

c. The expression 37 has one term, the constant 37. Its numerical coefficient is 37. ■

ORALS   *If $x = -2$ and $y = 3$, evaluate each expression.*

1. $x + y$ 2. $7x$ 3. $7x + y$ 4. $7(x + y)$
5. $4x^2$ 6. $(4x)^2$ 7. $-3x^2$ 8. $(-3x)^2$

E X E R C I S E   1.6

In Exercises 1–18, let x, y, and z represent three real numbers. Write an algebraic expression to denote each quantity.

1. The sum of x and y.

2. The product of x and y.

3. The product of x and twice y.

4. The sum of twice x and twice y.

5. The difference obtained when x is subtracted from y.

6. The difference obtained when twice x is subtracted from y.

7. The quotient obtained when y is divided by x.

8. The quotient obtained when the sum of x and y is divided by z.

9. The sum obtained when the quotient of x divided by y is added to z.

10. y decreased by x.

11. z less the product of x and y.

12. z less than the product of x and y.

13. The product of 3, x, and y.

14. The quotient obtained when the product of 3 and z is divided by the product of 4 and x.

15. The quotient obtained when the sum of x and y is divided by the sum of y and z.

16. The quotient obtained when the product of x and y is divided by the sum of x and z.

17. The sum of the product xy and the quotient obtained when y is divided by z.

18. The number obtained when x decreased by 4 is divided by the product of 3 and y.

19. Course load A man enrolls in college for c hours of credit, and his sister enrolls for 4 more hours than her brother. Write an expression that represents the number of hours the sister is taking.

20. Antique cars An antique Ford has 25,000 more miles on its odometer than a new car. If the new car has traveled m miles, find an expression that represents the mileage on the Ford.

21. T-bills Write an expression that represents the value of t T-bills, each worth $9987.

22. Real estate Write an expression that represents the value of a vacant lots if each lot is worth $35,000.

23. Cutting rope A rope x feet long is cut into 5 equal pieces. Find the length of each piece.

24. Plumbing A plumber cuts a pipe that is 12 feet long into x equal pieces. Find the length of each piece.

25. Comparing assets A girl had d dollars, and her brother had $5 more than three times that amount. How much money did the brother have?

26. Comparing investments Wendy has x shares of stock. Her sister has 2 fewer shares than twice Wendy's shares. How many shares does her sister have?

In Exercises 27–38, write each algebraic expression as an appropriate English phrase.

27. $x + 3$

28. $y - 2$

29. $\dfrac{x}{y}$

30. xz

31. $2xy$

32. $\dfrac{x + y}{2}$

33. $\dfrac{5}{x + y}$

34. $\dfrac{3x}{y + z}$

35. $\dfrac{3 + x}{y}$

36. $3 + \dfrac{x}{y}$

37. $xy(x + y)$

38. $(x + y + z)(xyz)$

In Exercises 39–46, let $x = 8$, $y = 4$, and $z = 2$. Write each phrase as an algebraic expression and evaluate it.

39. the sum of x and z

40. the product of x, y, and z

41. z less than y

42. the quotient obtained when y is divided by z

43. 3 less than the product of y and z

44. 7 less than the sum of x and y

45. the quotient obtained when the product of x and y is divided by z

46. the quotient obtained when 10 greater than x is divided by z

In Exercises 47–56, give the number of terms in each algebraic expression and also give the numerical coefficient of the first term.

47. $6d$

48. $-4c + 3d$

49. $-xy - 4t + 35$

50. xy

51. $3ab + bc - cd - ef$

52. $-2xyz + cde - 14$

53. $-4xyz + 7xy - z$

54. $5uvw - 4uv + 8uw$

55. $3x + 4y + 2z + 2$

56. $7abc - 9ab + 2bc + a - 1$

In Exercises 57–60, consider the algebraic expression $29xyz + 23xy + 19x$.

57. What are the factors of the third term?

58. What are the factors of the second term?

59. What are the factors of the first term?

60. What factor is common to all three terms?

In Exercises 61–64, consider the algebraic expression $3xyz + 5xy + 17xz$.

61. What are the factors of the first term?

62. What are the factors of the second term?

63. What are the factors of the third term?

64. What factor is common to all three terms?

In Exercises 65–68, consider the algebraic expression $5xy + yt + 8xyt$.

65. Find the numerical coefficients of each term.

66. What factor is common to all three terms?

67. What factors are common to the first and third terms?

68. What factors are common to the second and third terms?

In Exercises 69–72, consider the algebraic expression $3xy + y + 25xyz$.

69. Find the numerical coefficient of each term and find their product.

70. Find the numerical coefficient of each term and find their sum.

71. What factors are common to the first and third terms?

72. What factor is common to all three terms?

Writing Exercises ■ *Write a paragraph using your own words.*

1. Distinguish between the meanings of these two phrases: "3 less than x" and "3 is less than x."

2. Distinguish between *factor* and *term*.

3. What is the purpose of using variables? Why aren't ordinary numbers enough?

4. In words, xy is "the product of x and y." However, $\frac{x}{y}$ is "the quotient obtained when x is divided by y." Explain why the extra words are needed.

Something to Think About ■

1. If the value of x were doubled, what would happen to the value of $37x$?

2. If the values of both x and y were doubled, what would happen to the value of $5x^2y$?

Review Exercises ■ *Evaluate each of the following.*

1. 14% of 3800
2. $\frac{3}{5}$ of 4765
3. $\frac{-4 + (7 - 9)}{(-9 - 7) + 4}$
4. $\frac{5}{4}\left(1 - \frac{3}{5}\right)$

1.7 Properties of Real Numbers

■ The Closure Properties ■ The Commutative Properties ■ The Associative Properties ■ The Distributive Property ■ The Identity Elements ■ Inverses for Addition and Multiplication

GETTING READY *Do the indicated operations.*

1. $3 + (5 + 9)$
2. $(3 + 5) + 9$
3. $23.7 + 14.9$
4. $14.9 + 23.7$
5. $7(5 + 3)$
6. $7 \cdot 5 + 7 \cdot 3$
7. $125.3 + (-125.3)$
8. $125.3\left(\dfrac{1}{125.3}\right)$
9. $777 + 0$
10. $777 \cdot 1$

In this section, we will discuss the properties governing addition, subtraction, multiplication, and division of real numbers.

■ The Closure Properties

The **closure properties** guarantee that the sum, difference, product, or quotient (except for division by zero) of any two real numbers is also a real number.

Closure Properties   If a and b are real numbers, then

$a + b$ is a real number $a - b$ is a real number

ab is a real number $\dfrac{a}{b}$ is a real number $(b \neq 0)$

EXAMPLE 1 Assume that $x = 8$ and $y = -4$. Find the real number answer to show that **a.** $x + y$, **b.** $x - y$, **c.** xy, and **d.** $\frac{x}{y}$ all represent real numbers.

Solution We substitute 8 for x and -4 for y in each expression and simplify.

a. $x + y = 8 + (-4)$
$\qquad = 4$

b. $x - y = 8 - (-4)$
$\qquad = 8 + 4$
$\qquad = 12$

c. $xy = 8(-4)$
$\qquad = -32$

d. $\dfrac{x}{y} = \dfrac{8}{-4}$
$\qquad = -2$ ■

■ The Commutative Properties

The **commutative properties** (from the word *commute,* which means to go back and forth) guarantee that addition or multiplication of two real numbers can be done in either order.

Commutative Properties | If a and b are real numbers, then
$a + b = b + a$ $\qquad$ commutative property of addition
$ab = ba$ $\qquad$ commutative property of multiplication

EXAMPLE 2 Assume that $x = -3$ and $y = 7$. Show that **a.** $x + y = y + x$ and **b.** $xy = yx$.

Solution **a.** We can show that the sum $x + y$ is the same as the sum $y + x$ by substituting -3 for x and 7 for y in each expression and simplifying.

$x + y = -3 + 7$ $\qquad$ and $\qquad$ $y + x = 7 + (-3)$
$\qquad = 4$ $\qquad\qquad\qquad\qquad = 4$

b. We can show that the product xy is the same as the product yx by substituting -3 for x and 7 for y in each expression and simplifying.

$xy = -3(7)$ $\qquad$ and $\qquad$ $yx = 7(-3)$
$\qquad = -21$ $\qquad\qquad\qquad\qquad = -21$ ■

■ The Associative Properties

The **associative properties** guarantee that three real numbers can be regrouped in an indicated addition or multiplication.

Associative Properties	If a, b, and c are real numbers, then

$$(a + b) + c = a + (b + c) \quad \text{associative property of addition}$$
$$(ab)c = a(bc) \quad \text{associative property of multiplication}$$

Because of the associative property of addition, we can group, or *associate*, the numbers in a sum in any way that we wish. For example,

$$(\mathbf{3 + 4}) + 5 = \mathbf{7} + 5 \quad \text{and} \quad 3 + (\mathbf{4 + 5}) = 3 + \mathbf{9}$$
$$= 12 \qquad\qquad\qquad\qquad = 12$$

The answer is 12 regardless of how we group the three numbers.

The associative property of multiplication permits us to group, or associate, the numbers in a product in any way that we wish. For example,

$$(\mathbf{3 \cdot 4}) \cdot 7 = \mathbf{12} \cdot 7 \quad \text{and} \quad 3 \cdot (\mathbf{4 \cdot 7}) = 3 \cdot \mathbf{28}$$
$$= 84 \qquad\qquad\qquad\qquad = 84$$

The answer is 84 regardless of how we group the three numbers.

■ The Distributive Property

The **distributive property** shows how to multiply the sum of two numbers by a third number. Because of this property, we can often add first and then multiply, or multiply first and then add.

For example, $2(3 + 7)$ can be calculated in two different ways. We can add and then multiply, or we can multiply each number within the parentheses by 2 and then add.

$$2(\mathbf{3 + 7}) = 2(\mathbf{10}) \quad \text{and} \quad 2(3 + 7) = \mathbf{2 \cdot 3 + 2 \cdot 7}$$
$$= 20 \qquad\qquad\qquad\qquad = 6 + 14$$
$$\qquad\qquad\qquad\qquad = 20$$

Either way, the result is 20.

In general, we have the following property.

Distributive Property	If a, b, and c are real numbers, then

$$a(b + c) = ab + ac$$

Because multiplication is commutative, the distributive property can be written in the form

$$(b + c)a = ba + ca$$

EXAMPLE 3   Use the distributive property to evaluate each expression in two different ways: **a.** $3(5 + 9)$ and   **b.** $-2(-7 + 3)$.

Solution   **a.** $3(5 + 9) = 3(14)$ and $3(5 + 9) = 3 \cdot 5 + 3 \cdot 9$
$= 42$ $= 15 + 27$
$= 42$

b. $-2(-7 + 3) = -2(-4)$ and $-2(-7 + 3) = -2(-7) + (-2)3$
$= 8$ $= 14 + (-6)$
$= 8$ ∎

EXAMPLE 4   Use the distributive property to write $3(x + 2)$ without using parentheses.

Solution $3(x + 2) = 3x + 3 \cdot 2$
$= 3x + 6$ ∎

The distributive property can be extended to three or more terms. For example, if a, b, c, and d are real numbers, then

$$a(b + c + d) = ab + ac + ad$$

■ The Identity Elements

The numbers 0 and 1 play special roles in the arithmetic of real numbers. The number 0 is the only number that can be added to another number (say, a) and give an answer of that same number a:

$$0 + a = a + 0 = a$$

The number 1 is the only number that can be multiplied by another number (say, a) and give an answer of that same number a:

$$1 \cdot a = a \cdot 1 = a$$

Because adding 0 to a number or multiplying a number by 1 leaves that number identically the same, the numbers 0 and 1 are called **identity elements.**

Identity Elements   The number 0 is the **identity element for addition.**

The number 1 is the **identity element for multiplication.**

■ Inverses for Addition and Multiplication

If the sum of two numbers is 0, the numbers are called **negatives,** or **additive inverses,** of each other. For example, because $3 + (-3) = 0$, the numbers 3 and -3 are negatives or additive inverses of each other. In general, because

$$a + (-a) = 0$$

the numbers represented by a and $-a$ are negatives or additive inverses of each other.

If the product of two numbers is 1, the numbers are called **reciprocals,** or **multiplicative inverses,** of each other. For example, because $7(\frac{1}{7}) = 1$, the numbers 7 and $\frac{1}{7}$ are reciprocals. Because $(-0.25)(-4) = 1$, the numbers -0.25 and -4 are reciprocals. In general, because

$$a\left(\frac{1}{a}\right) = 1 \qquad \text{provided } a \neq 0$$

the numbers represented by a and $\frac{1}{a}$ are reciprocals or multiplicative inverses of each other.

Additive and Multiplicative Inverses	Because $a + (-a) = 0$, the numbers a and $-a$ are called **negatives** or **additive inverses.**
	Because $a\left(\dfrac{1}{a}\right) = 1$ $(a \neq 0)$, the numbers a and $\dfrac{1}{a}$ are called **reciprocals** or **multiplicative inverses.**

EXAMPLE 5 The statement in the right column justifies the statement in the left column.

$3 + 4$ is a real number	closure property of addition
$\dfrac{8}{3}$ is a real number	closure property of division
$3 + 4 = 4 + 3$	commutative property of addition
$-3 + (2 + 7) = (-3 + 2) + 7$	associative property of addition
$(5)(-4) = (-4)(5)$	commutative property of multiplication
$(ab)c = a(bc)$	associative property of multiplication
$3(a + 2) = 3a + 3 \cdot 2$	distributive property
$3 + 0 = 3$	additive identity property
$3(1) = 3$	multiplicative identity property
$2 + (-2) = 0$	additive inverse property
$\left(\dfrac{2}{3}\right)\left(\dfrac{3}{2}\right) = 1$	multiplicative inverse property

The properties of the real numbers are summarized as follows.

Properties of Real Numbers For all real numbers a, b, and c,

Closure properties	$a + b$ is a real number $a - b$ is a real number	$a \cdot b$ is a real number $a \div b$ is a real number $(b \neq 0)$
Commutative properties	$a + b = b + a$	$a \cdot b = b \cdot a$
Associative properties	$(a + b) + c = a + (b + c)$	$(ab)c = a(bc)$
Identity properties	$a + 0 = a$	$a \cdot 1 = a$
Inverse properties	$a + (-a) = 0$	$a \cdot \left(\dfrac{1}{a}\right) = 1 \quad (a \neq 0)$
Distributive property	$a(b + c) = ab + ac$	

ORALS *State an example of each of these properties of real numbers.*

1. The associative property of multiplication

2. The additive identity property

3. The distributive property

4. The inverse for multiplication

Provide an example to illustrate each statement.

5. Subtraction is not commutative.

6. Division is not associative.

EXERCISE 1.7

In Exercises 1–8, assume that $x = 12$ and $y = -2$. Show that each expression represents a real number by finding the real-number answer.

1. $x + y$

2. $y - x$

3. xy

4. $\dfrac{x}{y}$

5. x^2

6. y^2

7. $\dfrac{x}{y^2}$

8. $\dfrac{2x}{3y}$

In Exercises 9–14, assume that $x = 5$ and $y = 7$. Show that both given expressions have the same value.

9. $x + y$; $y + x$ **10.** xy; yx **11.** $3x + 2y$; $2y + 3x$ **12.** $3xy$; $3yx$

13. $x(x + y)$; $(x + y)x$ **14.** $xy + y^2$; $y^2 + xy$

In Exercises 15–20, assume that $x = 2$, $y = -3$, and $z = 1$. Show that the expressions have the same value.

15. $(x + y) + z$; $x + (y + z)$ **16.** $(xy)z$; $x(yz)$

17. $(xz)y$; $x(yz)$ **18.** $(x + y) + z$; $y + (x + z)$

19. $x^2(yz^2)$; $(x^2y)z^2$ **20.** $x(y^2z^3)$; $(xy^2)z^3$

In Exercises 21–32, use the distributive property to write each expression without parentheses. Simplify each result if possible.

21. $3(x + y)$ **22.** $4(a + b)$ **23.** $x(x + 3)$ **24.** $y(y + z)$

25. $-x(a + b)$ **26.** $a(x + y)$ **27.** $4(x^2 + x)$ **28.** $-2(a^2 + 3)$

29. $-5(t + 2)$ **30.** $2x(a - x)$ **31.** $-2a(x + a)$ **32.** $-p(p - q)$

In Exercises 33–44, give the additive and multiplicative inverse of each number when possible.

33. 2   **34.** 3   **35.** $\dfrac{1}{3}$ **36.** $-\dfrac{1}{2}$

37. 0   **38.** -2 **39.** $-\dfrac{5}{2}$ **40.** 0.5

41. -0.2 **42.** 0.75   **43.** $\dfrac{4}{3}$ **44.** -1.25

In Exercises 45–56, state which property of real numbers justifies each statement.

45. $3 + x = x + 3$ **46.** $(3 + x) + y = 3 + (x + y)$

47. $xy = yx$ **48.** $(3)(2) = (2)(3)$

49. $-2(x + 3) = -2x + (-2)(3)$ **50.** $x(y + z) = (y + z)x$

51. $(x + y) + z = z + (x + y)$ **52.** $3(x + y) = 3x + 3y$

53. $5 \cdot 1 = 5$ **54.** $x + 0 = x$

55. $3 + (-3) = 0$ **56.** $9 \cdot \dfrac{1}{9} = 1$

In Exercises 57–66, use the given property to rewrite the expression in a different form.

57. $3(x + 2)$; distributive property   **58.** $x + y$; commutative property of addition

59. y^2x; commutative property of multiplication   **60.** $x + (y + z)$; associative property of addition

61. $(x + y)z$; commutative property of addition   **62.** $x(y + z)$; distributive property

63. $(xy)z$; associative property of multiplication

64. $1x$; multiplicative identity property

65. $0 + x$; additive identity property

66. $5 \cdot \dfrac{1}{5}$; multiplicative inverse property

Writing Exercises ■   *Write a paragraph using your own words.*

1. Explain why division is not commutative.

2. Describe two ways of calculating the value of $3(x + 7)$.

Something to Think About ■   **1.** Suppose there were no other numbers than the odd integers.

- Would the closure property for addition still be true?
- Would the closure property for multiplication still be true?
- Would there still be an identity for addition?
- Would there still be an identity for multiplication?

2. Suppose there were no other numbers than the even integers. Answer the four parts of Question 1 again.

Review Exercises ■   **1.** Write as a mathematical expression: The sum of x and the square of y is greater than or equal to z.

2. Write as an English phrase: $3(x + z)$.

In Review Exercises 3–6, fill each box with an appropriate symbol.

3. For any number x, $|x| \geq$ ▢

4. $x - y = x + ($ ▢ $)$

5. The product of two negative numbers is a ▢ number.

6. The sum of two negative numbers is a ▢ number.

MATHEMATICS FOR FUN

In this chapter, we learned that subtraction is not associative. For example, $(10 - 5) - 2 \neq 10 - (5 - 2)$, because

$$(10 - 5) - 2 = 3 \quad \text{but} \quad 10 - (5 - 2) = 7$$

However, in the story on page 1, we expected subtraction to be associative. After the $5 refund on the incorrect $120 room cost, we reasoned incorrectly that

$$
\begin{aligned}
40 + 40 + 40 - 5 &= 120 - 5 \\
&= 120 - (3 + 2) \\
&= (120 - 3) + 2 \qquad \text{This step is false.} \\
&= (40 - 1) + (40 - 1) + (40 - 1) + 2 \\
&= 39 + 39 + 39 + 2 \\
&= 119
\end{aligned}
$$

That $119 is not equal to $120 is irrelevant. The cost of the room was $115.

P R O J E C T ■ A Slice of Pi

1. The circumference of any circle (the distance around the circle) and the diameter of the circle (the distance across) are related. When you divide the circumference by the diameter, the quotient is always the same number, **pi,** denoted by the Greek letter π.

   • Carefully measure the circumference of several circles: a quarter, a dinner plate, a bicycle tire, whatever you can find that is round. Then calculate approximations of π by dividing (with a calculator) each circle's circumference by its diameter.

   • Press the $\boxed{\pi}$ button on the calculator to obtain a more accurate value of π. How close were your calculations?

2. **a.** The fraction $\frac{22}{7}$ is often used as an approximation of π. To how many decimal places is this approximation accurate?

   b. Experiment with your calculator and try to do better: Find another fraction (with no more than three digits in either its numerator or its denominator) that is closer to π. Who in your class has done best?

Chapter Summary

KEY WORDS

absolute value (1.1)	graph (1.1)
algebraic expression (1.6)	grouping symbols (1.3)
algebraic term (1.6)	improper fractions (1.2)
area (1.3)	integer (1.1)
base (1.3)	irrational number (1.1)
circumference (1.3)	lowest (or least) common denominator (1.2)
coefficient (1.6)	
composite number (1.1)	lowest terms (1.2)
constant (1.6)	mixed number (1.2)
coordinate (1.1)	natural number (1.1)
decimal fraction (1.2)	number line (1.1)
denominator (1.2)	numerator (1.2)
difference (1.6)	numerical coefficient (1.6)
element of a set (1.1)	odd number (1.1)
even number (1.1)	origin (1.1)
exponent (1.3)	percent (1.2)
exponential expression (1.3)	perimeter (1.3)
factors (1.2, 1.6)	prime-factored form (1.2)

prime number (1.1)
product (1.6)
proper fractions (1.2)
quotient (1.6)
radius (1.3)
rational number (1.1)

reciprocal (1.2)
repeating decimal (1.2)
set (1.1)
sum (1.6)
terminating decimal (1.2)
whole number (1.1)

KEY IDEAS

(1.1) Sets of numbers can be graphed on the number line.

The absolute value of a number x, denoted $|x|$, is the distance between x and 0 on the number line.

$$|x| \geq 0$$

(1.2) To simplify a fraction, first factor the numerator and the denominator and then divide out all common factors.

To multiply two fractions, multiply their numerators and multiply their denominators.

To divide one fraction by another, multiply the first fraction by the reciprocal of the second fraction.

To add (or subtract) two fractions with the same denominator, add (or subtract) their numerators and keep their common denominator.

To add (or subtract) two fractions with unlike denominators, write the fractions as equivalent fractions with the same denominator, add (or subtract) their numerators, and use the common denominator.

Before performing arithmetic with mixed numbers, first convert them to improper fractions.

(1.3) If n is a natural number, then

$$x^n = \overbrace{x \cdot x \cdot x \cdot x \cdots \cdots x}^{n \text{ factors of } x}$$

Order of mathematical operations
Within each pair of grouping symbols (working from the innermost pair to the outermost pair), do the following operations:

1. Find the values of any exponential expressions.
2. Do all multiplications and divisions as they are encountered while working from left to right.

3. Do all additions and subtractions as they are encountered while working from left to right.

   When all of the grouping symbols have been removed, repeat the above rules to finish the calculation.

   In a fraction, simplify the numerator and the denominator separately. Then simplify the fraction, if possible.

(1.4) To find the sum of two real numbers with the same sign, we add their absolute values and keep their common sign.

To add two real numbers that have unlike signs, we subtract their absolute values (the smaller from the larger) and use the sign of the number with the greatest absolute value.

If x and y are two real numbers, then

$$x - y = x + (-y)$$

(1.5) The product of two real numbers with like signs is the product of their absolute values.

The product of two real numbers with unlike signs is the negative of the product of their absolute values.

The quotient of two real numbers with like signs is the quotient of their absolute values.

The quotient of two real numbers with unlike signs is the negative of the quotient of their absolute values.

Division by zero is undefined.

(1.7) The closure properties:
  $x + y$ is a real number.
  $x - y$ is a real number.
  xy is a real number.
  $\dfrac{x}{y}$ is a real number   (provided $y \neq 0$).

The commutative properties:
$$x + y = y + x$$
$$xy = yx$$

The associative properties:
$$(x + y) + z = x + (y + z)$$
$$(xy)z = x(yz)$$

The distributive property:
$$x(y + z) = xy + xz$$

The identity elements:
0 is the identity for addition.
1 is the identity for multiplication.

The additive and multiplicative inverse properties:
$$x + (-x) = 0$$
$$x\left(\frac{1}{x}\right) = 1 \quad \text{provided } x \neq 0$$

■ Chapter 1 Review Exercises

(1.1) *In Review Exercises 1–4, consider the set of numbers {0, 1, 2, 3, 4, 5}.*

1. Which numbers are natural numbers?

2. Which numbers are prime numbers?

3. Which numbers are odd natural numbers?

4. Which numbers are composite numbers?

In Review Exercises 5–8, place one of the symbols =, <, or > in the box to make a true statement.

5. -5 ▢ $12 - 12$

6. $\dfrac{24}{6}$ ▢ 5

7. $13 - 13$ ▢ $5 - \dfrac{25}{5}$

8. $\dfrac{21}{7}$ ▢ -33

In Review Exercises 9–12, draw a number line and graph each set of numbers.

9. the composite numbers from 10 to 20

10. the whole numbers between 15 and 25

11. the real numbers less than or equal to -3, or greater than 2

12. the real numbers greater than -4 and less than 3

In Review Exercises 13–14, find each absolute value.

13. $|53 - 42|$

14. $|-31|$

(1.2) *In Review Exercises 15–16, simplify each fraction.*

15. $\dfrac{45}{27}$

16. $\dfrac{121}{11}$

In Review Exercises 17–20, do each operation. Simplify each answer, if possible.

17. $\dfrac{31}{15} \cdot \dfrac{10}{62}$

18. $\dfrac{18}{21} \div \dfrac{6}{7}$

19. $\dfrac{1}{3} + \dfrac{1}{7}$

20. $\dfrac{2}{3} - \dfrac{1}{7}$

In Review Exercises 21–24, do each operation.

21. $32.71 + 15.9$

22. $27.92 - 14.93$

23. $5.3 \cdot 3.5$

24. $21.83 \div 5.9$

In Review Exercises 25–28, do each operation. Round each answer to two decimal places.

25. $2.7(4.92 - 3.18)$ **26.** $\dfrac{3.3 + 2.5}{0.22}$ **27.** $\dfrac{12.5}{14.7 - 11.2}$ **28.** $(3 - 0.7)(3.63 - 2)$

29. Average study time   Four students recorded time spent working on a take-home exam: 5.2, 4.7, 9.5, and 8 hours. Find the average time spent. (*Hint:* Add the numbers and divide by 4.)

30. Absentees   During the height of the flu season, 15% of the 380 university faculty were sick. How many were ill?

(1.3) In Review Exercises 31–42, find the value of each expression.

31. 3^4 **32.** $\left(\dfrac{2}{3}\right)^2$ **33.** -5^2 **34.** $(-5)^2$

35. $5 + 3^3$ **36.** $7 \cdot 2 - 7$ **37.** $4 + (8 \div 4)$ **38.** $(4 + 8) \div 4$

39. $5^3 - \dfrac{81}{3}$ **40.** $(5 - 2)^2 + 5^2 + 2^2$ **41.** $\dfrac{4 \cdot 3 + 3^4}{31}$ **42.** $\dfrac{4}{3} \cdot \dfrac{9}{2} + \dfrac{1}{2} \cdot 18$

In Review Exercises 43–46, let x = 6 and y = 8. Evaluate each expression.

43. $y^2 - x$ **44.** $(y - x)^2$ **45.** $\dfrac{x + y}{x - 4}$ **46.** $\dfrac{xy - 12}{4 + y}$

In Review Exercises 47–50, let x = 2 and y = 3. Evaluate each expression.

47. y^4 **48.** x^y **49.** $x^2 + xy^2$ **50.** $\dfrac{x^2 + y}{x^3 - 1}$

In Review Exercises 51–52, use a calculator to find each value.

51. Packaging   Four steel bands surround the shipping crate in Illustration 1. Find the total length of strapping needed.

52. Petroleum storage   Find the volume of the cylindrical storage tank in Illustration 2. Round the answer to one decimal place.

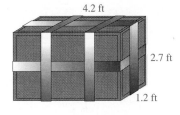

4.2 ft
2.7 ft
1.2 ft

ILLUSTRATION 1

32.1 ft
18.7 ft

ILLUSTRATION 2

(1.4–1.5) In Review Exercises 53–68, evaluate each expression.

53. $[-5 + (-5)] - (-5)$ **54.** $1 - [5 - (-3)]$ **55.** $\dfrac{5}{6} - \left(-\dfrac{2}{3}\right)$ **56.** $\dfrac{2}{3} - \left(\dfrac{1}{3} - \dfrac{2}{3}\right)$

57. $\left| \dfrac{3}{7} - \left(-\dfrac{4}{7} \right) \right|$

58. $\dfrac{3}{7} - \left| -\dfrac{4}{7} \right|$

59. $3.7 + (-2.5)$

60. $-5.6 - (-2.06)$

61. $\dfrac{-14}{-2}$

62. $\dfrac{(-2)(-7)}{4}$

63. $\left(-\dfrac{3}{14} \right)\left(-\dfrac{7}{6} \right)$

64. $\left(-\dfrac{1}{2} \right)\left(\dfrac{4}{3} \right)$

65. $\left(\dfrac{-3 + (-3)}{3} \right)\left(\dfrac{-15}{5} \right)$

66. $\dfrac{-2 - (-8)}{5 + (-1)}$

67. $\left(\dfrac{-10}{2} \right)^2 - (-1)^3$

68. $\dfrac{[-3 + (-4)]^2}{10 + (-3)}$

In Review Exercises 69–84, let $x = 2$, $y = -3$, and $z = -1$. Evaluate each expression.

69. $y + z$

70. $x + y$

71. $x + (y + z)$

72. $x - y$

73. $x - (y - z)$

74. $(x - y) - z$

75. xy

76. yz

77. $x(x + z)$

78. xyz

79. $y^2 z + x$

80. $yz^3 + (xy)^2$

81. $\dfrac{xy}{z}$

82. $\dfrac{|xy|}{3z}$

83. $\dfrac{3y^2 - x^2 + 1}{y|z|}$

84. $\dfrac{2y^2 - xyz}{x^2 |yz|}$

(1.6) *In Review Exercises 85–88, let x, y, and z represent three real numbers. Write an algebraic expression that represents each quantity.*

85. the product of x and z

86. the sum of x and twice y

87. twice the sum of x and y

88. x decreased by the product of y and z

In Review Exercises 89–92, write each algebraic expression as an appropriate English phrase.

89. $3xy$

90. $5 - yz$

91. $yz - 5$

92. $\dfrac{x + y + z}{2xyz}$

93. How many terms does the expression $3x + 4y + 9$ have?

94. What is the numerical coefficient of the term $7xy$?

95. What is the numerical coefficient of the term xy?

96. Find the sum of the numerical coefficients in the expression $2x^3 + 4x^2 + 3x$.

(1.7) *In Review Exercises 97–106, tell which property of real numbers justifies each statement. Assume that all variables represent real numbers.*

97. $x + y$ is a real number.

98. $3 \cdot (4 \cdot 5) = (4 \cdot 5) \cdot 3$

99. $3 + (4 + 5) = (3 + 4) + 5$

100. $3(x + 2) = 3 \cdot x + 3 \cdot 2$

101. $a + x = x + a$

102. $3(4 \cdot 5) = (3 \cdot 4) \cdot 5$

103. $3 + (x + 1) = (x + 1) + 3$

104. $x \cdot 1 = x$

105. $17 + (-17) = 0$

106. $x + 0 = x$

Chapter 1 Test

1. List the prime numbers between 30 and 50.

2. What is the only even prime number?

3. Graph the composite numbers less than 10 on a number line.

4. Graph the real numbers from 5 to 15 on a number line.

5. Evaluate $-|23|$.

6. Evaluate $-|7| + |-7|$.

In Problems 7–10, which of the symbols $=$, $<$, or $>$ placed in the box will make a true statement?

7. $3(4 - 2) \quad \boxed{} \quad -2(2 - 6)$

8. $1 + 4 \cdot 3 \quad \boxed{} \quad -2(-7)$

9. 25% of $136 \quad \boxed{} \quad \dfrac{1}{2}$ of 68

10. $-13.7 \quad \boxed{} \quad -|-13.7|$

In Problems 11–16, simplify each expression.

11. $\dfrac{26}{40}$

12. $\dfrac{7}{8} \cdot \dfrac{24}{21}$

13. $\dfrac{18}{35} \div \dfrac{9}{14}$

14. $\dfrac{24}{16} + 3$

15. $\dfrac{17 - 5}{36} - \dfrac{2(13 - 5)}{12}$

16. $\dfrac{|-7 - (-6)|}{-7 - |-6|}$

17. Find 17% of 457 and round the answer to one decimal place.

18. Find the area of a rectangle 12.8 feet wide and 23.56 feet long. Round the answer to two decimal places.

19. Find the area of the figure in Illustration 1.

20. Find the volume of the solid in Illustration 2. Use $\pi = \frac{22}{7}$.

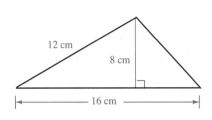

ILLUSTRATION 1

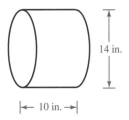

ILLUSTRATION 2

In Problems 21–26, let $x = -2$, $y = 3$, and $z = 4$. Evaluate each expression.

21. $xy + z$

22. $x(y + z)$

23. $\dfrac{z + 4y}{2x}$

24. $|x^y - z|$

25. $x^3 + y^2 + z$

26. $|x| - 3|y| - 4|z|$

27. Let x and y represent two real numbers. Write an algebraic expression to denote the quotient obtained when the product of the two numbers is divided by their sum.

28. Let x and y represent two real numbers. Write an algebraic expression to denote the difference obtained when the sum of x and y is subtracted from the product of 5 and y.

29. A man lives 12 miles from work and 7 miles from the grocery store. If he made x round trips to work and y round trips to the store, how many miles did he drive?

30. A baseball costs a dollars and a glove costs b dollars. How much will it cost the Community Center to buy 12 baseballs and 8 gloves?

31. What is the numerical coefficient of the term $3xy^2$?

32. How many terms are in the expression $3x^2y + 5xy^2 + x + 7$?

33. What is the identity element for addition?

34. What is the multiplicative inverse of $\frac{1}{5}$?

In Problems 35–38, state which property of the real numbers justifies each statement.

35. $(xy)z = z(xy)$

36. $3(x + y) = 3x + 3y$

37. $2 + x = x + 2$

38. $7 \cdot \dfrac{1}{7} = 1$

2

Equations and Inequalities

MATHEMATICS IN RETIREMENT

Employees of most corporations participate in retirement plans funded by contributions from both the company and the employee. Persons who are self-employed can also fund a retirement plan. One such plan, called an SEP (Simplified Employee Pension) allows an annual contribution that does not exceed 15% of the income available after deductible expenses.

That would seem to be an easy calculation—subtract deductible expenses from gross income and take 15% of what's left. However, the tax code is not so simple. The SEP contribution is considered a deductible expense. The SEP contribution is 15% of what is left after subtracting the amount of the SEP contribution. It would seem that to calculate the contribution, you must first know the contribution.

After reading this chapter, you will be able to calculate the maximum annual SEP contribution.

2.1 Introduction to Equations

■ The Addition Property of Equality ■ The Subtraction Property of Equality ■ The Division Property of Equality ■ The Multiplication Property of Equality ■ Percent ■ Applications Involving Percent

GETTING READY   *Find the missing number. Use the inverse properties on page 69.*

1. $3 + \rule{1cm}{0.4pt} = 0$ **2.** $(-7) + \rule{1cm}{0.4pt} = 0$

3. $(-4) + \rule{1cm}{0.4pt} = 0$ **4.** $7 - \rule{1cm}{0.4pt} = 0$

5. $\dfrac{17}{\rule{1cm}{0.4pt}} = 1$ **6.** $\dfrac{-23}{\rule{1cm}{0.4pt}} = 1$

7. $\rule{1cm}{0.4pt} \cdot \left(\dfrac{1}{3}\right) = 1$ **8.** $\rule{1cm}{0.4pt} \cdot \left(-\dfrac{1}{5}\right) = 1$

An **equation** is a statement indicating that two quantities are equal. Here are some examples of equations.

$$x + 5 = 21$$
$$3(3x + 4) = 3 + x$$
$$3x^2 - 4x + 5 = 0$$

In the equation $x + 5 = 21$, $x + 5$ is called the **left-hand side,** and 21 is called the **right-hand side.** The letter x is called the **variable** (or the **unknown**).

An equation can be true or false. The equation $16 + 5 = 21$ is a true equation, but the equation $10 + 5 = 21$ is a **false equation.** An equation such as $2x - 5 = 11$ might be true or false, depending upon the value of x. If $x = 8$, the equation is true, because

$$2(8) - 5 = 16 - 5$$
$$= 11$$

However, this equation is false for all other values of x. Any number that makes an equation true when substituted for its variable is said to **satisfy** the equation. All numbers that satisfy an equation are called its **solutions** or **roots.** Because 8 is the only number that satisfies the equation $2x - 5 = 11$, it is its only solution.

EXAMPLE 1 Is 6 a solution of $3x - 5 = 2x$?

Solution We substitute 6 for x in the equation and simplify.

$$3x - 5 = 2x$$
$$3 \cdot 6 - 5 \overset{?}{=} 2 \cdot 6 \qquad \text{Replace } x \text{ with 6.}$$
$$18 - 5 \overset{?}{=} 12$$
$$13 \neq 12$$

Since $13 \neq 12$, 6 is not a solution of $3x - 5 = 2x$.   ■

■ The Addition Property of Equality

To **solve an equation** means to find its solutions. To do so, we use the following properties of equality to get the variable by itself on one side of the equal sign and a single number on the other side. When either the addition or the subtraction property is applied, the resulting equation will have the same solutions as the original equation.

Addition Property of Equality	Suppose that a, b, and c are real numbers. Then If $a = b$, then $a + c = b + c$.

The **addition property of equality** can be stated in this way: *If the same quantity is added to equal quantities, the results will be equal quantities.*

EXAMPLE 2 Solve $x - 5 = 2$.

Solution To isolate x on one side of the equal sign, we undo the subtraction of 5 by adding 5 to both sides of the equation.

$$x - 5 = 2$$
$$x - 5 + 5 = 2 + 5 \qquad \text{Add 5 to both sides of the equation.}$$
$$x = 7$$

The solution of the given equation is 7. We check it by substituting 7 for x in the original equation and simplifying.

$$x - 5 = 2$$
$$7 - 5 \stackrel{?}{=} 2 \qquad \text{Replace } x \text{ with 7.}$$
$$2 = 2$$

Since $2 = 2$, the solution checks. ∎

■ The Subtraction Property of Equality

Subtraction Property of Equality | Suppose that a, b, and c are real numbers. Then
If $a = b$, then $a - c = b - c$.

The **subtraction property of equality** can be stated in this way: *If the same quantity is subtracted from equal quantities, the results will be equal quantities.*

EXAMPLE 3 Solve $x + 4 = 9$.

Solution To isolate x on one side of the equal sign, we undo the addition of 4 by subtracting 4 from both sides of the equation.

■ PERSPECTIVE

To find answers to such questions as How many? How far? How fast? and How heavy?, we often make use of mathematical statements called *equations*. The concept has a long history, and the techniques we will study in this chapter have been developed over many centuries.

The mathematical notation that we use today is the result of thousands of years of development.

The ancient Egyptians used a word for variables, best translated as *heap*. Others used the word *res*, which is Latin for *thing*. In the 15th century, the letters *p:* and *m:* were used for *plus* and *minus*. What we would now write as $2x + 3 = 5$ might have appeared to those early mathematicians as

2 *res p*: 3 *aequalis* 5

François Vieta (Viête) (1540–1603)
By using letters in place of unknown
numbers, Vieta simplified algebra
and brought its notation closer to
the notation that we use today. The
one symbol he didn't use was the
equal sign.

$$x + 4 = 9$$
$$x + 4 - 4 = 9 - 4 \qquad \text{Subtract 4 from both sides.}$$
$$x = 5$$

We can check the solution of 5 by substituting 5 for x in the original equation and simplifying.

$$x + 4 = 9$$
$$5 + 4 \stackrel{?}{=} 9 \qquad \text{Replace } x \text{ with 5.}$$
$$9 = 9$$

The solution checks. Instead of subtracting 4 from both sides, we could just as well have added -4 to both sides. ∎

EXAMPLE 4

Archeology An ancient piece of pottery is estimated to be 12,000 years older than bones found in a nearby burial site. If the pottery is known to be 75,000 years old, how old are the bones?

Solution We can let b represent the age of the bones and translate the words of the problem into an equation.

The pottery's age	is	12,000	more than	the bones' age.
75,000	=	12,000	+	b

We then solve the equation $75,000 = 12,000 + b$.

$$75,000 = 12,000 + b$$
$$75,000 - \mathbf{12,000} = 12,000 + b - \mathbf{12,000} \qquad \text{Subtract 12,000 from both sides.}$$
$$\mathbf{1.} \qquad 63,000 = b$$

Because 63,000 and b represent the same number, Equation 1 can be written as

$$b = 63,000$$

The bones are 63,000 years old. The pottery, at age 75,000, is 12,000 years older than the bones. The solution checks. ∎

To solve many equations, we must divide or multiply both sides of the equation by the same nonzero number. When either the division or the multiplication property is applied, the resulting equation will have the same solutions as the original equation.

■ The Division Property of Equality

Division Property of Equality	Suppose that a, b, and c are real numbers and that $c \neq 0$. Then If $a = b$, then $\dfrac{a}{c} = \dfrac{b}{c}$.

The **division property of equality** can be stated this way: *If equal quantities are divided by the same nonzero quantity, the results will be equal quantities.*

EXAMPLE 5   Solve $-5x = 15$.

Solution   To isolate x on one side of the equal sign, we undo the multiplication by -5 by dividing both sides of the equation by -5.

$$-5x = 15$$
$$\frac{-5x}{-5} = \frac{15}{-5} \qquad \text{Divide both sides by } -5.$$
$$x = -3$$

Verify that the solution of the equation is -3.   ■

■ The Multiplication Property of Equality

Multiplication Property of Equality	Suppose that a, b, and c are real numbers, and $c \neq 0$. Then If $a = b$, then $ca = cb$.

The **multiplication property of equality** can be stated this way: *If equal quantities are multiplied by the same nonzero quantity, the results will be equal quantities.*

EXAMPLE 6   Solve $\dfrac{x}{5} = 7$.

Solution   To find x, we undo the division by 5 by multiplying both sides of the equation by 5.

$$\frac{x}{5} = 7$$

$$5 \cdot \frac{x}{5} = 5 \cdot 7 \qquad \text{Multiply both sides by 5.}$$

$$x = 35$$

Verify that the solution checks. ■

■ Percent

Percent problems involve answering questions such as

- What is 30% of 1000?
- 405 is 45% of what amount?
- What percent of 400 is 60?

Our knowledge of equations unifies all percent problems and makes them easy to solve.

Since the word *of* often means *multiply,* 30% of 1000 means 30% times 1000.

$$
\begin{aligned}
30\% \text{ of } 1000 &= 30\% \cdot 1000 \\
&= (0.30)(1000) \qquad \text{Change 30\% to the decimal 0.30.} \\
&= 300 \qquad\qquad\;\; \text{Multiply.}
\end{aligned}
$$

In the statement 30% of 1000 is 300, the percent 30% is called the **rate,** 1000 is the **base,** and their product, 300, is called a **percentage.** Every percent problem is based on the equation **rate · base = percentage.**

Percentage	The product of a rate r and a base b is called a **percentage.** If p is the percentage, then
	$$rb = p$$

EXAMPLE 7   45% of what number is 405?

Solution   In this problem, the rate r is 45%, and the percentage p is 405. We can substitute these values into the percentage formula and solve for b.

Rate	·	**base**	=	**percentage**
r	·	b	=	p

$$45\% \cdot b = 405$$

$$0.45 \cdot b = 405 \qquad \text{Change 45\% to a decimal.}$$

$$b = \frac{405}{0.45} \qquad \text{Divide both sides by 0.45.}$$

$$b = 900 \qquad \text{Simplify.}$$

45% of 900 is 405. ■

EXAMPLE 8   What percent of 400 is 60?

Solution   In this problem, the base b is 400, and the percentage p is 60. We can substitute these values into the percentage formula and solve for r.

$$rb = p$$

$$r \cdot 400 = 60$$

$$r = \frac{60}{400} \qquad \text{Divide both sides by 400.}$$

$$r = 0.15 \qquad \text{Do the division.}$$

$$r = 15\% \qquad \text{To change the decimal into a percent, multiply by 100 and add a \% sign.}$$

60 is 15% of 400. ■

■ Applications Involving Percent

EXAMPLE 9   **Shares of stock**   At a recent stockholder's meeting, 4.5 million shares were voted in favor of a proposal for a mandatory retirement age for the board of directors. Since this represented 75% of the total number of shares, the proposal passed. How many shares were there?

Solution   Let b represent the total number of shares. Then 75% of b is 4.5 million. We can substitute 75% for r and 4.5 million for p in the formula for percentage and solve for b.

$$rb = p$$

$$75\% \cdot b = 4,500,000$$

$$0.75b = 4,500,000 \qquad \text{Change 75\% to a decimal.}$$

$$b = \frac{4,500,000}{0.75} \qquad \text{Divide both sides by 0.75.}$$

$$= 6,000,000 \qquad \text{Simplify.}$$

There are 6 million shares outstanding. ■

EXAMPLE 10   **Quality control**   After examining 240 wool sweaters, a quality control inspector found 5 with defective stitching, 8 with mismatched plaids, and 2 marked with incorrect size. What percent were defective?

Solution   Let r represent the percent that are defective. Then the base b is 240, the percentage p is $5 + 8 + 2 = 15$, and we find r by solving the equation

$$rb = p$$
$$r \cdot 240 = 15$$
$$r = \frac{15}{240} \qquad \text{Divide both sides by 240.}$$
$$r = 0.0625 \qquad \text{Use a calculator.}$$
$$r = 6.25\% \qquad \text{To change 0.0625 to a percent, multiply by 100 and add a \% sign.}$$

The defect rate is $6\frac{1}{4}\%$.   ∎

ORALS   *Solve each equation.*

1. $x - 9 = -11$

2. $13 = x - 3$

3. $-7x = 0$

4. $\dfrac{x}{5} = -5$

5. $w + 5 = 5$

6. $x - 345 = 345$

7. $\dfrac{1}{2}x = 222$

8. $2x = 222$

9. What number is 25% of 400?

10. What percent of 36 is 12?

EXERCISE 2.1

In Exercises 1–8, tell whether each statement is an equation.

1. $x = 2$

2. $y - 3$

3. $7x < 8$

4. $7 + x = 2$

5. $x + 7 = 0$

6. $3 - 3y > 2$

7. $1 + 1 = 3$

8. $5 = a + 2$

In Exercises 9–24, tell whether the indicated number is a solution of the equation.

9. $x + 2 = 3$; 1

10. $x - 2 = 4$; 6

11. $a - 7 = 0$; -7

12. $x + 4 = 4$; 0

13. $2x = 4$; 2

14. $3x = 6$; 3

15. $3x - 1 = 7$; 2

16. $2x + 1 = 7$; 3

17. $\dfrac{y}{7} = 4$; 28

18. $\dfrac{c}{-5} = -2$; -10

19. $\dfrac{x}{5} = x$; 0

20. $\dfrac{x}{7} = 7x$; 0

21. $3k + 5 = 5k - 1$; 3

22. $2s - 1 = s + 7$; 6

23. $\dfrac{5 + x}{10} - x = \dfrac{1}{2}$; 0

24. $\dfrac{x - 5}{6} = 12 - x$; 11

In Exercises 25–40, use the addition or the subtraction property of equality to solve each equation. Check all solutions.

25. $x - 7 = 3$

26. $y - 3 = 7$

27. $y + 7 = 12$

28. $c + 11 = 22$

29. $-37 + z = 37$

30. $-43 + a = -43$

31. $-57 = b - 29$

32. $-93 = 67 + y$

33. $\dfrac{4}{3} = -\dfrac{2}{3} + x$

34. $z + \dfrac{5}{7} = -\dfrac{2}{7}$

35. $d + \dfrac{2}{3} = \dfrac{3}{2}$

36. $s + \dfrac{2}{3} = \dfrac{1}{5}$

37. $-\dfrac{3}{5} = x - \dfrac{2}{5}$

38. $b + 7 = \dfrac{20}{3}$

39. $r - \dfrac{1}{5} = \dfrac{3}{10}$

40. $t + \dfrac{4}{7} = \dfrac{11}{14}$

In Exercises 41–56, use the division or the multiplication property of equality to solve each equation. Check each solution.

41. $3x = 3$

42. $5x = 5$

43. $\dfrac{x}{5} = 5$

44. $4x = 36$

45. $-32z = 64$

46. $15 = \dfrac{r}{-5}$

47. $18z = -9$

48. $-12z = 3$

49. $\dfrac{z}{7} = 14$

50. $-19x = -57$

51. $\dfrac{w}{7} = \dfrac{5}{7}$

52. $-17z = -51$

53. $\dfrac{s}{-3} = -\dfrac{5}{6}$

54. $1228 = \dfrac{x}{0.25}$

55. $0.25x = 1228$

56. $-255y = 51$

In Exercises 57–80, solve each equation and check each solution.

57. $x - 3 = 12$

58. $x + 2 = 20$

59. $-7y = 7$

60. $y - 11 = 9$

61. $4t = 108$

62. $-66 = -6t$

63. $11x = -121$

64. $x - 29 = -43$

65. $0 = 5 + x$

66. $3 = x + 7$

67. $-9 + y = -9$

68. $-9y = -9$

69. $\dfrac{b}{3} = 5$

70. $\dfrac{a}{5} = -3$

71. $-3 = \dfrac{s}{11}$

72. $\dfrac{s}{-12} = 4$

73. $\dfrac{b}{3} = \dfrac{1}{3}$

74. $\dfrac{a}{13} = \dfrac{1}{26}$

75. $-34w = -17$

76. $\dfrac{t}{-7} = \dfrac{1}{2}$

77. $\dfrac{u}{5} = -\dfrac{3}{10}$

78. $v - \dfrac{7}{3} = -\dfrac{5}{6}$

79. $x + 17 = \dfrac{33}{2}$

80. $z + \dfrac{7}{9} = \dfrac{2}{9}$

In Exercises 81–96, use the formula $rb = p$ to find each value.

81. What number is 40% of 200?

82. What number is 35% of 520?

83. What number is 50% of 38?

84. What number is 25% of 300?

85. 15% of what number is 48?

86. 26% of what number is 78?

87. 133 is 35% of what number?

88. 13.3 is 3.5% of what number?

89. 28% of what number is 42?

90. 44% of what number is 143?

91. What percent of 357.5 is 71.5?

92. What percent of 254 is 13.208?

93. 0.32 is what percent of 4?

95. 34 is what percent of 17?

94. 3.6 is what percent of 28.8?

96. 39 is what percent of 13?

In Exercises 97–108, solve each problem.

97. Buying real estate   The cost of a condominium is $57,595 less than the cost of a house. If the house costs $102,744, find the cost of the condominium.

98. Buying paint   After reading the ad in Illustration 1, a decorator bought one gallon of primer, one gallon of paint, and a brush. The total cost was $30.44. Find the cost of the brush.

Sale

Primer $10.99

Latex Flat Paint $14.50

ILLUSTRATION 1

99. Customer satisfaction   Two-thirds of the movie audience left the theater in disgust. If 78 angry patrons walked out, how many were there originally?

100. Stock split   After a three-for-two stock split, a shareholder will own 1.5 times as many shares as before. If 555 shares are owned after the split, how many were owned before?

101. Off-campus housing   Four-sevenths of the senior class is living in off-campus housing. If 868 students live off campus, how large is the senior class?

102. Union membership   The 2484 union members represent 90% of a factory's workforce. How many employees are there?

103. Shopper dissatisfaction   Refer to the survey results in Illustration 2. What percent of those surveyed were not satisfied with the service?

Shopper Survey Results	
First-time shoppers	1731
Major purchase today	539
Shopped within previous month	1823
Satisfied with service	4140
Seniors	2387
Total surveyed	9200

ILLUSTRATION 2

104. Charity overhead   Out of $237,000 donated to a certain charity, $5925 was used to pay for fundraising expenses. What percent of donations was overhead?

105. Selling price of a microwave oven   The 5% sales tax on a microwave oven amounts to $13.50. What is the microwave's selling price?

106. Sales tax rate   Sales tax on a $12 compact disk is $0.72. At what rate is sales tax computed?

107. Hospital occupancy   36% of hospital patients stay for less than 2 days. If 1008 patients in January stayed for less than 2 days, what total number of patients did the hospital treat in January?

108. Home prices   The average price of homes in one neighborhood decreased 8% since last year, a drop of $7800. What was the average price of a home last year?

Writing Exercises   ■   *Write a paragraph using your own words.*

1. Explain how you would decide whether a number is a solution of an equation.

2. Distinguish between *percent* and *percentage*.

Something to Think About ■ 1. The Ahmes papyrus mentioned at the beginning of Chapter 1 provides a window to the mathematical world of ancient Egypt. The papyrus contains this statement: *A circle nine units in diameter has the same area as a square eight units on a side.* From this, determine the ancient Egyptian value of π.

2. ▦ Calculate the Egyptians' **percent of error:** what percent of the actual value of π is the difference between the values?

Review Exercises ■ *Do the indicated operations and classify the result as an integer, a prime number, and a composite number.*

1. $3[2 - (-3)]$

2. $(2 - 4)^4$

3. $\dfrac{2^3 - 14}{3^2 - 3}$

4. $\dfrac{3 + 5}{3} - \dfrac{5}{7 - 4}$

Tell which property of real numbers or property of equality justifies each statement.

5. $3 + 31$ is a real number

6. $3(x + y) = 3x + 3y$

7. $a + (3 + b) = (3 + b) + a$

8. $a + (3 + b) = (a + 3) + b$

2.2 Solving More Equations

■ **Markup and Markdown**

GETTING READY *Do the indicated operations.*

1. $7 + 3 \cdot 5$

2. $3(5 + 7)$

3. $\dfrac{3 + 7}{2}$

4. $3 + \dfrac{7}{2}$

5. $\dfrac{3(5 - 8)}{9}$

6. $3 \cdot \dfrac{5 - 8}{9}$

7. $\dfrac{3 \cdot 5 - 8}{9}$

8. $3 \cdot \dfrac{5}{9} - 8$

We have solved equations by using just one of the addition, subtraction, multiplication, and division properties of equality. To solve more complicated equations, we need to use several of these properties in succession.

EXAMPLE 1 Solve $-12x + 5 = 17$.

Analysis The left-hand side of the equation indicates that a multiplication is done first, followed by an addition. To solve this equation, we must undo these operations in the opposite order.

- To undo the addition of 5, we subtract 5 from both sides.
- To undo the multiplication by -12, we divide both sides by -12.

Solution

$$-12x + 5 = 17$$

$$-12x + 5 - 5 = 17 - 5 \qquad \text{Subtract 5 from both sides.}$$

$$-12x = 12 \qquad \text{Simplify.}$$

$$\frac{-12x}{-12} = \frac{12}{-12} \qquad \text{Divide both sides by } -12.$$

$$x = -1$$

Check: $-12x + 5 = 17$

$$-12(-1) + 5 \stackrel{?}{=} 17 \qquad \text{Replace } x \text{ with } -1.$$

$$12 + 5 \stackrel{?}{=} 17 \qquad \text{Simplify.}$$

$$17 = 17$$

Because $17 = 17$, the solution checks. ∎

EXAMPLE 2 Solve $\dfrac{x}{3} - 7 = -3$.

Analysis The left-hand side of the equation indicates that a division is done first, followed by a subtraction. To solve this equation, we must undo these operations in the opposite order.

- To undo the subtraction of 7, we add 7 to both sides.
- To undo the division by 3, we multiply both sides by 3.

Solution

$$\frac{x}{3} - 7 = -3$$

$$\frac{x}{3} - 7 + 7 = -3 + 7 \qquad \text{Add 7 to both sides.}$$

$$\frac{x}{3} = 4 \qquad \text{Simplify.}$$

$$3\left(\frac{x}{3}\right) = 3 \cdot 4 \qquad \text{Multiply both sides by 3.}$$

$$x = 12 \qquad \text{Simplify.}$$

Check: $\dfrac{x}{3} - 7 = -3$

$$\frac{12}{3} - 7 \stackrel{?}{=} -3 \qquad \text{Replace } x \text{ with 12.}$$

$$4 - 7 \stackrel{?}{=} -3 \qquad \text{Simplify.}$$

$$-3 = -3$$

Since $-3 = -3$, the solution checks. ∎

EXAMPLE 3 Solve $\dfrac{x-7}{3} = 9$.

Analysis The left-hand side of the equation indicates that a subtraction is done first, followed by a division. To solve this equation, we must undo these operations in the opposite order.

- To undo the division by 3, we multiply both sides by 3.
- To undo the subtraction of 7, we add 7 to both sides.

Solution

$$\frac{x-7}{3} = 9$$

$$3\left(\frac{x-7}{3}\right) = 3(9) \qquad \text{Multiply both sides by 3.}$$

$$x - 7 = 27 \qquad \text{Simplify.}$$

$$x - 7 + 7 = 27 + 7 \qquad \text{Add 7 to both sides.}$$

$$x = 34 \qquad \text{Simplify.}$$

Verify that the solution checks. ∎

EXAMPLE 4 Solve $\dfrac{3x}{4} + 2 = -7$.

Analysis The left-hand side of the equation indicates that a multiplication (by 3) is done first, followed by a division (by 4) and then an addition. To solve this equation, we must undo these operations in the opposite order.

- To undo the addition of 2, we subtract 2 from both sides.
- To undo the division by 4, we multiply both sides by 4.
- To undo the multiplication by 3, we divide both sides by 3.

Solution

$$\frac{3x}{4} + 2 = -7$$

$$\frac{3x}{4} + 2 - 2 = -7 - 2 \qquad \text{Subtract 2 from both sides.}$$

$$\frac{3x}{4} = -9 \qquad \text{Simplify.}$$

$$4\left(\frac{3x}{4}\right) = 4(-9) \qquad \text{Multiply both sides by 4.}$$

$$3x = -36 \qquad \text{Simplify.}$$

$$\frac{3x}{3} = \frac{-36}{3} \qquad \text{Divide both sides by 3.}$$

$$x = -12 \qquad \text{Simplify.}$$

Verify that the solution checks. ∎

EXAMPLE 5   **Advertising**   The manager of a supermarket hires a woman to distribute advertising circulars door to door. She will be paid $5 a day plus $0.05 for every advertisement distributed. How many circulars must she distribute to earn $42.50?

Solution   We can let *a* represent the number of circulars that the woman must distribute. Her earnings can be expressed in two ways: as $5 more than the $0.05-apiece cost of distributing the circulars and as $42.50.

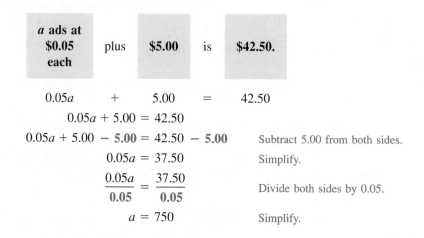

| *a* ads at $0.05 each | plus | $5.00 | is | $42.50. |

$$0.05a \quad + \quad 5.00 \quad = \quad 42.50$$

$$0.05a + 5.00 = 42.50$$

$$0.05a + 5.00 - \textbf{5.00} = 42.50 - \textbf{5.00} \qquad \text{Subtract 5.00 from both sides.}$$

$$0.05a = 37.50 \qquad \text{Simplify.}$$

$$\frac{0.05a}{\textbf{0.05}} = \frac{37.50}{\textbf{0.05}} \qquad \text{Divide both sides by 0.05.}$$

$$a = 750 \qquad \text{Simplify.}$$

The woman must distribute 750 advertisements. Check this result.   ■

■ Markup and Markdown

To make a profit, a merchant must sell a product for more than he paid for it. The retail price of the product is the sum of the cost and the markup.

| Retail price | = | cost | + | markup. |

Sometimes, markup is expressed as a **percent of cost.**

| Markup | = | percent of markup | · | cost. |

Suppose a store manager buys toasters for $21 and sells them at a 17% markup. To find the retail price, the manager begins with his cost and adds 17% of that cost.

$$\boxed{\text{Retail price}} \;=\; \boxed{\text{cost}} \;+\; \boxed{\text{markup.}}$$

$$=\; \boxed{\text{cost}} \;+\; \boxed{\text{percent of markup}} \;\cdot\; \boxed{\text{cost.}}$$

$$= 21 + (0.17)(21)$$
$$= 21 + 3.57$$
$$= 24.57$$

The retail price of a toaster is $24.57.

EXAMPLE 6 **Antique cars** In 1956, a Chevrolet BelAir automobile sold for $4000. Today, it sold for $28,600. Find the **percent of increase.**

Solution We let p represent the percent of increase, expressed as a decimal.

$$\boxed{\text{Current price}} \;=\; \boxed{\text{original price}} \;+\; \boxed{p \cdot \text{(original price).}}$$

$$28,600 \;=\; 4000 \;+\; p(4000)$$

$$28,600 - \mathbf{4000} = 4000 + 4000p - \mathbf{4000} \qquad \text{Subtract 4000 from both sides.}$$

$$24,600 = 4000p$$

$$\frac{24,600}{\mathbf{4000}} = \frac{4000p}{\mathbf{4000}} \qquad \text{Divide both sides by 4000.}$$

$$6.15 = p \qquad \text{Use a calculator.}$$

To convert the decimal 6.15 to a percent, we multiply by 100 and insert a % sign: The percent of increase is 615%. We can say that the car has appreciated 615%. ■

When the regular price of merchandise is reduced, the amount of reduction is the **markdown.**

$$\boxed{\text{Sale price}} \;=\; \boxed{\text{regular price}} \;-\; \boxed{\text{markdown.}}$$

Usually, markdown is expressed as a percent of the regular price.

$$\boxed{\text{Markdown}} \;=\; \boxed{\text{percent of markdown}} \;\cdot\; \boxed{\text{regular price.}}$$

The markdown is often called the **discount.**

Suppose that a television set that regularly sells for $570 has been marked down 25%. That means the customer will pay 25% less than the regular price. We find the sale price as follows:

$$\boxed{\textbf{Sale price}} \quad = \quad \boxed{\textbf{regular price}} \quad - \quad \boxed{\textbf{markdown.}}$$

$$= \$570 - 25\% \text{ of } \$570$$
$$= \$570 - (.25)(\$570)$$
$$= \$570 - \$142.50$$
$$= \$427.50$$

The television set is selling for $427.50.

EXAMPLE 7   **Buying a camera**   A camera on sale for $384.20 was originally priced at $452. Find the percent of discount.

Solution   We let r represent the percent of discount, expressed as a decimal. We substitute $384.20 for the sale price and $452 for the regular price.

$$\boxed{\textbf{Sale price}} \quad = \quad \boxed{\textbf{regular price}} \quad - \quad \boxed{\textbf{markdown.}}$$

$$384.20 = 452 - r(452)$$
$$384.20 - \mathbf{452} = 452 - r(452) - \mathbf{452} \qquad \text{Subtract 452 from both sides.}$$
$$-67.80 = -r(452)$$
$$\frac{-67.80}{-452} = \frac{-r(452)}{-452} \qquad \text{Divide both sides by } -452.$$
$$0.15 = r$$

The camera is on sale at a 15% discount.   ∎

WARNING!   When a price increases from $100 to $125, the percent of increase is 25%. When the price *decreases* from $125 to $100, the **percent of decrease** is 20%. These different results occur because the percent of increase is a percent of the original (smaller) price, $100. The percent of decrease is a percent of the original (larger) price, $125.

ORALS   *Tell what would be your first step in solving each equation.*

1. $5x - 7 = -12$

2. $15 = \dfrac{x}{5} + 3$

3. $\dfrac{x}{7} - 3 = 0$

4. $\dfrac{x - 3}{7} = -7$

5. $5w - 5 = 5$

6. $5w + 5 = 5$

7. $\dfrac{x - 7}{3} = 5$

8. $\dfrac{3x - 5}{2} + 2 = 0$

Find the value of the variable in each equation.

9. $7z - 7 = 14$

10. $\dfrac{t - 1}{2} = 6$

E X E R C I S E   2 . 2

In Exercises 1–52, solve each equation and check each solution.

1. $5x - 1 = 4$

2. $5x + 3 = 8$

3. $6x + 2 = -4$

4. $4x - 4 = 4$

5. $3x - 8 = 1$

6. $7x - 19 = 2$

7. $11x + 17 = -5$

8. $13x - 29 = -3$

9. $43t + 72 = 158$

10. $96t + 23 = -265$

11. $-47 - 21s = 58$

12. $-151 + 13s = -229$

13. $2y - \dfrac{5}{3} = \dfrac{4}{3}$

14. $9y + \dfrac{1}{2} = \dfrac{3}{2}$

15. $-4y - 12 = -20$

16. $-8y + 64 = -32$

17. $\dfrac{x}{3} - 3 = -2$

18. $\dfrac{x}{7} + 3 = 5$

19. $\dfrac{z}{9} + 5 = -1$

20. $\dfrac{y}{5} - 3 = 3$

21. $\dfrac{b}{3} + 5 = 2$

22. $\dfrac{a}{5} - 3 = -4$

23. $\dfrac{s}{11} + 9 = 6$

24. $\dfrac{r}{12} + 2 = 4$

25. $\dfrac{k}{5} - \dfrac{1}{2} = \dfrac{3}{2}$

26. $\dfrac{y}{5} - \dfrac{8}{7} = -\dfrac{1}{7}$

27. $\dfrac{w}{16} + \dfrac{5}{4} = 1$

28. $\dfrac{m}{7} - \dfrac{1}{14} = \dfrac{1}{14}$

29. $\dfrac{b + 5}{3} = 11$

30. $\dfrac{2 + a}{13} = 3$

31. $\dfrac{r + 7}{3} = 4$

32. $\dfrac{t - 2}{7} = -3$

33. $\dfrac{u - 2}{5} = 1$

34. $\dfrac{v - 7}{3} = -1$

35. $\dfrac{x - 4}{4} = -3$

36. $\dfrac{3 + y}{5} = -3$

37. $\dfrac{3x}{2} - 6 = 9$

38. $\dfrac{5x}{7} + 3 = 8$

39. $\dfrac{3y}{2} + 5 = 11$

40. $\dfrac{5z}{3} + 3 = -2$

41. $\dfrac{3x - 12}{2} = 9$

42. $\dfrac{5x + 10}{7} = 0$

43. $\dfrac{5k - 8}{9} = 1$

44. $\dfrac{2x - 1}{3} = -5$

45. $\dfrac{3z + 2}{17} = 0$ **46.** $\dfrac{10t - 4}{2} = 1$ **47.** $\dfrac{17k - 28}{21} + \dfrac{4}{3} = 0$ **48.** $\dfrac{5a - 2}{3} = \dfrac{1}{6}$

49. $-\dfrac{x}{3} - \dfrac{1}{2} = -\dfrac{5}{2}$ **50.** $\dfrac{17 - 7a}{8} = 2$ **51.** $\dfrac{9 - 5w}{15} = \dfrac{2}{5}$ **52.** $\dfrac{3t - 5}{5} + \dfrac{1}{2} = -\dfrac{19}{2}$

In Exercises 53–68, solve each problem.

53. Integer problem   Six less than 3 times a certain integer is 9. Find the integer.

54. Integer problem   If a certain integer is increased by 7 and that result divided by 2, the integer 5 is obtained. Find the original integer.

55. Apartment rental   A student moves in to a bigger apartment that rents for $400 per month. That rent is $100 less than twice what she is now paying. Find the current rent.

56. Auto repair   A mechanic charged $20 an hour to repair the water pump on a car, plus $95 for parts. If the total bill was $155, how many hours did the repair take?

57. Boarding dogs   A sportsman boarded his dog at the kennel for $16 plus $12 a day. If the stay cost $100, how many days was the owner gone?

58. Water billing   The city's water department charges $7 per month, plus 42 cents for every 100 gallons of water used. Last month, one homeowner used 1900 gallons and received a bill for $17.98. Was the billing correct?

59. Telephone charges   A call to Tucson from a pay phone in Chicago costs 85 cents for the first minute and 27 cents for each additional minute or portion of a minute. If a student has $8.50 in change, how long can she talk?

60. Monthly sales   A clerk's sales in February were $2000 less than three times her sales in January. If her February sales were $7000, by what amount did her sales increase?

61. Ticket sales   A music group charges $1500 for each performance, plus 20% of the total ticket sales. After a concert, the group received $2980. How much money did the ticket sales raise?

62. Getting an A   To receive a grade of A, the average of four 100-point exams must be 90 or better. If a student received scores of 88, 83, and 92 on the first three exams, what score does he need on the fourth exam to earn an A?

63. Getting an A   The grade in History class is based on the average of five 100-point exams. One student received scores of 85, 80, 95, and 78 on the first four exams. With an average of 90 needed, what chance does he have for an A?

64. Excess inventory   From the portion of the ad in Illustration 1, find the sale price of a shirt.

ILLUSTRATION 1

65. Clearance sales   Sweaters already on sale for 20% off the regular price cost $36 when purchased with a promotional coupon that allows an additional 10% discount off the sale price. Find the original price. (*Hint:* When you save 20%, you are paying 80%.)

66. Furniture sale   A $1250 sofa is marked down to $900. Find the percent of markdown.

67. Value of coupons The percent discount offered by the coupon in Illustration 2 depends on the amount purchased. Find the range of the percent discount.

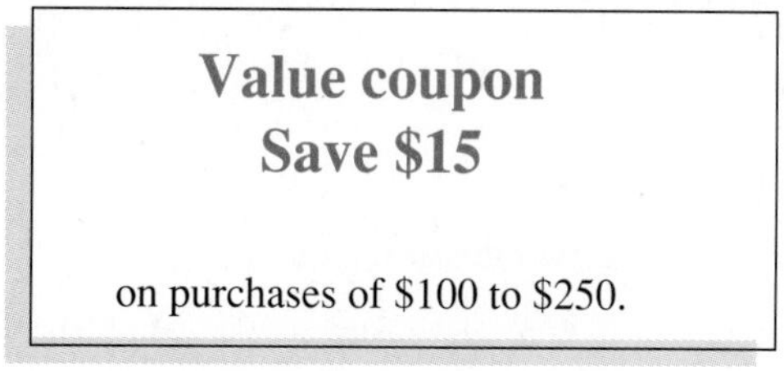

68. Furniture pricing A bedroom set selling for $1900 cost $1000 wholesale. Find the percent markup.

Value coupon
Save $15

on purchases of $100 to $250.

ILLUSTRATION 2

Writing Exercises ■ *Write a paragraph using your own words.*

1. In solving the equation $5x - 3 = 12$, explain why you would add 3 to both sides first, rather than divide by 5.

2. To solve the equation $\frac{3x-4}{7} = 2$, what operations would you perform, and in what order?

Something to Think About ■ **1.** Suppose you must solve the following equation but can't quite read one number. It reads $\frac{7x + ?}{22} = \frac{1}{2}$. The answer is $x = -2$.

What was the equation?

2. A store manager first increases his prices by 30%, and then advertises SALE!! 30% savings!! What is the real percent discount to the customer?

Review Exercises ■ *Refer to the formulas given in Section 1.3.*

1. Find the perimeter of a rectangle with sides measuring 8.5 cm and 16.5 cm.

2. Find the area of a rectangle with sides measuring 2.3 in. and 3.7 in.

3. Find the area of a trapezoid with a height of 8.5 in. and bases measuring 6.7 in. and 12.2 in.

4. Find the volume of a rectangular solid with dimensions of 8.2 cm by 7.6 cm by 10.2 cm.

2.3 Simplifying Expressions to Solve Equations

■ Combining Terms ■ Solving Equations ■ Identities and Impossible Equations ■ Problem Solving

GETTING READY   *Use the distributive property to remove parentheses.*

1. $(3 + 4)x$ **2.** $(7 + 2)x$ **3.** $(8 - 3)w$ **4.** $(10 - 4)y$

Simplify each expression by doing the operations within the parentheses.

5. $(3 + 4)x$ **6.** $(7 + 2)x$ **7.** $(8 - 3)w$ **8.** $(10 - 4)y$

Recall that a *term* is either a number or the product of numbers and variables. Some examples of terms are $7x$, $-3xy$, y^2, and 8. The **numerical coefficient** of the term $7x$ is 7, the numerical coefficient of the term $-3xy$ is -3, and the numerical coefficient of the term y^2 is the understood factor of 1. The number 8 is the numerical coefficient of the term 8.

Like Terms   **Like terms,** or **similar terms,** are terms with exactly the same variables and exponents.

The terms $3x$ and $5x$ are **like terms,** as are $9x^2$ and $-3x^2$. The terms $4xy$ and $3x^2$ are **unlike terms** because they have different variables. The terms $4x$ and $5x^2$ are unlike terms because the variables have different exponents.

■ Combining Terms

The distributive property can be used to combine terms of algebraic expressions that contain sums or differences of like terms. For example, the terms in the binomials $3x + 5x$ and $9xy^2 - 11xy^2$ can be combined as follows:

$$3x + 5x = (3 + 5)x \qquad 9xy^2 - 11xy^2 = (9 - 11)xy^2$$
$$= 8x \qquad\qquad\qquad = -2xy^2$$

These examples suggest the following rule.

Combining Like Terms   To combine like terms, add their numerical coefficients and keep the same variables and exponents.

WARNING! If the terms of an expression are unlike terms, they cannot be combined. Because the terms of the expression $9xy^2 - 11x^2y$, for example, have variables with different exponents, they are unlike terms and cannot be combined.

EXAMPLE 1 Simplify $3(x + 2) + 2(x - 8)$.

Solution

$$
\begin{aligned}
3(x + 2) + 2(x - 8) &= 3x + 3 \cdot 2 + 2x - 2 \cdot 8 && \text{Remove parentheses.} \\
&= 3x + 6 + 2x - 16 && \text{Simplify.} \\
&= 3x + 2x + 6 - 16 && \text{Use the associative and} \\
& && \text{commutative properties} \\
& && \text{of addition to rearrange} \\
& && \text{terms.} \\
&= 5x - 10 && \text{Combine like terms.} \quad \blacksquare
\end{aligned}
$$

EXAMPLE 2 Simplify $3(x - 3) - 5(x + 4)$.

Solution

$$
\begin{aligned}
3(x - 3) - 5(x + 4) &= 3(x - 3) + (-5)(x + 4) && a - b = a + (-b) \\
&= 3x - 3 \cdot 3 + (-5)x + (-5)4 && \text{Remove parentheses.} \\
&= 3x - 9 + (-5x) + (-20) && \text{Multiply.} \\
&= -2x - 29 && \text{Combine like} \\
& && \text{terms.} \quad \blacksquare
\end{aligned}
$$

◼ Solving Equations

To solve any equation, we must isolate the variable on one side. This is often a many-step process that may require combining like terms. As we solve equations, we will follow these steps.

Strategy for Solving Equations

1. Clear the equation of fractions.
2. Use the distributive property to remove parentheses.
3. Combine like terms if necessary.
4. Undo the operations of addition and subtraction to get the variables on one side and the constants on the other.
5. Combine like terms and undo the operations of multiplication and division to isolate the variable.

EXAMPLE 3   Solve $3(x + 2) - 5x = 0$.

Solution

$$3(x + 2) - 5x = 0$$
$$3x + 3 \cdot 2 - 5x = 0 \qquad \text{Remove parentheses.}$$
$$3x - 5x + 6 = 0 \qquad \text{Rearrange terms and simplify.}$$
$$-2x + 6 = 0 \qquad \text{Combine like terms.}$$
$$-2x + 6 - 6 = 0 - 6 \qquad \text{Subtract 6 from both sides.}$$
$$-2x = -6 \qquad \text{Combine like terms.}$$
$$\frac{-2x}{-2} = \frac{-6}{-2} \qquad \text{Divide both sides by } -2.$$
$$x = 3 \qquad \text{Simplify.}$$

Check: $3(x + 2) - 5x = 0$
$$3(3 + 2) - 5 \cdot 3 \overset{?}{=} 0 \qquad \text{Replace } x \text{ with 3.}$$
$$3 \cdot 5 - 5 \cdot 3 \overset{?}{=} 0$$
$$15 - 15 \overset{?}{=} 0$$
$$0 = 0$$

■

EXAMPLE 4   Solve $3(x - 5) = 4(x + 9)$.

Solution

$$3(x - 5) = 4(x + 9)$$
$$3x - 15 = 4x + 36 \qquad \text{Remove parentheses.}$$
$$3x - 15 - 3x = 4x + 36 - 3x \qquad \text{Subtract } 3x \text{ from both sides.}$$
$$-15 = x + 36 \qquad \text{Combine like terms.}$$
$$-15 - 36 = x + 36 - 36 \qquad \text{Subtract 36 from both sides.}$$
$$-51 = x \qquad \text{Combine like terms.}$$
$$x = -51$$

Check: $3(x - 5) = 4(x + 9)$
$$3(-51 - 5) \overset{?}{=} 4(-51 + 9) \qquad \text{Replace } x \text{ with } -51.$$
$$3(-56) \overset{?}{=} 4(-42)$$
$$-168 = -168$$

■

EXAMPLE 5   Solve $\dfrac{3x + 11}{5} = x + 3$.

Solution   We first multiply both sides by 5 to clear the equation of fractions. When we multiply the right-hand side of the equation by 5, we must multiply the *entire* right-hand side by 5.

$$\frac{3x + 11}{5} = x + 3$$

$5\left(\dfrac{3x + 11}{5}\right) = 5(x + 3)$	Multiply both sides by 5.
$3x + 11 = 5x + 15$	Remove parentheses.
$3x + 11 - 11 = 5x + 15 - 11$	Subtract 11 from both sides.
$3x = 5x + 4$	Combine like terms.
$3x - 5x = 5x + 4 - 5x$	Subtract $5x$ from both sides.
$-2x = 4$	Combine like terms.
$\dfrac{-2x}{-2} = \dfrac{4}{-2}$	Divide both sides by -2.
$x = -2$	Simplify.

$$Check: \quad \frac{3x + 11}{5} = x + 3$$

$\dfrac{3(-2) + 11}{5} \overset{?}{=} (-2) + 3$	Replace x with -2.
$\dfrac{-6 + 11}{5} \overset{?}{=} 1$	Siimplify.
$\dfrac{5}{5} \overset{?}{=} 1$	
$1 = 1$	∎

WARNING! Remember that when you multiply one side of an equation by a nonzero number, you must multiply the other side of the equation by the same number.

EXAMPLE 6 Solve $0.2x + 0.4(50 - x) = 19$.

Solution Because $0.2 = \frac{2}{10}$ and $0.4 = \frac{4}{10}$, this equation contains fractions. To clear the equation of these fractions, we multiply both sides by 10 and proceed as follows.

$0.2x + 0.4(50 - x) = 19$	
$10[0.2x + 0.4(50 - x)] = 10(19)$	Multiply both sides by 10.
$10[0.2x] + 10[0.4(50 - x)] = 10(19)$	Use the distributive property on the left-hand side.
$2x + 4(50 - x) = 190$	Do the multiplications.
$2x + 200 - 4x = 190$	Remove parentheses.
$-2x + 200 = 190$	Combine like terms.
$-2x = -10$	Subtract 200 from both sides.
$x = 5$	Divide both sides by -2.

Verify that the solution checks. ∎

EXAMPLE 7   Solve $x(x - 5) = x^2 + 15$.

Solution

$$x(x - 5) = x^2 + 15$$
$$x^2 - 5x = x^2 + 15 \qquad \text{Remove parentheses.}$$
$$x^2 - 5x - x^2 = x^2 + 15 - x^2 \qquad \text{Subtract } x^2 \text{ from both sides.}$$
$$-5x = 15 \qquad \text{Combine like terms.}$$
$$\frac{-5x}{-5} = \frac{15}{-5} \qquad \text{Divide both sides by } -5.$$
$$x = -3 \qquad \text{Simplify.}$$

Verify that the solution checks.   ∎

■ Identities and Impossible Equations

An equation that is true for all values of its variable is called an **identity.** For example, the equation

$$x + x = 2x$$

is an identity because it is true for all values of x.

Because no number can equal a number that is 1 larger than itself, the equation $x = x + 1$ is not true for any number x. Such equations are called **impossible equations** or **contradictions.**

The equations in Examples 3–7 have solutions, so they are not impossible equations. Those equations are not identities either, because they are false for some values of x. Such equations, called **conditional equations,** are true for some but not all real numbers.

EXAMPLE 8   Solve $3(x + 8) + 5x = 2(12 + 4x)$.

Solution

$$3(x + 8) + 5x = 2(12 + 4x)$$
$$3x + 24 + 5x = 24 + 8x \qquad \text{Remove parentheses.}$$
$$8x + 24 = 24 + 8x \qquad \text{Combine like terms.}$$
$$8x + 24 - 8x = 24 + 8x - 8x \qquad \text{Subtract } 8x \text{ from both sides.}$$
$$24 = 24 \qquad \text{Combine like terms.}$$

Since the result $24 = 24$ is true for every number x, every number x is a solution of the original equation. This equation is an identity.   ∎

EXAMPLE 9   Solve $3(x + 7) - x = 2(x + 10)$.

Solution
$$3(x + 7) - x = 2(x + 10)$$
$$3x + 21 - x = 2x + 20 \qquad \text{Remove parentheses.}$$
$$2x + 21 = 2x + 20 \qquad \text{Combine like terms.}$$
$$2x + 21 - \mathbf{2x} = 2x + 20 - \mathbf{2x} \qquad \text{Subtract } 2x \text{ from both sides.}$$
$$21 = 20 \qquad \text{Combine like terms.}$$

Since the result $21 = 20$ is false, the original equation has no solution. It is an impossible equation.   ■

■ Problem Solving

We can use equations to solve several types of problems. To set up those equations, we will follow these steps.

Strategy for Problem Solving

1. Read the problem several times and analyze the facts. What information is given? What are you asked to find? Occasionally a sketch, chart, or diagram will help you visualize the facts of the problem.
2. Pick a variable to represent the quantity to be found and write a sentence telling what the variable represents. Express all other important quantities mentioned in the problem as expressions involving this single variable.
3. Organize the data and find a way to express a quantity in two different ways.
4. Write an equation showing that the two quantities found in Step 3 are equal.
5. Solve the equation.
6. State the solution or solutions.
7. Check the answer in the words of the problem. Have all the questions been answered?

EXAMPLE 10   **Plumbing**   A plumber wants to cut a 17-foot pipe into three sections. The longest section is to be 3 times as long as the shortest, and the middle-sized section is to be 2 feet longer than the shortest. How long should each section be?

Analysis   The information in this problem is given in terms of the length of the shortest section of the pipe. Therefore, we let a variable represent the length of the shortest section. We then express the other facts in terms of that variable.

Solution   Let x represent the length of the shortest section. Then

$3x$ represents the length of the longest section, and

$x + 2$ represents the length of the middle-sized section.

We sketch the pipe as shown in Figure 2-1.

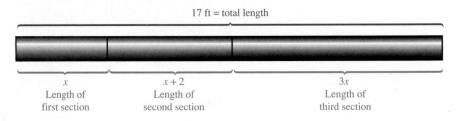

17 ft = total length

| x | $x + 2$ | $3x$ |
| Length of first section | Length of second section | Length of third section |

FIGURE 2-1

The sum of the lengths of these three sections must equal the total length of the pipe.

The length of section 1	plus	the length of section 2	plus	the length of section 3	equals	the total length.
x	$+$	$x + 2$	$+$	$3x$	$=$	17

$$x + x + 2 + 3x = 17 \qquad \text{The equation to solve.}$$
$$5x + 2 = 17 \qquad \text{Combine like terms.}$$
$$5x = 15 \qquad \text{Subtract 2 from both sides.}$$
$$x = 3 \qquad \text{Divide both sides by 5.}$$

The shortest section is 3 feet long. Because the middle-sized section is 2 feet longer than the shortest, it is 5 feet long. Because the longest section is 3 times the shortest, it is 9 feet long.

Check: Because 3 feet, 5 feet, and 9 feet total 17 feet, the solution checks.   ■

EXAMPLE 11   **Road construction**   A truck made five trips hauling asphalt to a road construction site. A larger truck made eight trips. When fully loaded, the larger truck carries 2 tons more asphalt than the smaller truck. If the two trucks hauled a total of 55 tons, how much asphalt can the smaller truck carry in one load?

Analysis   We let x represent the number of tons that the smaller truck can carry when it is fully loaded. Because the larger truck can carry 2 tons more, it can carry $(x + 2)$ tons when fully loaded. Since the smaller truck carries x tons in one load, it carries $5x$ tons in five loads. Since the larger truck carries $(x + 2)$ tons in one load, it carries $8(x + 2)$ tons in eight loads. The total number of tons hauled can be expressed two ways: as the sum of $5x$ and $8(x + 2)$ and as the number 55.

Solution   Let x represent the capacity of the smaller truck in tons. Then $x + 2$ represents the capacity of the larger truck in tons.

Total tonnage carried by small truck	plus	total tonnage carried by large truck	equals	total tonnage carried by both trucks.
$5x$	$+$	$8(x + 2)$	$=$	55

$$5x + 8(x + 2) = 55 \qquad \text{The equation to solve.}$$
$$5x + 8x + 16 = 55 \qquad \text{Remove parentheses.}$$
$$13x + 16 = 55 \qquad \text{Combine like terms.}$$
$$13x = 39 \qquad \text{Subtract 16 from both sides.}$$
$$x = 3 \qquad \text{Divide both sides by 13.}$$

The smaller truck can carry 3 tons of asphalt.

Check: Each load on the small truck is 3 tons. Because the larger truck carries 2 extra tons, it carries 5 tons. The small truck made five trips, carrying 3 tons on each trip, and the larger truck made eight trips, carrying 5 tons on each trip. The small truck hauled a total of $5 \cdot 3$ or 15 tons, while the larger truck hauled $8 \cdot 5$ or 40 tons. In total, they carried $15 + 40$, or 55 tons. The solution checks. ■

ORALS *Simplify by combining like terms.*

1. $3x + 5x - 7x$

2. $-2y + 3y - y$

3. $3x^2 + 2x^2 - 5x^2$

4. $3x + 2y - 5xy$

5. $3(x + 2) - 3x + 6$

6. $3(x + 2) + 3x - 6$

Solve each equation, when possible.

7. $5x = 4x + 3$

8. $2(x - 1) = 2(x + 1)$

9. $3x = 2(x + 1)$

10. $x + 2(x + 1) = 3$

E X E R C I S E   2.3

In Exercises 1–22, simplify each expression, when possible.

1. $3x + 17x$

2. $12y - 15y$

3. $8x^2 - 5x^2$

4. $17x^2 + 3x^2$

5. $9x + 3y$

6. $5x + 5y$

7. $3(x + 2) + 4x$

8. $9(y - 3) + 2y$

9. $5(z - 3) + 2z$

10. $4(y + 9) - 6y$

11. $12(x + 11) - 11$

12. $-3(3 + z) + 2z$

13. $8(y + 7) - 2(y - 3)$

14. $9(z + 2) + 5(3 - z)$

15. $2x + 4(y - x) + 3y$

16. $3y - 6(y + z) + y$

17. $(x + 2) - (x - y)$

18. $3z + 2(y - z) + y$

19. $2\left(4x + \dfrac{9}{2}\right) - 3\left(x + \dfrac{2}{3}\right)$

20. $7\left(3x - \dfrac{2}{7}\right) - 5\left(2x - \dfrac{3}{5}\right) + x$

21. $8x(x + 3) - 3x^2$

22. $2x + x(x + 3)$

In Exercises 23–70, solve each equation. Check all solutions.

23. $3x + 2 = 2x$

24. $5x + 7 = 4x$

25. $5x - 3 = 4x$

26. $4x + 3 = 5x$

27. $9y - 3 = 6y$

28. $8y + 4 = 4y$

29. $8y - 7 = y$

30. $9y - 8 = y$

31. $9 - 23w = 4w$

32. $y + 4 = -7y$

33. $22 - 3r = 8r$

34. $14 + 7s = s$

35. $3(a + 2) = 4a$

36. $4(a - 5) = 3a$

37. $5(b + 7) = 6b$

38. $8(b + 2) = 9b$

39. $2 + 3(x - 5) = 4(x - 1)$

40. $2 - (4x + 7) = 3 + 2(x + 2)$

41. $10x + 3(2 - x) = 5(x + 2) - 4$

42. $11x + 6(3 - x) = 3$

43. $3(a + 2) = 2(a - 7)$

44. $9(t - 1) = 6(t + 2) - t$

45. $9(x + 11) + 5(13 - x) = 0$

46. $3(x + 15) + 4(11 - x) = 0$

47. $\dfrac{3(t - 7)}{2} = t - 6$

48. $\dfrac{2(t + 9)}{3} = t - 8$

49. $\dfrac{5(2 - s)}{3} = s + 6$

50. $\dfrac{8(5 - s)}{5} = -2s$

51. $\dfrac{4(2x - 10)}{3} = 2(x - 4)$

52. $\dfrac{11(x - 12)}{2} = 9 - 2x$

53. $3.1(x - 2) = 1.3x + 2.8$

54. $0.6x - 0.8 = 0.8(2x - 1) - 0.7$

55. $2.7(y + 1) = 0.3(3y + 33)$

56. $1.5(5 - y) = 3y + 12$

57. $19.1x - 4(x + 0.3) = -46.5$

58. $18.6x + 7.2 = 1.5(48 - 2x)$

59. $14.3(x + 2) + 13.7(x - 3) = 15.5$

60. $1.25(x - 1) = 0.5(3x - 1) - 1$

61. $x(2x - 3) = 2x^2 + 15$

62. $2x(3x + 4) = 6x^2 + 32$

63. $a(a + 2) = a(a - 4) + 16$

64. $b(b - 1) + 18 = b(b + 5)$

65. $\dfrac{x(2x - 8)}{2} = x(x + 2)$

66. $\dfrac{3x(2x + 1)}{2} = 3x^2 - 6$

67. $2y^2 - 9 = y(y + 3) + y^2$

68. $y(3y - 4) - y^2 = 2y(y + 3) + 20$

69. $\dfrac{x(4x + 3) + 2(x^2 + 9)}{2} = 3x(x + 2)$

70. $x(x + 2) + 24 = \dfrac{x(x + 2) + x(2x - 8)}{3}$

In Exercises 71–82, solve each equation. If it is an identity or an impossible equation, so indicate.

71. $8x + 3(2 - x) = 5(x + 2) - 4$

72. $5(x + 2) = 5x - 2$

73. $s(s + 2) = s^2 + 2s + 1$

74. $21(b - 1) + 3 = 3(7b - 6)$

75. $\dfrac{2(t - 1)}{6} - 2 = \dfrac{t + 2}{6}$

76. $\dfrac{2(2r - 1)}{6} + 5 = \dfrac{3(r + 7)}{6}$

77. $2(3z + 4) = 2(3z - 2) + 13$

78. $x + 7 = \dfrac{2x + 6}{2} + 4$

79. $2(y - 3) - \dfrac{y}{2} = \dfrac{3}{2}(y - 4)$

80. $\dfrac{20 - a}{2} = \dfrac{3}{2}(a + 4)$

81. $\dfrac{3x + 14}{2} = x - 2 + \dfrac{x + 18}{2}$

82. $\dfrac{5(x + 3)}{3} - x = \dfrac{2(x + 8)}{3}$

83. Carpentry The 12-foot board in Illustration 1 has been cut into two sections, one twice as long as the other. How long is each section?

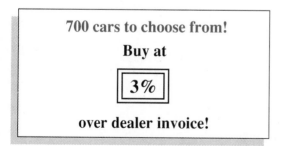

ILLUSTRATION 1

84. Plumbing A 20-foot pipe has been cut into two sections, one 3 times as long as the other. How long is each section?

85. Discount shopping A stereo system discounted 20% is selling for $969.20. What was its original price?

86. Auto sales An auto dealer's promotional ad appears in Illustration 2. One car is selling for $23,499. What was the dealer's invoice?

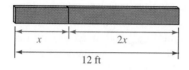

700 cars to choose from!

Buy at

3%

over dealer invoice!

ILLUSTRATION 2

87. Furniture pricing A sofa and a $300 chair are discounted 35%, and the price is $780 for both. Find the original price of the sofa.

88. Air purity standards A community has legislated that one of its industries decrease toxic emissions by 22%, to a level of 3.9 parts per billion. Find the industry's current emission level.

89. Hourly salary rate Tia worked 54 hours last week, but she is unsure of her hourly rate. She is paid $1\frac{1}{2}$ times her hourly rate for any hours beyond 40, and last week's pay was $332.45. Find the hourly rate.

90. Cost of a car The total cost of a new car, including an 8.5% sales tax, is $13,725.25. Find the cost before taxes.

91. Counting calories A slice of pie with a scoop of ice cream contains 850 calories. The number of calories in the pie alone is 100 greater than double the calories in the ice cream alone. How many calories are in the ice cream?

92. Publisher's inventories A novel can be purchased in a hardcover edition for $15.95 or in paperback for $4.95. The publisher printed 11 times as many paperbacks as hardcover books, a total of 114,000 copies. How many hardcover books were printed?

93. Manufacturing concrete Concrete contains 3 times as much gravel as cement. How much cement is in 500 pounds of dry concrete mix?

94. Building construction A 35-foot beam, 1 foot wide and 2 inches thick, is cut into three sections. One section is 14 feet long. Of the remaining two sections, one is twice as long as the other. Will the shortest section span an 8-foot-wide doorway?

95. Installing solar heating One solar panel in Illustration 3 is 3.4 feet wider than the other. Find the width of each.

← 18 ft →

ILLUSTRATION 3

96. Waste disposal   Two tanks hold a total of 45 gallons of a toxic solvent. One tank holds 6 gallons more than twice the amount in the other. Can the smaller tank be emptied into a 10-gallon waste disposal canister?

97. Buying vitamins   If you buy one bottle of vitamins, you can get a second bottle for half price. Two bottles cost $2.25. Find the usual price for a single bottle of vitamins.

98. Atomic weights   Water is made up of hydrogen and oxygen. In water, the mass of the oxygen is 8 times the mass of the hydrogen. How much hydrogen is in 2700 grams of water?

99. Choosing salary options   Carrie has her choice of two salary options. The first plan pays $600 per month plus a 2% commission on sales. The second plan is straight commission—5% of sales. To decide which plan to choose, Carrie needs to know the monthly sales that will produce equal income. Find the monthly sales.

100. Sales quotas   The sales force is hoping that last month's sales will exceed their one-million-dollar quota. Jennifer was last month's top sales representative, responsible for 45% of total sales. Jim came in second at 23%, with the remaining $317,920 credited to others. Did the sales team make their quota?

Writing Exercises ■ *Write a paragraph using your own words.*

1. Explain why $3x^2y$ and $5x^2y$ are like terms.

2. Explain why $3x^2y$ and $3xy^2$ are unlike terms.

3. Discuss whether $7xxy^3$ and $5x^2yyy$ are like terms.

4. Discuss whether $\dfrac{3}{2}x$ and $\dfrac{3x}{2}$ are like terms.

Something to Think About: ■ **1.** What number is equal to its own double?

2. What number is equal to one-half of itself?

Review Exercises ■ *Evaluate each expression when $x = -3$, $y = -5$, and $z = 0$.*

1. $x^2z(y^3 - z)$ **2.** $z - y^3$ **3.** $\dfrac{x - y^2}{2y - 1 + x}$ **4.** $\dfrac{2y + 1}{x} - x$

2.4 Literal Equations

GETTING READY   *Find the missing number.*

1. $\dfrac{3x}{\rule{1cm}{0.3cm}} = x$ **2.** $\dfrac{-5y}{\rule{1cm}{0.3cm}} = y$ **3.** $\dfrac{rx}{\rule{1cm}{0.3cm}} = x$ **4.** $\dfrac{-ay}{\rule{1cm}{0.3cm}} = y$

5. $\rule{1cm}{0.3cm} \cdot \dfrac{x}{7} = x$ **6.** $\rule{1cm}{0.3cm} \cdot \dfrac{y}{12} = y$ **7.** $\rule{1cm}{0.3cm} \cdot \dfrac{x}{d} = x$ **8.** $\rule{1cm}{0.3cm} \cdot \dfrac{y}{s} = y$

Equations with several variables are called **literal equations.** Often these equations are **formulas** such as $A = lw$, the formula for finding the area of a rectangle. Suppose that we wish to find the lengths of several rectangles whose areas and widths are known. It would be tedious to substitute values for A and w into the formula and then repeatedly solve the formula for l. It would be better to solve the formula $A = lw$ for l first and then substitute values for A and w and compute l directly.

To **solve an equation for a variable** means to isolate that variable on one side of the equation, with all other quantities on the opposite side.

EXAMPLE 1 Solve $A = lw$ for l.

Solution To isolate l on the left-hand side, we undo the multiplication by w by dividing both sides of the equation by w.

$$A = lw$$

$$\frac{A}{w} = \frac{lw}{w} \qquad \text{Divide both sides by } w.$$

$$\frac{A}{w} = l \qquad \text{Simplify.}$$

$$l = \frac{A}{w}$$ ∎

EXAMPLE 2 The formula $A = \frac{1}{2}bh$ gives the area of a triangle with base b and height h. Solve this formula for b.

Solution

$$A = \frac{1}{2}bh$$

$$2A = 2 \cdot \frac{1}{2}bh \qquad \text{Multiply both sides by 2.}$$

$$2A = bh \qquad \text{Simplify.}$$

$$\frac{2A}{h} = \frac{bh}{h} \qquad \text{Divide both sides by } h.$$

$$\frac{2A}{h} = b \qquad \text{Simplify.}$$

$$b = \frac{2A}{h}$$

If the area A and the height h of a triangle are known, the base b is given by the formula $b = \frac{2A}{h}$. ∎

EXAMPLE 3 The formula $C = \frac{5}{9}(F - 32)$ is used to convert Fahrenheit temperature readings into their Celsius equivalents. Solve the formula for F.

Solution
$$C = \frac{5}{9}(F - 32)$$

$$\frac{9}{5}C = \frac{9}{5} \cdot \frac{5}{9}(F - 32) \qquad \text{Multiply both sides by } \tfrac{9}{5}.$$

$$\frac{9}{5}C = 1(F - 32) \qquad \text{Simplify.}$$

$$\frac{9}{5}C = F - 32 \qquad \text{Remove parentheses.}$$

$$\frac{9}{5}C + 32 = F - 32 + 32 \qquad \text{Add 32 to both sides.}$$

$$\frac{9}{5}C + 32 = F \qquad \text{Combine terms.}$$

$$F = \frac{9}{5}C + 32$$

The formula $F = \frac{9}{5}C + 32$ is used to convert degrees Celsius to degrees Fahrenheit. ■

EXAMPLE 4 The area A of the trapezoid shown in Figure 2-2 is given by the formula

$$A = \frac{1}{2}(B + b)h$$

where B and b are its bases and h is its height. Solve the formula for b.

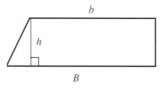

b

h

B

FIGURE 2-2

Solution *Method 1:* $\quad A = \frac{1}{2}(B + b)h$

$$2A = 2 \cdot \frac{1}{2}(B + b)h \qquad \text{Multiply both sides by 2.}$$

Albert Einstein (1879–1955)
Einstein was a theoretical physicist
best known for his theory of
relativity. Although Einstein was
born in Germany, he became a Swiss
citizen and earned his doctorate at
the University of Zurich in 1905. In
1910, he returned to Germany to
teach. He fled Germany because of
the Nazi government and became a
United States citizen in 1940. He is
famous for his equation, $E = mc^2$.

$$2A = Bh + bh$$ Simplify and remove parentheses.

$$2A - Bh = Bh + bh - Bh$$ Subtract Bh from both sides.

$$2A - Bh = bh$$ Combine like terms.

$$\frac{2A - Bh}{h} = \frac{bh}{h}$$ Divide both sides by h.

$$\frac{2A - Bh}{h} = b$$ Simplify.

Method 2: $A = \frac{1}{2}(B + b)h$

$$2A = 2 \cdot \frac{1}{2}(B + b)h$$ Multiply both sides by 2.

$$2A = (B + b)h$$ Simplify.

$$\frac{2A}{h} = \frac{(B + b)h}{h}$$ Divide both sides by h.

$$\frac{2A}{h} = B + b$$ Simplify.

$$\frac{2A}{h} - B = B + b - B$$ Subtract B from both sides.

$$\frac{2A}{h} - B = b$$ Combine like terms.

Although they look different, the results of Methods 1 and 2 are equivalent. ■

    EXAMPLE 5   Use the formula $P = 2l + 2w$ to find l when $P = 56$ and $w = 11$.

Solution   We first solve the formula $P = 2l + 2w$ for l.

$$P = 2l + 2w$$

$$P - 2w = 2l + 2w - 2w$$ Subtract $2w$ from both sides.

$$P - 2w = 2l$$ Combine like terms.

$$\frac{P - 2w}{2} = \frac{2l}{2}$$ Divide both sides by 2.

$$\frac{P - 2w}{2} = l$$ Simplify.

$$l = \frac{P - 2w}{2}$$

We then substitute 56 for P and 11 for w and simplify.

$$l = \frac{P - 2w}{2}$$

$$l = \frac{56 - 2(11)}{2}$$

$$= \frac{56 - 22}{2}$$

$$= \frac{34}{2}$$

$$= 17$$

Thus, $l = 17$. ∎

EXAMPLE 6 The volume V of the right-circular cone shown in Figure 2-3 is given by the formula

$$V = \frac{1}{3}Bh$$

where B is the area of its circular base and h is its height. Solve the formula for h and find the height of a right-circular cone with a volume of 64 cubic centimeters and a base area of 16 square centimeters.

Solution   We first solve the formula for h.

$$V = \frac{1}{3}Bh$$

$$3V = 3 \cdot \frac{1}{3}Bh \qquad \text{Multiply both sides by 3.}$$

$$3V = Bh \qquad \text{Simplify.}$$

$$\frac{3V}{B} = \frac{Bh}{B} \qquad \text{Divide both sides by } B.$$

$$\frac{3V}{B} = h \qquad \text{Simplify.}$$

$$h = \frac{3V}{b}$$

FIGURE 2-3

We then substitute 64 for V and 16 for B and simplify.

$$h = \frac{3V}{B}$$

$$h = \frac{3(64)}{16}$$
$$= 3(4)$$
$$= 12$$

The height of the right-circular cone is 12 centimeters. ■

ORALS *Solve the equation $ab + c - d = 0$,*

1. for a **2.** for b **3.** for c **4.** for d

Solve the equation $a + b = \dfrac{c}{d}$,

5. for a **6.** for b **7.** for c **8.** for d

EXERCISE 2.4

In Exercises 1–24, solve each formula for the variable indicated.

1. $E = IR$; for I

2. $i = prt$; for r

3. $V = lwh$; for w

4. $K = A + 32$; for A

5. $P = a + b + c$; for b

6. $P = 4s$; for s

7. $P = 2l + 2w$; for w

8. $d = rt$; for t

9. $A = P + Prt$; for t

10. $a = \dfrac{1}{2}(B + b)h$; for h

11. $C = 2\pi r$; for r

12. $I = \dfrac{E}{R}$; for R

13. $K = \dfrac{wv^2}{2g}$; for w

14. $V = \pi r^2 h$; for h

15. $P = I^2 R$; for R

16. $V = \dfrac{1}{3}\pi r^2 h$; for h

17. $K = \dfrac{wv^2}{2g}$; for g

18. $P = \dfrac{RT}{mV}$; for V

19. $F = \dfrac{GMm}{d^2}$; for M

20. $C = 1 - \dfrac{A}{a}$; for A

21. $F = \dfrac{GMm}{d^2}$; for d^2

22. $y = mx + b$; for x

23. $G = 2(r - 1)b$; for r

24. $F = f(1 - M)$; for M

In Exercises 25–32, solve each formula for the variable indicated. Then substitute numbers to find the variable's value.

25. $d = rt$ Find t if $d = 135$ and $r = 45$.

26. $d = rt$ Find r if $d = 275$ and $t = 5$.

27. $i = prt$ Find t if $i = 12$, $p = 100$, and $r = 0.06$.

28. $i = prt$ Find r if $i = 120$, $p = 500$, and $t = 6$.

29. $P = a + b + c$ Find c if $P = 37$, $a = 15$, and $b = 19$.

30. $y = mx + b$ Find x if $y = 30$, $m = 3$, and $b = 0$.

31. $K = \dfrac{1}{2}h(a + b)$ Find h if $K = 48$, $a = 7$, and $b = 5$.

32. $\dfrac{x}{2} + y = z^2$ Find x if $y = 3$ and $z = 3$.

33. Ohm's law   The formula $E = IR$, called **Ohm's law,** is used in electronics. Solve for I and then calculate the current I if the voltage E is 48 volts and the resistance R is 12 ohms. Current has units of *amperes.*

34. Volume of a cone   The volume V of a cone is given by the formula $V = \frac{1}{3}\pi r^2 h$. Solve the formula for h and then calculate the height h if V is 36π cubic inches and the radius r is 6 inches.

35. **Circumference of a circle**   The circumference C of a circle is given by $C = 2\pi r$, where r is the radius of the circle. Solve the formula for r and then calculate the radius of circle with a circumference of 14.32 feet. Round your answer to the nearest hundredth of a foot.

36. **Growth of money**   At a simple interest rate r, an amount of money P grows to an amount A in t years according to the formula $A = P(1 + rt)$. Solve the formula for P. After $t = 3$ years, a girl has an amount $A = \$4357$ on deposit. What amount P did she start with? Assume an interest rate of 6%.

37. **Power loss**   The power P lost when an electric current I passes through a resistance R is given by the formula $P = I^2 R$. Solve for R. If P is 2700 watts and I is 14 amperes, calculate R to the nearest hundredth of an ohm.

38. **Geometry**   The perimeter P of a rectangle with length l and width w is given by the formula $P = 2l + 2w$. Solve this formula for w. If the perimeter of a certain rectangle is 58.37 meters and its length is 17.23 meters, find its width. Round the answer to two decimal places.

39. Force of gravity   The masses of the two objects in Illustration 1 are m and M. The force of gravitation, F, between the masses is given by

$$F = \frac{GmM}{d^2}$$

where G is a constant and d is the distance between them. Solve for m.

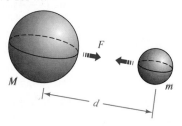

ILLUSTRATION 1

40. Thermodynamics   In thermodynamics, the Gibbs free-energy function is given by

$$G = U - TS + pV$$

Solve this equation for the pressure, p.

41. Pulleys The approximate length L of a belt joining two pulleys of radii r and R feet with centers D feet apart is given by the formula $L = 2D + 3.25(r + R)$. (See Illustration 2.) Solve the formula for D. If a 25-foot belt joins pulleys of radius 1 foot and 3 feet, how far apart are the centers of the pulleys?

42. Geometry The measure a of an interior angle of a regular polygon with n sides is given by the formula $a = 180°(1 - \frac{2}{n})$. (See Illustration 3.) Solve the formula for n. How many sides does a regular polygon have if an interior angle is 108°? (*Hint:* Distribute first.)

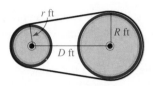

ILLUSTRATION 2

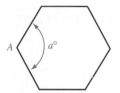

ILLUSTRATION 3

Writing Exercises ■ *Write a paragraph using your own words.*

1. The formula $P = 2l + 2w$ is also an equation, but an equation such as $2x + 3 = 5$ is not a formula. What equations do you think should be called formulas?

2. To solve the equation $s - A(s - 5) = r$ for the variable s, one student simply added $A(s - 5)$ to both sides to get $s = r + A(s - 5)$. Explain why this is not correct.

Something to Think About ■ **1.** The tremendous energy of the atomic bomb comes from the conversion of matter into energy, according to Einstein's formula $E = mc^2$. The constant c is the speed of light, about 300,000 meters per second. Find the energy in a mass, m, of 1 kilogram (which would weigh about 2 pounds). Energy has units of **joules.**

2. When a car of mass m collides with a wall, the energy of the collision is given by the formula $E = \frac{1}{2}mv^2$. Compare the energy of two collisions: a car striking a wall at 30 mph, and at 60 mph.

Review Exercises ■ *Simplify each expression, if possible.*

1. $2x - 5y + 3x$

2. $2x^2y + 5x^2y^2$

3. $\frac{3}{5}(x + 5) - \frac{8}{5}(10 + x)$

4. $\frac{2}{11}(22x - y^2) + \frac{9}{11}y^2$

2.5 Applications of Equations

■ Integer Problems ■ Geometric Problems ■ Quantity and Value Problems
■ Break-Even Analysis

GETTING READY

1. If 8 is the second of three consecutive integers, what are the other two?

2. If 8 is the second of three consecutive even integers, what are the other two?

3. If the base of a triangle is 6 and the height is k, find the expression that represents the area.

4. Find the value of s shares of stock, worth $72 each.

In this section, we solve more problems.

■ Integer Problems

EXAMPLE 1   **Integer problem**   The sum of two consecutive even integers is 22. Find the integers.

Analysis   Consecutive even integers differ by 2; numbers such as 2, 4, 6, 8, and 10 are consecutive even integers. Hence, if x is an even integer, then the expressions x and $x + 2$ represent two consecutive even integers. The sum of these two integers can be written in two different ways: as $x + (x + 2)$ and as 22.

Solution   Let x represent the first even integer. Then $x + 2$ represents the next consecutive even integer.
We can form the equation

The first even integer	plus	the second even integer	equals	their sum.
x	$+$	$x + 2$	$=$	22

$$x + x + 2 = 22 \qquad \text{The equation to solve.}$$
$$2x + 2 = 22 \qquad \text{Combine like terms.}$$
$$2x = 20 \qquad \text{Subtract 2 from both sides.}$$
$$x = 10 \qquad \text{Divide both sides by 2.}$$

Since the first even integer is 10, the next consecutive even integer is 12. The two consecutive even integers are 10 and 12.

Check: The numbers 10 and 12 are indeed consecutive even integers, and their sum is 22. The answers check.   ■

■ Geometric Problems

EXAMPLE 2   **Dimensions of a rectangle** The length of a rectangle is 4 meters longer than twice its width. If the perimeter of the rectangle is 26 meters, find its dimensions.

Analysis   We can draw a sketch like that in Figure 2-4. The formula for the perimeter of the rectangle is $p = 2l + 2w$. The perimeter of the rectangle in the figure is $2(4 + 2w) + 2w$, which is also 26.

Solution   Let w represent the width of the rectangle. Then $4 + 2w$ represents the length of the rectangle.
We can form the equation

$4 + 2w$

w

FIGURE 2-4

2 ·	the length	plus 2 ·	the width	equals	the perimeter.

$$2(4 + 2w) + 2w = 26 \qquad \text{The equation to solve.}$$
$$8 + 4w + 2w = 26 \qquad \text{Remove parentheses.}$$
$$6w + 8 = 26 \qquad \text{Combine like terms.}$$
$$6w = 18 \qquad \text{Subtract 8 from both sides.}$$
$$w = 3 \qquad \text{Divide both sides by 6.}$$

The width of the rectangle is 3 meters, and the length, $4 + 2w$, is 10 meters.

Check: If a rectangle has a width of 3 meters and a length of 10 meters, then the length is 4 meters longer than twice the width ($4 + 2 \cdot 3 = 10$). Furthermore, the perimeter is $2 \cdot 10 + 2 \cdot 3$, or 26 meters. The solution checks.   ■

EXAMPLE 3   **Angles in an isosceles triangle** The vertex angle of an isosceles triangle is 56°. Find the measure of each base angle.

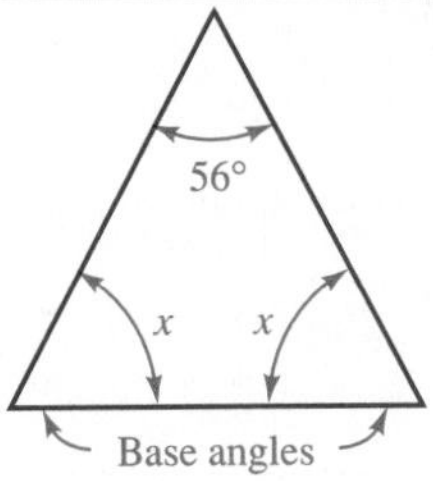

Analysis   An **isosceles triangle** has two equal sides, which meet to form the **vertex angle.** The angles opposite those sides, called **base angles,** are also equal (see Figure 2-5). If we let x represent the measure of one base angle, then the measure of the other base angle is also x. In any triangle the sum of the three angles is 180°. The sum of the angles is $x° + x° + 56°$, and it is also 180°.

56°

x x

Base angles

FIGURE 2-5

Solution   Let x represent the measure of one base angle. Then x also represents the measure of the other base angle.

We can form the equation

One base angle	plus	the other base angle	plus	the vertex angle	equals	**180°.**
x	$+$	x	$+$	56	$=$	180

$$x + x + 56 = 180 \qquad \text{The equation to solve.}$$
$$2x + 56 = 180 \qquad \text{Combine like terms.}$$
$$2x = 124 \qquad \text{Subtract 56 from both sides.}$$
$$x = 62 \qquad \text{Divide both sides by 2.}$$

The measure of each base angle is 62°.

Check: The measure of each base angle is 62°, and the vertex angle measures 56°. These three angles total 180°. The solution checks.   ■

■ Quantity and Value Problems

EXAMPLE 4   **Stock portfolios**   An investor has stock worth $240,000 in three segments of the economy. He owns five times as many shares in transportation as he owns in utilities, and 2000 more shares in pharmaceuticals than in transportation. Each share in transportation is worth $55, each utility share is worth $75, and each share in pharmaceuticals is worth $30. How many shares does he own in each segment?

Analysis   We let n represent the number of shares of utility stock. Then

$5n$ represents the number of shares of transportation stock, and

$5n + 2000$ represents the number of shares in pharmaceuticals.

It is important to distinguish between the *number* of shares and the *value* of those shares. Because the investor owns n shares of utility stock worth $75 each, the value of the stock is $75n$. Because he owns $5n$ shares of transportation stock worth $55 each, their value is $55(5n)$. Finally, his $(5n + 2000)$ shares of pharmaceutical stock, at $30 each, are worth $30(5n + 2000)$. The total value can be expressed in two ways: as

$$\$[75n + 55(5n) + 30(5n + 2000)] \qquad \text{and as} \qquad \$240,000$$

Solution

The value of the utility stock	$+$	the value of the transportation stock	$+$	the value of the pharmaceutical stock	$=$	the total value.
$75n$	$+$	$55(5n)$	$+$	$30(5n + 2000)$	$=$	$240,000$

$$75n + 275n + 150n + 60{,}000 = 240{,}000 \quad \text{Remove parentheses.}$$
$$500n + 60{,}000 = 240{,}000 \quad \text{Combine like terms.}$$
$$500n = 180{,}000 \quad \text{Subtract 60,000 from both sides.}$$
$$n = 360 \quad \text{Divide both sides by 500.}$$

The investor owns 360 utility stocks. His holdings are summarized as follows.

	Shares	Value
Utilities	$n = 360$	$ 27,000
Transportation	$5n = 1800$	$ 99,000
Pharmaceuticals	$5n + 2000 = 3800$	$114,000
Total		$240,000

Because the total value is $240,000, the solution checks. ∎

■ Break-Even Analysis

In manufacturing, there are two types of costs—**fixed costs** and **unit costs.** Fixed costs do not depend on the amount of product manufactured. Fixed costs would include the cost of plant rental, insurance, and machinery. Unit costs depend on the amount of product manufactured. Unit costs would include the cost of raw materials and labor.

Break-even analysis is used to find the production level where revenue will just offset the cost of production. When production exceeds the break-even point, the company will make a profit.

EXAMPLE 5

Manufacturing electronics An electronics company has fixed costs of $6405 a week and a unit cost of $75 for each compact disk player manufactured. If the company can sell all the CD players it can make at a wholesale price of $90, find the company's break-even point.

Analysis

Suppose the company manufactures x CD players each week. The cost of manufacturing these players is the sum of the fixed and unit costs. We are given that the fixed costs are $6405 each week. The unit cost is the product of x, the number of CD players manufactured each week, and $75, the cost of manufacturing a single player. Thus, the weekly cost is

Cost $= 75x + 6405$

Since the company can sell all the machines it can make, the weekly revenue is the product of x, the number of players manufactured (and sold) each week, and $90, the wholesale price of each CD player. Thus, the weekly revenue is

Revenue $= 90x$

The break-even point is the value of x for which the weekly revenue is equal to the weekly cost.

Solution Let x represent the number of CD players manufactured each week. Then

$90x$ represents the weekly revenue, and

$75x + 6405$ represents the weekly cost.

Because the break-even point occurs when revenue equals cost, we set up and solve the following equation:

The weekly revenue	=	the weekly cost.

$$90x = 75x + 6405$$
$$15x = 6405 \qquad \text{Subtract } 75x \text{ from both sides.}$$
$$x = 427 \qquad \text{Divide both sides by 15.}$$

If the company manufactures 427 CD players, the revenue will equal the cost, and the company will break even. ∎

ORALS *The sum of three consecutive even integers is 42. Find the integers.*

1. If x represents the smallest integer, how do you represent the other two?

2. What equation leads to the solution?

3. Solve the equation.

4. List the three consecutive integers.

E X E R C I S E 2.5

In Exercises 1–30, pick a variable to represent the unknown quantity, set up an equation involving the variable, solve the equation, and check it.

1. Finding consecutive even integers The sum of two consecutive even integers is 54. Find the integers.

2. Finding consecutive odd integers The sum of two consecutive odd integers is 88. Find the integers.

3. Finding consecutive integers The sum of three consecutive integers is 120. Find the integers.

4. Finding consecutive even integers The sum of three consecutive even integers is 72. Find the three even integers.

5. Finding integers The sum of an integer and twice the next integer is 23. Find the smaller integer.

6. Finding integers If 4 times the smallest of three consecutive integers is added to the largest, the result is 112. Find the three integers.

7. **Finding integers**   The larger of two integers is 10 greater than the smaller. The larger is 3 less than twice the smaller. Find the smaller integer.

8. **Finding integers**   The smaller of two integers is one-half of the larger, and 11 greater than one-third of the larger. Find the smaller integer.

9. **Triangular bracing**   The outside perimeter of the triangular brace in Illustration 1 is 57 feet. If all three sides are equal, find the length of each side.

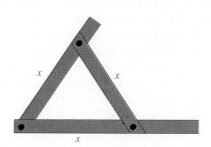

ILLUSTRATION 1

10. **Circuit boards**   The perimeter of the circuit board in Illustration 2 is 90 centimeters. Find the dimensions of the board.

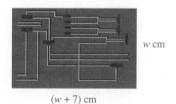

w cm

(*w* + 7) cm

ILLUSTRATION 2

11. **Swimming pools**   The width of a rectangular swimming pool is 11 meters less than the length. The perimeter is 94 meters. Find the dimensions of the pool.

12. **Wooden truss**   The truss in Illustration 3 is in the form of an isosceles triangle. Each of the two equal sides is 4 feet less than the third side. If the perimeter is 25 feet, find the lengths of the sides.

ILLUSTRATION 3

13. **Framing pictures**   The length of a rectangular picture frame is 5 inches greater than twice the width. If the perimeter is 112 inches, find the dimensions of the frame.

14. **Guy wires**   The two guy wires in Illustration 4 form an isosceles triangle. One of the two equal angles of the triangle is 4 times the third angle (the vertex angle). Find the measure of the vertex angle.

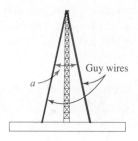

Guy wires

a

ILLUSTRATION 4

15. **Equilateral triangles**   Find the measure of each angle of an equilateral triangle. (*Hint:* The three angles of an equilateral triangle are all equal.)

16. **Land areas**   The perimeter of a square parcel of land is twice the perimeter of an equilateral (equal-sided) triangular plot. If one side of the square is 60 meters, find the length of a side of the triangle.

17. **Sorting hardware**   Liz has an assortment of bolts, washers, and locknuts—99 pieces in all. She has twice as many washers as bolts and one less locknut than washers. How many bolts does she have?

18. **CD collections**   A musician has a collection of 178 CDs. He has twice as many jazz as rock and 23 more classical than jazz. How many of each does he have?

19. Stock performance An investor owns twice as many shares of stock with average performance as he owns of shares that have far outpaced the market. He owns 200 shares of a stock that is now worthless. If he owns 3500 shares in all, how many shares have performed better than average?

20. Raising capital 30% of a corporation's stock offering is preferred stock worth $25 per share. The rest is common stock, selling at $15 per share. If the entire offering raised $306 million, how many common shares were sold?

21. Warehousing costs Monthly storage costs are $1.50 for each portable television, $4.00 for each console, and $7.50 for each wide-screen entertainment center. An appliance store warehouses 40 more portables than wide-screen sets, and 25 fewer consoles than portables. Television storage costs $276 for the month. How many wide-screen sets are in stock?

22. Apartment rental The designers of a new apartment complex will include one-, two-, and three-bedroom apartments. They feel they can rent equal numbers of each, with the monthly rents given in Table 1.

One bedroom	$550
Two bedroom	$700
Three bedroom	$900

TABLE 1

If the total monthly rental income will be $36,550, how many units of each kind are planned?

23. Software sales The three best-selling software applications at one computer store are priced as in Table 2.

Spreadsheet	$150
Database	$195
Word processing	$210

TABLE 2

Spreadsheet and database programs sold in equal numbers, but 15 more word processing applications were sold than the other two combined. The three applications generated sales of $72,000. How many spreadsheets were sold?

24. Appliance store inventory Stoves and refrigerators are expected to sell in equal numbers. With summer approaching, the number of air conditioners sold is expected to be double that of stoves and refrigerators combined. Stoves sell for $350, refrigerators for $450, and air conditioners for $500, and sales of $56,000 are expected. How many of each should be stocked?

25. Performance bond A construction company has agreed to finish a highway repair in 60 days, and pay a $1000 fine for each day the work takes beyond 60 days. To finish by the deadline, management expects to spend $12,000 per day. By cutting the work crew, expenses would be $9400 per day, but the job would take longer. How many days beyond deadline can the project last to make expenditures equal?

26. Tax-free investments An executive has $1000 to invest. One investment pays 11% interest, but the income is fully taxable. Income from a municipal bond fund is not taxable. If the executive is in a 32% tax bracket, what tax-free interest rate makes both investments equally attractive?

27. Manufacturing shoes A shoe company has fixed costs of $9600 per month and a unit cost of $20 per pair of shoes. The company can sell all the shoes it can make at a wholesale price of $30 per pair. Find the break-even point.

28. Manufacturing insulators A manufacturer of high-voltage insulators has fixed costs of $5400 per month and a unit cost of $12 per insulator. It can sell all the units it can make at a wholesale price of $15. Find the break-even point.

29. Scheduling machine usage A machine shop has two machines that can mill a certain brass plate. One machine has a setup cost of $500 and a cost of $2 per plate, while the other machine has a setup cost of $800 and a cost of $1 per plate. How many plates should be manufactured if the cost is to be the same using either machine?

30. Oriental rugs A rug manufacturer has two looms for weaving Oriental-style rugs. One loom has a setup cost of $750 and can produce a rug for $115. The other loom has a setup cost of $950 and can produce a rug for $95. How many rugs can be manufactured if the cost is to be the same using either loom?

In Exercises 31–34, a paint manufacturer can choose between two processes for manufacturing house paint, with monthly costs shown in Table 3. The paint can be sold for $21 per gallon.

Process	Fixed costs	Unit cost (per gallon)
A	$ 75,000	$11
B	$128,000	$ 5

TABLE 3

31. Find the break-even point for process A.

32. Find the break-even point for process B.

33. If expected sales will be 8800 gallons per month, which process should the company choose?

34. If expected sales will be 9000 gallons per month, which process should the company choose?

Writing Exercises ■ *Write a paragraph using your own words.*

1. Describe the steps you would use to analyze and solve problems.

2. Create a geometry problem that could be solved by using the equation $2w + 2(w + 5) = 26$.

Something to Think About ■

1. Is it possible for the equation of a problem to have a solution, but the problem to have no solution? For example, is it possible to find two consecutive even integers whose sum is 16?

2. Invent a geometry problem that leads to an equation that has a solution, although the problem does not.

Review Exercises ■ *Refer to the formulas in Section 1.3.*

1. Find the volume of a pyramid that has a height of 6 centimeters and a square base, 10 centimeters on each side.

2. Find the volume of a cone with a height of 6 centimeters and a circular base with radius 6 centimeters. Use $\pi \approx \frac{22}{7}$.

Simplify each expression.

3. $3(x + 2) + 4(x - 3)$

4. $4(x - 2) - 3(x + 1)$

5. $\frac{1}{2}(x + 1) - \frac{1}{2}(x + 4)$

6. $\frac{3}{2}\left(x + \frac{2}{3}\right) + \frac{1}{2}(x + 8)$

2.6 More Applications of Equations

■ Investment Problems ■ Uniform Motion Problems ■ Liquid Mixture Problems ■ Dry Mixture Problems

GETTING READY

1. Find 7% of $12,000.

2. At 55 miles per hour, how far would a car travel in 7 hours?

3. At 55 miles per hour, how long would it take to go 176 miles?

4. At $7.50 per pound, how many pounds of chocolate would be worth $71.25?

■ Investment Problems

EXAMPLE 1 **Investing money** A retired teacher invested part of $12,000 at 6% annual interest, and the rest at 9%. If the annual income from these investments was $945, how much did the teacher invest at each rate?

Analysis The interest i earned by an amount p invested at an annual rate r for t years is given by the formula $i = prt$. In this example, $t = 1$ year. Hence, if x dollars were invested at 6%, the interest earned would be $0.06x$. If x dollars were invested at 6%, then the rest of the money, $(12,000 - x)$, would be invested at 9%. The interest earned on that money would be $0.09(12,000 - x)$ dollars. The total interest earned in dollars can be expressed in two ways: as 945 and as the sum $0.06x + 0.09(12,000 - x)$.

Solution Let x represent the amount of money invested at 6%. Then $12,000 - x$ represents the amount of money invested at 9%.

We can form an equation as follows:

The interest earned at 6%	plus	the interest earned at 9%	equals	the total interest.
$0.06x$	$+$	$0.09(12,000 - x)$	$=$	945

$$0.06x + 0.09(12,000 - x) = 945 \qquad \text{The equation to solve.}$$
$$6x + 9(12,000 - x) = 94,500 \qquad \text{Multiply both sides by 100 to clear the equation of decimals.}$$
$$6x + 108,000 - 9x = 94,500 \qquad \text{Remove parentheses.}$$
$$-3x + 108,000 = 94,500 \qquad \text{Combine like terms.}$$
$$-3x = -13,500 \qquad \text{Subtract 108,000 from both sides.}$$
$$x = 4500 \qquad \text{Divide both sides by } -3.$$

The teacher invested $4500 at 6% and $12,000 − $4500, or $7500, at 9%.

Check: The first investment yielded 6% of $4500, or $270. The second investment yielded 9% of $7500, or $675. Because the total return was $270 + $675, or $945, the answers check.   ∎

■ Uniform Motion Problems

EXAMPLE 2   **Driving times**   Chicago, Illinois and Green Bay, Wisconsin are about 200 miles apart. A car leaves Chicago traveling toward Green Bay at 55 miles per hour. At the same time, a truck leaves Green Bay bound for Chicago at 45 miles per hour. How long will it take them to meet?

Analysis   Uniform motion problems are based on the formula $d = rt$, where d is the distance traveled, r is the rate, and t is the time. We can organize the information of this problem in chart form, as in Figure 2-6(a).

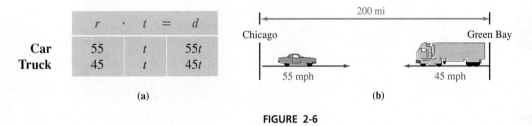

	r	$\cdot$	t	$=$	d
Car	55		t		$55t$
Truck	45		t		$45t$

(a)

(b)

FIGURE 2-6

We know that the two vehicles travel for the same amount of time—say, t hours. The faster car travels $55t$ miles, and the slower truck travels $45t$ miles. The total distance can be expressed in two ways: as the sum $55t + 45t$ and as 200 miles.

Solution   Let t represent the time that each vehicle travels until they meet. Then

$55t$ represents the distance traveled by the car, and

$45t$ represents the distance traveled by the truck.

After referring to Figure 2-6(b), we form the equation

The distance the car goes	plus	the distance the truck goes	equals	the total distance.
$55t$	+	$45t$	=	200

$$55t + 45t = 200 \qquad \text{The equation to solve.}$$
$$100t = 200 \qquad \text{Combine like terms.}$$
$$t = 2 \qquad \text{Divide both sides by 100.}$$

The vehicles will meet after 2 hours.

Check: During those 2 hours, the car travels $55 \cdot 2$, or 110 miles, while the truck travels $45 \cdot 2$, or 90 miles. The total distance traveled is $110 + 90$, or 200 miles. This is the total distance between Chicago and Green Bay. The answer checks. ■

■ Liquid Mixture Problems

EXAMPLE 3   **Mixing acids**   A chemistry instructor has one solution that is 50% sulfuric acid and another that is 20% sulfuric acid. How much of each should she use to make 12 liters of a solution that is 30% acid?

Analysis   The sulfuric acid present in the final mixture comes from the two solutions to be mixed. If x represents the numbers of liters of the 50% solution required for the mixture, then the rest of the mixture ($(12 - x)$ liters) must be the 20% solution. (See Figure 2-7.) Only 50% of the x liters, and only 20% of the $(12 - x)$ liters, is pure sulfuric acid. The total of these amounts is also the amount of acid in the final mixture, which is 30% of 12 liters.

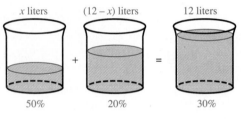

x liters	$(12 - x)$ liters	12 liters
50%	20%	30%

FIGURE 2-7

Solution   Let x represent the required number of liters of the 50% solution. Then $12 - x$ represents the required number of liters of the 20% solution.
We can form the equation

The acid in the 50% solution	plus	the acid in the 20% solution	equals	the acid in the final mixture.
50% of x	$+$	20% of $(12 - x)$	$=$	30% of 12

$$0.50x + 0.20(12 - x) = 0.30(12)$$ The equation to solve.

$$5x + 2(12 - x) = 3(12)$$ Multiply both sides by 10 to clear the equation of decimals.

$$5x + 24 - 2x = 36$$ Remove parentheses.

$$3x + 24 = 36$$ Combine like terms.

$$3x = 12$$ Subtract 24 from both sides.

$$x = 4$$ Divide both sides by 3.

The chemist must mix 4 liters of the 50% solution and 8 liters ($(12 - 4)$ liters) of the 20% solution. ■

■ Dry Mixture Problems

EXAMPLE 4 **Mixing nuts** Fancy cashews are not selling at $9 per pound, because they are too expensive. Filberts are selling at $6 per pound. How many pounds of filberts should be combined with 50 pounds of cashews to obtain a mixture that can be sold at $7 per pound?

Analysis Dry mixture problems are based on the formula $v = pn$, where v is the value of the mixture, p is the price per pound, and n is the number of pounds. Suppose x pounds of filberts are used in the mixture. At $6 per pound, they are worth $6x$. At $9 per pound, the 50 pounds of cashews are worth $9 \cdot 50$, or $450. The mixture will weigh $(50 + x)$ pounds, and at $7 per pound, it will be worth $7(50 + x)$. The value of the ingredients, $\$(6x + 450)$, is equal to the value of the mixture, $\$7(50 + x)$. (See Figure 2-8.)

Solution Let x represent the number of pounds of filberts in the mixture.

	v	$=$	p	$\cdot$	n
Filberts	$6x$		6		x
Cashews	$9(50)$		9		50
Mixture	$7(50 + x)$		7	$\cdot$	$50 + x$

FIGURE 2-8

We can form the equation

The value of the filberts	plus	the value of the cashews	equals	the value of the mixture.
$6x$	$+$	$9 \cdot 50$	$=$	$7(50 + x)$

$$6x + 9 \cdot 50 = 7(50 + x) \qquad \text{The equation to solve.}$$
$$6x + 450 = 350 + 7x \qquad \text{Remove parentheses and simplify.}$$
$$100 = x \qquad \text{Subtract } 6x \text{ and } 350 \text{ from both sides.}$$

The storekeeper should use 100 pounds of filberts in the mixture.

Check: The value of 100 pounds of filberts at $6 per pound is $600

The value of 50 pounds of cashews at $9 per pound is $\underline{450}$

The value of the mixture is $1050

The value of 150 pounds of mixture at $7 per pound is also $1050.   ■

ORALS

1. Find the value of 7 pounds of ground coffee worth d dollars per pound.

2. Find one year's interest on $18,000, invested at an annual rate r.

3. Find the length of a rectangle with area of A square feet and width of 6 feet.

4. Find the length of a rectangle with perimeter of P feet and width of 9 feet.

EXERCISE 2.6

In Exercises 1–28, pick a variable to represent the unknown quantity, set up an equation involving the variable, solve the equation, and check it.

1. **Investment problem** A broker has invested $24,000 in two accounts, one earning 9% annual interest and the other earning 8%. After 1 year, his combined interest is $1965. How much was invested at each rate?

2. **Investment problem** A woman's rollover IRA of $18,750 is invested in two accounts, one earning 7% interest and the other earning 10%. After 1 year, the combined interest income is $1512. How much has she invested at each rate?

3. **Investment problem** One investment pays 8% and another pays 11%. If equal amounts are invested in each, the combined interest income for 1 year is $712.50. How much is invested at each rate?

4. **Investment problem** When equal amounts are invested in each of three accounts paying 5%, 6%, and 7%, one year's combined interest income is $882. How much is invested in each account?

5. **Investment problem** A college professor wants to supplement her retirement income with investment interest. If she invests $15,000 at 6% annual interest, how much more would she have to invest at 7% to achieve a goal of $1250 in supplemental income?

6. **Investment problem** A retired teacher has a choice of two investment plans: an insured fund that pays 11% interest or a riskier investment that promises a 13% return. The same amount invested at the higher rate would generate an extra $150 per year. How much does the teacher have to invest?

7. **Investment problem** A financial adviser recommends investing twice as much in CDs as in a bond fund. A client follows his advice and invests $21,000 in CDs paying 1% more interest than the fund. The CDs generate $840 more interest than the fund. Find the two rates. (*Hint:* 1% = 0.01)

8. **Investment problem** The amount of annual interest earned by $8000 invested at a certain rate is $200 less than $12,000 would earn at a 1% lower rate. At what rate is the $8000 invested?

9. **Travel time** Ashford and Bartlett are 315 miles apart. A car leaves Ashford bound for Bartlett at 50 miles per hour. At the same time, another car leaves Bartlett and heads toward Ashford at 55 miles per hour. In how many hours will the two cars meet?

10. **Travel time** Granville and Preston are 535 miles apart. A car leaves Preston bound for Granville at 47 miles per hour. At the same time, another car leaves Granville and heads toward Preston at 60 miles per hour. How long will it take them to meet?

11. Travel time Two cars leave Peoria at the same time, one heading east at 60 miles per hour and the other west at 50 miles per hour. (See Illustration 1.) How long will it take them to be 715 miles apart?

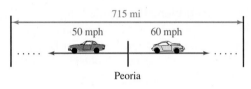

ILLUSTRATION 1

12. Boating Two boats leave port at the same time, one heading north at 35 knots (nautical miles per hour), the other south at 47 knots. How long will it take them to be 738 nautical miles apart?

13. Travel time Two cars start together and head east, one at 42 miles per hour and the other at 53 miles per hour. (See Illustration 2.) In how many hours will the cars be 82.5 miles apart?

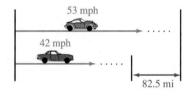

ILLUSTRATION 2

14. Speed of trains Two trains are 330 miles apart, and their speeds differ by 20 miles per hour. They travel toward each other and meet in 3 hours. Find the speed of each train.

15. Speed of an airplane Two planes are 6000 miles apart, and their speeds differ by 200 miles per hour. They travel toward each other and meet in 5 hours. Find the speed of the slower plane.

16. Average speed An automobile averaged 40 miles per hour for part of a trip and 50 miles per hour for the remainder. If the 5-hour trip covered 210 miles, for how long did the car average 40 miles per hour?

17. Mixing fuels How many gallons of fuel costing $1.15 per gallon must be mixed with 20 gallons of a fuel costing $.85 per gallon to obtain a mixture costing $1 per gallon? (See Illustration 3.)

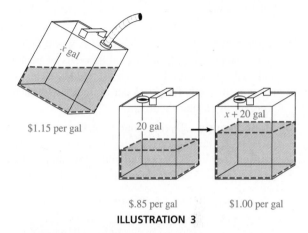

ILLUSTRATION 3

18. Mixing paint Paint costing $19 per gallon is to be mixed with 5 gallons of a $3 per gallon thinner to make a paint that can be sold for $14 per gallon. How much paint will be produced?

19. Brine solution How many gallons of a 3% salt solution must be mixed with 50 gallons of a 7% solution to obtain a 5% solution?

20. Making cottage cheese To make low-fat cottage cheese, milk containing 4% butterfat is mixed with 10 gallons of milk containing 1% butterfat to obtain a mixture containing 2% butterfat. How many gallons of the richer milk must be used?

21. Antiseptic solutions A nurse wishes to add water to 30 ounces of a 10% solution of benzalkonium chloride to dilute it to an 8% solution. How much water must she add?

22. Mixing photographic chemicals A photographer wishes to mix 2 liters of a 5% acetic acid solution with a 10% solution to get a 7% solution. How many liters of 10% solution must be added?

23. Mixing candy Lemon drops worth $1.90 per pound are to be mixed with jelly beans that cost $1.20 per pound to make 100 pounds of a mixture worth $1.48 per pound. How many pounds of each candy should be used?

24. Blending gourmet tea One grade of tea, worth $3.20 per pound, is to be mixed with another grade worth $2 per pound to make 20 pounds that will sell for $2.72 per pound. How much of each grade of tea must be used?

25. Mixing nuts A pound of peanuts is worth $.30 less than a pound of cashews. Equal amounts of peanuts and cashews are used to make 40 pounds of a mixture that sells for $1.05 per pound. How much is a pound of cashews worth?

26. Mixing candy Twenty pounds of lemon drops are to be mixed with cherry chews to make a mixture that will sell for $1.80 per pound. How much of the more expensive candy should be used? (See Table 1.)

27. Coffee blends A store sells regular coffee for $4 a pound and a gourmet coffee for $7 a pound. To get rid of 40 pounds of the gourmet coffee, the shopkeeper makes a gourmet blend that he will put on sale for $5 a pound. How many pounds of regular coffee should be used?

28. Lawn seed blends A garden store sells Kentucky bluegrass seed for $6 per pound and ryegrass seed for $3 per pound. How much rye must be mixed with 100 pounds of bluegrass to obtain a blend that will sell for $5 per pound?

	Price per pound
Peppermint patties	$1.35
Lemon drops	$1.70
Licorice lumps	$1.95
Cherry chews	$2.00

TABLE 1

Writing Exercises ■ *Write a paragraph using your own words.*

1. Create a mixture problem of your own and solve it.

2. In mixture problems, explain why it is important to distinguish between the quantity and the value of the materials being combined.

Something to Think About ■ **An impossible mixture** How many gallons of a 10% and a 20% solution should be mixed to obtain a 30% solution?

1. Without solving it, how do you know that the problem has no solution?

2. What happens if you try to solve it anyway?

Review Exercises ■ **1.** The amount, A, on deposit in a bank account bearing simple interest is given by the formula

$$A = P + Prt$$

Find A when $P = \$1200$, $r = 0.08$, and $t = 3$.

2. The distance, s, that a certain object falls in t seconds is given by the formula

$$s = 350 - 16t^2 + vt$$

Find s when $t = 4$ and $v = -3$.

2.7 Solving Inequalities

■ Double or Compound Inequalities ■ Problem Solving

GETTING READY   *Graph each set on the number line.*

1. All real numbers greater than -1.

2. All real numbers less than or equal to 5.

3. All real numbers between -2 and 4.

4. All real numbers less than -2 or greater than or equal to 4.

Recall the meaning of the following symbols.

Inequality Symbols

$<$ means	"is less than"
$>$ means	"is greater than"
$\leq$ means	"is less than or equal to"
$\geq$ means	"is greater than or equal to"

An **inequality** is a mathematical expression that indicates that two quantities are not necessarily equal. A **solution of an inequality** is any number that makes the inequality a true statement. The number 2 is a solution of the inequality

$$x \leq 3$$

because $2 \leq 3$.

The inequality $x \leq 3$ has many more solutions, because any real number that is less than or equal to 3 will satisfy the inequality. We can use a graph on the number line to exhibit the solutions of the inequality $x \leq 3$. The colored arrow in Figure 2-9 indicates all those points with coordinates that satisfy the inequality $x \leq 3$. The solid circle at the point with coordinate 3 indicates that the number 3 is a solution of the inequality $x \leq 3$.

FIGURE 2-9

The graph of the inequality $x > 1$ appears in Figure 2-10. The colored arrow indicates all those points whose coordinates satisfy the inequality $x > 1$. The open circle at the point with coordinate 1 indicates that 1 is not a solution of the inequality $x > 1$.

FIGURE 2-10

To solve more complicated inequalities, we need to use the addition, subtraction, multiplication, and division properties of inequalities. When we use any of these properties, the resulting inequality has the same solutions as the original inequality.

Addition Property of Inequality	If a, b, and c are real numbers, and If $a < b$, then $a + c < b + c$. Similar statements can be made for the symbols $>$, $\leq$, and $\geq$.

The **addition property of inequality** can be stated this way: *If any quantity is added to both sides of an inequality, the resulting inequality has the same direction as the original inequality.*

Subtraction Property of Inequality	If a, b, and c are real numbers, and If $a < b$, then $a - c < b - c$. Similar statements can be made for the symbols $>$, $\leq$, and $\geq$.

The **subtraction property of inequality** can be stated this way: *If any quantity is subtracted from both sides of an inequality, the resulting inequality has the same direction as the original inequality.*

The subtraction property of inequality is included in the addition property: To *subtract* a number a from both sides of an inequality, we could instead *add* the *negative* of a to both sides.

EXAMPLE 1   Solve the inequality $2x + 5 > x - 4$ and graph its solution on a number line.

Solution   To isolate the x on the left-hand side of the $>$ sign, we proceed as we would when solving equations.

$$2x + 5 > x - 4$$
$$2x + 5 - 5 > x - 4 - 5 \qquad \text{Subtract 5 from both sides.}$$
$$2x > x - 9 \qquad \text{Combine like terms.}$$
$$2x - x > x - 9 - x \qquad \text{Subtract } x \text{ from both sides.}$$
$$x > -9 \qquad \text{Combine like terms.}$$

The graph of this solution (see Figure 2-11) includes all points to the right of -9 but does not include -9 itself. For that reason, we use an open circle at -9.

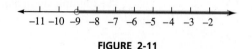

FIGURE 2-11   ■

If both sides of the true inequality $2 < 5$ are multiplied by a *positive* number, such as 3, another true inequality results.

$$2 < 5$$
$$3 \cdot 2 < 3 \cdot 5 \qquad \text{Multiply both sides by 3.}$$
$$6 < 15$$

The inequality $6 < 15$ is a true inequality. However, if both sides of the inequality $2 < 5$ are multiplied by a negative number, such as -3, the direction of the inequality symbol must be reversed to produce another true inequality.

$$2 < 5$$
$$-3 \cdot 2 > -3 \cdot 5 \qquad \text{Multiply both sides by the \textit{negative} number } -3$$
$$\text{and reverse the direction of the inequality.}$$
$$-6 > -15$$

The inequality $-6 > -15$ is a true inequality because -6 lies to the right of -15 on the number line.

Multiplication Property of Inequality

If a, b, and c are real numbers, and

If $a < b$ and $c > 0$, then $ac < bc$.

If $a < b$ and $c < 0$, then $ac > bc$.

There is a similar property for division.

Division Property of Inequality	If a, b, and c are real numbers, and

If $a < b$ and $c > 0$, then $\dfrac{a}{c} < \dfrac{b}{c}$.

If $a < b$ and $c < 0$, then $\dfrac{a}{c} > \dfrac{b}{c}$.

To *divide* both sides of an inequality by a nonzero number c, we could instead *multiply* both sides by $\frac{1}{c}$.

The multiplication and division properties of inequality are also true for $\leq$, $>$, and $\geq$.

 WARNING! If both sides of an inequality are multiplied by a *positive* number, the direction of the resulting inequality remains the same. However, if both sides of an inequality are multiplied by a *negative* number, the direction of the resulting inequality must be reversed.

EXAMPLE 2 Solve $3x + 7 \leq -5$ and graph the solution.

Solution

$$3x + 7 \leq -5$$
$$3x + 7 - 7 \leq -5 - 7 \qquad \text{Subtract 7 from both sides.}$$
$$3x \leq -12 \qquad \text{Combine like terms.}$$
$$\frac{3x}{3} \leq \frac{-12}{3} \qquad \text{Divide both sides by 3.}$$
$$x \leq -4$$

The solution of $3x + 7 \leq -5$ consists of all real numbers less than, and also including, -4. The solid circle at -4 in the graph of the solution in Figure 2-12 indicates that -4 is one of the solutions of the given inequality.

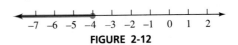

FIGURE 2-12

EXAMPLE 3 Solve $5 - 3x \leq 14$ and graph the solution.

Solution

$$5 - 3x \leq 14$$
$$5 - 3x - 5 \leq 14 - 5 \qquad \text{Subtract 5 from both sides.}$$
$$-3x \leq 9 \qquad \text{Combine like terms.}$$
$$\frac{-3x}{-3} \geq \frac{9}{-3} \qquad \text{Divide both sides by } -3 \text{ and reverse the direction of the } \leq \text{ symbol.}$$
$$x \geq -3$$

In the last step, both sides of the inequality were divided by -3. Because -3 is negative, the direction of the inequality was *reversed*. The graph of the solution appears in Figure 2-13. The solid circle at -3 indicates that -3 is one of the solutions.

FIGURE 2-13

EXAMPLE 4   Solve $x(x - 8) < x^2 + 3(x + 11)$ and graph the solution.

Solution

$$x(x - 8) < x^2 + 3(x + 11)$$

$x^2 - 8x < x^2 + 3x + 33$ Remove parentheses.

$-8x < 3x + 33$ Subtract x^2 from both sides.

$-11x < 33$ Subtract $3x$ from both sides.

$x > -3$ Divide both sides by -11 and reverse the direction of the inequality sign.

The graph of the solution appears in Figure 2-14. The open circle at -3 indicates that -3 is not a solution.

FIGURE 2-14

▪ Double or Compound Inequalities

Two inequalities can be combined into a **double inequality** or **compound inequality** to indicate that numbers lie *between* two fixed values. The inequality $2 < x < 5$, for example, indicates that x is greater than 2 and that x is *also* less than 5. The solution of the double inequality $2 < x < 5$ consists of all numbers that lie *between* 2 and 5. The graph of this set appears in Figure 2-15. These sets of numbers are called **intervals**.

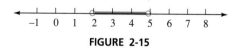

FIGURE 2-15

EXAMPLE 5   Solve $-4 < 2(x - 1) \leq 4$ and graph the solution.

Solution

$$-4 < 2(x - 1) \leq 4$$

$-4 < 2x - 2 \leq 4$ Remove parentheses.

$-2 < 2x \leq 6$ Add 2 to all three parts.

$-1 < x \leq 3$ Divide all three parts by 2.

The graph of the solution appears in Figure 2-16.

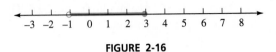

FIGURE 2-16 ■

■ Problem Solving

EXAMPLE 6 **Averaging grades** A student has scores of 72%, 74%, and 78% on three mathematics examinations. What score does he need on the last exam to earn a grade of at least a B (80% or better)?

Solution We can let x represent the score on the fourth (and last) exam. To find the average grade, we add the four scores and divide by 4. To earn a B, this average must be greater than or equal to 80%.

The average of the four grades	$\geq$	80.

$$\frac{72 + 74 + 78 + x}{4} \geq 80$$

We can solve this inequality for x.

$$\frac{224 + x}{4} \geq 80 \qquad 72 + 74 + 78 = 224.$$

$$224 + x \geq 320 \qquad \text{Multiply both sides by 4.}$$

$$x \geq 96 \qquad \text{Subtract 224 from both sides.}$$

To earn a B, the student must score 96% or better on the last exam. The graph of this solution appears in Figure 2-17.

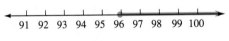

FIGURE 2-17 ■

EXAMPLE 7 **Geometry** If the perimeter of an equilateral triangle is less than 15 feet, how long could a side be?

Solution Recall that each side of an equilateral triangle is the same length and that the perimeter of a triangle is the sum of the lengths of its three sides.

Let x represent the length of one side of the triangle.

Then $x + x + x$ represents the perimeter.

The perimeter is to be less than 15 feet. We indicate this fact with the following inequality.

$$x + x + x < 15$$
$$3x < 15 \quad \text{Combine like terms.}$$
$$x < 5 \quad \text{Divide both sides by 3.}$$

Each side of the triangle must be less than 5 feet long.   ■

ORALS *Solve each inequality.*

1. $2x < 4$ **2.** $x + 5 > 6$
3. $-3x < -6$ **4.** $-x > 2$
5. $2x - 5 < 7$ **6.** $5 - 2x < 7$

EXERCISE 2.7

In Exercises 1–40, solve each inequality and graph the solution.

1. $x + 2 > 5$ **2.** $x + 5 \geq 2$ **3.** $-x - 3 \leq 7$ **4.** $-x - 9 > 3$

5. $3 + x < 2$ **6.** $5 + x \geq 3$ **7.** $2x - 3 \leq 5$ **8.** $-3x - 5 < 4$

9. $-3x - 7 > -1$ **10.** $-5x + 7 \leq 12$ **11.** $-4x + 1 > 17$ **12.** $7x - 9 > 5$

13. $2x + 9 \leq x + 8$ **14.** $3x + 7 \leq 4x - 2$ **15.** $9x + 13 \geq 8x$ **16.** $7x - 16 < 6x$

17. $8x + 4 > 6x - 2$ **18.** $7x + 6 \geq 4x$ **19.** $5x + 7 < 2x + 1$ **20.** $7x + 2 > 4x - 1$

21. $7 - x \leq 3x - 1$ **22.** $2 - 3x \geq 6 + x$

23. $9 - 2x > 24 - 7x$ **24.** $13 - 17x < 34 - 10x$

25. $3(x - 8) < 5x + 6$

26. $9(x - 11) > 13 + 7x$

27. $8(5 - x) \leq 10(8 - x)$

28. $17(3 - x) \geq 3 - 13x$

29. $x(5x - 5) > 5x^2 + 15$

30. $x(x - 7) < x^2 + 3x + 20$

31. $x(x + 8) \leq x^2 + 24$

32. $x(5 + 3x) \leq 15 + 3x^2$

33. $89x^2 - 178 > 89x(x - 1)$

34. $31x^2 + 124 > 3.1x(10x + 20)$

35. $\dfrac{5}{2}(7x - 15) + x \geq \dfrac{13}{2}x - \dfrac{3}{2}$

36. $\dfrac{5}{3}(x + 1) \leq -x + \dfrac{2}{3}$

37. $\dfrac{3x - 3}{2} < 2x + 2$

38. $\dfrac{x + 7}{3} \geq x - 3$

39. $\dfrac{2(x + 5)}{3} \leq 3x - 6$

40. $\dfrac{3(x - 1)}{4} > x + 1$

In Exercises 41–60, solve each inequality and graph the solution.

41. $2 < x - 5 < 5$

42. $3 < x - 2 < 7$

43. $-5 < x + 4 \leq 7$

44. $-9 \leq x + 8 < 1$

45. $0 \leq x + 10 \leq 10$

46. $-8 < x - 8 < 8$

47. $4 < -2x < 10$

48. $-4 \leq -4x < 12$

49. $-3 \leq \dfrac{x}{2} \leq 5$

50. $-12 \leq \dfrac{x}{3} < 0$

51. $3 \leq 2x - 1 < 5$

52. $4 < 3x - 5 \leq 7$

53. $0 < 10 - 5x \leq 15$

54. $1 \leq -7x + 8 \leq 15$

55. $-6 < 3(x + 2) < 9$

56. $-18 \leq 9(x - 5) < 27$

57. $x^2 + 3 \leq x(x - 3) \leq x^2 + 9$

58. $x^2 < x(x + 4) \leq x^2 + 8$

59. $3 - x < 5 < 7 - x$

60. $x + 1 < 2x + 3 < x + 5$

In Exercises 61–78, express each solution as an inequality.

61. Calculating grades A student has test scores of 68%, 75%, and 79%. What must she score on the last exam to earn a B (80% or better)?

62. Calculating grades A student has test scores of 70%, 74%, and 84%. What score does he need to maintain a C (70% or better)?

63. Fleet averages An auto manufacturer produces three sedan models in equal quantities. One model has an economy rating of 17 miles per gallon, and the second model is rated at 19 mpg. The manufacturer is required to have a fleet average of at least 21 mpg. What economy rating is required for the third model car?

64. Avoiding a service charge When the average daily balance of a customer's checking account falls below $500 in any week, the bank assesses a $5 service charge. Bill's account balances for the week were

Monday	$540.00
Tuesday	435.50
Wednesday	345.30
Thursday	310.00

What must Friday's balance be to avoid the service charge?

65. Geometry The perimeter of an equilateral triangle is at most 57 feet. What could be the length of a side? (*Hint:* All three sides of an equilateral triangle are equal.)

66. Geometry The perimeter of a square is no less than 68 centimeters. How long can a side be?

67. Land elevations The land elevations in Nevada range from the 13,143-foot height of Boundary Peak to the Colorado River at 470 feet. To the nearest tenth, what is the range of these elevations in miles? (*Hint:* 1 mile is 5280 feet.)

68. Doing homework A teacher requires that students do homework at least 2 hours a day. How many minutes should a student work each week?

69. Plane altitudes A pilot plans to fly at an altitude of between 17,500 and 21,700 feet. To the nearest tenth, what will be the range of altitudes in miles? (*Hint:* There are 5280 feet in 1 mile.)

70. Getting exercise Doctors advise exercising at least 15 minutes but less than 30 minutes per day. Find the range of exercise time for one week, in hours.

71. Comparing temperatures To hold the temperature of a room between 19° and 22° Celsius, what Fahrenheit temperatures must be maintained? (*Hint:* Fahrenheit temperature (F) and Celsius temperature (C) are related by the formula $C = \frac{5}{9}(F - 32)$.)

72. Melting iron To melt iron, the temperature of a furnace must be at least 1540° C but no more than 1650° C. What range of Fahrenheit temperatures must be maintained?

73. Phonograph records The radii of phonograph records must lie between 5.9 and 6.1 inches. What variation in circumference can occur? (*Hint:* The circumference of a circle is given by the formula $C = 2\pi r$, where r is the radius. Let $\pi = 3.14$.)

74. Pythons A large snake, the African Rock Python, can grow to a length of 25 feet. To the nearest hundredth, what is the range of lengths in meters? (*Hint:* There are about 3.281 feet in one meter.)

75. Comparing weights The normal weight of a 6-foot-2-inch man is between 150 and 190 pounds. To the nearest hundredth, what would such a person weigh in kilograms? (*Hint:* There are 2.2 pounds in one kilogram.)

76. Manufacturing stereos The time required to assemble a television set at the factory is 2 hours. A stereo receiver requires only 1 hour. The labor force at the factory can supply at least 640 and at most 810 hours of assembly time per week. When the factory is producing 3 times as many television sets as stereos, how many stereos could be manufactured in 1 week?

77. Geometry   A rectangle's length is 3 feet less than twice its width, and its perimeter is between 24 and 48 feet. What might be its width?

78. Geometry   A rectangle's width is 8 feet less than 3 times its length, and its perimeter is between 8 and 16 feet. What might be its length?

Writing Exercises ■ *Write a paragraph using your own words.*

1. Explain why multiplying both sides of an inequality by a negative constant reverses the direction of the inequality.

2. Explain the use of solid and open circles in the graphing of the solution of an inequality.

Something to Think About ■

1. To solve the inequality $1 < \frac{1}{x}$, one student multiplies both sides by x to get $x < 1$. Why is this not correct?

2. Find the solution of $1 < \frac{1}{x}$. (*Hint:* Will any negative values of x work?)

Review Exercises ■ *Simplify each expression*

1. $3x^2 - 2(y^2 - x^2)$

2. $5(xy + 2) - 3xy - 8$

3. $\frac{1}{3}(x + 6) - \frac{4}{3}(x - 9)$

4. $\frac{4}{5}x(y + 1) - \frac{9}{5}y(x - 1)$

P R O J E C T ■ The Math of Magic

Magicians don't really saw people in half, or pull rabbits out of empty hats. Most magic tricks are just clever illusions, fooling the audience into seeing what isn't really there. The most successful magicians are very believable liars—and it takes a lot of practice to be good.

Many magic tricks involve cutting ropes in various ways and then restoring them to their original lengths. For example, a magician holds up a long rope for all to see and then cuts it into three separate sections. He displays these to the audience; the sections are of three obviously different lengths, as in Illustration 1.

The magician then folds the ropes, twists them, coils them around his fist and arm, utters some magic words–all to distract and confuse the audience. When he holds up the three sections as in Illustration 2, they are now the same length!

The secret of the trick lies behind the magician's hand, hidden from the audience. What appears to be two equal lengths of rope in Illustration 3 is only one–the longest of the original three, folded in half. Those "two" sections are

ILLUSTRATION 1   ILLUSTRATION 2

equal to the "third," which is just the middle-sized of the original three. What happened to the shortest of the original three? The magician disposed of it when the audience was distracted.

To prepare for this trick, the magician places two marks on an 8-foot rope, so that the two cuts can be made quickly and accurately. A third mark of a different color is the center of the largest section, the point where that rope is to be folded in half.

ILLUSTRATION 3

- If the shortest section is to be 1 foot long, where does the magician make the marks?

- Get some rope, cut an 8-foot piece, mark it as you have determined, and practice. There are several ways to dispose of the shortest segment without being noticed. Try using a stretched rubber band to snap the rope up your sleeve, or fake a distracting sneeze while you slip it into your pocket. It is an easy trick to master, and an effective illusion.

MATHEMATICS IN RETIREMENT

To find the maximum allowable annual contribution to an SEP, we first find the net income N by subtracting deductible expenses from gross income. If C is the maximum contribution to the SEP, then

$$C = 0.15(N - C)$$

Solve this literal equation for C.

$C = 0.15(N - C)$	
$C = 0.15N - 0.15C$	Remove parentheses.
$C + 0.15C = 0.15N$	Add $0.15C$ to both sides.
$1.15C = 0.15N$	Combine terms.
$C = \dfrac{0.15}{1.15}N$	Divide both sides by 1.15.
$C = 0.1304N$	Simplify.

The maximum allowable annual contribution to an SEP plan is slightly greater than 13% of taxable income.

Since tax law covers many special cases and includes many exceptions, it is always wise to consult a tax professional for advice.

Chapter Summary

KEY WORDS

addition property of equality (2.1)

addition property of inequality (2.7)

conditional equations (2.3)

discount (2.2)

division property of equality (2.1)

division property of inequality (2.7)

double inequality (2.7)

equation (2.1)

formula (2.4)

identity (2.3)

impossible equation (2.3)

inequality (2.7)

interval (2.7)

like terms (2.3)

literal equation (2.4)

markdown (2.2)

markup (2.2)

multiplication property of equality (2.1)

multiplication property of inequality (2.7)

numerical coefficient (2.3)

percent of decrease (2.2)

percent of increase (2.2)

root of an equation (2.1)

solution of an equation (2.1)

subtraction property of equality (2.1)

subtraction property of inequality (2.7)

unknown (2.1)

unlike terms (2.3)

variable (2.1)

KEY IDEAS

(2.1) Any real number can be added to (or subtracted from) both sides of an equation to form another equation with the same solutions as the original equation.

Both sides of an equation can be multiplied (or divided) by any *nonzero* real number to form another equation with the same solutions as the original equation.

Percentage = rate · base

(2.3) Like terms can be combined by adding their numerical coefficients and using the same variables and exponents.

(2.4) A literal equation, or formula, can often be solved for any of its variables.

(2.5–2.6) Equations are useful in problem solving.

(2.7) Inequalities are solved by techniques similar to those used to solve equations, with this exception: *If both sides of an inequality are multiplied or divided by a negative number, the direction of the inequality must be reversed.*

The solution of an inequality can be graphed on the number line.

■ Chapter 2 Review Exercises

(2.1) *In Review Exercises 1–6, tell whether the indicated number is a solution of the given equation.*

1. $3x + 7 = 1$; -2

2. $5 - 2x = 3$; -1

3. $2(x + 3) = x$; -3

4. $5(3 - x) = 2 - 4x$; 13

5. $3(x + 5) = 2(x - 3)$; -21

6. $2(x - 7) = x + 14$; 0

In Review Exercises 7–18, solve each equation. Check all solutions.

7. $x - 7 = -6$

8. $x + \dfrac{3}{5} = \dfrac{3}{5}$

9. $y - \dfrac{7}{2} = \dfrac{1}{2}$

10. $z + \dfrac{5}{3} = -\dfrac{1}{3}$

11. $3x = 15$

12. $8r = -16$

13. $10z = 5$

14. $14s = 21$

15. $\dfrac{y}{3} = 6$

16. $\dfrac{w}{7} = -5$

17. $\dfrac{a}{-7} = \dfrac{1}{14}$

18. $\dfrac{t}{12} = \dfrac{1}{2}$

19. What number is 35% of 700?

20. 72% of what number is 936?

21. What percent of 2300 is 851?

22. 72 is what percent of 576?

(2.2) *In Review Exercises 23–42, solve each equation. Check all solutions.*

23. $5y + 6 = 21$

24. $5y - 9 = 1$

25. $12z + 4 = -8$

26. $17z + 3 = 20$

27. $13 - 13t = 0$

28. $10 + 7t = -4$

29. $23a - 43 = 3$

30. $84 - 21a = -63$

31. $3x + 7 = 1$

32. $7 - 9x = 16$

33. $\dfrac{b + 3}{4} = 2$

34. $\dfrac{b - 7}{2} = -2$

35. $\dfrac{x - 8}{5} = 1$

36. $\dfrac{x + 10}{2} = -1$

37. $\dfrac{2(y - 1)}{4} = 2$

38. $\dfrac{3(y + 4)}{11} = 3$

39. $\dfrac{x}{2} + \dfrac{7}{2} = 11$ **40.** $\dfrac{r}{3} - 3 = 7$ **41.** $\dfrac{a}{2} + \dfrac{9}{4} = 6$ **42.** $\dfrac{x}{8} - 2.3 = 3.2$

43. Discount shopping A compact disk player is on sale for $240, a 25% savings from the regular price. Find the regular price.

44. Sales tax rate A $38 dictionary costs $40.47, with sales tax. Find the tax rate.

45. Oriental rugs A Turkish rug was purchased for $560. If it is now worth $1100, find the percent of increase.

46. Percent of discount A clock on sale for $215 was regularly priced at $465. Find the percent of discount.

(2.3) *In Review Exercises 47–56, simplify each expression, if possible.*

47. $5x + 9x$ **48.** $7a + 12a$ **49.** $18b - 13b$ **50.** $21x - 23x$

51. $5y - 7y$ **52.** $19x - 19$ **53.** $y^2 + 3(y^2 - 2)$ **54.** $2x^2 - 2(x^2 - 2)$

55. $7(x + 2) + 2(x - 7)$ **56.** $2(3 - x) + x - 6x$

In Review Exercises 57–80, solve each equation. Check all solutions.

57. $2x - 19 = 2 - x$ **58.** $5b - 19 = 2b + 20$ **59.** $3x + 20 = 5 - 2x$ **60.** $9x + 100 = 7x + 18$

61. $10(t - 3) = 3(t + 11)$ **62.** $2(5x - 7) = 2(x - 35)$ **63.** $x(x + 6) = x^2 + 6$ **64.** $x(x - 5) = x^2 + 10$

65. $\dfrac{3(u - 2)}{5} = 3$ **66.** $\dfrac{5(v - 7)}{3} = -5$ **67.** $\dfrac{7(x - 4)}{4} = -21$ **68.** $\dfrac{9(3 + y)}{5} = -27$

69. $\dfrac{2}{3}(5x - 3) = 38$ **70.** $\dfrac{3}{2}(3x + 2) = 9$ **71.** $\dfrac{3}{4}(7k - 5) = 12$ **72.** $\dfrac{5}{3}(6 + 5b) = 10$

73. $\dfrac{1}{3}(4n - 3) = 5$ **74.** $\dfrac{3}{5}(4x + 4) = -6$ **75.** $\dfrac{3}{5}(2x - 1) = -1$ **76.** $\dfrac{3}{7}(3m - 7) = -2$

77. $\dfrac{1}{4}\left(\dfrac{q}{2} + 1\right) = 1$ **78.** $\dfrac{1}{2}\left(\dfrac{a}{3} - 4\right) = -3$ **79.** $\dfrac{2r - 10}{6} + 7 = 17$ **80.** $\dfrac{5a + 10}{10} + 7 = 0$

(2.4) *In Review Exercises 81–92, solve each equation for the variable indicated.*

81. $E = IR$; for R **82.** $i = prt$; for t **83.** $P = I^2 R$; for R **84.** $d = rt$; for r

85. $V = lwh$; for h **86.** $y = mx + b$; for m **87.** $V = \pi r^2 h$; for h **88.** $a = 2\pi rh$; for r

89. $F = \dfrac{GMm}{d^2}$; for G **90.** $P = \dfrac{RT}{mV}$; for m **91.** $T = n(V - 3)$; for V **92.** $T = n(V - 3)$; for n

(2.5–2.6) In Review Exercises 93–102, solve each problem and check the solution.

93. Integer problem   If twice an integer is decreased by 7, the result is 9. Find the original integer.

94. Utility bills   The electric company charges $17.50 per month, plus 18 cents for every kilowatt-hour of energy used. One resident's bill was $43.96. How many kilowatt-hours were used that month?

95. Installing rain gutters   A contractor charges $35 for the installation of rain gutters, plus $1.50 per foot. One installation cost $162.50. How many feet of gutter are required?

96. Cutting a rope   A 45-foot rope is to be cut into three sections. One section is to be 15 feet long. Of the remaining sections, one must be 2 feet less than 3 times the length of the other. Find the length of the shortest section.

97. Width of a rectangle   If the length of the rectangular painting in Illustration 1 is 3 inches more than twice the width, how wide is the rectangle?

84 in.

ILLUSTRATION 1

98. Manufacturing auto parts   Costs for machining automotive parts run $668.50 per week, plus variable costs of $1.25 per unit. The company can sell as many as it can make for $3 each. Find the break-even point.

99. Making baseball caps   A company can manufacture baseball caps on either of two machines, with costs as shown in Table 1. At the projected sales level, they find the costs on the two machines are equal. Find the expected sales.

Machine	Startup cost	Unit cost (per cap)
1	$ 85	$3
2	$105	$2.50

TABLE 1

100. Investment income   A woman has $27,000. Part is invested for 1 year in a certificate of deposit paying 7% interest and the remaining amount in a cash management fund paying 9%. The total interest on the two investments is $2110. How much did she invest at each rate?

101. Walking and bicycling   A bicycle path is 5 miles long. A man walks from one end at the rate of 3 miles per hour. At the same time, a friend bicycles from the other end, traveling 12 miles per hour. In how many minutes will they meet?

102. Mixtures   A store manager mixes candy worth 90 cents per pound with gumdrops worth $1.50 per pound to make 20 pounds of a mixture worth $1.20 per pound. How many pounds of each kind of candy must he use?

(2.7) In Review Exercises 103–112, solve each inequality and graph its solution.

103. $3x + 2 < 5$

104. $-5x - 8 > 7$

105. $5x - 3 \geq 2x + 9$

106. $7x + 1 \leq 8x - 5$

107. $5(3 - x) \leq 3(x - 3)$

108. $3(5 - x) \geq 2x$

109. $8 < x + 2 < 13$

110. $0 \le 2 - 2x < 4$

111. $x^2 < x(x + 1) \le x^2 + 9$

112. $x^2 + 4 \ge x^2 + x > x^2 + 3$

Chapter 2 Test

In Problems 1–4, tell whether the indicated number is a solution of the given equation.

1. $5x + 3 = -2; -1$

2. $3(x + 2) = 2x; -6$

3. $-3(2 - x) = 0; -2$

4. $x(x + 1) = x^2 + 1; 1$

In Problems 5–14, solve each equation.

5. $x + 17 = -19$

6. $12x = -144$

7. $\dfrac{x}{7} = -1$

8. $8x + 2 = -14$

9. $3 = 5 - 2x$

10. $23 - 5(x + 10) = -12$

11. $x(x + 5) = x^2 + 3x - 8$

12. $\dfrac{3(x - 6)}{2} = 6x$

13. $\dfrac{5}{3}(x - 7) = 15(x + 1)$

14. $\dfrac{7}{8}(x - 4) = 5x - \dfrac{7}{2}$

In Problems 15–20, simplify each expression.

15. $x + 5(x - 3)$

16. $3x - 5(2 - x)$

17. $x(x - 3) + 2x^2 - 3x$

18. $4(x^2 - 3^2) - x(x + 36)$

19. $-3x(x + 3) + 3x(x - 3)$

20. $-4x(2x - 5) - 7x(4x + 1)$

In Problems 21–26, solve each equation for the variable indicated.

21. $d = rt$; for t

22. $P = 2l + 2w$; for l

23. $A = 2\pi rh$; for h

24. $A = P + Prt$; for r

25. $P = \dfrac{RT}{v}$; for v

26. $A = \dfrac{1}{3}\pi r^2 h$; for h

27. Integer problem The sum of two consecutive odd integers is 36. Find the integers.

28. Investment problem Part of $13,750 is invested at 9% annual interest and the rest at 8%. After 1 year, the accounts paid $1185 in interest. How much was invested at the lower rate?

29. Travel times   A car leaves Rockford, Illinois at the rate of 65 miles per hour, bound for Madison, Wisconsin. At the same time, a truck leaves Madison at the rate of 55 miles per hour, bound for Rockford. If the cities are 72 miles apart, how long will it take for the car and the truck to meet?

30. Mixture problem   How many liters of water must be added to 30 liters of a 10% brine solution to dilute it to an 8% solution?

In Problems 31–34, solve each inequality and graph its solution.

31. $8x - 20 \geq 4$

32. $x^2 - x(x + 7) > 14$

33. $-4 \leq 2(x + 1) < 10$

34. $-2 < 5(x - 1) \leq 10$

3

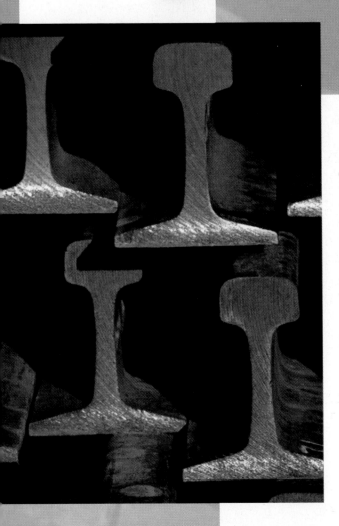

Polynomials

MATHEMATICS IN MEDICINE

The red cells of our blood pick up oxygen in the lungs and carry it to all parts of the body. Each red cell is a tiny disk with an approximate radius of 0.00015 inch. Because the amount of oxygen carried depends on the surface area of the cells, and the cells are so tiny, a very great number is needed—25 trillion in an average adult.

What is the total surface area of all of the red blood cells in the body?

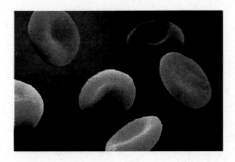

■ ■ ■ ■ ■ ■ ■ ■   After reading this chapter, you will be able to answer this question.

In this chapter, we shall develop rules for finding products, powers, and quotients of certain exponential expressions and then discuss how to add, subtract, multiply, and divide polynomials.

3.1 Natural-Number Exponents

■ **Properties of Natural-Number Exponents**

GETTING READY   *Evaluate each expression.*

1. 2^3 **2.** 3^2 **3.** $3(2)$ **4.** $2(3)$

5. $2^3 + 2^2$ **6.** $2^3 \cdot 2^2$ **7.** $3^3 - 3^2$ **8.** $\dfrac{3^3}{3^2}$

We have used natural-number exponents to indicate repeated multiplication. For example,

$$2^5 = 2 \cdot 2 \cdot 2 \cdot 2 \cdot 2 = 32 \qquad (-7)^3 = (-7)(-7)(-7) = -343$$
$$x^4 = x \cdot x \cdot x \cdot x \qquad \qquad -y^5 = -y \cdot y \cdot y \cdot y \cdot y$$

These examples suggest a definition for x^n, where n is a natural number.

Natural-Number Exponents	If n is a natural number, then

$$\overset{n \text{ factors of } x}{\overbrace{x^n = x \cdot x \cdot x \cdots \cdots x}}$$

In the exponential expression x^n, x is called the **base**, and n is called the **exponent**. The entire expression is called a **power of x.** If an exponent is a natural number, it tells how many times its base is to be used as a factor in a product. An exponent of 1 indicates that its base is to be used as a factor one time, an exponent of 2 indicates that its base is to be used as a factor two times, and so on.

$$3^1 = 3 \quad (-y)^1 = -y \quad (-4z)^2 = (-4z)(-4z) \quad \text{and} \quad (t^2)^3 = t^2 \cdot t^2 \cdot t^2$$

EXAMPLE 1 Show that -2^4 and $(-2)^4$ are different numbers.

Solution We write each expression without exponents and observe different results.

$$-2^4 = -(2^4) \qquad\qquad (-2)^4 = (-2)(-2)(-2)(-2)$$
$$= -(2 \cdot 2 \cdot 2 \cdot 2) \qquad\qquad = 16$$
$$= -16$$

Since $-16 \neq 16$, it follows that $-2^4 \neq (-2)^4$. ■

EXAMPLE 2 Write each expression without exponents: **a.** r^3, **b.** $(-2s)^4$, and **c.** $\left(\frac{1}{3}ab\right)^5$.

Solution **a.** $r^3 = r \cdot r \cdot r$

b. $(-2s)^4 = (-2s)(-2s)(-2s)(-2s)$

c. $\left(\frac{1}{3}ab\right)^5 = \left(\frac{1}{3}ab\right)\left(\frac{1}{3}ab\right)\left(\frac{1}{3}ab\right)\left(\frac{1}{3}ab\right)\left(\frac{1}{3}ab\right)$ ■

■ Properties of Natural-Number Exponents

Because of the definition of natural-number exponents, we can develop a rule for multiplying exponential expressions. To multiply x^3 by x^2, for example, we note that the expression x^2 means that x is to be used as a factor two times and the expression x^3 means that x is to be used as a factor three times. Thus,

$$\overset{\text{2 factors of } x \qquad \text{3 factors of } x}{x^2 x^3 = \qquad \overbrace{x \cdot x} \qquad \cdot \qquad \overbrace{x \cdot x \cdot x}}$$

$$= \overbrace{x \cdot x \cdot x \cdot x \cdot x}^{\text{5 factors of } x}$$

$$= x^5$$

In general,

$$x^m \cdot x^n = \overbrace{x \cdot x \cdot x \cdots \cdots x}^{m \text{ factors of } x} \cdot \overbrace{x \cdot x \cdot x \cdots \cdots x}^{n \text{ factors of } x}$$

$$= \overbrace{x \cdot x \cdot x \cdot x \cdot x \cdot x \cdots \cdots x \cdot x \cdot x}^{m + n \text{ factors of } x}$$

$$= x^{m+n}$$

This discussion justifies the rule for multiplying exponential expressions: *To multiply two exponential expressions with the same base, we keep the same base and add the exponents.*

Product Rule for Exponents If m and n are natural numbers, then
$$x^m x^n = x^{m+n}$$

EXAMPLE 3 (a, d, f)

a. $x^3 x^4 = x^{3+4}$
    $= x^7$

b. $y^2 y^4 = y^{2+4}$
    $= y^6$

c. $zz^3 = z^1 z^3$
    $= z^{1+3}$
    $= z^4$

d. $x^2 x^3 x^6 = (x^2 x^3)x^6$
    $= (x^{2+3})x^6$
    $= x^5 x^6$
    $= x^{5+6}$
    $= x^{11}$

e. $(2y^3)(3y^2) = 2(3)y^3 y^2$
    $= 6y^{3+2}$
    $= 6y^5$

f. $(4x)(-3x^2) = 4(-3)xx^2$
    $= -12x^{1+2}$
    $= -12x^3$

∎

WARNING! The product rule for exponents applies only to exponential expressions with the same base. An expression such as $x^2 y^3$ cannot be simplified, because x^2 and y^3 have different bases.

To find another rule of exponents, we consider the expression $(x^3)^4$, which can be written as $x^3 \cdot x^3 \cdot x^3 \cdot x^3$. Because each of the four factors of x^3 contains three factors of x, there are $4 \cdot 3$, or 12, factors of x. Thus, the product can be written as x^{12}.

$$(x^3)^4 = x^3 \cdot x^3 \cdot x^3 \cdot x^3$$

$$= \overbrace{x \cdot x \cdot x \; \cdot \; x \cdot x \cdot x \; \cdot \; x \cdot x \cdot x \; \cdot \; x \cdot x \cdot x}^{\text{12 factors of } x}$$

$$\underbrace{\quad}_{x^3} \underbrace{\quad}_{x^3} \underbrace{\quad}_{x^3} \underbrace{\quad}_{x^3}$$

$$= x^{12}$$

In general,

$$(x^m)^n = \overbrace{x^m \cdot x^m \cdot x^m \cdots \cdots x^m}^{n \text{ factors of } x^m}$$

$$= \overbrace{x \cdot x \cdot x \cdot x \cdot x \cdot x \cdot x \cdots \cdots x}^{mn \text{ factors of } x}$$

$$= x^{mn}$$

This discussion justifies a rule for raising an exponential expression to a power: *To raise an exponential expression to a power, we keep the same base and multiply the exponents.*

First Power Rule for Exponents	If m and n are natural numbers, then $(x^m)^n = x^{mn}$

EXAMPLE 4 (a)   **a.** $(2^3)^7 = 2^{3 \cdot 7}$ **b.** $(y^5)^2 = y^{5 \cdot 2}$
$\qquad\qquad\quad = 2^{21}$ $\qquad\qquad\qquad\quad = y^{10}$

c. $(z^7)^7 = z^{7 \cdot 7}$ **d.** $(u^x)^y = u^{x \cdot y}$
$\qquad\quad = z^{49}$ $\qquad\qquad\quad = u^{xy}$ ∎

In Example 5, both the product and power rules of exponents are applied.

EXAMPLE 5 (a, d)   **a.** $(x^2x^5)^2 = (x^7)^2$ **b.** $(yy^6y^2)^3 = (y^9)^3$
$\qquad\qquad\qquad = x^{14}$ $\qquad\qquad\qquad\quad = y^{27}$

c. $(z^2)^4(z^3)^3 = z^8z^9$ **d.** $(x^3)^2(x^5x^2)^3 = x^6(x^7)^3$
$\qquad\qquad\qquad = z^{17}$ $\qquad\qquad\qquad\qquad = x^6x^{21}$
$\qquad\qquad\qquad\qquad\qquad\qquad\qquad = x^{27}$ ∎

To find two more power rules for exponents, we consider the expressions $(2x)^3$ and $\left(\frac{2}{x}\right)^3$.

$$(2x)^3 = (2x)(2x)(2x)$$
$$= (2 \cdot 2 \cdot 2)(x \cdot x \cdot x)$$
$$= 2^3 x^3$$
$$= 8x^3$$

$$\left(\frac{2}{x}\right)^3 = \left(\frac{2}{x}\right)\left(\frac{2}{x}\right)\left(\frac{2}{x}\right) \qquad (x \neq 0)$$
$$= \frac{2 \cdot 2 \cdot 2}{x \cdot x \cdot x}$$
$$= \frac{2^3}{x^3}$$
$$= \frac{8}{x^3}$$

These examples suggest that *to raise a product to a power, we raise each factor of the product to that power*, and *to raise a fraction to a power, we raise both the numerator and denominator to that power*.

More Power Rules for Exponents	If n is a natural number, then

$$(xy)^n = x^n y^n \qquad \text{and if } y \neq 0, \text{ then} \qquad \left(\frac{x}{y}\right)^n = \frac{x^n}{y^n}$$

 (c, e)

EXAMPLE 6

a. $(ab)^4 = a^4 b^4$

b. $(3c)^3 = 3^3 c^3$
$$= 27c^3$$

c. $(x^2 y^3)^5 = (x^2)^5 (y^3)^5$
$$= x^{10} y^{15}$$

d. $(-2x^3 y)^2 = (-2)^2 (x^3)^2 y^2$
$$= 4x^6 y^2$$

e. $\left(\frac{4}{k}\right)^3 = \frac{4^3}{k^3}$
$$= \frac{64}{k^3}$$

f. $\left(\frac{3x^2}{2y^3}\right)^5 = \frac{3^5 (x^2)^5}{2^5 (y^3)^5}$
$$= \frac{243x^{10}}{32y^{15}}$$ ∎

To find a rule for dividing exponential expressions, we consider the fraction

$$\frac{4^5}{4^2}$$

where the exponent in the numerator is greater than the exponent in the denominator. We can simplify the fraction as follows:

$$\frac{4^5}{4^2} = \frac{4 \cdot 4 \cdot 4 \cdot 4 \cdot 4}{4 \cdot 4}$$

$$= \frac{\overset{1}{4} \cdot \overset{1}{4} \cdot 4 \cdot 4 \cdot 4}{\underset{1}{4} \cdot \underset{1}{4}}$$

$$= 4^3$$

The result of 4^3 has a base of 4 and an exponent of $5 - 2$, or 3. This suggests that *to divide exponential expressions with the same base, we keep the base and subtract the exponents.*

Quotient Rule for Exponents   If m and n are natural numbers, $m > n$ and $x \neq 0$, then

$$\frac{x^m}{x^n} = x^{m-n}$$

(a, d)   **EXAMPLE 7**   If there are no divisions by 0, then

a. $\dfrac{x^4}{x^3} = x^{4-3}$
    $= x^1$
    $= x$

b. $\dfrac{8y^2y^6}{4y^3} = \dfrac{8y^8}{4y^3}$
    $= \dfrac{8}{4}y^{8-3}$
    $= 2y^5$

c. $\dfrac{a^3a^5a^7}{a^4a} = \dfrac{a^{15}}{a^5}$
    $= a^{15-5}$
    $= a^{10}$

d. $\dfrac{(a^3b^4)^2}{ab^5} = \dfrac{a^6b^8}{ab^5}$
    $= a^{6-1}b^{8-5}$
    $= a^5b^3$ ■

The rules for positive exponents are summarized as follows:

Properties of Exponents   If n is a natural number, then

$$x^n = \overbrace{x \cdot x \cdot x \cdot \cdots \cdot x}^{n \text{ factors of } x}$$

If m and n are natural numbers and there are no divisions by 0, then

$$x^m x^n = x^{m+n} \qquad (x^m)^n = x^{mn} \qquad (xy)^n = x^n y^n \qquad \left(\frac{x}{y}\right)^n = \frac{x^n}{y^n}$$

$$\frac{x^m}{x^n} = x^{m-n} \quad \text{provided } m > n$$

ORALS *Find the base and the exponent in each expression.*

1. x^3 **2.** 3^x **3.** ab^c **4.** $(ab)^c$

Evaluate each expression.

5. 6^2 **6.** $(-6)^2$ **7.** $2^3 + 1^3$ **8.** $(2 + 1)^3$

E X E R C I S E   3.1

In Exercises 1–12, identify the base and the exponent in each expression.

1. 4^3 **2.** $(-5)^2$ **3.** x^5 **4.** y^8

5. $(2y)^3$ **6.** $(-x)^2$ **7.** $-x^4$ **8.** $(-x)^4$

9. x **10.** (xy) **11.** $2x^3$ **12.** $-3y^6$

In Exercises 13–20, write each expression without using exponents.

13. 5^3 **14.** -4^5 **15.** x^7 **16.** $3x^3$

17. $-4x^5$ **18.** $(-2y)^4$ **19.** $(3t)^5$ **20.** a^3b^2

In Exercises 21–28, write each expression using exponents.

21. $2 \cdot 2 \cdot 2$ **22.** $5 \cdot 5$ **23.** $x \cdot x \cdot x \cdot x$ **24.** $y \cdot y \cdot y \cdot y \cdot y \cdot y$

25. $(2x)(2x)(2x)$ **26.** $(-4y)(-4y)$ **27.** $-4t \cdot t \cdot t \cdot t$ **28.** $5 \cdot u \cdot u$

In Exercises 29–36, evaluate each expression.

29. 5^4 **30.** $(-3)^3$ **31.** $2^2 + 3^2$ **32.** $2^3 - 2^2$

33. $5^4 - 4^3$ **34.** $2(4^3 + 3^2)$ **35.** $-5(3^4 + 4^3)$ **36.** $-5^2(4^3 - 2^6)$

In Exercises 37–52, write each expression as an expression involving only one exponent.

37. x^4x^3 **38.** y^5y^2 **39.** x^5x^5 **40.** yy^3

41. tt^2 **42.** w^3w^5 **43.** $a^3a^4a^5$ **44.** $b^2b^3b^5$

45. $y^3(y^2y^4)$ **46.** $(y^4y)y^6$ **47.** $4x^2(3x^5)$ **48.** $-2y(y^3)$

49. $(-y^2)(4y^3)$ **50.** $(-4x^3)(-5x)$ **51.** $6x^3(-x^2)(-x^4)$ **52.** $-2x(-x^2)(-3x)$

In Exercises 53–68, write each expression as an expression involving only one exponent.

53. $(3^2)^4$ **54.** $(4^3)^3$ **55.** $(y^5)^3$ **56.** $(b^3)^6$

57. $(a^3)^7$ **58.** $(b^2)^3$ **59.** $(x^2x^3)^5$ **60.** $(y^3y^4)^4$

61. $(3zz^2z^3)^5$ **62.** $(4t^3t^6t^2)^2$ **63.** $(x^5)^2(x^7)^3$ **64.** $(y^3y)^2(y^2)^2$

65. $(r^3r^2)^4(r^3r^5)^2$ **66.** $(s^2)^3(s^3)^2(s^4)^4$ **67.** $(s^3)^3(s^2)^2(s^5)^4$ **68.** $(yy^3)^3(y^2y^3)^4(y^3y^3)^2$

In Exercises 69–84, write each expression without using parentheses.

69. $(xy)^3$

70. $(uv^2)^4$

71. $(r^3s^2)^2$

72. $(a^3b^2)^3$

73. $(4ab^2)^2$

74. $(3x^2y)^3$

75. $(-2r^2s^3t)^3$

76. $(-3x^2y^4z)^2$

77. $\left(\dfrac{a}{b}\right)^3$

78. $\left(\dfrac{r^2}{s}\right)^4$

79. $\left(\dfrac{x^2}{y^3}\right)^5$

80. $\left(\dfrac{u^4}{v^2}\right)^6$

81. $\left(\dfrac{-2a}{b}\right)^5$

82. $\left(\dfrac{2t}{3}\right)^4$

83. $\left(\dfrac{b^2}{3a}\right)^3$

84. $\left(\dfrac{a^3b}{c^4}\right)^5$

In Exercises 85–100, simplify each expression.

85. $\dfrac{x^5}{x^3}$

86. $\dfrac{a^6}{a^3}$

87. $\dfrac{y^3y^4}{yy^2}$

88. $\dfrac{b^4b^5}{b^2b^3}$

89. $\dfrac{12a^2a^3a^4}{4(a^4)^2}$

90. $\dfrac{16(aa^2)^3}{2a^2a^3}$

91. $\dfrac{(ab^2)^3}{(ab)^2}$

92. $\dfrac{(m^3n^4)^3}{(mn^2)^3}$

93. $\dfrac{20(r^4s^3)^4}{6(rs^3)^3}$

94. $\dfrac{15(x^2y^5)^5}{21(x^3y)^2}$

95. $\dfrac{17(x^4y^3)^8}{34(x^5y^2)^4}$

96. $\dfrac{35(r^3s^2)^2}{49r^2s^4}$

97. $\left(\dfrac{y^3y}{2yy^2}\right)^3$

98. $\left(\dfrac{3t^3t^4t^5}{4t^2t^6}\right)^3$

99. $\left(\dfrac{-2r^3r^3}{3r^4r}\right)^3$

100. $\left(\dfrac{-6y^4y^5}{5y^3y^5}\right)^2$

Writing Exercises ■ *Write a paragraph using your own words.*

1. Describe how you would multiply two exponential expressions with like bases.

2. Describe how you would divide two exponential expressions with like bases.

Something to Think About ■

1. Is the operation of raising to a power commutative? That is, is $a^b = b^a$? Explain.

2. Is the operation of raising to a power associative? That is, is $(a^b)^c = a^{(bc)}$? Explain.

Review Exercises ■

1. Graph the real numbers $-3, 0, 2,$ and $-\frac{3}{2}$ on a number line.

2. Graph the *negatives* of 0, 5, and -3 on a number line.

Write each algebraic expression as an English phrase.

3. $3(x + y)$

4. $3x + y$

Write each English phrase as an algebraic expression.

5. Three greater than the absolute value of twice x.

6. The sum of the numbers y and z decreased by the sum of their squares.

3.2 Zero and Negative Integral Exponents

GETTING READY

Simplify by dividing out common factors.

1. $\dfrac{3 \cdot 3 \cdot 3}{3 \cdot 3 \cdot 3 \cdot 3}$ **2.** $\dfrac{2yy}{2yyy}$ **3.** $\dfrac{3xx}{3xx}$ **4.** $\dfrac{xxy}{xxxyy}$

When we developed the quotient rule for exponents in Section 3.1, the exponent in the numerator was greater than the exponent in the denominator. We now consider what happens when the exponents are equal, and when the denominator exponent is greater than the numerator exponent.

If we apply the quotient rule to the fraction $\dfrac{5^3}{5^3}$, where the exponents in the numerator and denominator are equal, we obtain

$$\frac{5^3}{5^3} = 5^{3-3} = 5^0$$

However, because any nonzero number divided by itself is equal to 1, we have

$$\frac{5^3}{5^3} = 1$$

To make the results of 5^0 and 1 consistent, we shall define 5^0 to be equal to 1. In general, we have

Zero Exponent

If x is any nonzero real number, then
$$x^0 = 1$$
Since $x \neq 0$, 0^0 is undefined.

[oo] (a, e, f)   **EXAMPLE 1**

a. $\left(\dfrac{1}{13}\right)^0 = 1$ **b.** $(-0.115)^0 = 1$

c. $\dfrac{4^2}{4^2} = 4^{2-2}$ **d.** $\dfrac{x^5}{x^5} = x^{5-5}$ $(x \neq 0)$

        $= 4^0$ $= x^0$

        $= 1$ $= 1$

e. $3x^0 = 3(1)$ **f.** $(3x)^0 = 1$

        $= 3$

g. $\dfrac{6^n}{6^n} = 6^{n-n}$

$= 6^0$

$= 1$

h. $\dfrac{y^m}{y^m} = y^{m-m} \quad (y \neq 0)$

$= y^0$

$= 1$

It is clear from parts **e** and **f** that $3x^0 \neq (3x)^0$. ∎

If we apply the quotient rule to the fraction $\dfrac{6^2}{6^5}$, where the exponent in the numerator is less than the exponent in the denominator, we obtain

$$\frac{6^2}{6^5} = 6^{2-5} = 6^{-3}$$

However, we know that

$$\frac{6^2}{6^5} = \frac{\overset{1}{\cancel{6}} \cdot \overset{1}{\cancel{6}}}{\underset{1}{\cancel{6}} \cdot \underset{1}{\cancel{6}} \cdot 6 \cdot 6 \cdot 6} = \frac{1}{6^3}$$

To make the results of 6^{-3} and $\dfrac{1}{6^3}$ consistent, we define 6^{-3} to be equal to $\dfrac{1}{6^3}$. In general,

Negative Exponent   If x is any nonzero number and n is a natural number, then

$$x^{-n} = \frac{1}{x^n}$$

⊙⊙ (b, c, d)   **EXAMPLE 2**   Express each quantity without using negative exponents or parentheses. Assume that no denominators are zero.

a. $3^{-5} = \dfrac{1}{3^5}$

$= \dfrac{1}{243}$

b. $x^{-4} = \dfrac{1}{x^4}$

c. $(2x)^{-2} = \dfrac{1}{(2x)^2}$

$= \dfrac{1}{4x^2}$

d. $2x^{-2} = 2\left(\dfrac{1}{x^2}\right)$

$= \dfrac{2}{x^2}$

e. $(-3a)^{-4} = \dfrac{1}{(-3a)^4}$

$= \dfrac{1}{81a^4}$

f. $(x^3x^2)^{-3} = (x^5)^{-3}$

$= \dfrac{1}{(x^5)^3}$

$= \dfrac{1}{x^{15}}$ ■

Because of the definition of negative and zero exponents, the product, power, and quotient rules are also true for all integral exponents. We restate the properties of exponents for integral exponents.

Properties of Exponents   If m and n are integers and there are no divisions by 0, then

$$x^m x^n = x^{m+n} \qquad (x^m)^n = x^{mn} \qquad (xy)^n = x^n y^n \qquad \left(\dfrac{x}{y}\right)^n = \dfrac{x^n}{y^n}$$

$$x^0 = 1 \quad (x \neq 0) \qquad x^{-n} = \dfrac{1}{x^n} \qquad \dfrac{x^m}{x^n} = x^{m-n}$$

(a, d)   **EXAMPLE 3**   Simplify each quantity and express the result without using negative exponents. Assume that no denominators are zero.

a. $(x^{-3})^2 = x^{-6}$

$= \dfrac{1}{x^6}$

b. $\dfrac{x^3}{x^7} = x^{3-7}$

$= x^{-4}$

$= \dfrac{1}{x^4}$

c. $\dfrac{y^{-4}y^{-3}}{y^{-20}} = \dfrac{y^{-7}}{y^{-20}}$

$= y^{-7-(-20)}$

$= y^{-7+20}$

$= y^{13}$

d. $\dfrac{12a^3b^4}{4a^5b^2} = 3a^{3-5}b^{4-2}$

$= 3a^{-2}b^2$

$= \dfrac{3b^2}{a^2}$

e. $\left(-\dfrac{x^3y^2}{xy^{-3}}\right)^{-2} = (-x^{3-1}y^{2-(-3)})^{-2}$

$= (-x^2y^5)^{-2}$

$= \dfrac{1}{(-x^2y^5)^2}$

$= \dfrac{1}{x^4y^{10}}$ ■

These properties of exponents are also true when the exponents are algebraic expressions.

EXAMPLE 4

a. $x^{2m}x^{3m} = x^{2m+3m}$
$\qquad\qquad = x^{5m}$

b. $\dfrac{y^{2m}}{y^{4m}} = y^{2m-4m} \quad (y \neq 0)$
$\qquad\quad = y^{-2m}$
$\qquad\quad = \dfrac{1}{y^{2m}}$

c. $a^{2m-1}a^{2m} = a^{2m-1+2m}$
$\qquad\qquad = a^{4m-1}$

d. $(b^{m+1})^{2m} = b^{(m+1)2m}$
$\qquad\qquad = b^{2m^2+2m}$ ∎

ORALS *Simplify each quantity.*

1. 2^{-1}

2. 2^{-2}

3. $\left(\dfrac{1}{2}\right)^{-1}$

4. $\left(\dfrac{7}{9}\right)^{0}$

5. $x^{-1}x^{2}$

6. $y^{-2}y^{-5}$

7. $\dfrac{x^{5}x^{2}}{x^{7}}$

8. $\left(\dfrac{x}{y}\right)^{-1}$

E X E R C I S E 3.2

In Exercises 1–64, simplify each expression. Write each answer without using parentheses or negative exponents.

1. $2^{5} \cdot 2^{-2}$

2. $10^{2} \cdot 10^{-4} \cdot 10^{5}$

3. $4^{-3} \cdot 4^{-2} \cdot 4^{5}$

4. $3^{-4} \cdot 3^{5} \cdot 3^{-3}$

5. $\dfrac{3^{5} \cdot 3^{-2}}{3^{3}}$

6. $\dfrac{6^{2} \cdot 6^{-3}}{6^{-2}}$

7. $\dfrac{2^{5} \cdot 2^{7}}{2^{6} \cdot 2^{-3}}$

8. $\dfrac{5^{-2} \cdot 5^{-4}}{5^{-6}}$

9. $2x^{0}$

10. $(2x)^{0}$

11. $(-x)^{0}$

12. $-x^{0}$

13. $\left(\dfrac{a^{2}b^{3}}{ab^{4}}\right)^{0}$

14. $\dfrac{2}{3}\left(\dfrac{xyz}{x^{2}y}\right)^{0}$

15. $\dfrac{x^{0} - 5x^{0}}{2x^{0}}$

16. $\dfrac{4a^{0} + 2a^{0}}{3a^{0}}$

17. x^{-2}

18. y^{-3}

19. b^{-5}

20. c^{-4}

21. $(2y)^{-4}$

22. $(-3x)^{-1}$

23. $(ab^{2})^{-3}$

24. $(m^{2}n^{3})^{-2}$

25. $\dfrac{y^{4}}{y^{5}}$

26. $\dfrac{t^{7}}{t^{10}}$

27. $\dfrac{(r^{2})^{3}}{(r^{3})^{4}}$

28. $\dfrac{(b^{3})^{4}}{(b^{5})^{4}}$

29. $\dfrac{y^{4}y^{3}}{y^{4}y^{-2}}$

30. $\dfrac{x^{12}x^{-7}}{x^{3}x^{4}}$

31. $\dfrac{a^{4}a^{-2}}{a^{2}a^{0}}$

32. $\dfrac{b^{0}b^{3}}{b^{-3}b^{4}}$

33. $(ab^{2})^{-2}$

34. $(c^{2}d^{3})^{-2}$

35. $(x^{2}y)^{-3}$

36. $(-xy^{2})^{-4}$

37. $(x^{-4}x^{3})^{3}$

38. $(y^{-2}y)^{3}$

39. $(y^{3}y^{-2})^{-2}$

40. $(x^{-3}x^{-2})^{2}$

41. $(a^{-2}b^{-3})^{-4}$

42. $(y^{-3}z^{5})^{-6}$

43. $(-2x^{3}y^{-2})^{-5}$

44. $(-3u^{-2}v^{3})^{-3}$

45. $\left(\dfrac{a^3}{a^{-4}}\right)^2$

46. $\left(\dfrac{a^4}{a^{-3}}\right)^3$

47. $\left(\dfrac{b^5}{b^{-2}}\right)^{-2}$

48. $\left(\dfrac{b^{-2}}{b^3}\right)^{-3}$

49. $\left(\dfrac{4x^2}{3x^{-5}}\right)^4$

50. $\left(\dfrac{-3r^4r^{-3}}{r^{-3}r^7}\right)^3$

51. $\left(\dfrac{12y^3z^{-2}}{3y^{-4}z^3}\right)^2$

52. $\left(\dfrac{6xy^3}{3x^{-1}y}\right)^3$

53. $\left(\dfrac{2x^3y^{-2}}{4xy^2}\right)^7$

54. $\left(\dfrac{9u^2v^3}{18u^{-3}v}\right)^4$

55. $\left(\dfrac{14u^{-2}v^3}{21u^{-3}v}\right)^4$

56. $\left(\dfrac{-27u^{-5}v^{-3}w}{18u^3v^{-2}}\right)^4$

57. $\left(\dfrac{6a^2b^3}{2ab^2}\right)^{-2}$

58. $\left(\dfrac{15r^2s^{-2}t}{3r^{-3}s^3}\right)^{-3}$

59. $\left(\dfrac{18a^2b^3c^{-4}}{3a^{-1}b^2c}\right)^{-3}$

60. $\left(\dfrac{21x^{-2}y^2z^{-2}}{7x^3y^{-1}}\right)^{-2}$

61. $\dfrac{(2x^{-2}y)^{-3}}{(4x^2y^{-1})^3}$

62. $\dfrac{(ab^{-2}c)^2}{(a^{-2}b)^{-3}}$

63. $\dfrac{(17x^5y^{-5}z)^{-3}}{(17x^{-5}y^3z^2)^{-4}}$

64. $\dfrac{16(x^{-2}yz)^{-2}}{(2x^{-3}z^0)^4}$

In Exercises 65–80, write each expression with a single exponent.

65. $x^{2m}x^m$

66. $y^{3m}y^{2m}$

67. $u^{2m}v^{3n}u^{3m}v^{-3n}$

68. $r^{2m}s^{-3}r^{3m}s^3$

69. $y^{3m+2}y^{-m}$

70. $x^{m+1}x^m$

71. $\dfrac{y^{3m}}{y^{2m}}$

72. $\dfrac{z^{4m}}{z^{2m}}$

73. $\dfrac{x^{3n}}{x^{6n}}$

74. $\dfrac{x^m}{x^{5m}}$

75. $(x^{m+1})^2$

76. $(y^2)^{m+1}$

77. $(x^{3-2n})^{-4}$

78. $(y^{1-n})^{-3}$

79. $(y^{2-n})^{-4}$

80. $(x^{3-4n})^{-2}$

Writing Exercises ■ *Write a paragraph using your own words.*

1. Tell how you would help a friend understand that 2^{-3} is not equal to -8.

2. Describe how you would use a calculator to verify that

$$2^{-3} = \dfrac{1}{2^3}$$

Something to Think About ■

1. If a positive number, x, is raised to a negative power, is the result greater than, equal to, or less than x? Explore the possibilities.

2. We know that $x^{-n} = \dfrac{1}{x^n}$. Is it also true that $x^n = \dfrac{1}{x^{-n}}$? Explain.

Review Exercises ■

1. If $a = -2$ and $b = 3$, evaluate $\dfrac{3a^2 + 4b + 8}{a + 2b^2}$.

2. Evaluate $|-3 + 5 \cdot 2|$.

Solve each equation.

3. $5\left(x - \dfrac{1}{2}\right) = \dfrac{7}{2}$

4. $\dfrac{5(2 - x)}{6} = \dfrac{x + 6}{2}$

5. Solve $P = L + \dfrac{s}{f}i$ for s.

6. Solve $P = L + \dfrac{s}{f}i$ for i.

3.3 Scientific Notation

■ Using Scientific Notation to Simplify Computations

Evaluate each expression.

1. 10^2 **2.** 10^3 **3.** 10^1 **4.** 10^{-2}

5. $5(10)^2$ **6.** $8(10^3)$ **7.** $3(10^1)$ **8.** $7(10^{-2})$

Scientists often deal with extremely large and extremely small numbers. For example, the distance from the earth to the sun is approximately 150,000,000 kilometers, and ultraviolet light emitted from a mercury arc has a wavelength of approximately 0.000025 centimeter. The large number of zeros in these numbers makes them difficult to read and hard to remember. To make such numbers easier to work with, scientists use a compact form of notation called **scientific notation.**

Scientific Notation   A number is written in **scientific notation** if it is written as the product of a number between 1 (including 1) and 10 and an integer power of 10.

EXAMPLE 1   Change 150,000,000 to scientific notation.

Solution   To write the number 150,000,000 in scientific notation, we must write it as a product of a number between 1 and 10 and some power of 10. Note that the number 1.5 lies between 1 and 10.

To obtain the number 150,000,000, the decimal point in the number 1.5 must be moved eight places to the right. Because multiplying a number by 10 moves the decimal place one place to the right, we can accomplish this by multiplying 1.5 by 10 eight times. To see this, we count from the decimal point in 1.5 to where the decimal point should be in 150,000,000:

$$1.50000000.$$

8 places to the right

Thus, the number 150,000,000 written in scientific notation is 1.5×10^8.   ■

EXAMPLE 2   Change 0.000025 to scientific notation.

Solution   To write the number 0.000025 in scientific notation, we write it as a product of a number between 1 and 10 and some power of 10. To obtain the number 0.000025, the decimal point in the number 2.5 must be moved five places to the left. We can accomplish this by dividing 2.5 by 10^5, which is equivalent to multiplying 2.5 by

$\frac{1}{10^5}$, or by 10^{-5}. To do this, we count from the decimal point in 2.5 to where the decimal point should be in 0.000025:

.00002.5

5 places to the left

Thus, the number 0.000025 written in scientific notation is 2.5×10^{-5}. ∎

EXAMPLE 3   Write   **a.** 235,000   and   **b.** 0.00000235   in scientific notation.

Solution   **a.** $235{,}000 = 2.35 \times 10^5$ because $2.35 \times 10^5 = 235{,}000$ and 2.35 is between 1 and 10.

         b. $0.00000235 = 2.35 \times 10^{-6}$ because $2.35 \times 10^{-6} = 0.00000235$ and 2.35 is between 1 and 10. ∎

■ PERSPECTIVE

The Metric System

A common metric unit of length is the kilometer, which is 1000 meters. Because 1000 is 10^3, we can write 1 km = 10^3 m. Similarly, 1 centimeter is one-hundredth of a meter: 1 cm = 10^{-2} m. In the metric system, prefixes such as *kilo* and *centi* refer to powers of 10. Other prefixes are used in the metric system, as shown in the table.

Prefix	Symbol	Meaning	
peta	P	10^{15} =	1,000,000,000,000,000.
tera	T	10^{12} =	1,000,000,000,000.
giga	G	10^9 =	1,000,000,000.
mega	M	10^6 =	1,000,000.
kilo	k	10^3 =	1,000.
deci	d	10^{-1} =	0.1
centi	c	10^{-2} =	0.01
milli	m	10^{-3} =	0.001
micro	μ	10^{-6} =	0.000 001
nano	n	10^{-9} =	0.000 000 001
pico	p	10^{-12} =	0.000 000 000 001
femto	f	10^{-15} =	0.000 000 000 000 001
atto	a	10^{-18} =	0.000 000 000 000 000 001

To appreciate the magnitudes involved, consider these facts: Light, which travels 186,000 miles every second, will travel about one foot in one nanosecond. The distance to the nearest star is 43 petameters, and the diameter of an atom is about 10 nanometers. To measure some quantities, however, even these units are inadequate. The sun, for example, radiates 5×10^{26} watts. That's a lot of light bulbs!

We can change a number written in scientific notation to **standard notation.** For example, to write the number 9.3×10^7 in standard notation, we multiply 9.3 by 10^7.

$$9.3 \times 10^7 = 9.3 \times 10,000,000$$
$$= 93,000,000$$

EXAMPLE 4 Write **a.** 3.4×10^5 and **b.** 2.1×10^{-4} in standard notation.

Solution **a.** $3.4 \times 10^5 = 3.4 \times 100,000$ **b.** $2.1 \times 10^{-4} = 2.1 \times \dfrac{1}{10^4}$
$$= 340,000$$
$$= 2.1 \times \dfrac{1}{10,000}$$
$$= 0.00021 \quad\blacksquare$$

Each of the following numbers is written in both scientific and standard notation. In each case, the exponent gives the number of places that the decimal point moves and the sign of the exponent indicates the direction that it moves.

$5.32 \times 10^5 = 5\,3\,2\,0\,0\,0.$ 5 places to the right
$2.37 \times 10^6 = 2\,3\,7\,0\,0\,0\,0.$ 6 places to the right
$8.95 \times 10^{-4} = 0.\,0\,0\,0\,8\,9\,5$ 4 places to the left
$8.375 \times 10^{-3} = 0.\,0\,0\,8\,3\,7\,5$ 3 places to the left
$9.77 \times 10^0 = 9.77$ no movement of the decimal point

EXAMPLE 5 Write 432.0×10^5 in scientific notation.

Solution The number 432.0×10^5 is not written in scientific notation, because 432.0 is not a number between 1 and 10. To write the number in scientific notation, we proceed as follows:

$$432.0 \times 10^5 = 4.32 \times 10^2 \times 10^5 \quad \text{Write 432.0 in scientific notation.}$$
$$= 4.32 \times 10^7 \qquad\qquad 10^2 \times 10^5 = 10^7. \quad\blacksquare$$

■ Using Scientific Notation to Simplify Computations

Another advantage of scientific notation becomes apparent when we must simplify fractions such as

$$\frac{(320)(25,000)}{0.000040}$$

that contain very large or very small numbers. Although we can simplify this fraction by using ordinary arithmetic, scientific notation provides an easier way. First we write each number in scientific notation and then do the arithmetic on the numbers and the exponential expressions separately. Then we write the answer in standard form, if desired.

$$\frac{(320)(25,000)}{0.000040} = \frac{(3.2 \times 10^2)(2.5 \times 10^4)}{4.0 \times 10^{-5}}$$

$$= \frac{(3.2)(2.5)}{4.0} \times \frac{10^2 10^4}{10^{-5}}$$

$$= \frac{8.0}{4.0} \times 10^{2+4-(-5)}$$

$$= 2.0 \times 10^{11}$$

$$= 200,000,000,000$$

If we use a scientific calculator to simplify the fraction, the final display will look like

$$2. \quad {}^{11}$$

which means 2×10^{11}. A scientific calculator cannot give the final result in standard notation, because its display can show only ten digits.

EXAMPLE 6   **Speed of light**   In a vacuum, light travels 1 meter in approximately 0.000000003 second. How long does it take for light to travel 500 kilometers?

Solution   Because 1 kilometer is equal to 1000 meters, the length of time for light to travel 500 kilometers (500 · 1000 meters) is given by

$$(0.000000003)(500)(1000) = (3 \times 10^{-9})(5 \times 10^2)(1 \times 10^3)$$

$$= 3(5) \times 10^{-9+2+3}$$

$$= 15 \times 10^{-4}$$

$$= 1.5 \times 10^1 \times 10^{-4}$$

$$= 1.5 \times 10^{-3}$$

$$= 0.0015$$

Light travels 500 kilometers in approximately 0.0015 second.   ■

ORALS   *Indicate which number of each pair is the larger.*

1. 37.2 or 3.72×10^2 **2.** 37.2 or 3.72×10^{-1}

3. 3.72×10^3 or 4.72×10^3 **4.** 3.72×10^3 or 4.72×10^2

5. 3.72×10^{-1} or 4.72×10^{-2} **6.** 3.72×10^{-3} or 2.72×10^{-2}

E X E R C I S E 3.3

In Exercises 1–12, write each number in scientific notation.

1. 23,000 **2.** 4750 **3.** 1,700,000 **4.** 290,000

5. 0.062 **6.** 0.00073 **7.** 0.0000051 **8.** 0.04

9. 42.5×10^2 **10.** 0.3×10^3 **11.** 0.25×10^{-2} **12.** 25.2×10^{-3}

In Exercises 13–24, write each number in standard notation.

13. 2.3×10^2 **14.** 3.75×10^4 **15.** 8.12×10^5 **16.** 1.2×10^3

17. 1.15×10^{-3} **18.** 4.9×10^{-2} **19.** 9.76×10^{-4} **20.** 7.63×10^{-5}

21. 25×10^6 **22.** 0.07×10^3 **23.** 0.51×10^{-3} **24.** 617×10^{-2}

25. Distance to Alpha Centauri The distance from the earth to the nearest star outside our solar system is approximately 25,700,000,000,000 miles. Express this number in scientific notation.

26. Speed of sound The speed of sound in air is 33,100 centimeters per second. Express this number in scientific notation.

27. Distance to Mars The distance from Mars to the sun is approximately 1.14×10^8 miles. Express this number in standard notation.

28. Distance to Venus The distance from Venus to the sun is approximately 6.7×10^7 miles. Express this number in standard notation.

29. Length of a meter One meter is approximately 0.00622 mile. Use scientific notation to express this number.

30. Angstrom One angstrom is 1×10^{-7} millimeter. Express this number in standard notation.

In Exercises 31–38, use scientific notation to simplify each expression. Check your work with a scientific calculator. Give answers in standard notation when possible.

31. $(3.4 \times 10^2)(2.1 \times 10^3)$

32. $(4.1 \times 10^{-3})(3.4 \times 10^4)$

33. $\dfrac{9.3 \times 10^2}{3.1 \times 10^{-2}}$

34. $\dfrac{7.2 \times 10^6}{1.2 \times 10^8}$

35. $\dfrac{96,000}{(12,000)(0.00004)}$

36. $\dfrac{(0.48)(14,400,000)}{96,000,000}$

37. $\dfrac{(12,000)(3600)}{0.0003}$

38. $\dfrac{(0.0004)(0.0012)}{80,000}$

39. Distance between Mercury and the sun The distance from Mercury to the sun is approximately 3.6×10^7 miles. Use scientific notation to express this distance in feet. (*Hint:* 5280 feet = 1 mile.)

40. Mass of a proton The mass of one proton is approximately 1.7×10^{-24} gram. Use scientific notation to express the mass of 1 million protons.

41. Speed of sound The speed of sound in air is approximately 3.3×10^4 centimeters per second. Use scientific notation to express this speed in kilometers per second. (*Hint:* 100 centimeters = 1 meter and 1000 meters = 1 kilometer.)

42. Light year One light year is approximately 5.87×10^{12} miles. Use scientific notation to express this distance in feet. (*Hint:* 5280 feet = 1 mile.)

Writing Exercises ■ *Write a paragraph using your own words.*

1. In what situations would scientific notation be more convenient than standard notation?

2. To multiply a number by a power of 10, we move the decimal point. Which way, and how far? Explain.

Something to Think About ■

1. Two positive numbers are written in scientific notation. How could you decide which is larger, without converting either to standard notation?

2. ▦ The product $1 \cdot 2 \cdot 3 \cdot 4 \cdot 5$, or 120, is called **5 factorial,** written 5!. Similarly, $6! = 1 \cdot 2 \cdot 3 \cdot 4 \cdot 5 \cdot 6 = 620$. Factorials get large very quickly. Calculate 30! and write the number in standard notation. (*Hint:* Experiment with the $x!$ button.)

Review Exercises ■

1. If $y = -1$, find the value of $-5y^{55}$.

2. Evaluate $\dfrac{3a^2 - 2b}{2a + 2b}$ if $a = 4$ and $b = 3$.

Tell which property of real numbers justifies each statement.

3. $5 + z = z + 5$

4. $7(u + 3) = 7u + 7 \cdot 3$

Solve each equation.

5. $3(x - 4) - 6 = 0$

6. $8(3x - 5) - 4(2x + 3) = 12$

3.4 Polynomials

■ $P(x)$ Notation

GETTING READY *Write each expression using exponents.*

1. $2xxyyy$

2. $3xyyy$

3. $2xx + 3yy$

4. $xxx + yyy$

5. $(3xxy)(2xyy)$

6. $(5xyzzz)(xyz)$

7. $3(5xy)\left(\dfrac{1}{3}xy\right)$

8. $(xy)(xz)(yz)(xyz)$

Expressions such as

$$3x \qquad 4y^2 \qquad -8x^2y^3 \qquad \text{and} \qquad 25$$

with constant and/or variable factors are called **algebraic terms.** The numerical coefficients of the first three of these terms are 3, 4, and -8, respectively. Because

$25 = 25x^0$, the number 25 is considered to be the numerical coefficient of the term 25.

An algebraic expression that is the sum of one or more terms containing whole number exponents on its variables is called a **polynomial.** The expressions

$$8xy^2t \qquad 3x + 2 \qquad 4y^2 - 2y + 3 \qquad \text{and} \qquad 3a - 4b - 4c + 8d$$

are examples of polynomials.

 WARNING! An expression such as $2x^3 - 3y^{-2}$ is not a polynomial, because the second term contains a negative exponent on a variable base.

A polynomial with exactly one term is called a **monomial.** A polynomial with exactly two terms is called a **binomial.** A polynomial with exactly three terms is called a **trinomial.** Here are some examples of each:

Monomials	*Binomials*	*Trinomials*
$5x^2y$	$3u^3 - 4u^2$	$-5t^2 + 4t + 3$
$-6x$	$18a^2b + 4ab$	$27x^3 - 6x - 2$
29	$-29z^{17} - 1$	$-32r^6 + 7y^3 - z$

The monomial $7x^6$ is called a **monomial of sixth degree** or a **monomial of degree 6,** because the variable x occurs as a factor six times. The monomial $3x^3y^4$ is a monomial of the seventh degree, because the variables x and y occur as factors a total of seven times. Other examples are:

$-2x^3$ is a monomial of degree 3

$47x^2y^3$ is a monomial of degree 5

$18x^4y^2z^8$ is a monomial of degree 14

8 is a monomial of degree 0, because $8 = 8x^0$

These examples illustrate the following definition.

Degree of a Monomial   If a is a nonzero constant, the **degree of the monomial** ax^n is n. The degree of a monomial with several variables is the sum of the exponents of those variables.

We define the degree of a polynomial by considering the degrees of each of its terms.

Degree of a Polynomial   The **degree of a polynomial** is the same as the degree of its term with largest degree.

Amalie Noether (1882–1935)
Albert Einstein described Noether as the most creative female mathematical genius since the beginning of higher education for women. Her work was in the area of abstract algebra. Although she received a doctoral degree in mathematics, she was denied a mathematics position in Germany because she was a woman.

For example,

- $x^2 + 2x$ is a binomial of degree 2, because the degree of its first term is 2 and the degree of its other term is less than 2.
- $3x^3y^2 + 4x^4y^4 - 3x^3$ is a trinomial of degree 8, because the degree of its second term is 8 and the degree of each of its other terms is less than 8.
- $25x^4y^3z^7 - 15xy^8z^{10} - 32x^8y^8z^3 + 4$ is a polynomial of degree 19, because its second and third terms are of degree 19. Its other terms have degrees less than 19.

■ *P(x)* Notation

Polynomials that contain a single variable are often denoted by symbols such as

$P(x)$	Read as "P of x."
$Q(t)$	Read as "Q of t."
$R(z)$	Read as "R of z."

where the letter within the parentheses represents the variable of the polynomial.

 WARNING!   The symbol $P(x)$ does not indicate the product of P and x. Instead, it represents a polynomial with the variable x.

The symbols $P(x)$, $Q(t)$, and $R(z)$ could represent the polynomials

$$P(x) = 3x + 4, \qquad Q(t) = 3t^2 + 4t - 5, \qquad \text{or} \qquad R(z) = -z^3 - 2z + 3$$

The symbol $P(x)$ is convenient, because it provides a way to indicate the value of a polynomial in x at different values of x. If $P(x) = 3x + 4$, for example, then $P(1)$ represents the value of the polynomial $P(x) = 3x + 4$ when $x = 1$.

$$P(x) = 3x + 4$$
$$P(1) = 3(1) + 4$$
$$= 7$$

Likewise, if $Q(t) = 3t^2 + 4t - 5$, then $Q(-2)$ represents the value of the polynomial $Q(t) = 3t^2 + 4t - 5$ when $t = -2$.

$$Q(t) = 3t^2 + 4t - 5$$
$$Q(-2) = 3(-2)^2 + 4(-2) - 5$$
$$= 3(4) - 8 - 5$$
$$= 12 - 8 - 5$$
$$= -1$$

(a, c)

EXAMPLE 1 Consider the polynomial $P(z)$ where $P(z) = 3z^2 + 2$. Find **a.** $P(0)$, **b.** $P(2)$, **c.** $P(-3)$, and **d.** $P(s)$.

Solution

a. $P(z) = 3z^2 + 2$
$P(0) = 3(0)^2 + 2$
$= 2$

b. $P(z) = 3z^2 + 2$
$P(2) = 3(2)^2 + 2$
$= 3(4) + 2$
$= 12 + 2$
$= 14$

c. $P(z) = 3z^2 + 2$
$P(-3) = 3(-3)^2 + 2$
$= 3(9) + 2$
$= 27 + 2$
$= 29$

d. $P(z) = 3z^2 + 2$
$P(s) = 3s^2 + 2$

EXAMPLE 2 **Trajectory** The polynomial $h(t) = -16t^2 + 28t + 8$ gives the height (in feet) of an object t seconds after it has been thrown upward. Find the height of the object at **a.** 1 second and **b.** 2 seconds.

Solution

a. To find the height at 1 second, we find $h(1)$.

$h(t) = -16t^2 + 28t + 8$
$h(1) = -16(1)^2 + 28(1) + 8$
$= -16 + 28 + 8$
$= 20$

At 1 second, the object is 20 feet above the ground.

b. To find the height at 2 seconds, we find $h(2)$.

$h(t) = -16t^2 + 28t + 8$
$h(2) = -16(2)^2 + 28(2) + 8$
$= -16(4) + 28(2) + 8$
$= -64 + 56 + 8$
$= 0$

At 2 seconds, the height is 0 feet. That is when the object strikes the ground.

(b)

EXAMPLE 3 Consider $P(x)$ where $P(x) = x^3 + 1$. Find **a.** $P(2t)$, **b.** $P(-3y)$, **c.** $P(s^4)$, and **d.** $P(x) + P(a)$.

Solution

a. $P(x) = x^3 + 1$
$P(2t) = (2t)^3 + 1$
$= 8t^3 + 1$

b. $P(x) = x^3 + 1$
$P(-3y) = (-3y)^3 + 1$
$= -27y^3 + 1$

c. $P(x) = x^3 + 1$
$P(s^4) = (s^4)^3 + 1$
$= s^{12} + 1$

d. $P(x) + P(a) = x^3 + 1 + a^3 + 1$
$= x^3 + a^3 + 2$

ORALS *Give an example of a polynomial that is . . .*

1. a binomial   **2.** a monomial   **3.** a trinomial   **4.** not a monomial, binomial, or trinomial

5. of degree 3   **6.** of degree 1   **7.** of degree 0   **8.** has no defined degree

EXERCISE 3.4

In Exercises 1–12, classify each polynomial as a monomial, a binomial, or a trinomial, if possible.

1. $3x + 7$

2. $3y - 5$

3. $3y^2 + 4y + 3$

4. $3xy$

5. $3z^2$

6. $3x^4 - 2x^3 + 3x - 1$

7. $5t - 32$

8. $9x^2y^3z^4$

9. $s^2 - 23s + 31$

10. $12x^3 - 12x^2 + 36x - 3$

11. $3x^5 - 2x^4 - 3x^3 + 17$

12. x^3

In Exercises 13–24, give the degree of each polynomial.

13. $3x^4$

14. $3x^5 - 4x^2$

15. $-2x^2 + 3x^3$

16. $-5x^5 + 3x^2 - 3x$

17. $3x^2y^3 + 5x^3y^5$

18. $-2x^2y^3 + 4x^3y^2z$

19. $-5r^2s^2t - 3r^3st^2$

20. $4r^2s^3t^3 - 5r^2s^8 + 3$

21. $x^{12} + 3x^2y^3z^4$

22. 17^2x

23. 38

24. -25

In Exercises 25–32, let $P(x) = 5x - 3$. Find each value.

25. $P(2)$

26. $P(0)$

27. $P(-1)$

28. $P(-2)$

29. $P(w)$

30. $P(t)$

31. $P(-y)$

32. $P(2t)$

In Exercises 33–40, let $Q(z) = -z^2 - 4$. Find each value.

33. $Q(0)$

34. $Q(1)$

35. $Q(-1)$

36. $Q(-2)$

37. $Q(r)$

38. $Q(-u)$

39. $Q(3s)$

40. $Q(-2x)$

In Exercises 41–48, let $R(y) = y^2 - 2y + 3$. Find each value.

41. $R(0)$

42. $R(3)$

43. $R(-2)$

44. $R(-1)$

45. $R(-b)$

46. $R(t)$

47. $R\left(-\dfrac{1}{4}w\right)$

48. $R\left(\dfrac{1}{2}u\right)$

In Exercises 49–64, let $P(x) = 5x - 2$. Find each value.

49. $P\left(\dfrac{1}{5}\right)$ **50.** $P\left(\dfrac{1}{10}\right)$ **51.** $P(u^2)$ **52.** $P(-v^4)$

53. $P(-4z^6)$ **54.** $P(10x^7)$ **55.** $P(x^2y^2)$ **56.** $P(x^3y^3)$

57. $P(x + h)$ **58.** $P(x - h)$ **59.** $P(x) + P(h)$ **60.** $P(x) - P(h)$

61. $P(2y + z)$ **62.** $P(-3r + 2s)$ **63.** $P(2y) + P(z)$ **64.** $P(-3r) + P(2s)$

In Exercises 65–68, assume that the height in feet of an object t seconds after it is thrown is given by $h(t) = -16t^2 + 120t + 16$. Find each value.

65. The height from which the object was thrown: find $h(0)$.

66. The height at 2 seconds.

67. The height at 6 seconds.

68. The height at 8 seconds.

Writing Exercises ■ *Write a paragraph using your own words.*

1. Describe how to find the degree of a polynomial.

2. Describe how to classify a polynomial as a monomial, a binomial, a trinomial, or none of these.

Something to Think About ■

1. Find a polynomial $P(x)$ for which $P\left(\frac{3}{2}\right) = 1$.

2. If $P(x)$ is a polynomial, must $P(2)$ be less than $P(3)$? Can you invent a polynomial for which $P(2) > P(3)$?

Review Exercises ■ *Solve each equation.*

1. $5(u - 5) + 9 = 2(u + 4)$

2. $8(3a - 5) - 12 = 4(2a + 3)$

Solve each inequality and graph the solution set.

3. $-4(3y + 2) \le 20$

4. $-5 < 3t + 4 \le 13$

Write each expression without using parentheses or negative exponents.

5. $(x^2x^4)^3$ **6.** $(a^2)^3(a^3)^2$ **7.** $\left(\dfrac{y^2y^5}{y^4}\right)^3$ **8.** $\left(\dfrac{2t^3}{t}\right)^{-4}$

3.5 Adding and Subtracting Polynomials

■ Adding and Subtracting Multiples of Polynomials

Combine terms and simplify, if possible.

1. $3x + 2x$ **2.** $5y - 3y$ **3.** $19x + 6x$ **4.** $8z - 3z$

5. $9r + 3r$ **6.** $4r - 3s$ **7.** $7r - 7r$ **8.** $17r - 17r^2$

Like terms have the same variables with the same exponents. For example,

$$3xyz^2 \quad \text{and} \quad -2xyz^2$$

are like terms and

$$\frac{1}{2}ab^2c \quad \text{and} \quad \frac{1}{3}a^2bd^2$$

are unlike terms.

    Because of the distributive property, we can combine like terms by adding their coefficients and using the same variables and exponents. For example,

$$2y + 5y = (2 + 5)y \quad \text{and} \quad -3x^2 + 7x^2 = (-3 + 7)x^2$$
$$= 7y \qquad\qquad\qquad = 4x^2$$

Likewise,

$$4x^3y^2 + 9x^3y^2 = 13x^3y^2 \quad \text{and} \quad 4r^2s^3t^4 + 7r^2s^3t^4 = 11r^2s^3t^4$$

Thus, to add like monomials, we simply combine like terms.

⬭⬭ (b, d)   **EXAMPLE 1**   **a.** $5xy^3 + 7xy^3 = 12xy^3$

b. $-7x^2y^2 + 6x^2y^2 + 3x^2y^2 = -x^2y^2 + 3x^2y^2$
$$= 2x^2y^2$$

c. $(2x^2)^2 + (3x)^4 = 4x^4 + 81x^4$
$$= 85x^4$$

d. $2(x + y) + 3(x + y) = 5(x + y)$
$$= 5x + 5y \qquad\qquad\qquad ■$$

    Recall that to subtract one monomial from another, we add the negative of the monomial that is to be subtracted. In symbols, $x - y = x + (-y)$.

(c)

EXAMPLE 2

a. $8x^2 - 3x^2 = 8x^2 + (-3x^2)$
$$= 5x^2$$

b. $6x^3y^2 - 9x^3y^2 = 6x^3y^2 + (-9x^3y^2)$
$$= -3x^3y^2$$

c. $-3r^2st^3 - 5r^2st^3 = -3r^2st^3 + (-5r^2st^3)$
$$= -8r^2st^3 \qquad \blacksquare$$

Because of the distributive property, we can remove parentheses enclosing several terms when the sign preceding the parentheses is a + sign. We simply drop the parentheses.

$$+(3x^2 + 3x - 2) = +1(3x^2 + 3x - 2)$$
$$= 1(3x^2) + 1(3x) + 1(-2)$$
$$= 3x^2 + 3x + (-2)$$
$$= 3x^2 + 3x - 2$$

Polynomials are added by removing parentheses, if necessary, and then combining any like terms that are contained within the polynomials.

EXAMPLE 3

$$(3x^2 - 3x + 2) + (2x^2 + 7x - 4) = 3x^2 - 3x + 2 + 2x^2 + 7x - 4$$
$$= 3x^2 + 2x^2 - 3x + 7x + 2 + (-4)$$
$$= 5x^2 + 4x - 2 \qquad \blacksquare$$

Problems such as Example 3 are often written with the terms aligned vertically.

$$3x^2 - 3x + 2$$
$$\underline{2x^2 + 7x - 4}$$
$$5x^2 + 4x - 2$$

EXAMPLE 4

Add: $4x^2y + 8x^2y^2 - 3x^2y^3$
$$\underline{3x^2y - 8x^2y^2 + 8x^2y^3}$$
$$7x^2y \qquad\quad + 5x^2y^3 \qquad \blacksquare$$

Because of the distributive property, we can also remove parentheses enclosing several terms when the sign preceding the parentheses is a − sign. We simply drop the minus sign and the parentheses and *change the sign of every term within the parentheses.*

$$-(3x^2 + 3x - 2) = -1(3x^2 + 3x - 2)$$
$$= -1(3x^2) + (-1)(3x) + (-1)(-2)$$
$$= -3x^2 + (-3x) + 2$$
$$= -3x^2 - 3x + 2$$

This suggests that the way to subtract polynomials is to remove parentheses and combine like terms.

EXAMPLE 5   **a.** $(3x - 4) - (5x + 7) = 3x - 4 - 5x - 7$
$$= -2x - 11$$

b. $(3x^2 - 4x - 6) - (2x^2 - 6x + 12) = 3x^2 - 4x - 6 - 2x^2 + 6x - 12$
$$= x^2 + 2x - 18$$

c. $(-4rt^3 + 2r^2t^2) - (-3rt^3 + 2r^2t^2) = -4rt^3 + 2r^2t^2 + 3rt^3 - 2r^2t^2$
$$= -rt^3$$ ∎

To subtract polynomials in vertical form, we add the negative of the **subtrahend** (the bottom polynomial) to the **minuend** (the top polynomial).

EXAMPLE 6   Subtract $3x^2y - 2xy^2$ from $2x^2y + 4xy^2$.

Solution   We write the subtraction in vertical form, change the signs of the terms of the subtrahend, and add:

$$
\begin{array}{r}
2x^2y + 4xy^2 \\
-\ \underline{3x^2y - 2xy^2}
\end{array}
\quad\longrightarrow\quad
\begin{array}{r}
2x^2y + 4xy^2 \\
+\ \underline{-3x^2y + 2xy^2} \\
-\ x^2y + 6xy^2
\end{array}
$$

In horizontal form, the solution is

$$2x^2y + 4xy^2 - (3x^2y - 2xy^2) = 2x^2y + 4xy^2 - 3x^2y + 2xy^2$$
$$= -x^2y + 6xy^2$$ ∎

EXAMPLE 7   Subtract $6xy^2 + 4x^2y^2 - x^3y^2$ from $-2xy^2 - 3x^3y^2$.

Solution
$$
\begin{array}{r}
-2xy^2 \qquad\quad - 3x^3y^2 \\
-\ \underline{6xy^2 + 4x^2y^2 - x^3y^2}
\end{array}
\quad\longrightarrow\quad
\begin{array}{r}
-2xy^2 \qquad\quad -3x^3y^2 \\
+\ \underline{-6xy^2 - 4x^2y^2 + x^3y^2} \\
-8xy^2 - 4x^2y^2 - 2x^3y^2
\end{array}
$$

In horizontal form, the solution is

$$-2xy^2 - 3x^3y^2 - (6xy^2 + 4x^2y^2 - x^3y^2) = -2xy^2 - 3x^3y^2 - 6xy^2 - 4x^2y^2 + x^3y^2$$
$$= -8xy^2 - 4x^2y^2 - 2x^3y^2$$ ∎

■ Adding and Subtracting Multiples of Polynomials

Because of the distributive property, we can remove parentheses enclosing several terms when a monomial precedes the parentheses. We simply multiply every term

within the parentheses by that monomial. For example, to add $3(2x + 5)$ and $2(4x - 3)$, we proceed as follows:

$$3(2x + 5) + 2(4x - 3) = 6x + 15 + 8x - 6$$
$$= 6x + 8x + 15 - 6$$
$$= 14x + 9$$

EXAMPLE 8   (a, b)

a. $3(x^2 + 4x) + 2(x^2 - 4) = 3x^2 + 12x + 2x^2 - 8$
$$= 5x^2 + 12x - 8$$

b. $8(y^2 - 2y + 3) - 4(2y^2 + y - 3) = 8y^2 - 16y + 24 - 8y^2 - 4y + 12$
$$= -20y + 36$$

c. $-4x(xy^2 - xy + 3) - x(xy^2 - 2) + 3(x^2y^2 + 2x^2y)$
$$= -4x^2y^2 + 4x^2y - 12x - x^2y^2 + 2x + 3x^2y^2 + 6x^2y$$
$$= -2x^2y^2 + 10x^2y - 10x$$

ORALS   *Simplify.*

1. $x^3 + 3x^3$ **2.** $3xy + xy$
3. $(x + 3y) - (x + y)$ **4.** $5(1 - x) + 3(x - 1)$
5. $(2x - y^2) - (2x + y^2)$ **6.** $5(x^2 + y) + (x^2 - y)$
7. $3x^2 + 2y + x^2 - y$ **8.** $2x^2y + y - (2x^2y - y)$

E X E R C I S E   3.5

In Exercises 1–12, tell whether the terms are like or unlike terms. If they are like terms, add them.

1. $3y, 4y$ **2.** $3x^2, 5x^2$ **3.** $3x, 3y$ **4.** $3x^2, 6x$

5. $3x^3, 4x^3, 6x^3$ **6.** $-2y^4, -6y^4, 10y^4$ **7.** $-5x^3y^2, 13x^3y^2$ **8.** $23, 12x$

9. $-23t^6, 32t^6, 56t^6$ **10.** $32x^5y^3, -21x^5y^3, -11x^5y^3$
11. $-x^2y, xy, 3xy^2$ **12.** $4x^3y^2z, -6x^3y^2z, 2x^3y^2z$

In Exercises 13–30, simplify each expression if possible.

13. $4y + 5y$ **14.** $-2x + 3x$ **15.** $-8t^2 - 4t^2$ **16.** $15x^2 + 10x^2$
17. $32u^3 - 16u^3$ **18.** $25xy^2 - 7xy^2$ **19.** $18x^5y^2 - 11x^5y^2$ **20.** $17x^6y - 22x^6y$
21. $3rst + 4rst + 7rst$ **22.** $-2ab + 7ab - 3ab$
23. $-4a^2bc + 5a^2bc - 7a^2bc$ **24.** $(xy)^2 + 4x^2y^2 - 2x^2y^2$
25. $(3x)^2 - 4x^2 + 10x^2$ **26.** $(2x)^4 - (3x^2)^2$

27. $5x^2y^2 + 2(xy)^2 - (3x^2)y^2$

28. $-3x^3y^6 + 2(xy^2)^3 - (3x)^3y^6$

29. $(-3x^2y)^4 + (4x^4y^2)^2 - 2x^8y^4$

30. $5x^5y^{10} - (2xy^2)^5 + (3x)^5y^{10}$

In Exercises 31–62, do the indicated operations and simplify.

31. $(3x + 7) + (4x - 3)$

32. $(2y - 3) + (4y + 7)$

33. $(4a + 3) - (2a - 4)$

34. $(5b - 7) - (3b + 5)$

35. $(2x + 3y) + (5x - 10y)$

36. $(5x - 8y) - (2x + 5y)$

37. $(-8x - 3y) - (11x + y)$

38. $(-4a + b) + (5a - b)$

39. $(3x^2 - 3x - 2) + (3x^2 + 4x - 3)$

40. $(3a^2 - 2a + 4) - (a^2 - 3a + 7)$

41. $(2b^2 + 3b - 5) - (2b^2 - 4b - 9)$

42. $(4c^2 + 3c - 2) + (3c^2 + 4c + 2)$

43. $(2x^2 - 3x + 1) - (4x^2 - 3x + 2) + (2x^2 + 3x + 2)$

44. $(-3z^2 - 4z + 7) + (2z^2 + 2z - 1) - (2z^2 - 3z + 7)$

45. $2(x + 3) + 3(x + 3)$

46. $5(x + y) + 7(x + y)$

47. $-8(x - y) + 11(x - y)$

48. $-4(a - b) - 5(a - b)$

49. $2(x^2 - 5x - 4) - 3(x^2 - 5x - 4) + 6(x^2 - 5x - 4)$

50. $7(x^2 + 3x + 1) + 9(x^2 + 3x + 1) - 5(x^2 + 3x + 1)$

51. *Add:* $3x^2 + 4x + 5$
$\underline{2x^2 - 3x + 6}$

52. *Add:* $2x^3 + 2x^2 - 3x + 5$
$\underline{3x^3 - 4x^2 - x - 7}$

53. *Add:* $2x^3 - 3x^2 + 4x - 7$
$\underline{-9x^3 - 4x^2 - 5x + 6}$

54. *Add:* $-3x^3 + 4x^2 - 4x + 9$
$\underline{2x^3 + 9x - 3}$

55. *Add:* $-3x^2y + 4xy + 25y^2$
$\underline{5x^2y - 3xy - 12y^2}$

56. *Add:* $-6x^3z - 4x^2z^2 + 7z^3$
$\underline{-7x^3z + 9x^2z^2 - 21z^3}$

57. *Subtract:* $3x^2 + 4x - 5$
$\underline{-2x^2 - 2x + 3}$

58. *Subtract:* $3y^2 - 4y + 7$
$\underline{6y^2 - 6y - 13}$

59. *Subtract:* $4x^3 + 4x^2 - 3x + 10$
$\underline{5x^3 - 2x^2 - 4x - 4}$

60. *Subtract:* $3x^3 + 4x^2 + 7x + 12$
$\underline{-4x^3 + 6x^2 + 9x - 3}$

61. *Subtract:* $-2x^2y^2 - 4xy + 12y^2$
$\underline{10x^2y^2 + 9xy - 24y^2}$

62. *Subtract:* $25x^3 - 45x^2z + 31xz^2$
$\underline{12x^3 + 27x^2z - 17xz^2}$

63. Find the sum when $x^2 + x - 3$ is added to the sum of $2x^2 - 3x + 4$ and $3x^2 - 2$.

64. Find the sum when $3y^2 - 5y + 7$ is added to the sum of $-3y^2 - 7y + 4$ and $5y^2 + 5y - 7$.

65. Find the difference when $t^3 - 2t^2 + 2$ is subtracted from the sum of $3t^3 + t^2$ and $-t^3 + 6t - 3$.

66. Find the difference when $-3z^3 - 4z + 7$ is subtracted from the sum of $2z^2 + 3z - 7$ and $-4z^3 - 2z - 3$.

67. Find the sum when $3x^2 + 4x - 7$ is added to the sum of $-2x^2 - 7x + 1$ and $-4x^2 + 8x - 1$.

68. Find the difference when $32x^2 - 17x + 45$ is subtracted from the sum of $23x^2 - 12x - 7$ and $-11x^2 + 12x + 7$.

In Exercises 69–78, simplify each expression.

69. $2(x + 3) + 4(x - 2)$

70. $3(y - 4) - 5(y + 3)$

71. $-2(x^2 + 7x - 1) - 3(x^2 - 2x + 7)$

72. $-5(y^2 - 2y - 6) + 6(2y^2 + 2y - 5)$

73. $2(2y^2 - 2y + 2) - 4(3y^2 - 4y - 1) + 4y(y^2 - y - 1)$

74. $-4(z^2 - 5z) - 5(4z^2 - 1) + 6(2z - 3)$

75. $2a(ab^2 - b) - 3b(a + 2ab) + b(b - a + a^2b)$

76. $3y(xy + y) - 2y^2(x - 4 + y) + 2(y^3 + y^2)$

77. $-4xy^2(x + y + z) - 2x(xy^2 - 4y^2z) - 2y(8xy^2 - 1)$

78. $-3uv(u - v^2 + w) + 4w(uv + w) - 3w(w + uv)$

In Exercises 79–80, let $P(x) = 3x - 5$. Find each value.

79. $P(x + h) + P(x)$

80. $P(x + h) - P(x)$

Writing Exercises ■ *Write a paragraph using your own words.*

1. How do you recognize like terms?   **2.** How do you add like terms?

Something to Think About ■ **1.** If $P(x) = x^{23} + 5x^2 + 73$ and $Q(x) = x^{23} + 4x^2 + 73$, find $P(7) - Q(7)$.

2. If two numbers written in scientific notation have the same power of 10, they can be added as similar terms:

$$2 \times 10^3 + 3 \times 10^3 = 5 \times 10^3$$

Without converting to standard form, tell how you would add

$$2 \times 10^3 + 3 \times 10^4$$

Review Exercises ■ *Let $a = 3$, $b = -2$, $c = -1$, and $d = 2$. Evaluate each expression.*

1. $ab + cd$ **2.** $ad + bc$ **3.** $a(b + c)$ **4.** $d(b + a)$

5. Solve the inequality $-4(2x - 9) \geq 12$ and graph the solution set.

6. The **kinetic energy** of a moving object is given by the formula

$$K = \frac{mv^2}{2}$$

Solve the formula for m.

3.6 Multiplying Polynomials

■ The FOIL Method ■ Multiplying Binomials to Solve Equations

GETTING READY *Simplify.*

1. $(2x)(3)$ **2.** $(3xxx)(x)$ **3.** $5x^2 \cdot x$ **4.** $8x^2x^3$

Use the distributive property to remove parentheses.

5. $3(x + 5)$ **6.** $x(x + 5)$ **7.** $4(y - 3)$ **8.** $2y(y - 3)$

In Section 3.1, we multiplied certain monomials by other monomials. To multiply $4x^2$ by $-2x^3$, for example, we use the commutative and associative properties of multiplication to group the numerical factors and the variable factors together and multiply.

$$4x^2(-2x^3) = 4(-2)x^2x^3$$
$$= -8x^5$$

 (a, c)   **EXAMPLE 1** **a.** $3x^5(2x^5) = 3(2)x^5x^5$ **b.** $-2a^2b^3(5ab^2) = -2(5)a^2ab^3b^2$
$$= 6x^{10} \qquad\qquad\qquad\qquad = -10a^3b^5$$

c. $-4y^5z^2(2y^3z^3)(3yz) = -4(2)(3)y^5y^3yz^2z^3z$
$$= -24y^9z^6$$ ■

The previous examples suggest the following rule.

Multiplying Two Monomials To multiply two monomials, first multiply the numerical factors and then multiply the variable factors.

To find the product of a monomial and a polynomial with more than one term, we use the distributive property. To multiply $x + 4$ by $3x$, for example, we proceed as follows:

$$3x(x + 4) = 3x \cdot x + 3x \cdot 4$$
$$= 3x^2 + 12x$$

OO (b)

EXAMPLE 2

a. $2a^2(3a^2 - 4a) = 2a^2 \cdot 3a^2 - 2a^2 \cdot 4a$
$$= 6a^4 - 8a^3$$

b. $-2xz^2(2x - 3z + 2x^2z^2) = -2xz^2(2x) - (-2xz^2)(3z) + (-2xz^2)(2x^2z^2)$
$$= -4x^2z^2 + 6xz^3 - 4x^3z^4 \qquad \blacksquare$$

The results of Example 2 suggest the following rule.

Multiplying Polynomials by Monomials To multiply a polynomial with more than one term by a monomial, use the distributive property to remove parentheses and simplify.

We must use the distributive property more than once to multiply a polynomial by a binomial. For example, to multiply $3x^2 + 3x - 5$ by $2x + 3$, we proceed as follows:

$$(2x + 3)(3x^2 + 3x - 5) = (2x + 3)3x^2 + (2x + 3)3x - (2x + 3)5$$
$$= 3x^2(2x + 3) + 3x(2x + 3) - 5(2x + 3)$$
$$= 6x^3 + 9x^2 + 6x^2 + 9x - 10x - 15$$
$$= 6x^3 + 15x^2 - x - 15$$

EXAMPLE 3

a. $(2x - 4)(3x + 5) = (2x - 4)3x + (2x - 4)5$
$$= 2x \cdot 3x - 4 \cdot 3x + 2x \cdot 5 - 4 \cdot 5$$
$$= 6x^2 - 12x + 10x - 20$$
$$= 6x^2 - 2x - 20$$

b. $(3x - 2y)(2x + 3y) = (3x - 2y)2x + (3x - 2y)3y$
$$= 3x \cdot 2x - 2y \cdot 2x + 3x \cdot 3y - 2y \cdot 3y$$
$$= 6x^2 - 4xy + 9xy - 6y^2$$
$$= 6x^2 + 5xy - 6y^2$$

c. $(3y + 1)(3y^2 + 2y + 2) = (3y + 1)3y^2 + (3y + 1)2y + (3y + 1)2$
$$= 3y \cdot 3y^2 + 1 \cdot 3y^2 + 3y \cdot 2y + 1 \cdot 2y + 3y \cdot 2 + 1 \cdot 2$$
$$= 9y^3 + 3y^2 + 6y^2 + 2y + 6y + 2$$
$$= 9y^3 + 9y^2 + 8y + 2 \qquad \blacksquare$$

The results of Example 3 suggest the following rule.

Multiplying Polynomials To multiply one polynomial by another, multiply each term of one polynomial by each term of the other polynomial and combine like terms.

It is often convenient to organize the work vertically.

EXAMPLE 4   **a.** Multiply:

$$
\begin{array}{r}
2x - 4 \\
3x + 2 \\
\hline
\end{array}
$$

$3x(2x - 4) \rightarrow$ $\quad 6x^2 - 12x$

$2(2x - 4) \rightarrow$ $\qquad\quad + \; 4x - 8$

$$
\begin{array}{r}
\hline
6x^2 - \; 8x - 8
\end{array}
$$

b. Multiply:

$$
\begin{array}{r}
3a^2 - 4a + 7 \\
2a + 5 \\
\hline
\end{array}
$$

$2a(3a^2 - 4a + 7) \rightarrow$ $\quad 6a^3 - \; 8a^2 + 14a$

$5(3a^2 - 4a + 7) \rightarrow$ $\qquad\quad + \; 15a^2 - 20a + 35$

$$
\begin{array}{r}
\hline
6a^3 + \; 7a^2 - \; 6a + 35
\end{array}
$$

c. Multiply:

$$
\begin{array}{r}
3y^2 - \; 5y \; + \; 4 \\
- \; 4y^2 - \; 3 \\
\hline
\end{array}
$$

$-4y^2(3y^2 - 5y + 4) \rightarrow$ $\; -12y^4 + 20y^3 - 16y^2$

$-3(3y^2 - 5y + 4) \rightarrow$ $\qquad\qquad\quad - \; 9y^2 + 15y - 12$

$$
\begin{array}{r}
\hline
-12y^4 + 20y^3 - 25y^2 + 15y - 12
\end{array}
$$

■

■ The FOIL Method

We can use the FOIL method to multiply one binomial by another. In this method we multiply each term of one binomial by each term of the other binomial. **FOIL** is an acronym for **F**irst terms, **O**uter terms, **I**nner terms, and **L**ast terms. To use the FOIL method to multiply $2a - 4$ by $3a + 5$, we

1. multiply the **F**irst terms $2a$ and $3a$ to obtain $6a^2$,
2. multiply the **O**uter terms $2a$ and 5 to obtain $10a$,
3. multiply the **I**nner terms -4 and $3a$ to obtain $-12a$, and
4. multiply the **L**ast terms -4 and 5 to obtain -20.

Then we simplify the resulting polynomial, if possible.

$$
\begin{aligned}
(2a - 4)(3a + 5) &= 2a(3a) + 2a(5) + (-4)(3a) + (-4)(5) \\
&= 6a^2 + 10a - 12a - 20 \qquad \text{Simplify.} \\
&= 6a^2 - 2a - 20 \qquad\qquad \text{Combine like terms.}
\end{aligned}
$$

Last terms

First terms

Inner terms

Outer terms

(b)

EXAMPLE 5 Use the FOIL method to find each product.

$$\begin{array}{ccc} F & L \\ \end{array}$$

a. $(3x + 4)(2x - 3) = 3x(2x) + 3x(-3) + 4(2x) + 4(-3)$
$$= 6x^2 - 9x + 8x - 12$$
$$= 6x^2 - x - 12$$

$$\begin{array}{ccc} F & L \\ \end{array}$$

b. $(2y - 7)(5y - 4) = 2y(5y) + 2y(-4) + (-7)(5y) + (-7)(-4)$
$$= 10y^2 - 8y - 35y + 28$$
$$= 10y^2 - 43y + 28$$

$$\begin{array}{ccc} F & L \\ \end{array}$$

c. $(2r - 3s)(2r + t) = 2r(2r) + 2r(t) - 3s(2r) - 3s(t)$
$$= 4r^2 + 2rt - 6rs - 3st$$

EXAMPLE 6 Simplify each expression.

a. $3(2x - 3)(x + 1) = 3(2x^2 + 2x - 3x - 3)$ Use FOIL to multiply the binomials.

$$= 3(2x^2 - x - 3)$$ Combine like terms.

$$= 6x^2 - 3x - 9$$ Use the distributive property to remove parentheses.

b. $(x + 1)(x - 2) - 3x(x + 3) = x^2 - 2x + x - 2 - 3x^2 - 9x$
$$= -2x^2 - 10x - 2$$ Combine like terms.

The products discussed in Example 7 are called **special products.**

EXAMPLE 7 Use the FOIL method to find each special product.

a. $(x + y)^2 = (x + y)(x + y)$
$$= x^2 + xy + xy + y^2$$
$$= x^2 + 2xy + y^2$$

The square of the sum of two quantities such as $x + y$ has three terms: the square of the first quantity, plus twice the product of the quantities, and the square of the second quantity.

b. $(x - y)^2 = (x - y)(x - y)$
$$= x^2 - xy - xy + y^2$$
$$= x^2 - 2xy + y^2$$

The square of the difference of two quantities such as $x - y$ has three terms: the square of the first quantity, minus twice the product of the quantities, and the square of the second quantity.

c. $(x + y)(x - y) = x^2 - xy + xy - y^2$

$$= x^2 - y^2$$

The product of a sum and a difference of two quantities such as $x + y$ and $x - y$ is a binomial. It is the product of the first quantities minus the product of the second quantities.

Binomials that have the same terms but different signs between them are often called **conjugate binomials.** For example, the conjugate of $x + y$ is $x - y$, and the conjugate of $ab - c$ is $ab + c$. ■

Because the special products discussed in Example 7 occur so often, it is wise to learn their forms.

Special Products

$$(x + y)^2 = x^2 + 2xy + y^2$$
$$(x - y)^2 = x^2 - 2xy + y^2$$
$$(x + y)(x - y) = x^2 - y^2$$

 WARNING! Note that $(x + y)^2 \neq x^2 + y^2$ and $(x - y)^2 \neq x^2 - y^2$.

■ Multiplying Binomials to Solve Equations

To solve an equation such as $(x + 2)(x + 3) = x(x + 7)$, we use the FOIL method to remove the parentheses on the left-hand side, use the distributive property to remove parentheses on the right-hand side, and proceed as follows:

$$(x + 2)(x + 3) = x(x + 7)$$

$x^2 + 3x + 2x + 6 = x^2 + 7x$

$5x + 6 = 7x$ Subtract x^2 from both sides and combine terms.

$6 = 2x$ Subtract $5x$ from both sides.

$3 = x$ Divide both sides by 2.

Check: $(x + 2)(x + 3) = x(x + 7)$

$(3 + 2)(3 + 3) \stackrel{?}{=} 3(3 + 7)$ Replace x with 3.

$5(6) \stackrel{?}{=} 3(10)$ Do the additions within parentheses.

$30 = 30$

EXAMPLE 8 Solve $(x+5)(x+4)=(x+9)(x+10)$.

Solution We use the FOIL method to remove parentheses on both sides of the equation. Then we proceed as follows:

$$\begin{aligned}
(x+5)(x+4) &= (x+9)(x+10) \\
x^2+4x+5x+20 &= x^2+10x+9x+90 \\
9x+20 &= 19x+90 && \text{Subtract } x^2 \text{ from both sides and combine terms.} \\
20 &= 10x+90 && \text{Subtract } 9x \text{ from both sides.} \\
-70 &= 10x && \text{Subtract 90 from both sides.} \\
-7 &= x && \text{Divide both sides by 10.}
\end{aligned}$$

Check:

$$\begin{aligned}
(x+5)(x+4) &= (x+9)(x+10) \\
(-7+5)(-7+4) &\stackrel{?}{=} (-7+9)(-7+10) && \text{Replace } x \text{ with } -7. \\
(-2)(-3) &\stackrel{?}{=} (2)(3) && \text{Do the additions within parentheses.} \\
6 &= 6
\end{aligned}$$

■

EXAMPLE 9 **Dimensions of a painting** The square painting in Figure 3-1 is surrounded by a border 2 inches wide. If the area of the border is 96 square inches, find the dimensions of the painting.

Solution Let x represent the length of each side of the square painting. Then the outer rectangle is also a square, and its dimensions are $(x+4)$ by $(x+4)$ inches. If we subtract the area of the painting from the area of the larger square, the difference is 96.

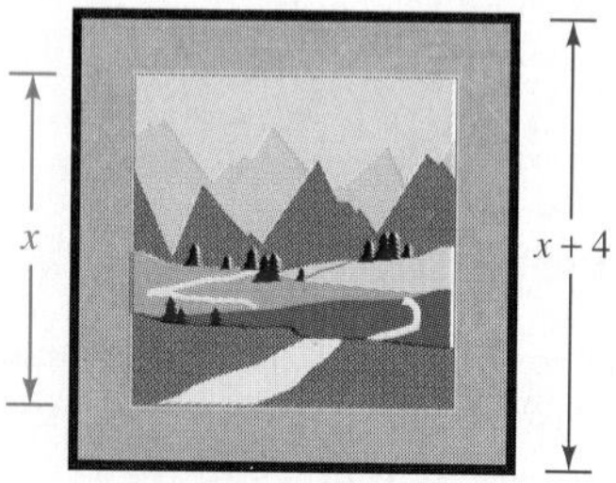

FIGURE 3-1

$$\begin{aligned}
(x+4)(x+4)-x^2 &= 96 \\
x^2+8x+16-x^2 &= 96 && (x+4)(x+4)=x^2+8x+16. \\
8x+16 &= 96 && \text{Combine like terms.} \\
8x &= 80 && \text{Subtract 16 from both sides.} \\
x &= 10 && \text{Divide both sides by 8.}
\end{aligned}$$

The dimensions of the painting are 10 inches by 10 inches. ■

ORALS *Find each product.*

1. $2x^2(3x-1)$
2. $5y(2y^2-3)$
3. $7xy(x+y)$
4. $-2y(2x-3y)$
5. $(x+3)(x+2)$
6. $(x-3)(x+2)$
7. $(2x+3)(x+2)$
8. $(3x-1)(3x+1)$
9. $(x+3)^2$
10. $(x-5)^2$

EXERCISE 3.6

In Exercises 1–12, find each product.

1. $(3x^2)(4x^3)$

2. $(-2a^3)(3a^2)$

3. $(3b^2)(-2b)(4b^3)$

4. $(3y)(2y^2)(-y^4)$

5. $(2x^2y^3)(3x^3y^2)$

6. $(-x^3y^6z)(x^2y^2z^7)$

7. $(x^2y^5)(x^2z^5)(-3y^2z^3)$

8. $(-r^4st^2)(2r^2st)(rst)$

9. $(x^2y^3)^5$

10. $(a^3b^2c)^4$

11. $(a^3b^2c)(abc^3)^2$

12. $(xyz^3)(xy^2z^2)^3$

In Exercises 13–30, find each product.

13. $3(x + 4)$

14. $-3(a - 2)$

15. $-4(t + 7)$

16. $6(s^2 - 3)$

17. $3x(x - 2)$

18. $4y(y + 5)$

19. $-2x^2(3x^2 - x)$

20. $4b^3(2b^2 - 2b)$

21. $3xy(x + y)$

22. $-4x^2(3x^2 - x)$

23. $2x^2(3x^2 + 4x - 7)$

24. $3y^3(2y^2 - 7y - 8)$

25. $\frac{1}{4}x^2(8x^5 - 4)$

26. $\frac{4}{3}a^2b(6a - 5b)$

27. $-\frac{2}{3}r^2t^2(9r - 3t)$

28. $-\frac{4}{5}p^2q(10p + 15q)$

29. $(3xy)(-2x^2y^3)(x + y)$

30. $(-2a^2b)(-3a^3b^2)(3a - 2b)$

In Exercises 31–50, use the FOIL method to find each product.

31. $(a + 4)(a + 5)$

32. $(y - 3)(y + 5)$

33. $(3x - 2)(x + 4)$

34. $(t + 4)(2t - 3)$

35. $(2a + 4)(3a - 5)$

36. $(2b - 1)(3b + 4)$

37. $(3x - 5)(2x + 1)$

38. $(2y - 5)(3y + 7)$

39. $(x + 3)(2x - 3)$

40. $(2x + 3)(2x - 5)$

41. $(2t + 3s)(3t - s)$

42. $(3a - 2b)(4a + b)$

43. $(x + y)(x + z)$

44. $(a - b)(x + y)$

45. $(u + v)(u + 2t)$

46. $(x - 5y)(a + 2y)$

47. $(-2r - 3s)(2r + 7s)$

48. $(-4a + 3)(-2a - 3)$

49. $(4t - u)(-3t + u)$

50. $(-3t + 2s)(2t - 3s)$

In Exercises 51–56, find each product.

51. $4x + 3$
$\underline{x + 2}$

52. $5r + 6$
$\underline{2r - 1}$

53. $4x - 2y$
$\underline{3x + 5y}$

54. $5r + 6s$
$\underline{2r - s}$

55. $x^2 + x + 1$
　　　$\underline{x - 1}$

56. $4x^2 - 2x + 1$
　　　$\underline{2x + 1}$

In Exercises 57–74, find each special product.

57. $(x + 4)(x + 4)$

58. $(a + 3)(a + 3)$

59. $(t - 3)(t - 3)$

60. $(z - 5)(z - 5)$

61. $(r + 4)(r - 4)$

62. $(b + 2)(b - 2)$

63. $(x + 5)^2$

64. $(y - 6)^2$

65. $(2s + 1)(2s + 1)$

66. $(3t - 2)(3t - 2)$

67. $(4x + 5)(4x - 5)$

68. $(5z + 1)(5z - 1)$

69. $(x - 2y)^2$

70. $(3a + 2b)^2$

71. $(2a - 3b)^2$

72. $(2x + 5y)^2$

73. $(4x + 5y)(4x - 5y)$

74. $(6p + 5q)(6p - 5q)$

In Exercises 75–84, find each product.

75. $2(x - 4)(x + 1)$

76. $-3(2x + 3y)(3x - 4y)$

77. $3a(a + b)(a - b)$

78. $-2r(r + s)(r + s)$

79. $(4t + 3)(t^2 + 2t + 3)$

80. $(3x + y)(2x^2 - 3xy + y^2)$

81. $(-3x + y)(x^2 - 8xy + 16y^2)$

82. $(3x - y)(x^2 + 3xy - y^2)$

83. $(x - 2y)(x^2 + 2xy + 4y^2)$

84. $(2m + n)(4m^2 - 2mn + n^2)$

In Exercises 85–94, simplify each expression.

85. $2t(t + 2) + 3t(t - 5)$

86. $3y(y + 2) + (y + 1)(y - 1)$

87. $3xy(x + y) - 2x(xy - x)$

88. $(a + b)(a - b) - (a + b)(a + b)$

89. $(x + y)(x - y) + x(x + y)$

90. $(2x - 1)(2x + 1) + x(2x + 1)$

91. $(x + 2)^2 - (x - 2)^2$

92. $(x - 3)^2 - (x + 3)^2$

93. $(2s - 3)(s + 2) + (3s + 1)(s - 3)$

94. $(3x + 4)(2x - 2) - (2x + 1)(x + 3)$

In Exercises 95–104, solve each equation.

95. $(s - 4)(s + 1) = s^2 + 5$

96. $(y - 5)(y - 2) = y^2 - 4$

97. $z(z + 2) = (z + 4)(z - 4)$

98. $(z + 3)(z - 3) = z(z - 3)$

99. $(x + 4)(x - 4) = (x - 2)(x + 6)$

100. $(y - 1)(y + 6) = (y - 3)(y - 2) + 8$

101. $(a - 3)^2 = (a + 3)^2$

102. $(b + 2)^2 = (b - 1)^2$

103. $4 + (2y - 3)^2 = (2y - 1)(2y + 3)$

104. $7s^2 + (s - 3)(2s + 1) = (3s - 1)^2$

105. Integer problem The difference between the squares of two consecutive integers is 11. Find the integers.

106. Integer problem If 3 less than a certain integer is multiplied by 4 more than the integer, the product is 6 less than the square of the integer. Find the integer.

107. Baseball In major league baseball, the distance between bases is 30 feet greater than it is in softball. The bases in major league baseball mark the corners of a square that has an area 4500 square feet greater than for softball. Find the distance between the bases in baseball.

108. Bookbinding Two square sheets of cardboard used for making book covers differ in area by 44 square inches. An edge of the larger square is 2 inches greater than an edge of the smaller square. Find the length of an edge of the smaller square.

109. Pulley design The radius of one pulley in Illustration 1 is 1 inch greater than the radius of the second pulley, and their areas differ by 4π square inches. Find the radius of the smaller pulley.

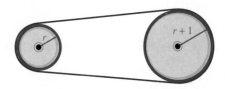

ILLUSTRATION 1

110. Stone-ground flour The radius of one millstone in Illustration 2 is 3 meters greater than the radius of another, and their areas differ by 15π square meters. Find the radius of the larger millstone.

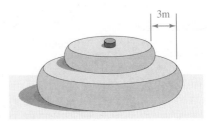

ILLUSTRATION 2

Writing Exercises ■ *Write a paragraph using your own words.*

1. Describe the steps involved in finding the product of a binomial and its conjugate.

2. Writing the expression $(x + y)^2$ as $x^2 + y^2$ illustrates a common error. Explain.

Something to Think About ■

1. Use the FOIL method to find the product: $(x^{-1} + 1)(x + 1)$

2. Simplify: $(x^{-1} + x)^2$

Review Exercises ■ *Tell which property of real numbers justifies each statement.*

1. $3(x + 5) = 3x + 3 \cdot 5$

2. $(x + 3) + y = x + (3 + y)$

3. $3(ab) = (ab)3$

4. $a + 0 = a$

5. Solve the equation $\frac{5}{3}(5y + 6) - 10 = 0$

6. Solve the equation $F = \dfrac{GMm}{d^2}$ for m.

3.7 Dividing Polynomials by Monomials

■ Dividing a Polynomial by a Monomial

GETTING READY *Simplify each fraction.*

1. $\dfrac{4x^2y^3}{2xy}$ 2. $\dfrac{9xyz}{9xz}$ 3. $\dfrac{15x^2y}{10x}$ 4. $\dfrac{6x^2y}{6xy^2}$

5. $\dfrac{(2x^2)(5y^2)}{10xy}$ 6. $\dfrac{(5x^3y)(6xy^3)}{10x^4y^4}$

Dividing by a number is equivalent to multiplying by its reciprocal. For example, dividing the number 8 by 2 gives the same answer as multiplying 8 by $\frac{1}{2}$.

$$\frac{8}{2} = 4 \qquad \text{and} \qquad \frac{1}{2} \cdot 8 = 4$$

In general, the following is true.

Division $\dfrac{a}{b} = \dfrac{1}{b} \cdot a \qquad (b \neq 0)$

Recall that to simplify a fraction, we write both its numerator and denominator as the product of several factors and then divide out all common factors. For example, to simplify $\frac{4}{6}$, we can write

$$\frac{4}{6} = \frac{2 \cdot 2}{2 \cdot 3} = \frac{\overset{1}{\cancel{2}} \cdot 2}{\underset{1}{\cancel{2}} \cdot 3} = \frac{2}{3}$$

To simplify the fraction $\frac{20}{25}$, we can write

$$\frac{20}{25} = \frac{4 \cdot 5}{5 \cdot 5} = \frac{4 \cdot \overset{1}{\cancel{5}}}{\underset{1}{\cancel{5}} \cdot 5} = \frac{4}{5}$$

To simplify algebraic fractions, we can either use the method just used for simplifying arithmetic fractions or use the rules of exponents.

EXAMPLE 1   Simplify   **a.** $\dfrac{x^2y}{xy^2}$ and   **b.** $\dfrac{-8a^3b^2}{4ab^3}$.

Solution   *Method for arithmetic fractions*

a. $\dfrac{x^2 y}{xy^2} = \dfrac{xxy}{xyy}$

$= \dfrac{\cancel{x}x\cancel{y}}{\cancel{x}y\cancel{y}}$

$= \dfrac{x}{y}$

b. $\dfrac{-8a^3 b^2}{4ab^3} = \dfrac{-2 \cdot 4aaabb}{4abbb}$

$= \dfrac{-2 \cdot \cancel{4}\cancel{a}aa\cancel{b}\cancel{b}}{\cancel{4}\cancel{a}b\cancel{b}\cancel{b}}$

$= \dfrac{-2a^2}{b}$

Using the rules of exponents

a. $\dfrac{x^2 y}{xy^2} = x^{2-1}y^{1-2}$

$= x^1 y^{-1}$

$= \dfrac{x}{y}$

b. $\dfrac{-8a^3 b^2}{4ab^3} = \dfrac{-2^3 a^3 b^2}{2^2 ab^3}$

$= -2^{3-2}a^{3-1}b^{2-3}$

$= -2^1 a^2 b^{-1}$

$= \dfrac{-2a^2}{b}$

∎

■ Dividing a Polynomial by a Monomial

To divide a polynomial with more than one term by a monomial, we rewrite the division as a product, use the distributive law to remove parentheses, and simplify each resulting fraction.

EXAMPLE 2   Simplify $\dfrac{9x + 6y}{3xy}$.

Solution $\dfrac{9x + 6y}{3xy} = \dfrac{1}{3xy}(9x + 6y)$

$= \dfrac{9x}{3xy} + \dfrac{6y}{3xy}$ Remove parentheses.

$= \dfrac{3}{y} + \dfrac{2}{x}$ Simplify each fraction.   ∎

EXAMPLE 3   Simplify $\dfrac{6x^2 y^2 + 4x^2 y - 2xy}{2xy}$.

Solution $\dfrac{6x^2 y^2 + 4x^2 y - 2xy}{2xy} = \dfrac{1}{2xy}(6x^2 y^2 + 4x^2 y - 2xy)$

$= \dfrac{6x^2 y^2}{2xy} + \dfrac{4x^2 y}{2xy} - \dfrac{2xy}{2xy}$ Remove parentheses.

$= 3xy + 2x - 1$ Simplify each fraction.   ∎

EXAMPLE 4 Simplify $\dfrac{12a^3b^2 - 4a^2b + a}{6a^2b^2}$.

Solution

$$\dfrac{12a^3b^2 - 4a^2b + a}{6a^2b^2} = \dfrac{1}{6a^2b^2}(12a^3b^2 - 4a^2b + a)$$

$$= \dfrac{12a^3b^2}{6a^2b^2} - \dfrac{4a^2b}{6a^2b^2} + \dfrac{a}{6a^2b^2} \qquad \text{Remove parentheses.}$$

$$= 2a - \dfrac{2}{3b} + \dfrac{1}{6ab^2} \qquad \text{Simplify each fraction.} \qquad ■$$

EXAMPLE 5 Simplify $\dfrac{(x-y)^2 - (x+y)^2}{xy}$.

Solution

$$\dfrac{(x-y)^2 - (x+y)^2}{xy} = \dfrac{x^2 - 2xy + y^2 - (x^2 + 2xy + y^2)}{xy} \qquad \begin{array}{l}\text{Multiply the binomials}\\\text{in the numerator.}\end{array}$$

$$= \dfrac{x^2 - 2xy + y^2 - x^2 - 2xy - y^2}{xy} \qquad \text{Remove parentheses.}$$

$$= \dfrac{-4xy}{xy} \qquad \text{Combine like terms.}$$

$$= -4 \qquad \text{Divide out } xy. \qquad ■$$

ORALS *Simplify each fraction.*

1. $\dfrac{4x^3y}{2xy}$ 2. $\dfrac{6x^3y^2}{3x^3y}$ 3. $\dfrac{35ab^2c^3}{7abc}$ 4. $\dfrac{-14p^2q^5}{7pq^4}$

5. $\dfrac{(x+y) + (x-y)}{2x}$ 6. $\dfrac{(2x^2 - z) + (x^2 + z)}{x}$

E X E R C I S E   3.7

In Exercises 1–12, simplify each fraction.

1. $\dfrac{5}{15}$ 2. $\dfrac{64}{128}$ 3. $\dfrac{-125}{75}$ 4. $\dfrac{-98}{21}$

5. $\dfrac{120}{160}$ 6. $\dfrac{70}{420}$ 7. $\dfrac{-3612}{-3612}$ 8. $\dfrac{-288}{-112}$

9. $\dfrac{-90}{360}$ 10. $\dfrac{8423}{-8423}$ 11. $\dfrac{5880}{2660}$ 12. $\dfrac{-762}{366}$

In Exercises 13–40, do each division by simplifying each fraction. Write all answers without using negative or zero exponents.

13. $\dfrac{xy}{yz}$

14. $\dfrac{a^2b}{ab^2}$

15. $\dfrac{r^3s^2}{rs^3}$

16. $\dfrac{y^4z^3}{y^2z^2}$

17. $\dfrac{8x^3y^2}{4xy^3}$

18. $\dfrac{-3y^3z}{6yz^2}$

19. $\dfrac{12u^5v}{-4u^2v^3}$

20. $\dfrac{16rst^2}{-8rst^3}$

21. $\dfrac{-16r^3y^2}{-4r^2y^4}$

22. $\dfrac{35xyz^2}{-7x^2yz}$

23. $\dfrac{-65rs^2t}{15r^2s^3t}$

24. $\dfrac{112u^3z^6}{-42u^3z^6}$

25. $\dfrac{x^2x^3}{xy^6}$

26. $\dfrac{(xy)^2}{x^2y^3}$

27. $\dfrac{(a^3b^4)^3}{ab^4}$

28. $\dfrac{(a^2b^3)^3}{a^6b^6}$

29. $\dfrac{15(r^2s^3)^2}{-5(rs^5)^3}$

30. $\dfrac{-5(a^2b)^3}{10(ab^2)^3}$

31. $\dfrac{-32(x^3y)^3}{128(x^2y^2)^3}$

32. $\dfrac{68(a^6b^7)^2}{-96(abc^2)^3}$

33. $\dfrac{(5a^2b)^3}{(2a^2b^2)^3}$

34. $\dfrac{-(4x^3y^3)^2}{(x^2y^4)^8}$

35. $\dfrac{-(3x^3y^4)^3}{-(9x^4y^5)^2}$

36. $\dfrac{(2r^3s^2t)^2}{-(4r^2s^2t^2)^2}$

37. $\dfrac{(a^2a^3)^4}{(a^4)^3}$

38. $\dfrac{(b^3b^4)^5}{(bb^2)^2}$

39. $\dfrac{(z^3z^{-4})^3}{(z^{-3})^2}$

40. $\dfrac{(t^{-3}t^5)}{(t^2)^{-3}}$

In Exercises 41–54, do each division.

41. $\dfrac{6x + 9y}{3xy}$

42. $\dfrac{8x + 12y}{4xy}$

43. $\dfrac{5x - 10y}{25xy}$

44. $\dfrac{2x - 32}{16x}$

45. $\dfrac{3x^2 + 6y^3}{3x^2y^2}$

46. $\dfrac{4a^2 - 9b^2}{12ab}$

47. $\dfrac{15a^3b^2 - 10a^2b^3}{5a^2b^2}$

48. $\dfrac{9a^4b^3 - 16a^3b^4}{12a^2b}$

49. $\dfrac{4x - 2y + 8z}{4xy}$

50. $\dfrac{5a^2 + 10b^2 - 15ab}{5ab}$

51. $\dfrac{12x^3y^2 - 8x^2y - 4x}{4xy}$

52. $\dfrac{12a^2b^2 - 8a^2b - 4ab}{4ab}$

53. $\dfrac{-25x^2y + 30xy^2 - 5xy}{-5xy}$

54. $\dfrac{-30a^2b^2 - 15a^2b - 10ab^2}{-10ab}$

In Exercises 55–64, simplify each numerator and do the division.

55. $\dfrac{5x(4x - 2y)}{2y}$

56. $\dfrac{9y^2(x^2 - 3xy)}{3x^2}$

57. $\dfrac{(-2x)^3 + (3x^2)^2}{6x^2}$

58. $\dfrac{(-3x^2y)^3 + (3xy^2)^3}{27x^3y^4}$

59. $\dfrac{4x^2y^2 - 2(x^2y^2 + xy)}{2xy}$

60. $\dfrac{-5a^3b - 5a(ab^2 - a^2b)}{10a^2b^2}$

61. $\dfrac{(3x - y)(2x - 3y)}{6xy}$

62. $\dfrac{(2m - n)(3m - 2n)}{-3m^2n^2}$

63. $\dfrac{(a + b)^2 - (a - b)^2}{2ab}$

64. $\dfrac{(x - y)^2 + (x + y)^2}{2x^2y^2}$

Writing Exercises ■ *Write a paragraph using your own words.*

1. Describe how you would simplify the fraction

$$\dfrac{4x^2y + 8xy^2}{4xy}$$

2. What would you say to another student who attempts to simplify the fraction $\dfrac{3x + 5}{x + 5}$ by dividing out the $x + 5$, as follows?

$$\dfrac{3x + 5}{x + 5} = \dfrac{3x + \cancel{5}}{\cancel{x + 5}} = 3$$

Something to Think About ■ **1.** If $x = 501$, evaluate $\dfrac{x^{500} - x^{499}}{x^{499}}$.

2. An exercise, *Simplify* $\dfrac{3x^3y + 6xy^2}{3xy^3}$, contains a misprint: one mistyped letter or digit. The correct answer is $\dfrac{x^2}{y} + 2$. Fix the exercise.

Review Exercises ■ *Let P(x) = 3x² + x.*

1. Find $P(4)$.

2. Find $P(-2)$.

3. Write 0.000265 in scientific notation.

4. Write 5.67×10^3 in standard notation.

Simplify each expression.

5. $(3x^2y^3z)^0$

6. $(a^2b^3a^4b^5)^3$

3.8 Dividing Polynomials by Polynomials

GETTING READY *Divide.*

1. $12\overline{)156}$ **2.** $17\overline{)357}$ **3.** $13\overline{)247}$ **4.** $19\overline{)247}$

To divide one polynomial by another, we can use a process similar to the long division of one number by another. We illustrate the process with several examples.

EXAMPLE 1   Divide $x^2 + 5x + 6$ by $x + 2$.

Solution   *Step 1:*

$$
\begin{array}{r}
x \\
x + 2 \overline{)\,x^2 + 5x + 6\,}
\end{array}
$$

How many times does x divide x^2? $x^2/x = x$.
Place the x above the division symbol.

Step 2:

$$
\begin{array}{r}
x \\
x + 2 \overline{)\,x^2 + 5x + 6\,} \\
x^2 + 2x
\end{array}
$$

Multiply each term in the divisor by x. Place the product under $x^2 + 5x$ as indicated and draw a line.

Step 3:

$$
\begin{array}{r}
x \\
x + 2 \overline{)\,x^2 + 5x + 6\,} \\
\underline{x^2 + 2x} \\
3x + 6
\end{array}
$$

Subtract $x^2 + 2x$ from $x^2 + 5x$ by adding the negative of $x^2 + 2x$ to $x^2 + 5x$.

Bring down the next term.

Step 4:

$$
\begin{array}{r}
x \;\; + 3 \\
x + 2 \overline{)\,x^2 + 5x + 6\,} \\
\underline{x^2 + 2x} \\
3x + 6
\end{array}
$$

How many times does x divide $3x$? $3x/x = +3$.
Place the $+3$ above the division symbol.

Step 5:

$$
\begin{array}{r}
x \;\; + 3 \\
x + 2 \overline{)\,x^2 + 5x + 6\,} \\
\underline{x^2 + 2x} \\
3x + 6 \\
3x + 6
\end{array}
$$

Multiply each term in the divisor by 3. Place the product under the $3x + 6$ as indicated and draw a line.

Step 6:

$$
\begin{array}{r}
x \;\; + 3 \\
x + 2 \overline{)\,x^2 + 5x + 6\,} \\
\underline{x^2 + 2x} \\
3x + 6 \\
\underline{3x + 6} \\
0
\end{array}
$$

Subtract $3x + 6$ from $3x + 6$ by adding the negative of $3x + 6$.

The quotient is $x + 3$ and the remainder is 0.

Step 7:   Check the work by verifying that the product of $x + 2$ and $x + 3$ is $x^2 + 5x + 6$.

$$(x + 2)(x + 3) = x^2 + 3x + 2x + 6$$
$$= x^2 + 5x + 6$$

The answer checks. ■

EXAMPLE 2 Divide: $\dfrac{6x^2 - 7x - 2}{2x - 1}$.

Solution *Step 1:*

$$
\begin{array}{r}
3x \\
2x - 1\overline{)6x^2 - 7x - 2}
\end{array}
$$

How many times does $2x$ divide $6x^2$? $6x^2/2x = 3x$. Place the $3x$ above the division symbol.

Step 2:

$$
\begin{array}{r}
3x \\
2x - 1\overline{)6x^2 - 7x - 2} \\
6x^2 - 3x
\end{array}
$$

Multiply each term in the divisor by $3x$. Place the product under $6x^2 - 7x$ as indicated and draw a line.

Step 3:

$$
\begin{array}{r}
3x \\
2x - 1\overline{)6x^2 - 7x - 2} \\
6x^2 - 3x \\
\hline
- 4x - 2
\end{array}
$$

Subtract $6x^2 - 3x$ from $6x^2 - 7x$ by adding the negative of $6x^2 - 3x$ to $6x^2 - 7x$.

Bring down the next term.

Step 4:

$$
\begin{array}{r}
3x - 2 \\
2x - 1\overline{)6x^2 - 7x - 2} \\
6x^2 - 3x \\
\hline
- 4x - 2
\end{array}
$$

How many times does $2x$ divide $- 4x$? $-4x/2x = -2$. Place the -2 above the division symbol.

Step 5:

$$
\begin{array}{r}
3x - 2 \\
2x - 1\overline{)6x^2 - 7x - 2} \\
6x^2 - 3x \\
\hline
- 4x - 2 \\
- 4x + 2
\end{array}
$$

Multiply each term in the divisor by -2. Place the product under the $-4x - 2$ as indicated and draw a line.

Step 6:

$$
\begin{array}{r}
3x - 2 \\
2x - 1\overline{)6x^2 - 7x - 2} \\
6x^2 - 3x \\
\hline
-4x - 2 \\
-4x + 2 \\
\hline
- 4
\end{array}
$$

Subtract $-4x + 2$ from $-4x - 2$ by adding the negative of $-4x + 2$.

In this example, the quotient is $3x - 2$ and the remainder is -4. It is common to write the answer in quotient $+ \frac{\text{remainder}}{\text{divisor}}$ form:

$$
3x - 2 + \dfrac{-4}{2x - 1}
$$

where the fraction $\dfrac{-4}{2x - 1}$ is formed by dividing the remainder by the divisor.

Step 7: To check the answer, we multiply $3x - 2 + \dfrac{-4}{2x - 1}$ by $2x - 1$. The product should be the dividend.

$$(2x - 1)\left(3x - 2 + \frac{-4}{2x - 1}\right) = (2x - 1)(3x - 2) + (2x - 1)\left(\frac{-4}{2x - 1}\right)$$
$$= (2x - 1)(3x - 2) - 4$$
$$= 6x^2 - 4x - 3x + 2 - 4$$
$$= 6x^2 - 7x - 2$$

Because the result is the dividend, the answer checks. ■

EXAMPLE 3 Divide $4x^2 + 2x^3 + 12 - 2x$ by $x + 3$.

Solution The division process works most efficiently if the terms of both the divisor and the dividend are written in descending order. This means that the term involving the highest power of x appears first, the term involving the second-highest power of x appears second, and so on. We can use the commutative property to rearrange the terms of the dividend in descending order and divide as follows:

$$
\begin{array}{r}
2x^2 - 2x + 4 \\
x + 3\overline{)2x^3 + 4x^2 - 2x + 12} \\
2x^3 + 6x^2 \\
\hline
-2x^2 - 2x \\
-2x^2 - 6x \\
\hline
+4x + 12 \\
+4x + 12 \\
\hline
\end{array}
$$

Check: $(x + 3)(2x^2 - 2x + 4) = 2x^3 - 2x^2 + 4x + 6x^2 - 6x + 12$
$$= 2x^3 + 4x^2 - 2x + 12$$ ■

EXAMPLE 4 Divide: $\dfrac{x^2 - 4}{x + 2}$.

Solution Since the binomial $x^2 - 4$ does not have a term involving x, we must either include the term $0x$ or leave a space for it. After this adjustment, the division is routine.

$$
\begin{array}{r}
x - 2 \\
x + 2\overline{)x^2 + 0x - 4} \\
x^2 + 2x \\
\hline
-2x - 4 \\
-2x - 4 \\
\hline
\end{array}
$$

Check: $(x + 2)(x - 2) = x^2 - 2x + 2x - 4$
$$= x^2 - 4$$ ■

EXAMPLE 5   Divide $x^3 + y^3$ by $x + y$.

Solution   We write $x^3 + y^3$ leaving spaces for the missing terms and proceed as follows:

$$
\begin{array}{r}
x^2 - xy + y^2 \\
x + y \overline{)x^3 \qquad\qquad + y^3} \\
\underline{x^3 + x^2 y} \\
- x^2 y \\
\underline{- x^2 y - xy^2} \\
+ xy^2 + y^3 \\
\underline{xy^2 + y^3}
\end{array}
$$

Check: $(x + y)(x^2 - xy + y^2) = x^3 - x^2 y + xy^2 + x^2 y - xy^2 + y^3$
$$= x^3 + y^3 \qquad \blacksquare$$

ORALS   *Divide, and give the answer in quotient $+ \frac{remainder}{divisor}$ form.*

1. $x \overline{)2x + 3}$ 2. $x \overline{)3x - 5}$

3. $x + 1 \overline{)2x + 3}$ 4. $x + 1 \overline{)3x + 5}$

5. $x + 1 \overline{)x^2 + x}$ 6. $x + 2 \overline{)x^2 + 2x}$

E X E R C I S E   3.8

1. Divide $x^2 + 4x + 4$ by $x + 2$.   2. Divide $x^2 - 5x + 6$ by $x - 2$.
3. Divide $y^2 + 13y + 12$ by $y + 1$.   4. Divide $z^2 - 7z + 12$ by $z - 3$.
5. Divide $a^2 + 2ab + b^2$ by $a + b$.   6. Divide $a^2 - 2ab + b^2$ by $a - b$.

In Exercises 7–12, do each division.

7. $\dfrac{6a^2 + 5a - 6}{2a + 3}$ 8. $\dfrac{8a^2 + 2a - 3}{2a - 1}$ 9. $\dfrac{3b^2 + 11b + 6}{3b + 2}$ 10. $\dfrac{3b^2 - 5b + 2}{3b - 2}$

11. $\dfrac{2x^2 - 7xy + 3y^2}{2x - y}$ 12. $\dfrac{3x^2 + 5xy - 2y^2}{x + 2y}$

In Exercises 13–24, rearrange the terms so that the powers of x are in descending order. Then do each division.

13. $5x + 3 \overline{)11x + 10x^2 + 3}$ 14. $2x - 7 \overline{)-x - 21 + 2x^2}$

15. $4 + 2x \overline{)-10x - 28 + 2x^2}$ 16. $1 + 3x \overline{)9x^2 + 1 + 6x}$

17. $2x - y \overline{)xy - 2y^2 + 6x^2}$ 18. $2y + x \overline{)3xy + 2x^2 - 2y^2}$

19. $x + 3y \overline{)2x^2 - 3y^2 + 5xy}$ 20. $2x - 3y \overline{)2x^2 - 3y^2 - xy}$

21. $3x - 2y)\overline{-10y^2 + 13xy + 3x^2}$

22. $2x + 3y)\overline{-12y^2 + 10x^2 + 7xy}$

23. $4x + y)\overline{-19xy + 4x^2 - 5y^2}$

24. $x - 4y)\overline{5x^2 - 4y^2 - 19xy}$

In Exercises 25–30, do each division.

25. $2x + 3)\overline{2x^3 + 7x^2 + 4x - 3}$

26. $2x - 1)\overline{2x^3 - 3x^2 + 5x - 2}$

27. $3x + 2)\overline{6x^3 + 10x^2 + 7x + 2}$

28. $4x + 3)\overline{4x^3 - 5x^2 - 2x + 3}$

29. $2x + y)\overline{2x^3 + 3x^2y + 3xy^2 + y^3}$

30. $3x - 2y)\overline{6x^3 - x^2y + 4xy^2 - 4y^3}$

In Exercises 31–40, do each division. If there is a remainder, leave the answer in quotient $+ \frac{remainder}{divisor}$ form.

31. $\dfrac{2x^2 + 5x + 2}{2x + 3}$

32. $\dfrac{3x^2 - 8x + 3}{3x - 2}$

33. $\dfrac{4x^2 + 6x - 1}{2x + 1}$

34. $\dfrac{6x^2 - 11x + 2}{3x - 1}$

35. $\dfrac{x^3 + 3x^2 + 3x + 1}{x + 1}$

36. $\dfrac{x^3 + 6x^2 + 12x + 8}{x + 2}$

37. $\dfrac{2x^3 + 7x^2 + 4x + 3}{2x + 3}$

38. $\dfrac{6x^3 + x^2 + 2x + 1}{3x - 1}$

39. $\dfrac{2x^3 + 4x^2 - 2x + 3}{x - 2}$

40. $\dfrac{3y^3 - 4y^2 + 2y + 3}{y + 3}$

In Exercises 41–50, do each division.

41. $\dfrac{x^2 - 1}{x - 1}$

42. $\dfrac{x^2 - 9}{x + 3}$

43. $\dfrac{4x^2 - 9}{2x + 3}$

44. $\dfrac{25x^2 - 16}{5x - 4}$

45. $\dfrac{x^3 + 1}{x + 1}$

46. $\dfrac{x^3 - 8}{x - 2}$

47. $\dfrac{a^3 + a}{a + 3}$

48. $\dfrac{y^3 - 50}{y - 5}$

49. $3x - 4)\overline{15x^3 - 23x^2 + 16x}$

50. $2y + 3)\overline{21y^2 + 6y^3 - 20}$

Writing Exercises ■ *Write a paragraph using your own words.*

1. Distinguish between *dividend, divisor, quotient,* and *remainder.*

2. How would you check the results of a division?

Something to Think About ■ **1.** What's wrong here?

$$x - 2)\overline{x^2 + 3x - 2} \quad \begin{array}{l} x + 1 \\ \underline{x^2 - 2x} \\ x - 2 \\ \underline{x - 2} \\ 0 \end{array}$$

2. What's wrong here?

$$x + 3)\overline{3x^2 + 10x + 7} \quad \begin{array}{l} 3x \\ \underline{3x^2 + 9x} \\ x + 7 \end{array}$$

The quotient is $3x$ and the remainder is $x + 7$.

Review Exercises ■

1. List the composite numbers from 20 to 30.

2. Graph the set of prime numbers between 10 and 20 on a number line.

Let a = −2 and b = 3. Evaluate each expression.

3. $|a - b|$ **4.** $|a + b|$ **5.** $-|a^2 - b^2|$ **6.** $a - |-b|$

Simplify each expression.

7. $3(2x^2 - 4x + 5) + 2(x^2 + 3x - 7)$ **8.** $-2(y^3 + 2y^2 - y) - 3(3y^3 + y)$

MATHEMATICS IN MEDICINE

The area of a circle is given by the formula $A = \pi r^2$. The bulk of the surface area of the red blood cell in Illustration 1 is contained on its top and bottom. That area is $2\pi r^2$, twice the area of one circle. If there are N disks, their total surface area, T, will be N times the surface area of a single disk: $T = 2N\pi r^2$.

To find the total surface area of the oxygen-carrying red cells, we first express the given quantities in scientific notation.

Radius $= r = 0.00015$ in. $= 1.5 \times 10^{-4}$ in.

Quantity $= N = 25$ trillion

$= 2.5 \times 10^{13}$

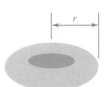

ILLUSTRATION 1

Then we substitute these values into the formula for total surface area.

$T = 2N\pi r^2$

$T = 2(2.5 \times 10^{13})(3.14)(1.5 \times 10^{-4})^2$

$\quad = 2(2.5)(3.14)(1.5)^2 \times 10^{13} \times 10^{-8}$

$\quad \approx 35.3 \times 10^5$

$\quad \approx 3.53 \times 10^6$

$\quad \approx 3,530,000$

The total surface area of the red blood cells is over $3\frac{1}{2}$ million square inches, or approximately 24,500 square feet—almost one-half the area of a football field!

■ ■ ■ ■ ■ ■ ■ ■

P R O J E C T ■ **A Polynomial Pattern**

There is a pattern in the behavior of polynomials. To discover it, let $P(x) = 2x^2 - 3x - 5$.

1. Find $P(1)$ and $P(3)$.

2. Divide $P(x)$ by $x - 1$ and again by $x - 3$.

- What do you notice about the remainders of these divisions?
- Try others. For example, find $P(2)$, and divide by $x - 2$.
- How can you make the pattern hold for $P(-2)$?
- Does the pattern hold for other polynomials? Invent some polynomials of your own, experiment, and report your conclusions.

Chapter Summary

KEY WORDS

algebraic terms (3.4)
base (3.1)
binomial (3.4)
conjugate binomials (3.6)
degree of a monomial (3.4)
degree of a polynomial (3.4)
dividend (3.8)
divisor (3.8)
exponent (3.1)
FOIL method (3.6)
like terms (3.5)

minuend (3.5)
monomial (3.4)
polynomial (3.4)
power (3.1)
quotient (3.8)
scientific notation (3.3)
special products (3.6)
standard notation (3.3)
subtrahend (3.5)
trinomial (3.4)

KEY IDEAS

(3.1–3.2) Properties of exponents. If n is a natural number, then

$$\overbrace{x^n = x \cdot x \cdot x \cdot x \cdots \cdot x}^{n \text{ factors of } x}$$

If m and n are integers, then

$$x^m x^n = x^{m+n}$$
$$(x^m)^n = x^{mn}$$

$$(xy)^n = x^n y^n$$

$$\left(\frac{x}{y}\right)^n = \frac{x^n}{y^n} \quad \text{provided } y \neq 0$$

$$\frac{x^m}{x^n} = x^{m-n} \quad \text{provided } x \neq 0$$

$$x^0 = 1 \quad \text{provided } x \neq 0$$

$$x^{-n} = \frac{1}{x^n} \quad \text{provided } x \neq 0$$

(3.3) A number is written in scientific notation if it is written as the product of a number between 1 (including 1) and 10 and an integer power of 10.

(3.4) If $P(x)$ is a polynomial in x, then $P(r)$ is the value of the polynomial when $x = r$.

(3.5) When adding or subtracting polynomials, combine like terms by adding or subtracting the numerical coefficients and using the same variables and the same exponents.

(3.6) To multiply two monomials, first multiply the numerical factors and then multiply the variable factors.

To multiply a polynomial with more than one term by a monomial, multiply each term of the polynomial by the monomial and simplify.

To multiply one polynomial by another, multiply each term of one polynomial by each term of the other polynomial and simplify.

To multiply two binomials, use the FOIL method.

Special products:

$$(x + y)^2 = x^2 + 2xy + y^2$$
$$(x - y)^2 = x^2 - 2xy + y^2$$
$$(x + y)(x - y) = x^2 - y^2$$

(3.7) To simplify a fraction, divide out all factors common to the numerator and the denominator of the fraction.

To divide a polynomial by a monomial, rewrite the division as a product, use the distributive law to remove parentheses, and simplify each resulting fraction.

(3.8) Use long division to divide one polynomial by another.

■ Chapter 3 Review Exercises

(3.1–3.2) In Review Exercises 1–8, evaluate each expression.

1. 5^3

2. 3^5

3. $(-8)^2$

4. -8^2

5. $3^2 + 2^2$

6. $(3 + 2)^2$

7. $3(3^3 + 3^3)$

8. $1^{17} + 17^1$

In Review Exercises 9–24, do the operations and simplify.

9. $x^3 x^2$

10. $x(x^2 y)$

11. $y^7 y^3$

12. $x^0 y^5$

13. $2b^3 b^4 b^5$

14. $(-z^2)(z^3 y^2)$

15. $(4^4 s)s^2$

16. $-3y(y^5)$

17. $(x^2 x^3)^3$

18. $(2x^2 y)^2$

19. $(3x^0)^2$

20. $(3x^2 y^2)^0$

21. $\dfrac{x^7}{x^3}$

22. $\left(\dfrac{x^2 y}{xy^2}\right)^2$

23. $\dfrac{8(y^2 x)^2}{2^3 (yx^2)^2}$

24. $\dfrac{(5x^0 y^2 z^3)^3}{25(yz)^5}$

In Review Exercises 25–32, write each expression without using negative exponents or parentheses.

25. $x^{-2} x^3$

26. $y^4 y^{-3}$

27. $\dfrac{x^3}{x^{-7}}$

28. $(x^{-3} x^{-4})^{-2}$

29. $\dfrac{x^3}{x^7}$

30. $\left(\dfrac{x^2}{x}\right)^{-5}$

31. $\left(\dfrac{3s}{6s^2}\right)^3$

32. $\left(\dfrac{15z^4}{5z^3}\right)^{-2}$

(3.3) *In Review Exercises 33–40, write each number in scientific notation.*

33. 728   **34.** 9370   **35.** 0.0136   **36.** 0.00942

37. 7.73   **38.** 753×10^3 **39.** 0.018×10^{-2} **40.** 600×10^2

In Review Exercises 41–48, write each number in standard notation.

41. 7.26×10^5 **42.** 3.91×10^{-4} **43.** 2.68×10^0 **44.** 5.76×10^1

45. 739×10^{-2} **46.** 0.437×10^{-3} **47.** $\dfrac{(0.00012)(0.00004)}{0.00000016}$ **48.** $\dfrac{(4800)(20,000)}{600,000}$

(3.4) *In Review Exercises 49–52, give the degree of each polynomial and classify the polynomial as a monomial, a binomial, or a trinomial.*

49. $13x^7$ **50.** $5^3x + x^2$ **51.** $-3x^5 + x - 1$ **52.** $9x + 21x^3$

In Review Exercises 53–56, let $P(x) = 3x + 2$. Find each value.

53. $P(3)$ **54.** $P(0)$ **55.** $P(-2)$ **56.** $P(2t)$

In Review Exercises 57–60, let $P(x) = 5x^4 - x$. Find each value.

57. $P(3)$ **58.** $P(0)$ **59.** $P(-2)$ **60.** $P(2t)$

(3.5) *In Review Exercises 61–70, simplify each expression.*

61. $3x + 5x - x$ **62.** $3x + 2y$ **63.** $(xy)^2 + 3x^2y^2$ **64.** $-2x^2yz + 3yx^2z$

65. $3x^2y^0 + 2x^2$ **66.** $2(x + 7) + 3(x + 7)$

67. $(3x^2 + 2x) + (5x^2 - 8x)$ **68.** $(7a^2 + 2a - 5) - (3a^2 - 2a + 1)$

69. $3(9x^2 + 3x + 7) + 2(2x^2 - 8x + 3) - 2(11x^2 - 5x + 9)$

70. $4(4x^3 + 2x^2 - 3x - 8) - 5(2x^3 - 3x + 8)$

(3.6) *In Review Exercises 71–94, find each product.*

71. $(2x^2y^3)(5xy^2)$ **72.** $(xyz^3)(x^3z)^2$ **73.** $5(x + 3)$ **74.** $3(2x + 4)$

75. $x^2(3x^2 - 5)$ **76.** $2y^2(y^2 + 5y)$ **77.** $-x^2y(y^2 - xy)$ **78.** $-3xy(xy - x)$

79. $(x + 3)(x + 2)$ **80.** $(2x + 1)(x - 1)$ **81.** $(3a - 3)(2a + 2)$ **82.** $6(a - 1)(a + 1)$

83. $(a - b)(2a + b)$ **84.** $(3x - y)(2x + y)$ **85.** $(-3a - b)(3a - b)$ **86.** $(x + 5)(x - 5)$

87. $(y - 2)(y + 2)$ **88.** $(x + 4)^2$ **89.** $(x - 3)^2$ **90.** $y(y + 1)^2$

91. $(2y + 1)^2$ **92.** $(y^2 + 1)(y^2 - 1)$ **93.** $(3x + 1)(x^2 + 2x + 1)$ **94.** $(2a - 3)(4a^2 + 6a + 9)$

In Review Exercises 95–100, solve each equation.

95. $x^2 + 3 = x(x + 3)$

96. $x^2 + x = (x + 1)(x + 2)$

97. $(x + 2)(x - 5) = (x - 4)(x - 1)$

98. $(x - 1)(x - 2) = (x - 3)(x + 1)$

99. $x^2 + x(x + 2) = x(2x + 1) + 1$

100. $(x + 5)(3x + 1) = x^2 + (2x - 1)(x - 5)$

(3.7–3.8) *In Review Exercises 101–110, do each division.*

101. $\dfrac{3x + 6y}{2xy}$

102. $\dfrac{14xy - 21x}{7xy}$

103. $\dfrac{15a^2bc + 20ab^2c - 25abc^2}{-5abc}$

104. $\dfrac{(x + y)^2 + (x - y)^2}{-2xy}$

105. $x + 2 \overline{)x^2 + 3x + 5}$

106. $x - 1 \overline{)x^2 - 6x + 5}$

107. $x + 3 \overline{)2x^2 + 7x + 3}$

108. $3x - 1 \overline{)3x^2 + 14x - 2}$

109. $2x - 1 \overline{)6x^3 + x^2 + 1}$

110. $3x + 1 \overline{)-13x - 4 + 9x^3}$

Chapter 3 Test

1. Use exponents to rewrite $2xxxyyy$.

2. Evaluate $3^2 + 5^3$.

In Problems 3–6, write each expression as an expression containing only one exponent.

3. $y^2(yy^3)$

4. $(-3b^2)(2b^3)(-b^2)$

5. $(2x^3)^5(x^2)^3$

6. $(2rr^2r^3)^3$

In Problems 7–10, simplify each expression. Write answers without using parentheses or negative exponents.

7. $3x^0$

8. $2y^{-5}y^2$

9. $\dfrac{y^2}{yy^{-2}}$

10. $\left(\dfrac{a^2b^{-1}}{4a^3b^{-2}}\right)^{-3}$

11. Write 28,000 in scientific notation.

12. Write 0.0025 in scientific notation.

13. Write 7.4×10^3 in standard notation.

14. Write 9.3×10^{-5} in standard notation.

15. Identify $3x^2 + 2$ as a monomial, a binomial, or a trinomial.

16. Find the degree of the polynomial $3x^2y^3z^4 + 2x^3y^2z - 5x^2y^3z^5$.

17. If $P(x) = x^2 + x - 2$, find $P(-2)$.

18. Simplify: $(xy)^2 + 5x^2y^2 - (3x)^2y^2$.

19. Simplify: $-6(x - y) + 2(x + y) - 3(x + 2y)$

20. Simplify:
$-2(x^2 + 3x - 1) - 3(x^2 - x + 2) + 5(x^2 + 2)$

21. Add: $3x^3 + 4x^2 - x - 7$
$\underline{2x^3 - 2x^2 + 3x + 2}$

22. Subtract: $2x^2 - 7x + 3$
$\underline{3x^2 - 2x - 1}$

In Problems 23–26, find each product.

23. $(-2x^3)(2x^2y)$ **24.** $3y^2(y^2 - 2y + 3)$ **25.** $(2x - 5)(3x + 4)$ **26.** $(2x - 3)(x^2 - 2x + 4)$

27. Solve the equation $(a + 2)^2 = (a - 3)^2$.

28. Simplify the fraction $\dfrac{8x^2y^3z^4}{16x^3y^2z^4}$.

29. Do the division $\dfrac{6a^2 - 12b^2}{24ab}$.

30. Divide: $2x + 3\overline{\smash{\big)}\,2x^2 - x - 6}$.

Cumulative Review Exercises

In Exercises 1–2, classify each number as an integer, a rational number, an irrational number, a real number, a positive number, or a negative number. Each number may be in several classifications.

1. $\dfrac{27}{9}$

2. -0.25

In Exercises 3–4, graph each set of numbers on the number line.

3. The natural numbers between 2 and 7.

4. The real numbers between 2 and 7.

In Exercises 5–8, simplify each expression.

5. $\dfrac{|-3| - |3|}{|-3 - 3|}$ **6.** $\dfrac{5}{7} \cdot \dfrac{14}{3}$ **7.** $2\dfrac{3}{5} + 5\dfrac{1}{2}$ **8.** $35.7 - 0.05$

In Exercises 9–12, let $x = -5$, $y = 3$, and $z = 0$. Evaluate each expression.

9. $(3x - 2y)z$ **10.** $\dfrac{x - 3y + |z|}{2 - x}$ **11.** $x^2 - y^2 + z^2$ **12.** $\dfrac{x}{y} + \dfrac{y + 2}{3 - z}$

13. What is $7\frac{1}{2}\%$ of 330?   **14.** 1688 is 32% of what number?

In Exercises 15–16, consider the algebraic expression $3x^3 + 5x^2y + 37y$.

15. Find the coefficient of the second term.   **16.** List the factors of the third term.

In Exercises 17–20, simplify each expression.

17. $3x - 5x + 2y$

18. $3(x - 7) + 2(8 - x)$

19. $2x^2y^3 - xy(xy^2)$

20. $x^2(3 - y) + x(xy + x)$

In Exercises 21–24, solve each equation.

21. $3(x - 5) + 2 = 2x$

22. $\dfrac{x - 5}{3} - 5 = 7$

23. $\dfrac{2x - 1}{5} = \dfrac{1}{2}$

24. $x(x - 7) + 2x = x^2 - 10$

In Exercises 25–26, solve each formula for the variable indicated.

25. $A = \dfrac{1}{2}h(b + B)$; for h

26. $y = mx + b$; for x

In Exercises 27–30, evaluate each expression.

27. $4^2 - 5^2$

28. $(4 - 5)^2$

29. $5(4^3 - 2^3)$

30. $2(5^4 - 7^3)^0$

In Exercises 31–32, solve each inequality and graph the solution set.

31. $8(4 + x) > 10(6 + x)$

32. $-9 < 3(x + 2) \le 3$

In Exercises 33–36, write each expression as an expression using only one exponent.

33. $(y^3y^5)y^6$

34. $\dfrac{x^3y^4}{x^2y^3}$

35. $\dfrac{a^4b^{-3}}{a^{-3}b^3}$

36. $\left(\dfrac{-x^{-2}y^3}{x^{-3}y^2}\right)^2$

In Exercises 37–40, do the indicated operations.

37. $(3x^2 + 2x - 7) - (2x^2 - 2x + 7)$

38. $(3x - 7)(2x + 8)$

39. $(x - 2)(x^2 + 2x + 4)$

40. $x - 3\overline{)2x^2 - 5x - 3}$

41. Astronomy The **parsec,** a unit of distance used in astronomy, is 3×10^{16} meters. The distance to Betelgeuse, a star in the constellation Orion, is 1.6×10^2 parsecs. Use scientific notation to express this distance in meters.

42. Surface area The total surface area, A, of a box with dimensions l, w, and d (see Illustration 1) is given by the formula

$$A = 2lw + 2wd + 2ld$$

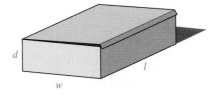

ILLUSTRATION 1

If $A = 202$ square inches, $l = 9$ inches, and $w = 5$ inches, find d.

43. **Concentric circles**   The area of the ring between the two concentric circles of radius r and R (see Illustration 2) is given by the formula

$$A = \pi(R + r)(R - r)$$

If $r = 3$ inches and $R = 17$ inches, find A to the nearest tenth.

ILLUSTRATION 2

44. Employee discount   Employees at an appliance store can purchase merchandise at 25% less than the regular price. An employee buys a TV set for $414.72, including an 8% sales tax. Find the regular price of the TV.

4

Factoring Polynomials

MATHEMATICS IN ECOLOGY

Many types of bacteria cannot survive in air. In one of the steps in waste treatment, sewage is exposed to the air by placing it in large, shallow, circular aeration pools.

One sewage processing plant has two such pools, with diameters of 40 and 42 meters. To meet new clean water standards, the plant must double its capacity, which includes building another aeration pool. How large a pool should the design engineers specify to double the capacity of this phase of sewage treatment?

After reading this chapter, you will be able to answer this question.

In this chapter, we shall reverse the operation of multiplication and show how to find the factors of a known product. The process of finding the individual factors of a product is called **factoring.** In this chapter, we will use factoring to solve a new type of equation, called a **quadratic equation;** in the next chapter, we will use factoring to simplify algebraic fractions.

4.1 Factoring Out the Greatest Common Factor

■ Factoring Out a Negative Factor ■ Quadratic Equations

GETTING READY *Simplify each expression by removing parentheses.*

1. $5(x + 3)$ **2.** $7(y - 8)$ **3.** $x(3x - 2)$ **4.** $y(5y + 9)$

5. $a(b + 9)$ **6.** $x(3 + x + y)$ **7.** $xy(x - 4)$ **8.** $xy^2(2x - 5y)$

Because the natural number 4 divides the natural number 12 exactly, 4 is called a **factor** of 12. The natural-number factors of 8, for example, are 1, 2, 4, and 8, because each of these numbers divides 8. The only natural number factors of 17 are 1 and 17. Recall that a natural number greater than 1 whose only factors are 1 and the number itself is called a **prime number.** Thus, 17 is a prime number.

To **factor** a natural number means to write the number as the product of prime numbers.

Prime-Factored Form A natural number is said to be in **prime-factored form** if it is written as the product of factors that are prime numbers.

The right-hand sides of the equations

$42 = 2 \cdot 3 \cdot 7$
$60 = 2^2 \cdot 3 \cdot 5$
$90 = 2 \cdot 3^2 \cdot 5$

show the prime-factored forms or **prime factorizations** of 42, 60, and 90. A theorem called the **fundamental theorem of arithmetic** points out that there is *exactly one* prime factorization for every natural number greater than 1.

The largest natural number that divides each of several natural numbers is called the **greatest common factor** or the **greatest common divisor** of these numbers. For example, 6 is the greatest common factor of 42, 60, and 90 because

$$\frac{42}{6} = 7 \qquad \frac{60}{6} = 10 \qquad \text{and} \qquad \frac{90}{6} = 15$$

and no natural number greater than 6 divides 42, 60, and 90.

Algebraic monomials also have a greatest common factor. The right-hand sides of the equations

$6a^2b^3 = 2 \cdot 3 \cdot a \cdot a \cdot b \cdot b \cdot b$
$4a^3b^2 = 2 \cdot 2 \cdot a \cdot a \cdot a \cdot b \cdot b$
$18a^2b = 2 \cdot 3 \cdot 3 \cdot a \cdot a \cdot b$

show the prime factorizations of $6a^2b^3$, $4a^3b^2$, and $18a^2b$. Because all three of these monomials have one factor of 2, two factors of a, and one factor of b in common, their greatest common factor is

$2 \cdot a \cdot a \cdot b \qquad \text{or} \qquad 2a^2b$

To find the greatest common factor of several monomials, we follow these steps:

Steps for Finding the Greatest Common Factor	**1.** Find the prime factorization of each monomial.
	2. Use each common factor the least number of times it appears in any one monomial.
	3. Find the product of the factors found in Step 2 to obtain the greatest common factor.

The distributive property provides a way to multiply a polynomial by a monomial. For example,

$$3x^2(2x - 3y) = 3x^2 \cdot 2x - 3x^2 \cdot 3y$$
$$= 6x^3 - 9x^2y$$

To factor a polynomial such as $6x^3 - 9x^2y$, we can find the greatest common factor of each monomial and then use the distributive property in reverse.

$$6x^3 - 9x^2y = 3x^2 \cdot 2x - 3x^2 \cdot 3y$$
$$= 3x^2(2x - 3y)$$

Because $3x^2$ is the greatest common factor of the terms $6x^3$ and $-9x^2y$, this process is called **factoring out the greatest common factor.**

EXAMPLE 1   Factor $12y^2 + 20y$ by factoring out the greatest common factor.

Solution   To find the greatest common factor, we find the prime factorizations of $12y^2$ and $20y$.

$$12y^2 = 2 \cdot 2 \cdot 3 \cdot y \cdot y$$
$$20y = 2 \cdot 2 \cdot 5 \cdot y$$

■ **PERSPECTIVE**

Much of the mathematics we have inherited from the ancients is the result of teamwork. In a battle early in the 12th century, control of the Spanish city of Toledo was taken from the Mohammedans, who had ruled for four centuries. Libraries in this great city contained many books written in Arabic, full of knowledge that was unknown in Europe.

The Archbishop of Toledo wanted to share this knowledge with the rest of the world. He knew that these books should be translated into Latin, the universal language of scholarship. But what

European scholar could read Arabic? The archbishop wasn't concerned. The citizens of Toledo could read both Arabic and Spanish, and most scholars of Europe could understand Spanish.

Teamwork saved the day. A citizen of Toledo read an Arabic text aloud, in Spanish. The scholar listened to the Spanish version and wrote in Latin. One of these scholars was an Englishman, Robert of Chester. It was he who translated al-Khowarazmi's book, *Ihm al-jabr wa'l muqabalah,* the beginning of the subject we now know as *algebra.*

The greatest common factor in $12y^2$ and $20y$ is $2 \cdot 2 \cdot y$, or $4y$, and we can use the distributive property to factor it out.

$$12y^2 + 20y = 4y \cdot 3y + 4y \cdot 5$$
$$= 4y(3y + 5)$$

Check this result by verifying that $4y(3y + 5) = 12y^2 + 20y$. ■

EXAMPLE 2   Factor $35a^3b^2 - 14a^2b^3$.

Solution   To find the greatest common factor, we find the prime factorizations of $35a^3b^2$ and $-14a^2b^3$.

$$35a^3b^2 = 5 \cdot 7 \cdot a \cdot a \cdot a \cdot b \cdot b$$
$$-14a^2b^3 = -2 \cdot 7 \cdot a \cdot a \cdot b \cdot b \cdot b$$

The greatest factor common in $35a^3b^2$ and $-14a^2b^3$ is $7 \cdot a \cdot a \cdot b \cdot b$, or $7a^2b^2$, and we can use the distributive property to factor it out.

$$35a^3b^2 - 14a^2b^3 = 7a^2b^2 \cdot 5a - 7a^2b^2 \cdot 2b$$
$$= 7a^2b^2(5a - 2b)$$

Check this result by verifying that $7a^2b^2(5a - 2b) = 35a^3b^2 - 14a^2b^3$. ■

    EXAMPLE 3   Factor $a^2b^2 - ab$.

Solution   We factor out the greatest common factor, which is ab.

$$a^2b^2 - ab = ab \cdot ab - ab \cdot 1$$
$$= ab(ab - 1)$$

WARNING!   It is important to understand where the 1 comes from. The last term of the binomial $a^2b^2 - ab$ has an implied coefficient of 1. When ab is factored out, this coefficient of 1 must be written.

Check the result by verifying that $ab(ab - 1) = a^2b^2 - ab$. ■

EXAMPLE 4   Factor $12x^3y^2z + 6x^2yz - 3xz$.

Solution   We factor out the greatest common factor, which is $3xz$.

$$12x^3y^2z + 6x^2yz - 3xz = 3xz \cdot 4x^2y^2 + 3xz \cdot 2xy - 3xz \cdot 1$$
$$= 3xz(4x^2y^2 + 2xy - 1)$$

Check the result by verifying that $3xz(4x^2y^2 + 2xy - 1) = 12x^3y^2z + 6x^2yz - 3xz$. ■

■ Factoring Out a Negative Factor

EXAMPLE 5 Factor -1 out of $-a^3 + 2a^2 - 4$.

Solution

$$-a^3 + 2a^2 - 4 = (-1)a^3 + (-1)(-2a^2) + (-1)4 \qquad (-1)(-2a^2) = +2a^2.$$
$$= -1(a^3 - 2a^2 + 4)$$
$$= -(a^3 - 2a^2 + 4)$$

Check the result by verifying that $-(a^3 - 2a^2 + 4) = -a^3 + 2a^2 - 4$. ■

EXAMPLE 6 Factor out the negative of the greatest common factor in $-18a^2b + 6ab^2 - 12a^2b^2$.

Solution The greatest common factor in the trinomial is $6ab$. Since we want to factor out the negative of this greatest common factor, we factor out $-6ab$ as follows:

$$-18a^2b + 6ab^2 - 12a^2b^2 = (-6ab)3a - (-6ab)b + (-6ab)2ab$$
$$= -6ab(3a - b + 2ab)$$

Check by verifying that $-6ab(3a - b + 2ab) = -18a^2b + 6ab^2 - 12a^2b^2$. ■

■ Quadratic Equations

Equations such as $9x - 6 = 0$ that involve first-degree polynomials are called **linear equations.** Equations such as $9x^2 - 6x = 0$ that involve second-degree polynomials are called **quadratic equations.**

Quadratic Equations A **quadratic equation** is an equation of the form
$$ax^2 + bx + c = 0$$
where a, b, and c are real numbers, and $a \neq 0$.

The techniques we have used to solve linear equations cannot be used to solve a quadratic equation such as $9x^2 - 6x = 0$, because none of those techniques can be used to isolate x on one side of the equation. However, we can often solve quadratic equations by factoring and by using the following property of real numbers.

The Zero-Factor Property of Real Numbers Suppose a and b represent two real numbers. Then
If $ab = 0$, then $a = 0$ or $b = 0$.

We already know that if either of two numbers is 0, then their product is 0. The **zero-factor property** says that if the product of two numbers is 0, then at least one of the numbers must be 0.

For example, the equation $(x - 4)(x + 5) = 0$ indicates that a product is equal to 0. By the zero-factor property, one of the factors must be 0:

$$x - 4 = 0 \quad \text{or} \quad x + 5 = 0$$

We can solve each of these linear equations to get

$$x = 4 \quad \text{or} \quad x = -5$$

The equation $(x - 4)(x + 5) = 0$ has two solutions: 4 and -5.

EXAMPLE 7   Solve the quadratic equation $9x^2 - 6x = 0$.

Solution   We begin by factoring the left-hand side of the equation.

$$9x^2 - 6x = 0$$
$$3x(3x - 2) = 0$$

By the zero-factor property, we have

$$3x = 0 \quad \text{or} \quad 3x - 2 = 0$$

We can solve these equations to get

$$x = 0 \quad \text{or} \quad x = \frac{2}{3}$$

Check: To check, we substitute these answers for x in the original equation and simplify.

For $x = 0$	For $x = \dfrac{2}{3}$
$9x^2 - 6x = 0$	$9x^2 - 6x = 0$
$9(0)^2 - 6(0) \stackrel{?}{=} 0$	$9\left(\dfrac{2}{3}\right)^2 - 6\left(\dfrac{2}{3}\right) \stackrel{?}{=} 0$
$0 - 0 \stackrel{?}{=} 0$	$9\left(\dfrac{4}{9}\right) - 6\left(\dfrac{2}{3}\right) \stackrel{?}{=} 0$
$0 = 0$	$4 - 4 \stackrel{?}{=} 0$
	$0 = 0$

Both solutions check.   ■

ORALS   *Find the prime factorization of each number.*

1. 36   **2.** 27   **3.** 81   **4.** 45

Find the greatest common factor.

5. 3, 6, and 9   **6.** $3a^2b$, $6ab$, and $9ab^2$

Factor out the greatest common factor.

7. $15xy + 10$ **8.** $15xy + 10xy^2$

EXERCISE 4.1

In Exercises 1–12, find the prime factorization of each number.

1. 12   **2.** 24   **3.** 15   **4.** 20

5. 40   **6.** 62   **7.** 98   **8.** 112

9. 225   **10.** 144   **11.** 288   **12.** 968

In Exercises 13–40, factor out the greatest common factor.

13. $3x + 6$ **14.** $2y - 10$ **15.** $xy - xz$ **16.** $uv + ut$

17. $t^3 + 2t^2$ **18.** $b^3 - 3b^2$ **19.** $r^4 - r^2$ **20.** $a^3 + a^2$

21. $a^3b^3z^3 - a^2b^3z^2$ **22.** $r^3s^6t^9 + r^2s^2t^2$ **23.** $24x^2y^3z^4 + 8xy^2z^3$ **24.** $3x^2y^3 - 9x^4y^3z$

25. $12uvw^3 - 18uv^2w^2$ **26.** $14xyz - 16x^2y^2z$ **27.** $3x + 3y - 6z$ **28.** $2x - 4y + 8z$

29. $ab + ac - ad$ **30.** $rs - rt + ru$ **31.** $4y^2 + 8y - 2xy$ **32.** $3x^2 - 6xy + 9xy^2$

33. $12r^2 - 3rs + 9r^2s^2$ **34.** $6a^2 - 12a^3b + 36ab$

35. $abx - ab^2x + abx^2$ **36.** $a^2b^2x^2 + a^3b^2x^2 - a^3b^3x^3$

37. $4x^2y^2z^2 - 6xy^2z^2 + 12xyz^2$ **38.** $32xyz + 48x^2yz + 36xy^2z$

39. $70a^3b^2c^2 + 49a^2b^3c^3 - 21a^2b^2c^2$ **40.** $8a^2b^2 - 24ab^2c + 9b^2c^2$

In Exercises 41–52, factor out -1 from each polynomial.

41. $-a - b$ **42.** $-x - 2y$ **43.** $-2x + 5y$ **44.** $-3x + 8z$

45. $-2a + 3b$ **46.** $-2x + 5y$ **47.** $-3m - 4n + 1$ **48.** $-3r + 2s - 3$

49. $-3xy + 2z + 5w$ **50.** $-4ab + 3c - 5d$

51. $-3ab - 5ac + 9bc$ **52.** $-6yz + 12xz - 5xy$

In Exercises 53–62, factor each polynomial by factoring out the greatest common factor, including -1.

53. $-3x^2y - 6xy^2$ **54.** $-4a^2b^2 + 6ab^2$

55. $-4a^2b^3 + 12a^3b^2$

56. $-25x^4y^3z^2 + 30x^2y^3z^4$

57. $-4a^2b^2c^2 + 14a^2b^2c - 10ab^2c^2$

58. $-10x^4x^3z^2 + 8x^3y^2z - 20x^2y$

59. $-14a^6b^6 + 49a^2b^3 - 21ab$

60. $-35r^9s^9t^9 + 25r^6s^6t^6 + 75r^3s^3t^3$

61. $-5a^2b^3c + 15a^3b^4c^2 - 25a^4b^3c$

62. $-7x^5y^4z^3 + 49x^5y^5z^4 - 21x^6y^4z^3$

In Exercises 63–70, solve each equation.

63. $(x - 2)(x + 3) = 0$ **64.** $(x - 3)(x - 2) = 0$

65. $(x - 4)(x + 1) = 0$ **66.** $(x + 5)(x + 2) = 0$

67. $(2x - 5)(3x + 6) = 0$ **68.** $(3x - 4)(x + 1) = 0$

69. $(x - 1)(x + 2)(x - 3) = 0$ **70.** $(x + 2)(x + 3)(x - 4) = 0$

In Exercises 71–82, solve each equation.

71. $x(x - 3) = 0$ **72.** $x(x + 5) = 0$ **73.** $x(2x - 5) = 0$ **74.** $x(5x + 7) = 0$

75. $x^2 - 7x = 0$ **76.** $x^2 + 5x = 0$ **77.** $3x^2 + 8x = 0$ **78.** $5x^2 - x = 0$

79. $8x^2 - 16x = 0$ **80.** $15x^2 - 20x = 0$ **81.** $10x^2 - 2x = 0$ **82.** $5x^2 + x = 0$

Writing Exercises ■ *Write a paragraph using your own words.*

1. When we add $5x$ and $7x$, we combine like terms: $5x + 7x = 12x$. Explain how this is related to factoring out a common factor.

2. One student summarized the zero-factor property of real numbers by saying "Anything times zero is zero." This answer is true, but it does not describe the zero-factor property. Explain.

Something to Think About ■

1. Think of two positive integers. Divide their product by their greatest common factor. Why do you think the result is called the **lowest common multiple** of the two integers? (*Hint:* The **multiples** of an integer such as 5 are 5, 10, 15, 20, 25, 30, and so on.)

2. Two integers are **relatively prime** if their greatest common factor is 1. For example, 6 and 25 are relatively prime, but 6 and 15 are not. If the greatest common factor of three integers is 1, must any two of them be relatively prime? Explain.

Review Exercises ■ *Solve each equation and check the solution.*

1. $3x - 2(x + 1) = 5$

2. $5(y - 1) + 1 = y$

3. $\dfrac{2x - 7}{5} = 3$

4. $2x - \dfrac{x}{2} = 5x$

4.2 Factoring by Grouping

■ Factoring Out a Polynomial ■ Factoring by Grouping

GETTING READY *Remove parentheses and simplify.*

1. $3(x + y) + a(x + y)$

2. $x(y + 1) + 5(y + 1)$

3. $5(x + 1) - y(x + 1)$

4. $x(x + 2) - y(x + 2)$

5. $(3x - y)x + (3x - y)y$

6. $5(y - 7) - y(y - 7)$

■ Factoring Out a Polynomial

Sometimes the greatest common factor of several terms is a polynomial, and we can factor out that common factor. Because the binomial $a + b$ is a common factor of both $(a + b)x$ and $(a + b)y$, for example, we can use the techniques of the previous section to factor $a + b$ out of the expression $(a + b)x + (a + b)y$:

$$(a + b)x + (a + b)y = (a + b)(x + y)$$

We can check the result by verifying that $(a + b)(x + y) = (a + b)x + (a + b)y$.

EXAMPLE 1 Factor $a + 3$ out of the expression $(a + 3) + (a + 3)^2$.

Solution We recall that $a + 3$ is equal to $(a + 3)1$ and that $(a + 3)^2$ is equal to $(a + 3)(a + 3)$. Then we factor out $a + 3$ and simplify.

$$\begin{aligned}(a + 3) + (a + 3)^2 &= (a + 3)1 + (a + 3)(a + 3) \\ &= (a + 3)[1 + (a + 3)] \\ &= (a + 3)(a + 4)\end{aligned}$$ ■

EXAMPLE 2 Factor $6a^2b^2(x + 2y) - 9ab(x + 2y)$.

Solution $\begin{aligned}6a^2b^2(x + 2y) - 9ab(x + 2y) &= (x + 2y)(6a^2b^2 - 9ab) \\ &= (x + 2y)3ab(2ab - 3) \\[1em] &= 3ab(x + 2y)(2ab - 3)\end{aligned}$

Factor out $(x + 2y)$.

Factor out $3ab$ from $6a^2b^2 - 9ab$.

Use the commutative property of multiplication.   ■

◼ Factoring by Grouping

Suppose we wish to factor the expression

$$ax + ay + cx + cy$$

Although no factor is common to all four terms, there is a common factor of a in $ax + ay$ and a common factor of c in $cx + cy$. We can factor out the a and c to obtain

$$ax + ay + cx + cy = a(x + y) + c(x + y)$$
$$= (x + y)(a + c)$$

We can check the result by multiplication.

$$(x + y)(a + c) = ax + cx + ay + cy$$
$$= ax + ay + cx + cy$$

Thus, $ax + ay + cx + cy$ factors as $(x + y)(a + c)$. This type of factoring is called **factoring by grouping.**

EXAMPLE 3   Factor $2c + 2d - cd - d^2$.

Solution $2c + 2d - cd - d^2 = 2(c + d) - d(c + d)$ Factor out 2 from $2c + 2d$ and factor out $-d$ from $-cd - d^2$.

$$= (c + d)(2 - d)$$ Factor out $c + d$.

Check: $(c + d)(2 - d) = 2c - cd + 2d - d^2$
$$= 2c + 2d - cd - d^2$$ ◼

EXAMPLE 4   Factor $x^2y - ax - xy + a$.

Solution $x^2y - ax - xy + a = x(xy - a) - 1(xy - a)$ Factor out x from $x^2y - ax$ and factor out -1 from $-xy + a$.

$$= (xy - a)(x - 1)$$ Factor out $xy - a$.

Check by multiplication.   ◼

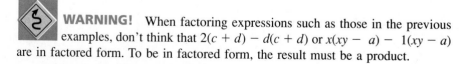 **WARNING!** When factoring expressions such as those in the previous examples, don't think that $2(c + d) - d(c + d)$ or $x(xy - a) - 1(xy - a)$ are in factored form. To be in factored form, the result must be a product.

The method of factoring by grouping often works on polynomials that contain more than four terms.

EXAMPLE 5 Factor $6am - 6bm + 6cm + 3an - 3bn + 3cn$.

Solution Factor $6m$ from the first three terms and $3n$ from the last three terms to obtain

$$6am - 6bm + 6cm + 3an - 3bn + 3cn = 6m(a - b + c) + 3n(a - b + c)$$

Then factor out the common factor of $(a - b + c)$.

$$6am - 6bm + 6cm + 3an - 3bn + 3cn = (a - b + c)(6m + 3n)$$

Check by multiplication. ∎

EXAMPLE 6 Factor **a.** $a(c - d) + b(d - c)$ and **b.** $ac + bd - ad - bc$.

Solution **a.**
$$
\begin{aligned}
a(c - d) + b(d - c) &= a(c - d) - b(-d + c) &&\text{Factor } -1 \text{ from } d - c. \\
&= a(c - d) - b(c - d) &&-d + c = c - d. \\
&= (c - d)(a - b) &&\text{Factor out } (c - d).
\end{aligned}
$$

b. In this example, we cannot factor anything from the first two terms or the last two terms. However, if we rearrange the terms, the factoring is routine:

$$
\begin{aligned}
ac + bd - ad - bc &= ac - ad + bd - bc &&bd - ad = -ad + bd. \\
&= a(c - d) - b(-d + c) &&\text{Factor } a \text{ from } ac - ad \text{ and} \\
& &&-b \text{ from } bd - bc. \\
&= a(c - d) - b(c - d) &&-d + c = c - d. \\
&= (c - d)(a - b) &&\text{Factor out } c - d.
\end{aligned}
$$ ∎

ORALS *Find the common factor of the given terms.*

1. $a(x + 3)$ and $3(x + 3)$
2. $5(a - 1)$ and $xy(a - 1)$
3. $b(x - 2)$ and $(x - 2)^2$
4. $(y + 5)$ and $(y + 5)^2$
5. $a(x - 7)$, $9(x - 7)$, and $x(x - 7)$
6. $5(2y + 9)$, $y(2y + 9)$, and $y^2(2y + 9)$

E X E R C I S E   4.2

In Exercises 1–20, factor each expression.

1. $(x + y)2 + (x + y)b$
2. $(a - b)c + (a - b)d$
3. $3(x + y) - a(x + y)$
4. $x(y + 1) - 5(y + 1)$
5. $3(r - 2s) - x(r - 2s)$
6. $x(a + 2b) + y(a + 2b)$
7. $(x - 3)^2 + (x - 3)$
8. $(3t + 5)^2 - (3t + 5)$
9. $2x(a^2 + b) + 2y(a^2 + b)$
10. $3x(c - 3d) + 6y(c - 3d)$
11. $3x^2(r + 3s) - 6y^2(r + 3s)$
12. $9a^2b^2(3x - 2y) - 6ab(3x - 2y)$

13. $3x(a + b + c) - 2y(a + b + c)$

14. $2m(a - 2b + 3c) - 21xy(a - 2b + 3c)$

15. $14x^2y(r + 2s - t) - 21xy(r + 2s - t)$

16. $15xy^3(2x - y + 3z) + 25xy^2(2x - y + 3z)$

17. $(x + 3)(x + 1) - y(x + 1)$

18. $x(x^2 + 2) - y(x^2 + 2)$

19. $(3x - y)(x^2 - 2) + (x^2 - 2)$

20. $(x - 5y)(a + 2) - (x - 5y)$

In Exercises 21–40, factor each expression.

21. $2x + 2y + ax + ay$

22. $bx + bz + 5x + 5z$

23. $7r + 7s - kr - ks$

24. $9p - 9q + mp - mq$

25. $xr + xs + yr + ys$

26. $pm - pn + qm - qn$

27. $2ax + 2bx + 3a + 3b$

28. $3xy + 3xz - 5y - 5z$

29. $2ab + 2ac + 3b + 3c$

30. $3ac + a + 3bc + b$

31. $2x^2 + 2xy - 3x - 3y$

32. $3ab + 9a - 2b - 6$

33. $3tv - 9tw + uv - 3uw$

34. $ce - 2cf + 3de - 6df$

35. $9mp + 3mq - 3np - nq$

36. $ax + bx - a - b$

37. $mp - np - m + n$

38. $6x^2u - 3x^2v + 2yu - yv$

39. $x(a - b) + y(b - a)$

40. $p(m - n) - q(n - m)$

In Exercises 41–48, factor each expression. Factor out all common factors first if they exist.

41. $ax^3 + bx^3 + 2ax^2y + 2bx^2y$

42. $x^3y^2 - 2x^2y^2 + 3xy^2 - 6y^2$

43. $4a^2b + 12a^2 - 8ab - 24a$

44. $-4abc - 4ac^2 + 2bc + 2c^2$

45. $x^3 + 2x^2 + x + 2$

46. $y^3 - 3y^2 - 5y + 15$

47. $x^3y - x^2y - xy^2 + y^2$

48. $2x^3z - 4x^2z + 32xz - 64z$

In Exercises 49–56, factor each expression completely.

49. $x^2 + xy + x + 2x + 2y + 2$

50. $ax + ay + az + bx + by + bz$

51. $am + bm + cm - an - bn - cn$

52. $x^2 + xz - x - xy - yz + y$

53. $ad - bd - cd + 3a - 3b - 3c$

54. $ab + ac - ad - b - c + d$

55. $ax^2 - ay + bx^2 - by + cx^2 - cy$

56. $a^2x - bx - a^2y + by + a^2z - bz$

In Exercises 57–68, factor each expression completely. You may have to rearrange some terms first.

57. $2r - bs - 2s + br$

58. $5x + ry + rx + 5y$

59. $ax + by + bx + ay$

60. $mr + ns + ms + nr$

61. $ac + bd - ad - bc$

62. $sx - ry + rx - sy$

63. $ar^2 - brs + ars - br^2$

64. $a^2bc + a^2c + abc + ac$

65. $ba + 3 + a + 3b$

66. $xy + 7 + y + 7x$

67. $pr + qs - ps - qr$

68. $ac - bd - ad + bc$

Writing Exercises ■ *Write a paragraph using your own words.*

1. Explain why $a - b$ and $b - a$ are negatives of each other.

2. Explain how you would factor $x(a - b) + y(b - a)$.

Something to Think About ■ **1.** Factor $ax + ay + bx + by$ by grouping the first two terms and the last two terms. Then rearrange the terms as $ax + bx + ay + by$, and factor again by grouping the first two and the last two. Do the results agree?

2. Factor $2xy + 2xz - 3y - 3z$ by grouping in two different ways.

Review Exercises ■ *Simplify each expression. Write all results without using negative exponents.*

1. $u^3u^2u^4$ **2.** $\dfrac{y^6}{y^8}$ **3.** $\dfrac{a^3b^4}{a^2b^5}$ **4.** $(3x^5)^0$

4.3 Factoring the Difference of Two Squares

■ Solving Equations

GETTING READY   *Multiply the binomials.*

1. $(a + b)(a - b)$ **2.** $(2r + s)(2r - s)$
3. $(3x + 2y)(3x - 2y)$ **4.** $(4x^2 + 3)(4x^2 - 3)$

Whenever we multiply a binomial of the form $x + y$ by a binomial of the form $x - y$, we obtain another binomial:

$$(x + y)(x - y) = x^2 - y^2$$

The binomial $x^2 - y^2$ is called the **difference of two squares,** because x^2 is the square of x and y^2 is the square of y. The difference of the squares of two quantities such as x and y always factors into the sum of those two quantities multiplied by the difference of those two quantities.

Factoring the Difference of Two Squares $x^2 - y^2 = (x + y)(x - y)$

To factor $x^2 - 9$, for example, we note that $x^2 - 9$ can be written in the form $x^2 - 3^2$ and that $x^2 - 3^2$ is the difference of the square of x and the square of 3. Thus, it factors into the product of (x plus 3) and (x minus 3).

$$x^2 - 9 = x^2 - 3^2$$
$$= (x + 3)(x - 3)$$

We can check this result by verifying that $(x + 3)(x - 3) = x^2 - 9$.

To factor the difference of two squares, it is useful to know the integers that are perfect squares. The number 400, for example, is a perfect square integer because $20^2 = 400$. The perfect square integers less than 400 are

1, 4, 9, 16, 25, 36, 49, 64, 81, 100, 121, 144, 169, 196, 225, 256, 289, 324, 361

Expressions containing variables such as x^4y^2 are also perfect squares because they can be written as the square of a quantity:

$$x^4y^2 = (x^2y)^2$$

EXAMPLE 1   Factor $4y^4 - 25z^2$.

Solution   Because the binomial $4y^4 - 25z^2$ can be written in the form $(2y^2)^2 - (5z)^2$, it represents the difference of the squares of $2y^2$ and $5z$. Thus, it factors into the sum of these two quantities times the difference of these two quantities.

$$4y^4 - 25z^2 = (2y^2)^2 - (5z)^2$$
$$= (2y^2 + 5z)(2y^2 - 5z)$$

Check this result by multiplication.   ■

We can often factor out a greatest common factor before factoring the difference of two squares. For example, to factor $8x^2 - 32$, we begin by factoring out the greatest common factor of 8 and then factor the resulting difference of two squares.

$$8x^2 - 32 = 8(x^2 - 4) \qquad \text{Factor out 8.}$$
$$= 8(x^2 - 2^2) \qquad \text{Write 4 as } 2^2.$$
$$= 8(x + 2)(x - 2) \qquad \text{Factor the difference of two squares.}$$

We can verify this result by multiplication:

$$8(x + 2)(x - 2) = 8(x^2 - 4)$$
$$= 8x^2 - 32$$

EXAMPLE 2   Factor $2a^2x^3y - 8b^2xy$.

Solution
$$2a^2x^3y - 8b^2xy = 2xy(a^2x^2 - 4b^2) \qquad \text{Factor out } 2xy.$$
$$= 2xy[(ax)^2 - (2b)^2]$$
$$= 2xy(ax + 2b)(ax - 2b) \qquad \begin{array}{l}\text{Factor the difference}\\ \text{of two squares.}\end{array}$$

Check this result by multiplication. ■

Sometimes we must factor a difference of two squares more than once to completely factor a polynomial. For example, the binomial $625a^4 - 81b^4$ can be written in the form $(25a^2)^2 - (9b^2)^2$. The difference of the squares of $25a^2$ and $9b^2$ factors as

$$625a^4 - 81b^4 = (25a^2)^2 - (9b^2)^2$$
$$= (25a^2 + 9b^2)(25a^2 - 9b^2)$$

The factor $25a^2 - 9b^2$, however, can be written in the form $(5a)^2 - (3b)^2$ and can be factored as $(5a + 3b)(5a - 3b)$. Thus, the complete factorization of $625a^4 - 81b^4$ is

$$625a^4 - 81b^4 = (25a^2 + 9b^2)(5a + 3b)(5a - 3b)$$

WARNING!   The binomial $25a^2 + 9b^2$ is called the **sum of two squares** because it can be written in the form $(5a)^2 + (3b)^2$. Such binomials cannot be factored if we are limited to integer coefficients. Polynomials that do not factor over the integers are called **irreducible** or **prime polynomials.**

EXAMPLE 3   Factor $2x^4y - 32y$.

Solution
$$2x^4y - 32y = 2y \cdot x^4 - 2y \cdot 16$$
$$= 2y(x^4 - 16) \qquad \text{Factor out } 2y.$$
$$= 2y(x^2 + 4)(x^2 - 4) \qquad \text{Factor } x^4 - 16.$$
$$= 2y(x^2 + 4)(x + 2)(x - 2) \qquad \begin{array}{l}\text{Factor } x^2 - 4. \text{ Note that } x^2 + 4\\ \text{does not factor.}\end{array} \quad ■$$

Example 4 requires the techniques of factoring out a common factor, factoring by grouping, and factoring the difference of two squares.

EXAMPLE 4   Factor $2x^3 - 8x + 2yx^2 - 8y$.

Solution
$$
\begin{aligned}
2x^3 - 8x + 2yx^2 - 8y &= 2(x^3 - 4x + yx^2 - 4y) \\
&= 2[x(x^2 - 4) + y(x^2 - 4)] \\
&= 2[(x^2 - 4)(x + y)] \\
&= 2(x + 2)(x - 2)(x + y)
\end{aligned}
$$

Factor out 2.

Factor out x from $x^3 - 4x$ and factor out y from $yx^2 - 4y$.

Factor out $x^2 - 4$.

Factor $x^2 - 4$.

Check by multiplication.   ∎

 WARNING!   To *factor* an expression means to factor the expression *completely*.

Solving Equations

We can use factoring the difference of two squares to solve many quadratic equations.

EXAMPLE 5   Solve the equation $4x^2 = 36$.

Solution   Before we can use the zero-factor property, we must subtract 36 from both sides to make the right-hand side 0.

$$
\begin{aligned}
4x^2 &= 36 \\
4x^2 - 36 &= 0 \\
x^2 - 9 &= 0 \\
(x + 3)(x - 3) &= 0
\end{aligned}
$$

Subtract 36 from both sides.

Divide both sides by 4.

Factor $x^2 - 9$.

$$
\begin{array}{ccc}
x + 3 = 0 & \text{or} & x - 3 = 0 \\
x = -3 & | & x = 3
\end{array}
$$

Set each factor equal to 0.

Solve each linear equation.

Check each possible solution.

For $x = -3$	For $x = 3$
$4x^2 = 36$	$4x^2 = 36$
$4(-3)^2 \overset{?}{=} 36$	$4(3)^2 \overset{?}{=} 36$
$4(9) \overset{?}{=} 36$	$4(9) \overset{?}{=} 36$
$36 = 36$	$36 = 36$

Both solutions check.   ∎

ORALS *Factor each binomial.*

1. $x^2 - 9$

2. $y^2 - 36$

3. $z^2 - 4$

4. $p^2 - q^2$

5. $25 - t^2$

6. $36 - r^2$

7. $100 - y^2$

8. $100 - y^4$

E X E R C I S E 4.3

In Exercises 1–20, factor each expression, if possible.

1. $x^2 - 16$

2. $x^2 - 25$

3. $y^2 - 49$

4. $y^2 - 81$

5. $4y^2 - 49$

6. $9z^2 - 4$

7. $9x^2 - y^2$

8. $4x^2 - z^2$

9. $25t^2 - 36u^2$

10. $49u^2 - 64v^2$

11. $16a^2 - 25b^2$

12. $36a^2 - 121b^2$

13. $a^2 + b^2$

14. $121a^2 - 144b^2$

15. $a^4 - 4b^2$

16. $9y^2 + 16z^2$

17. $49y^2 - 225z^4$

18. $25x^2 + 36y^2$

19. $196x^4 - 169y^2$

20. $144a^4 + 169b^4$

In Exercises 21–36, factor each expression.

21. $8x^2 - 32y^2$

22. $2a^2 - 200b^2$

23. $2a^2 - 8y^2$

24. $32x^2 - 8y^2$

25. $3r^2 - 12s^2$

26. $45u^2 - 20v^2$

27. $x^3 - xy^2$

28. $a^2b - b^3$

29. $4a^2x - 9b^2x$

30. $4b^2y - 16c^2y$

31. $3m^3 - 3mn^2$

32. $2p^2q - 2q^3$

33. $4x^4 - x^2y^2$

34. $9xy^2 - 4xy^4$

35. $2a^3b - 242ab^3$

36. $50c^4d^2 - 8c^2d^4$

In Exercises 37–48, factor each expression.

37. $x^4 - 81$

38. $y^4 - 625$

39. $a^4 - 16$

40. $b^4 - 256$

41. $a^4 - b^4$

42. $m^4 - 16n^4$

43. $81r^4 - 256s^4$

44. $x^8 - y^4$

45. $a^4 - b^8$

46. $16y^8 - 81z^4$

47. $x^8 - y^8$

48. $x^8y^8 - 1$

In Exercises 49–68, factor each expression.

49. $2x^4 - 2y^4$

50. $a^5 - ab^4$

51. $a^4b - b^5$

52. $m^5 - 16mn^4$

53. $48m^4n - 243n^5$

54. $2x^4y - 512y^5$

55. $3a^5y + 6ay^5$

56. $2p^{10}q - 32p^2q^5$

57. $3a^{10} - 3a^2b^4$

58. $2x^9y + 2xy^9$

59. $2x^8y^2 - 32y^6$

60. $3a^8 - 243a^4b^8$

61. $a^6b^2 - a^2b^6c^4$

62. $a^2b^3c^4 - a^2b^3d^4$

63. $a^2b^7 - 625a^2b^3$

64. $16x^3y^4z - 81x^3y^4z^5$

65. $243r^5s - 48rs^5$

66. $1024m^5n - 324mn^5$

67. $16(x - y)^2 - 9$

68. $9(x + 1)^2 - y^2$

In Exercises 69–78, factor each expression.

69. $a^3 - 9a + 3a^2 - 27$

70. $b^3 - 25b - 2b^2 + 50$

71. $y^3 - 16y - 3y^2 + 48$

72. $a^3 - 49a + 2a^2 - 98$

73. $3x^3 - 12x + 3x^2 - 12$

74. $2x^3 - 18x - 6x^2 + 54$

75. $3m^3 - 3mn^2 + 3am^2 - 3an^2$

76. $ax^3 - axy^2 - bx^3 + bxy^2$

77. $2m^3n^2 - 32mn^2 + 8m^2 - 128$

78. $2x^3y + 4x^2y - 98xy - 196y$

In Exercises 79–90, solve each equation.

79. $x^2 - 25 = 0$

80. $x^2 - 36 = 0$

81. $y^2 - 49 = 0$

82. $z^2 - 121 = 0$

83. $4x^2 - 1 = 0$

84. $9y^2 - 1 = 0$

85. $9y^2 - 4 = 0$

86. $16z^2 - 25 = 0$

87. $x^2 = 49$

88. $z^2 = 25$

89. $4x^2 = 81$

90. $9y^2 = 64$

Writing Exercises ■ *Write a paragraph using your own words.*

1. Explain how to factor the difference of two squares.

2. Explain why $x^4 - y^4$ is not completely factored as $(x^2 + y^2)(x^2 - y^2)$.

Something to Think About ■

1. It is easy to multiply 399 by 401 without a calculator: The product is $400^2 - 1$, or 159,999. Explain.

2. Use the method in the previous exercise to find $498 \cdot 502$ without a calculator.

Review Exercises ■

1. In the study of the flow of fluids, Bernoulli's law is given by the equation

$$\frac{p}{w} + \frac{v^2}{2g} + h = k$$

Solve the equation for p.

2. Solve Bernoulli's law for h. (See Review Exercise 1.)

4.4 Factoring Trinomials with Lead Coefficients of 1

■ Prime Polynomials ■ Solving Equations

GETTING READY   *Multiply the binomials.*

1. $(x + 6)(x + 6)$ **2.** $(y - 7)(y - 7)$
3. $(a - 3)(a - 3)$ **4.** $(x + 4)(x + 5)$
5. $(r - 2)(r - 5)$ **6.** $(m + 3)(m - 7)$
7. $(a - 3b)(a + 4b)$ **8.** $(u - 3v)(u - 5v)$
9. $(x + 4y)(x - 6y)$ **10.** $(2a + b)(a - b)$

The product of two binomials is often a trinomial. For example,

$$(x + 3)(x + 3) = x^2 + 6x + 9$$
$$(x - 4y)(x - 4y) = x^2 - 8xy + 16y^2$$

and

$$(3x - 4)(2x + 3) = 6x^2 + x - 12$$

Because the product of two binomials can be a trinomial, we are not surprised that many trinomials factor as the product of two binomials.

Many trinomials can be factored by using the following two special product formulas, first discussed in Section 3.6.

Special Product Formulas

$$(x + y)(x + y) = x^2 + 2xy + y^2$$
$$(x - y)(x - y) = x^2 - 2xy + y^2$$

The trinomials $x^2 + 2xy + y^2$ and $x^2 - 2xy + y^2$ are called **perfect square trinomials,** because each one can be written as the square of a binomial.

1. $x^2 + 2xy + y^2 = (x + y)(x + y)$ **2.** $x^2 - 2xy + y^2 = (x - y)(x - y)$
$\qquad\qquad\quad = (x + y)^2$ $\qquad\qquad\quad = (x - y)^2$

To factor a perfect square trinomial such as $x^2 + 8x + 16$, we note that the trinomial can be written in the form

$$x^2 + 2(x)(4) + 4^2$$

and that if $y = 4$, this form matches the left-hand side of Equation 1. Thus, $x^2 + 8x + 16$ factors as

$$x^2 + 8x + 16 = x^2 + 2(x)(4) + 4^2$$
$$= (x + 4)(x + 4)$$

This result can be verified by multiplication by using the FOIL method.

$$(x + 4)(x + 4) = x^2 + 4x + 4x + 16$$
$$= x^2 + 8x + 16 \qquad \text{Combine like terms.}$$

Likewise, the perfect square trinomial $a^2 - 4ab + 4b^2$ can be written in the form

$$a^2 - 2(a)(2b) + (2b)^2$$

If $x = a$ and $y = 2b$, this form matches the left-hand side of Equation 2. Thus, $a^2 - 4ab + 4b^2$ factors as

$$a^2 - 4ab + 4b^2 = a^2 - 2(a)(2b) + (2b)^2$$
$$= (a - 2b)(a - 2b)$$

This result can also be verified by multiplication.

Because the trinomial $x^2 + 5x + 6$ is not a perfect square trinomial, it cannot be factored by using a special product formula. However, it can be factored into the product of two binomials. To find those binomial factors, we note that the product of their first terms must be x^2. Thus, the first term of each binomial must be x.

$$\overset{x^2}{(x \qquad)(x \qquad)}$$

Because the product of their last terms must be 6, and the sum of the products of the outer and inner terms must be $5x$, we must find two numbers whose product is 6 and whose sum is 5.

$$\overset{6}{(x + ?)(x + ?)}$$
$$\underset{O + I = 5x}{}$$

Two such numbers are $+3$ and $+2$. Thus, we have

3. $\quad x^2 + 5x + 6 = (x + 3)(x + 2)$

We can verify this factorization by multiplying $x + 3$ and $x + 2$ and observing that the product is $x^2 + 5x + 6$.

$$(x + 3)(x + 2) = x^2 + 2x + 3x + 6$$
$$= x^2 + 5x + 6 \qquad \text{Combine like terms.}$$

Because of the commutative property of multiplication, the order of the factors listed in Equation 3 is not important. Equation 3 can be written as

$$x^2 + 5x + 6 = (x + 2)(x + 3)$$

Carl Friedrich Gauss (1777–1855)
Many people consider Gauss to be the greatest mathematician of all time. He made contributions in the areas of number theory, solutions of equations, geometry of curved surfaces, and statistics. For his efforts, he has earned the title "Prince of the Mathematicians."

EXAMPLE 1   Factor $y^2 - 7y + 12$.

Solution   If this trinomial is to be the product of two binomials, the product of their first terms must be y^2. Thus, the first term of each binomial must be y.

$$\overset{y^2}{(y \quad)(y \quad)}$$

Because the product of the last terms must be $+12$, and the sum of the products of the outer and inner terms must be $-7y$, we must find two negative numbers whose product is $+12$ and whose sum is -7.

$$\overset{12}{(y - ?)(y - ?)}$$
$$\text{O} + \text{I} = -7y$$

Two such numbers are -4 and -3. Thus,

$$y^2 - 7y + 12 = (y - 4)(y - 3)$$

We can check by verifying that the product of $y - 4$ and $y - 3$ is $y^2 - 7y + 12$.

$$(y - 4)(y - 3) = y^2 - 3y - 4y + 12$$
$$= y^2 - 7y + 12 \qquad \blacksquare$$

EXAMPLE 2   Factor $a^2 + 2a - 15$.

Solution   Because the first term of this trinomial is a^2, the first term of each binomial factor must be a.

$$\overset{a^2}{(a \quad)(a \quad)}$$

Because the product of the last terms must be -15, and the sum of the products of the outer terms and inner terms must be $+2a$, we must find two numbers whose product is -15 and whose sum is $+2$.

$$\overset{-15}{(a \quad ?)(a \quad ?)}$$
$$\text{O} + \text{I} = 2a$$

Two such numbers are $+5$ and -3. Thus,

$$a^2 + 2a - 15 = (a + 5)(a - 3)$$

We can check by verifying that the product of $a + 5$ and $a - 3$ is $a^2 + 2a - 15$.

$$(a + 5)(a - 3) = a^2 - 3a + 5a - 15$$
$$= a^2 + 2a - 15 \qquad \blacksquare$$

EXAMPLE 3   Factor $z^2 - 4z - 21$.

Solution   Because the first term of this trinomial is z^2, the first term of each binomial factor must be z.

$$z^2$$
$$(z\quad)(z\quad)$$

Because the product of the last terms must be -21, and the sum of the products of the outer terms and inner terms must be $-4z$, we must find two numbers whose product is -21 and whose sum is -4.

$$-21$$
$$(z\quad?)(z\quad?)$$
$$O + I = -4z$$

Two such numbers are -7 and $+3$. Thus,

$$z^2 - 4z - 21 = (z - 7)(z + 3)$$

We can check by verifying that the product of $z - 7$ and $z + 3$ is $z^2 - 4z - 21$.

$$(z - 7)(z + 3) = z^2 + 3z - 7z - 21$$
$$= z^2 - 4z - 21$$ ∎

EXAMPLE 4   Factor $x^2 + xy - 6y^2$.

Solution   Because the first term of this trinomial is x^2, the first term of each binomial factor must be x.

$$x^2$$
$$(x\quad)(x\quad)$$

Because the product of the last terms must be $-6y^2$, and the sum of the products of the outer terms and inner terms must be xy, we must find two numbers whose product is $-6y^2$ that will give a middle term of xy.

$$-6y^2$$
$$(x\quad?)(x\quad?)$$
$$O + I = xy$$

Two such numbers are $3y$ and $-2y$. Thus,

$$x^2 + xy - 6y^2 = (x + 3y)(x - 2y)$$

We can check by verifying that the product of $x + 3y$ and $x - 2y$ is $x^2 + xy - 6y^2$.

$$(x + 3y)(x - 2y) = x^2 - 2xy + 3xy - 6y^2$$
$$= x^2 + xy - 6y^2$$ ∎

When the coefficient of the first term of a trinomial is -1, we begin by factoring out -1.

EXAMPLE 5   Factor $-x^2 + 11x - 18$.

Solution
$$-x^2 + 11x - 18 = -(x^2 - 11x + 18) \qquad \text{Factor out } -1.$$
$$= -(x - 9)(x - 2) \qquad \text{Factor } x^2 - 11x + 18.$$

We can check by verifying that the product of -1, $x - 9$, and $x - 2$ is $-x^2 + 11x - 18$.

$$-(x - 9)(x - 2) = -(x^2 - 2x - 9x + 18)$$
$$= -(x^2 - 11x + 18)$$
$$= -x^2 + 11x - 18$$
■

EXAMPLE 6   Factor $-x^2 + 2x + 15$.

Solution
$$-x^2 + 2x + 15 = -(x^2 - 2x - 15) \qquad \text{Factor out } -1.$$
$$= -(x - 5)(x + 3) \qquad \text{Factor } x^2 - 2x - 15.$$

We can check the work by verifying that the product of -1, $x - 5$, and $x + 3$ is $-x^2 + 2x + 15$.

$$-(x - 5)(x + 3) = -(x^2 + 3x - 5x - 15)$$
$$= -(x^2 - 2x - 15)$$
$$= -x^2 + 2x + 15$$
■

■ Prime Polynomials

Not all trinomials are factorable. To attempt to factor the trinomial $x^2 + 2x + 3$, for example, we would begin by noting that the product of the first terms of the binomial factors is x^2. Thus, the first term of each binomial must be x.

$$\overset{\displaystyle x^2}{(x\quad)(x\quad)}$$

Because the last term of the trinomial is 3 and the middle term is $2x$, we must find two factors of 3 whose sum is 2 so that

$$\overset{\displaystyle 3}{(x\quad?)(x\quad?)}$$
$$\underset{\text{O} + \text{I} = 2x}{}$$

Because 3 factors only as $(1)(3)$ and $(-1)(-3)$, it has no factors whose sum is 2. Thus, $x^2 + 2x + 3$ cannot be factored over the integers. It is a prime polynomial.

EXAMPLE 7 Factor $-3ax^2 + 9a - 6ax$.

Solution We write the trinomial in descending powers of x and factor out the common factor of $-3a$.

$$-3ax^2 + 9a - 6ax = -3ax^2 - 6ax + 9a$$
$$= -3a(x^2 + 2x - 3)$$

Finally, we factor the trinomial $x^2 + 2x - 3$.

$$-3ax^2 + 9a - 6ax = -3a(x + 3)(x - 1)$$

We can check the result by multiplying.

$$-3a(x + 3)(x - 1) = -3a(x^2 + 2x - 3)$$
$$= -3ax^2 - 6ax + 9a$$
$$= -3ax^2 + 9a - 6ax \qquad \blacksquare$$

The next example requires factoring a trinomial and factoring a difference of two squares.

EXAMPLE 8 Factor $m^2 - 2mn + n^2 - 64a^2$.

Solution We group the first three terms together and factor the resulting trinomial:

$$m^2 - 2mn + n^2 - 64a^2 = (m - n)(m - n) - 64a^2$$
$$= (m - n)^2 - (8a)^2$$

Then we factor the resulting difference of two squares:

$$m^2 - 2mn + n^2 - 64a^2 = (m - n)^2 - (8a)^2$$
$$= (m - n + 8a)(m - n - 8a) \qquad \blacksquare$$

■ Solving Equations

We can use the factoring of trinomials and the zero factor property to solve more equations.

EXAMPLE 9 Solve $x^3 - 2x^2 - 63x = 0$.

Solution
$$x^3 - 2x^2 - 63x = 0$$
$$x(x^2 - 2x - 63) = 0 \qquad \text{Factor out the common factor of } x.$$
$$x(x + 7)(x - 9) = 0 \qquad \text{Factor the trinomial.}$$

$x = 0$	$x + 7 = 0$	$x - 9 = 0$	Set each factor equal to 0.
	$x = -7$	$x = 9$	Solve each linear equation.

The equation has three solutions: 0, -7, and 9. Check each of them. $\qquad \blacksquare$

ORALS *Finish each factoring problem.*

1. $x^2 + 5x + 4 = (x + 1)(x + \quad)$ 2. $x^2 - 5x + 6 = (x \quad 2)(x \quad 3)$
3. $x^2 + x - 6 = (x \quad 2)(x + \quad)$ 4. $x^2 - x - 6 = (x \quad 3)(x + \quad)$
5. $x^2 + 5x - 6 = (x + \quad)(x - \quad)$ 6. $x^2 - 7x + 6 = (x - \quad)(x - \quad)$

E X E R C I S E   4.4

In Exercises 1–12, factor each perfect square trinomial.

1. $x^2 + 6x + 9$ 2. $x^2 + 10x + 25$ 3. $y^2 - 8y + 16$ 4. $z^2 - 2z + 1$

5. $t^2 + 20t + 100$ 6. $r^2 + 24r + 144$ 7. $u^2 - 18u + 81$ 8. $v^2 - 14v + 49$

9. $x^2 + 4xy + 4y^2$ 10. $a^2 + 6ab + 9b^2$ 11. $r^2 - 10rs + 25s^2$ 12. $m^2 - 12mn + 36n^2$

In Exercises 13–40, factor each trinomial, if possible. Use the FOIL method to check each result.

13. $x^2 + 3x + 2$ 14. $y^2 + 4y + 3$ 15. $a^2 - 4a - 5$ 16. $b^2 + 6b - 7$

17. $z^2 + 12z + 11$ 18. $x^2 + 7x + 10$ 19. $t^2 - 9t + 14$ 20. $c^2 - 9c + 8$

21. $u^2 + 10u + 15$ 22. $v^2 + 9v + 15$ 23. $y^2 - y - 30$ 24. $x^2 - 3x - 40$

25. $a^2 + 6a - 16$ 26. $x^2 + 5x - 24$ 27. $t^2 - 5t - 50$ 28. $a^2 - 10a - 39$

29. $r^2 - 9r - 12$ 30. $s^2 + 11s - 26$ 31. $y^2 + 2yz + z^2$ 32. $r^2 - 2rs + 4s^2$

33. $x^2 + 4xy + 4y^2$ 34. $a^2 + 10ab + 9b^2$ 35. $m^2 + 3mn - 10n^2$ 36. $m^2 - mn - 12n^2$

37. $a^2 - 4ab - 12b^2$ 38. $p^2 + pq - 6q^2$ 39. $u^2 + 2uv - 15v^2$ 40. $m^2 + 3mn - 10n^2$

In Exercises 41–52, factor each trinomial. Factor out −1 first.

41. $-x^2 - 7x - 10$ 42. $-x^2 + 9x - 20$ 43. $-y^2 - 2y + 15$ 44. $-y^2 - 3y + 18$

45. $-t^2 - 15t + 34$ 46. $-t^2 - t + 30$ 47. $-r^2 + 14r - 40$ 48. $-r^2 + 14r - 45$

49. $-a^2 - 4ab - 3b^2$ 50. $-a^2 - 6ab - 5b^2$ 51. $-x^2 + 6xy + 7y^2$ 52. $-x^2 - 10xy + 11y^2$

In Exercises 53–64, write each trinomial in descending powers of one variable and then factor.

53. $4 - 5x + x^2$

54. $y^2 + 5 + 6y$

55. $10y + 9 + y^2$

56. $x^2 - 13 - 12x$

57. $c^2 - 5 + 4c$

58. $b^2 - 6 - 5b$

59. $-r^2 + 2s^2 + rs$

60. $u^2 - 3v^2 + 2uv$

61. $4rx + r^2 + 3x^2$

62. $-a^2 + 5b^2 + 4ab$

63. $-3ab + a^2 + 2b^2$

64. $-13yz + y^2 - 14z^2$

In Exercises 65–76, completely factor each trinomial. Factor out any common monomials first (including -1 if necessary).

65. $2x^2 + 10x + 12$

66. $3y^2 - 21y + 18$

67. $3y^3 + 6y^2 + 3y$

68. $4x^4 + 16x^3 + 16x^2$

69. $-5a^2 + 25a - 30$

70. $-2b^2 + 20b - 18$

71. $3z^2 - 15tz + 12t^2$

72. $5m^2 + 45mn - 50n^2$

73. $12xy + 4x^2y - 72y$

74. $48xy + 6xy^2 + 96x$

75. $-4x^2y - 4x^3 + 24xy^2$

76. $3x^2y^3 + 3x^3y^2 - 6xy^4$

In Exercises 77–84, completely factor each expression.

77. $ax^2 + 4ax + 4a + bx + 2b$

78. $mx^2 + mx - 6m + nx - 2n$

79. $a^2 + 8a + 15 + ab + 5b$

80. $x^2 + 2xy + y^2 + 2x + 2y$

81. $a^2 + 2ab + b^2 - 4$

82. $a^2 + 6a + 9 - b^2$

83. $b^2 - y^2 - 4y - 4$

84. $c^2 - a^2 + 8a - 16$

In Exercises 85–102, solve each equation.

85. $x^2 - 13x + 12 = 0$

86. $x^2 + 7x + 6 = 0$

87. $x^2 - 2x - 15 = 0$

88. $x^2 - x - 20 = 0$

89. $-4x - 21 + x^2 = 0$

90. $2x + x^2 - 15 = 0$

91. $x^2 + 8 - 9x = 0$

92. $45 + x^2 - 14x = 0$

93. $a^2 + 8a = -15$

94. $a^2 - a = 56$

95. $2y - 8 = -y^2$

96. $-3y + 18 = y^2$

97. $x^3 + 3x^2 + 2x = 0$

98. $x^3 - 7x^2 + 10x = 0$

99. $x^3 - 27x - 6x^2 = 0$

100. $x^3 - 22x - 9x^2 = 0$

101. $(x - 1)(x^2 + 5x + 6) = 0$

102. $(x - 2)(x^2 - 8x + 7) = 0$

Writing Exercises ■ *Write a paragraph using your own words.*

1. Explain how you would write a trinomial in descending order.

2. Explain how to use the FOIL method to check the factoring of a trinomial.

Something to Think About ■

1. Two students factor $2x^2 + 20x + 42$ and get two different answers: $(2x + 6)(x + 7)$, and $(x + 3)(2x + 14)$. Do both answers check? Why don't they agree? Is either one completely correct?

2. Find the error:

$$x = y$$
$$x^2 = xy \qquad \text{Multiply both sides by } x.$$
$$x^2 - y^2 = xy - y^2 \qquad \text{Subtract } y^2 \text{ from both sides.}$$
$$(x + y)(x - y) = y(x - y) \qquad \text{Factor.}$$
$$x + y = y \qquad \text{Divide both sides by } (x - y).$$
$$y + y = y \qquad \text{Substitute } y \text{ for its equal, } x.$$
$$2y = y \qquad \text{Combine terms.}$$
$$2 = 1 \qquad \text{Divide both sides by } y.$$

Review Exercises ■ *Graph the solution of each inequality on a number line.*

1. $x - 3 > 5$ **2.** $x + 4 \leq 3$ **3.** $-3x - 5 \geq 4$ **4.** $2x - 3 < 7$

5. $\dfrac{3(x - 1)}{4} < 12$ **6.** $\dfrac{-2(x + 3)}{3} \geq 9$ **7.** $-2 < x \leq 4$ **8.** $-5 \leq x + 1 < 0$

4.5 Factoring General Trinomials

■ The Key Number Method ■ Solving Equations

GETTING READY *Multiply and combine terms.*

1. $(2x + 1)(3x + 2)$ **2.** $(3y - 2)(2y - 5)$

3. $(4t - 3)(2t + 3)$ **4.** $(2r + 5)(2r - 3)$

5. $(2m - 3)(3m - 2)$ **6.** $(4a + 3)(4a + 1)$

We must consider more combinations when we factor trinomials with lead coefficients other than 1. To factor $2x^2 - 7x + 3$, for example, we must find binomials of the form $ax + b$ and $cx + d$ such that

$$2x^2 - 7x + 3 = (ax + b)(cx + d)$$

Because the first term of the trinomial $2x^2 - 7x + 3$ is $2x^2$, the first terms of the binomial factors must be $2x$ and x.

$$2x^2$$
$$(2x \quad)(x \quad)$$

Because the product of the last terms is 3, and the sum of the products of the outer terms and inner terms is $-7x$, we must find the two numbers with a product of 3 that will give a middle term of $-7x$.

$$+3$$
$$(2x \quad ?)(x \quad ?)$$
$$O + I = -7x$$

Because both $(3)(1)$ and $(-3)(-1)$ give a product of 3, there are four possible combinations to consider:

$$(2x + 3)(x + 1) \qquad (2x + 1)(x + 3)$$
$$(2x - 3)(x - 1) \qquad (2x - 1)(x - 3)$$

Of these possibilities, only the last one gives the required middle term of $-7x$. Thus,

$$2x^2 - 7x + 3 = (2x - 1)(x - 3)$$

We can check this result by multiplication.

$$(2x - 1)(x - 3) = 2x^2 - 6x - x + 3$$
$$= 2x^2 - 7x + 3$$

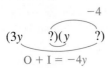

EXAMPLE 1 Factor $3y^2 - 4y - 4$.

Solution Because the first term of this trinomial is $3y^2$, the first terms of the binomial factors must be $3y$ and y.

$$3y^2$$
$$(3y \quad)(y \quad)$$

The product of the last terms must be -4, and the sum of the products of the outer terms and inner terms must be $-4y$.

$$-4$$
$$(3y \quad ?)(y \quad ?)$$
$$O + I = -4y$$

Because $(1)(-4)$, $(-1)(4)$, and $(-2)(2)$ all give a product of -4, there are six possible combinations to consider:

$$(3y + 1)(y - 4) \qquad (3y - 4)(y + 1)$$
$$(3y - 1)(y + 4) \qquad (3y + 4)(y - 1)$$
$$(3y - 2)(y + 2) \qquad (3y + 2)(y - 2)$$

Again, only the last possibility gives the required middle term of $-4y$. Thus,

$$3y^2 - 4y - 4 = (3y + 2)(y - 2)$$

We can check by multiplication.

$$(3y + 2)(y - 2) = 3y^2 - 6y + 2y - 4$$
$$= 3y^2 - 4y - 4$$

EXAMPLE 2   Factor $2x^2 + 7xy + 6y^2$.

Solution   The first terms of the two binomial factors must be $2x$ and x.

$$2x^2$$
$$(2x \quad)(x \quad)$$

The product of the last terms must be $+6y^2$, and the sum of the products of the outer terms and inner terms must be $+7xy$.

$$6y^2$$
$$(2x \quad ?)(x \quad ?)$$
$$O + I = 7xy$$

The products $(6y)(y)$, $(3y)(2y)$, $(-6y)(-y)$, and $(-3y)(-2y)$ all give a last term of $6y^2$. However, only the products $(6y)(y)$ and $(3y)(2y)$ can lead to a middle term that is preceded by a $+$ sign. Thus, there are only four possibilities to consider:

$$(2x + 6y)(x + y) \qquad (2x + y)(x + 6y)$$
$$(2x + 3y)(x + 2y) \qquad (2x + 2y)(x + 3y)$$

Of these possibilities, only $(2x + 3y)(x + 2y)$ gives the correct middle term of $7xy$. Thus,

$$2x^2 + 7xy + 6y^2 = (2x + 3y)(x + 2y)$$

We can check by multiplication.

$$(2x + 3y)(x + 2y) = 2x^2 + 4xy + 3xy + 6y^2$$
$$= 2x^2 + 7xy + 6y^2$$

EXAMPLE 3   Factor $6b^2 + 7b - 20$.

Solution   This time there are many possible combinations for the first terms of the binomial factors. They are

$$(b \quad)(6b \quad) \qquad (6b \quad)(b \quad)$$
$$(3b \quad)(2b \quad) \qquad (2b \quad)(3b \quad)$$

There are also many combinations for the last terms of the binomial factors. We must try to find one that will (in combination with our choice of first terms) give a last term of -20 and a sum of the products of the outer terms and inner terms of $+7b$.

We begin, for example, by picking factors of b and $6b$ for the first terms and $+4$ and -5 for the last terms. The possible factorization

$$\underbrace{(b + 4)(6b - 5)}_{O + I = 19b}$$

gives a middle term of $19b$, so it is incorrect.

We then try, for example, factors of $3b$ and $2b$ for the first terms and $+4$ and -5 for the last terms. The possible factorization

$$\underbrace{(3b + 4)(2b - 5)}_{O + I = -7b}$$

gives a middle term of $-7b$, so it is incorrect.

The possible factorization

$$(3b - 4)(2b + 5)$$

does give a middle term of $+7b$ and a last term of -20, so it is correct. Thus, we have

$$6b^2 + 7b - 20 = (3b - 4)(2b + 5)$$

We can check by multiplication.

$$(3b - 4)(2b + 5) = 6b^2 + 15b - 8b - 20$$
$$= 6b^2 + 7b - 20 \qquad \blacksquare$$

EXAMPLE 4   Factor $4x^2 + 4xy - 3y^2$.

Solution   Again, there are many combinations for the first terms of the binomial factors:

$$(4x \quad)(x \quad) \qquad (x \quad)(4x \quad) \qquad (2x \quad)(2x \quad)$$

We must try to find last terms that will give a third term of $-3y^2$ and a middle term of $+4xy$.

We begin by trying, for example, factors of $4x$ and x for the first terms and factors of $3y$ and $-y$ for the last terms. The possible factorization

$$(4x + 3y)(x - y)$$
$$\underbrace{}_{O + I = -xy}$$

gives a middle term of $-xy$, so it is incorrect.

We then try, for example, factors of $2x$ and $2x$ for the first terms and factors of $3y$ and $-y$ for the last terms. The possible factorization

$$(2x + 3y)(2x - y)$$
$$\underbrace{}_{O + I = +4xy}$$

gives a middle term of $+4xy$ and a last term of $-3y^2$, so it is correct. Thus,

$$4x^2 + 4xy - 3y^2 = (2x + 3y)(2x - y)$$

We can check by multiplication.

$$(2x + 3y)(2x - y) = 4x^2 - 2xy + 6xy - 3y^2$$
$$= 4x^2 + 4xy - 3y^2 \qquad \blacksquare$$

EXAMPLE 5 Factor $-8x^3 + 22x^2 - 12x$.

Solution We factor out the common monomial factor of $-2x$.

$$-8x^3 + 22x^2 - 12x = -2x(4x^2 - 11x + 6)$$

Then we find the binomial factors of $4x^2 - 11x + 6$.

$$-8x^3 + 22x^2 - 12x = -2x(x - 2)(4x - 3)$$

We can check by multiplication.

$$-2x(x - 2)(4x - 3) = -2x(4x^2 - 3x - 8x + 6)$$
$$= -2x(4x^2 - 11x + 6)$$
$$= -8x^3 + 22x^2 - 12x \qquad \blacksquare$$

It is not easy to give specific rules for factoring trinomials, because some guess-work is often necessary. However, the following hints are often helpful.

Factoring a General Trinomial

1. Write the trinomial in descending powers of one variable.
2. Factor out any greatest common factor (including -1 if that is necessary to make the coefficient of the first term positive).
3. If the sign of the third term is plus $(+)$, the signs between the terms of the binomial factors are the same as the sign of the middle term of the trinomial. If the sign of the third term is minus $(-)$, the signs between the terms of the binomial factors are opposite.
4. Mentally, try various combinations of first terms and last terms until you find one that works, or until you exhaust all the possibilities. In that case, the trinomial does not factor using only integer coefficients.
5. Check the factorization by multiplication.

EXAMPLE 6 Factor $2x^2y - 8x^3 + 3xy^2$.

Solution *Step 1:* We rewrite the trinomial in descending powers of x.

$$-8x^3 + 2x^2y + 3xy^2$$

Step 2: We factor out $-x$.

$$-8x^3 + 2x^2y + 3xy^2 = -x(8x^2 - 2xy - 3y^2)$$

Step 3: Because the sign of the third term of the trinomial factor is minus $(-)$, the signs within its binomial factors must be different. Thus, the sign between the terms in one binomial must be plus $(+)$, and the sign between the terms of the other binomial must be minus $(-)$.

Step 4: We find the binomial factors of the trinomial.

$$-8x^3 + 2x^2y + 3xy^2 = -x(8x^2 - 2xy - 3y^2)$$
$$= -x(2x + y)(4x - 3y)$$

Step 5: We can check by multiplication.

$$-x(2x + y)(4x - 3y) = -x(8x^2 - 6xy + 4xy - 3y^2)$$
$$= -x(8x^2 - 2xy - 3y^2)$$
$$= -8x^3 + 2x^2y + 3xy^2$$
$$= 2x^2y - 8x^3 + 3xy^2 \qquad \blacksquare$$

The next example combines the techniques of factoring by grouping, factoring the difference of two squares, and factoring trinomials.

EXAMPLE 7   Factor $4x^2 - 4xy + y^2 - 9$.

Solution $4x^2 - 4xy + y^2 - 9 = (2x - y)^2 - 9$ Factor the first three terms as a perfect square.

$\qquad\qquad\qquad = [(2x - y) + 3][(2x - y) - 3]$ Factor the difference of two squares.

$\qquad\qquad\qquad = (2x - y + 3)(2x - y - 3)$ Remove parentheses.

We can check by multiplication.   ■

EXAMPLE 8   Factor $9 - 4x^2 - 4xy - y^2$.

Solution $9 - 4x^2 - 4xy - y^2 = 9 - (4x^2 + 4xy + y^2)$ Factor -1 from the trinomial.

$\qquad\qquad\qquad = 9 - (2x + y)(2x + y)$ Factor the trinomial.

$\qquad\qquad\qquad = 9 - (2x + y)^2$ $(2x + y)(2x + y) = (2x + y)^2$.

$\qquad\qquad\qquad = [3 + (2x + y)][3 - (2x + y)]$ Factor the difference of two squares.

$\qquad\qquad\qquad = (3 + 2x + y)(3 - 2x - y)$ Remove parentheses.

We can check by multiplication.   ■

■ The Key Number Method

The method of factoring by grouping can be used to help factor trinomials of the form $ax^2 + bx + c$, where $a > 0$. For example, to factor the trinomial $4x^2 - 4x - 3$, where $a = 4$, $b = -4$, and $c = -3$, we proceed as follows:

1. Find the product of a and c. This is the **key number:** $ac = 4(-3) = -12$.
2. Find two factors of the key number whose sum is b. In this example, the key number is -12 and $b = -4$.

$$2(-6) = -12 \qquad \text{and} \qquad 2 + (-6) = -4$$

3. Use the factors 2 and -6 as coefficients of terms to be placed between $4x^2$ and -3:

$$4x^2 + 2x - 6x - 3$$

4. Factor by grouping:

$$4x^2 + 2x - 6x - 3 = 2x(2x + 1) - 3(2x + 1)$$
$$= (2x + 1)(2x - 3)$$

We can verify this result by multiplication.

■ **Solving Equations**

EXAMPLE 9 Solve the equation $2x^2 + 3x = 2$.

Solution We write the equation in the form $ax^2 + bx + c = 0$, and then solve for x as follows.

$$2x^2 + 3x = 2$$
$$2x^2 + 3x - 2 = 0 \qquad \text{Add } -2 \text{ to both sides.}$$
$$(2x - 1)(x + 2) = 0 \qquad \text{Factor } 2x^2 + 3x - 2.$$

$$2x - 1 = 0 \quad \text{or} \quad x + 2 = 0 \qquad \text{Set each factor equal to 0.}$$
$$2x = 1 \qquad\qquad\quad x = -2 \qquad \text{Solve each linear equation.}$$
$$x = \frac{1}{2}$$

Check each solution. ■

EXAMPLE 10 Solve the equation $6x^3 + 12x = 17x^2$.

Solution
$$6x^3 + 12x = 17x^2$$
$$6x^3 - 17x^2 + 12x = 0 \qquad \text{Add } -17x^2 \text{ to both sides.}$$
$$x(6x^2 - 17x + 12) = 0 \qquad \text{Factor out } x.$$
$$x(2x - 3)(3x - 4) = 0 \qquad \text{Factor } 6x^2 - 17x + 12.$$

$$x = 0 \quad \text{or} \quad 2x - 3 = 0 \quad \text{or} \quad 3x - 4 = 0 \qquad \text{Set each factor equal to 0.}$$
$$x = 0 \qquad\qquad 2x = 3 \qquad\qquad 3x = 4 \qquad \text{Solve the linear equations.}$$
$$x = \frac{3}{2} \qquad\qquad x = \frac{4}{3}$$

Verify that all three solutions check. ■

ORALS *Finish each factoring problem.*

1. $2x^2 + 5x + 3 = (x + \quad)(x + 1)$ **2.** $6x^2 + 5x + 1 = (x + 1)(x + 1)$

3. $6x^2 + 5x - 1 = (x \quad 1)(6x \quad 1)$ **4.** $6x^2 + x - 1 = (2x \quad 1)(3x \quad 1)$

5. $4x^2 + 4x - 3 = (2x + \quad)(2x - \quad)$ **6.** $4x^2 - x - 3 = (4x + \quad)(x - \quad)$

EXERCISE 4.5

In Exercises 1–24, factor each trinomial.

1. $2x^2 - 3x + 1$

2. $2y^2 - 7y + 3$

3. $3a^2 + 13a + 4$

4. $2b^2 + 7b + 6$

5. $4z^2 + 13z + 3$

6. $4t^2 - 4t + 1$

7. $6y^2 + 7y + 2$

8. $4x^2 + 8x + 3$

9. $6x^2 - 7x + 2$

10. $4z^2 - 9z + 2$

11. $3a^2 - 4a - 4$

12. $8u^2 - 2u - 15$

13. $2x^2 - 3x - 2$

14. $12y^2 - y - 1$

15. $2m^2 + 5m - 12$

16. $10u^2 - 13u - 3$

17. $10y^2 - 3y - 1$

18. $6m^2 + 19m + 3$

19. $12y^2 - 5y - 2$

20. $10x^2 + 21x - 10$

21. $5t^2 + 13t + 6$

22. $16y^2 + 10y + 1$

23. $16m^2 - 14m + 3$

24. $16x^2 + 16x + 3$

In Exercises 25–36, factor each trinomial.

25. $3x^2 - 4xy + y^2$

26. $2x^2 + 3xy + y^2$

27. $2u^2 + uv - 3v^2$

28. $2u^2 + 3uv - 2v^2$

29. $4a^2 - 4ab + b^2$

30. $2b^2 - 5bc + 2c^2$

31. $6r^2 + rs - 2s^2$

32. $3m^2 + 5mn + 2n^2$

33. $4x^2 + 8xy + 3y^2$

34. $4b^2 + 15bc - 4c^2$

35. $4a^2 - 15ab + 9b^2$

36. $12x^2 + 5xy - 3y^2$

In Exercises 37–52, write the terms of each trinomial in descending powers of one variable. Then factor the trinomial, if possible.

37. $-13x + 3x^2 - 10$

38. $-14 + 3a^2 - a$

39. $15 + 8a^2 - 26a$

40. $16 - 40a + 25a^2$

41. $12y^2 + 12 - 25y$

42. $12t^2 - 1 - 4t$

43. $3x^2 + 6 + x$

44. $25 + 2u^2 + 3u$

45. $2a^2 + 3b^2 + 5ab$

46. $11uv + 3u^2 + 6v^2$

47. $pq + 6p^2 - q^2$

48. $-11mn + 12m^2 + 2n^2$

49. $b^2 + 4a^2 + 16ab$

50. $3b^2 + 3a^2 - ab$

51. $12x^2 + 10y^2 - 23xy$

52. $5ab + 25a^2 - 2b^2$

In Exercises 53–68, factor each polynomial.

53. $4x^2 + 10x - 6$

54. $9x^2 + 21x - 18$

55. $y^3 + 13y^2 + 12y$

56. $2xy^2 + 8xy - 24x$

57. $6x^3 - 15x^2 - 9x$

58. $9y^3 + 3y^2 - 6y$

59. $30r^5 + 63r^4 - 30r^3$

60. $6s^5 - 26s^4 - 20s^3$

61. $4a^2 - 4ab - 8b^2$ **62.** $6x^2 + 3xy - 18y^2$ **63.** $8x^2 - 12xy - 8y^2$ **64.** $24a^2 + 14ab + 2b^2$

65. $-16m^3n - 20m^2n^2 - 6mn^3$ **66.** $-84x^4 - 100x^3y - 24x^2y^2$

67. $-28u^3v^3 + 26u^2v^4 - 6uv^5$ **68.** $-16x^4y^3 + 30x^3y^4 + 4x^2y^5$

In Exercises 69–76, factor each polynomial.

69. $4x^2 + 4xy + y^2 - 16$ **70.** $9x^2 - 6x + 1 - d^2$

71. $9 - a^2 - 4ab - 4b^2$ **72.** $25 - 9a^2 + 6ac - c^2$

73. $4x^2 + 4xy + y^2 - a^2 - 2ab - b^2$ **74.** $a^2 - 2ab + b^2 - x^2 + 2x - 1$

75. $2x^2z - 4xyz + 2y^2z - 18z^3$ **76.** $9s - r^2s + 2rs^2 - s^3$

In Exercises 77–84, use factoring by grouping to factor each trinomial.

77. $x^2 + 9x + 20$ **78.** $y^2 - 8y + 15$ **79.** $2r^2 + 9r + 10$ **80.** $2v^2 + 5v - 12$

81. $6x^2 - 7x - 5$ **82.** $2y^2 + 5y - 12$ **83.** $12t^2 + 13t - 4$ **84.** $2m^2 + 7m - 15$

In Exercises 85–100, solve each equation.

85. $2x^2 - 5x + 2 = 0$ **86.** $2x^2 + x - 3 = 0$ **87.** $5x^2 - 6x + 1 = 0$ **88.** $6x^2 - 5x + 1 = 0$

89. $3x^2 - 8x = 3$ **90.** $2x^2 - 11x = 21$ **91.** $15x^2 - 2 = 7x$ **92.** $8x^2 + 10x = 3$

93. $x(6x + 5) = 6$ **94.** $x(2x - 3) = 14$ **95.** $(x + 1)(8x + 1) = 18x$ **96.** $4x(3x + 2) = x + 12$

97. $2x(3x^2 + 10x) = -6x$ **98.** $2x^3 = 2x(x + 2)$ **99.** $x^3 + 7x^2 = x^2 - 9x$ **100.** $x^2(x + 10) = 2x(x - 8)$

Writing Exercises ■ *Write a paragraph using your own words.*

1. Describe an organized approach to finding all of the possibilities when you attempt to factor $12x^2 - 4x + 9$.

2. Explain how to determine if a trinomial is prime.

Something to Think About ■

1. For what values of b will the trinomial $6x^2 + bx + 6$ be factorable?

2. Create a quadratic equation with the two solutions $x = 3$ and $x = \frac{3}{2}$.

Review Exercises ■

1. The nth term, l, of an arithmetic sequence is

$$l = f + (n - 1)d$$

where f is the first term and d is the common difference. Remove the parentheses and solve the equation for n.

2. The sum, S, of n consecutive terms of an arithmetic sequence is

$$S = \frac{n}{2}(f + l)$$

where f is the first term of the sequence and l is the nth term. Solve for f.

4.6 Factoring the Sum and Difference of Two Cubes

GETTING READY   *Find each product.*

1. $(x - 3)(x^2 + 3x + 9)$

2. $(x + 2)(x^2 - 2x + 4)$

3. $(y + 4)(y^2 - 4y + 16)$

4. $(r - 5)(r^2 + 5r + 25)$

5. $(a - b)(a^2 + ab + b^2)$

6. $(a + b)(a^2 - ab + b^2)$

The difference of the squares of two quantities factors as the product of two binomials. One binomial is the sum of those two quantities, and the other is the difference of those quantities.

$$x^2 - y^2 = (x + y)(x - y)$$

There are similar formulas for factoring the sum of the cubes of two quantities and the difference of the cubes of two quantities. To discover these formulas, we need to find the following two products:

$$
\begin{aligned}
(x + y)(x^2 - xy + y^2) &= (x + y)x^2 - (x + y)xy + (x + y)y^2 \\
&= x^3 + x^2y - x^2y - xy^2 + xy^2 + y^3 \\
&= x^3 + y^3
\end{aligned}
$$

$$
\begin{aligned}
(x - y)(x^2 + xy + y^2) &= (x - y)x^2 + (x - y)xy + (x - y)y^2 \\
&= x^3 - x^2y + x^2y - xy^2 + xy^2 - y^3 \\
&= x^3 - y^3
\end{aligned}
$$

These results justify the formulas for factoring the **sum and difference of two cubes.**

Factoring the Sum and Difference of Two Cubes

$$x^3 + y^3 = (x + y)(x^2 - xy + y^2)$$
$$x^3 - y^3 = (x - y)(x^2 + xy + y^2)$$

The factorization of $x^3 + y^3$ has a first factor of $x + y$. The second factor has three terms: the square of x, the *negative* of the product of x and y, and the square of y.

The factorization of $x^3 - y^3$ has a first factor of $x - y$. The second factor has three terms: the square of x, the product of x and y, and the square of y.

To factor the sum or difference of two cubes, it is helpful to know the cubes of the numbers from 1 to 10:

1, 8, 27, 64, 125, 216, 343, 512, 729, 1000

Expressions containing variables such as x^6y^3 are also perfect cubes because they can be written as the cubes of a quantity:

$$x^6y^3 = (x^2y)^3$$

EXAMPLE 1   Factor $x^3 + 8$.

Solution   The binomial $x^3 + 8$ is the sum of two cubes.

$$x^3 + 8 = x^3 + 2^3$$

Thus, $x^3 + 8$ factors as the product of the sum of x and 2 and the trinomial $x^2 - 2x + 2^2$.

$$x^3 + 8 = x^3 + \mathbf{2}^3$$
$$= (x + \mathbf{2})(x^2 - \mathbf{2}x + \mathbf{2}^2)$$
$$= (x + 2)(x^2 - 2x + 4)$$

We can check by multiplication.

$$(x + 2)(x^2 - 2x + 4) = (x + 2)x^2 - (x + 2)2x + (x + 2)4$$
$$= x^3 + 2x^2 - 2x^2 - 4x + 4x + 8$$
$$= x^3 + 8 \qquad \blacksquare$$

EXAMPLE 2   Factor $a^3 - 64b^3$.

Solution   The binomial $a^3 - 64b^3$ is the difference of two cubes.

$$a^3 - 64b^3 = a^3 - (4b)^3$$

Thus, its factors are the difference $a - 4b$ and the trinomial $a^2 + a(4b) + (4b)^2$.

$$a^3 - 64b^3 = a^3 - (\mathbf{4b})^3$$
$$= (a - \mathbf{4b})[a^2 + a(\mathbf{4b}) + (\mathbf{4b})^2]$$
$$= (a - 4b)(a^2 + 4ab + 16b^2)$$

Check by multiplication.

$$(a - 4b)(a^2 + 4ab + 16b^2) = (a - 4b)a^2 + (a - 4b)4ab + (a - 4b)16b^2$$
$$= a^3 - 4a^2b + 4a^2b - 16ab^2 + 16ab^2 - 64b^3$$
$$= a^3 - 64b^3 \qquad \blacksquare$$

Sometimes we must factor out a greatest common factor before factoring a sum or difference of two cubes.

EXAMPLE 3   Factor $-2t^5 + 128t^2$.

Solution $$-2t^5 + 128t^2 = -2t^2(t^3 - 64) \qquad \text{Factor out } -2t^2.$$
$$= -2t^2(t - 4)(t^2 + 4t + 16) \qquad \text{Factor } t^3 - 64.$$

Verify this factorization by multiplication. $\qquad \blacksquare$

EXAMPLE 4   Factor $x^6 - 64$.

Solution   The binomial $x^6 - 64$ is both the difference of two squares and the difference of two cubes. Since it is easier to factor the difference of two squares first, the expression factors into the product of a sum and a difference.

$$x^6 - 64 = (x^3)^2 - 8^2$$
$$= (x^3 + 8)(x^3 - 8)$$

Because $x^3 + 8$ is the sum of two cubes and $x^3 - 8$ is the difference of two cubes, each of these binomials can be factored.

$$x^6 - 64 = (x^3 + 8)(x^3 - 8)$$
$$= (x + 2)(x^2 - 2x + 4)(x - 2)(x^2 + 2x + 4)$$

Verify this factorization by multiplication. $\qquad \blacksquare$

ORALS   *Factor each sum or difference of two cubes.*

1. $x^3 - y^3$ 2. $x^3 + y^3$

3. $a^3 + 8$ 4. $b^3 - 27$

5. $1 + 8x^3$ 6. $8 - r^3$

7. $x^3y^3 + 1$ 8. $125 - 8t^3$

EXERCISE 4.6

In Exercises 1–20, factor each expression.

1. $y^3 + 1$

2. $x^3 - 8$

3. $a^3 - 27$

4. $b^3 + 125$

5. $8 + x^3$

6. $27 - y^3$

7. $s^3 - t^3$

8. $8u^3 + w^3$

9. $27x^3 + y^3$

10. $x^3 - 27y^3$

11. $a^3 + 8b^3$

12. $27a^3 - b^3$

13. $64x^3 - 27$

14. $27x^3 + 125$

15. $27x^3 - 125y^3$

16. $64x^3 + 27y^3$

17. $a^6 - b^3$

18. $a^3 + b^6$

19. $x^9 + y^6$

20. $x^3 - y^9$

In Exercises 21–36, factor each expression. Factor out any greatest common factors first.

21. $2x^3 + 54$

22. $2x^3 - 2$

23. $-x^3 + 216$

24. $-x^3 - 125$

25. $64m^3x - 8n^3x$

26. $16r^4 + 128rs^3$

27. $x^4y + 216xy^4$

28. $16a^5 - 54a^2b^3$

29. $81r^4s^2 - 24rs^5$

30. $4m^5n + 500m^2n^4$

31. $125a^6b^2 + 64a^3b^5$

32. $216a^4b^4 - 1000ab^7$

33. $y^7z - yz^4$

34. $x^{10}y^2 - xy^5$

35. $2mp^4 + 16mpq^3$

36. $24m^5n - 3m^2n^4$

In Exercises 37–40, factor each expression completely. Factor a difference of two squares first.

37. $x^6 - 1$

38. $x^6 - y^6$

39. $x^{12} - y^6$

40. $a^{12} - 64$

In Exercises 41–54, factor each expression completely. Some exercises do not involve the sum or difference of two cubes.

41. $3(x^3 + y^3) - z(x^3 + y^3)$

42. $x(8a^3 - b^3) + 4(8a^3 - b^3)$

43. $(m^3 + 8n^3) + (m^3x + 8n^3x)$

44. $(a^3x + b^3x) - (a^3y + b^3y)$

45. $(a^4 + 27a) - (a^3b + 27b)$

46. $(x^4 + xy^3) - (x^3y + y^4)$

47. $x^2(y + z) - 4(y + z)$

48. $z^2(x + 1) - 9(x + 1)$

49. $r^2(x - a) - s^2(x - a)$

50. $pq^2(r + s) - p(r + s)$

51. $(x - 1)^2 + 2(x - 1)$

52. $(z + 3)^2 - 5(z + 3)$

53. $y^3(y^2 - 1) - 27(y^2 - 1)$

54. $z^3(y^2 - 4) + 8(y^2 - 4)$

Writing Exercises ■ *Write a paragraph using your own words.*

1. Explain how to factor $a^3 + b^3$.

2. Explain the difference between $x^3 - y^3$ and $(x - y)^3$.

Something to Think About ■

1. Use a calculator to verify that $a^3 - b^3 = (a - b)(a^2 + ab + b^2)$ when $a = 11$ and $b = 7$.

2. What difficulty do you encounter when you solve $x^3 - 8 = 0$ by factoring?

Review Exercises ■

1. A length of one Fermi is 1×10^{-13} centimeter, approximately the radius of a proton. Express this number in standard notation.

2. In the 14th century, the Black Plague killed about 25,000,000 people, which was 25% of the population of Europe. Find the population at that time, expressed in scientific notation.

4.7 Summary of Factoring Techniques

GETTING READY *Factor each polynomial.*

1. $3ax^2 + 3a^2x$

2. $x^2 - 9y^2$

3. $x^3 - 8$

4. $2x^2 - 8$

5. $x^2 - 3x - 10$

6. $6x^2 - 13x + 6$

7. $6x^2 - 14x + 4$

8. $ax^2 + bx^2 - ay^2 - by^2$

In this brief section, we will discuss ways to approach a randomly chosen factoring problem. For example, suppose we wish to factor the trinomial

$$x^4y + 7x^3y - 18x^2y$$

We begin by attempting to identify the problem type. The first type we look for is **factoring out a common factor.** Because the trinomial has a common factor of x^2y, we factor it out:

$$x^4y + 7x^3y - 18x^2y = x^2y(x^2 + 7x - 18)$$

We can factor the remaining trinomial $x^2 + 7x - 18$ as $(x + 9)(x - 2)$. Thus,

$$x^4y + 7x^3y - 18x^2y = x^2y(x^2 + 7x - 18)$$
$$= x^2y(x + 9)(x - 2)$$

To identify the type of factoring problem, we follow these steps.

Steps for Factoring a Polynomial

1. Factor out all common factors.
2. If an expression has two terms, check to see if the problem type is
   a. the **difference of two squares:** $a^2 - b^2 = (a + b)(a - b)$
   b. the **sum of two cubes:** $a^3 + b^3 = (a + b)(a^2 - ab + b^2)$
   c. the **difference of two cubes:** $a^3 - b^3 = (a - b)(a^2 + ab + b^2)$
3. If an expression has three terms, check to see if the problem type is a **perfect trinomial square:** $a^2 + 2ab + b^2 = (a + b)(a + b)$
   $$a^2 - 2ab + b^2 = (a - b)(a - b)$$
   If the trinomial is not a trinomial square, attempt to factor the trinomial as a **general trinomial.**
4. If an expression has four or more terms, try to factor it by **grouping.**
5. Continue factoring until each individual factor is prime.
6. Check the results by multiplying.

EXAMPLE 1   Factor $x^5y^2 - xy^6$.

Solution   We begin by factoring out the common factor of xy^2:

$$x^5y^2 - xy^6 = xy^2(x^4 - y^4)$$

Because the expression $x^4 - y^4$ has two terms, we check to see if it is the difference of two squares, which it is. As the difference of two squares, it factors as $(x^2 + y^2)(x^2 - y^2)$. Thus,

$$x^5y^2 - xy^6 = xy^2(x^4 - y^4)$$
$$= xy^2(x^2 + y^2)(x^2 - y^2)$$

The binomial $x^2 + y^2$ is the sum of two squares and cannot be factored. However, the binomial $x^2 - y^2$ is the difference of two squares and factors as $(x + y)(x - y)$. Thus,

$$x^5y^2 - xy^6 = xy^2(x^4 - y^4)$$
$$= xy^2(x^2 + y^2)(x^2 - y^2)$$
$$= xy^2(x^2 + y^2)(x + y)(x - y)$$

Because each of the individual factors is prime, the given expression is in completely factored form. ∎

EXAMPLE 2 Factor $x^6 - x^4y^2 - x^3y^3 + xy^5$.

Solution We begin by factoring out the common factor of x.

$$x^6 - x^4y^2 - x^3y^3 + xy^5 = x(x^5 - x^3y^2 - x^2y^3 + y^5)$$

Because the expression $x^5 - x^3y^2 - x^2y^3 + y^5$ has four terms, we try factoring it by grouping:

$$x^6 - x^4y^2 - x^3y^3 + xy^5 = x(x^5 - x^3y^2 - x^2y^3 + y^5)$$
$$= x[x^3(x^2 - y^2) - y^3(x^2 - y^2)]$$
$$= x(x^2 - y^2)(x^3 - y^3) \qquad \text{Factor out } x^2 - y^2.$$

Finally, we factor the difference of two squares and the difference of two cubes:

$$x^6 - x^4y^2 - x^3y^3 + xy^5 = x(x + y)(x - y)(x - y)(x^2 + xy + y^2)$$

Because each of the factors is prime, the given expression is in completely factored form. ∎

ORALS *Indicate which factoring technique you would use first.*

1. $2x^2 - 4x$
2. $16 - 25y^2$
3. $125 + r^3s^3$
4. $ax + ay - x - y$
5. $x^2 + 4$
6. $8x^2 - 50$
7. $25r^2 - s^4$
8. $8a^3 - 27b^3$

E X E R C I S E 4.7

In Exercises 1–50, factor each expression, if possible.

1. $6x + 3$
2. $x^2 - 9$
3. $x^2 - 6x - 7$
4. $a^3 + b^3$

5. $6t^2 + 7t - 3$
6. $3rs^2 - 6r^2st$
7. $4x^2 - 25$
8. $ac + ad + bc + bd$

9. $t^2 - 2t + 1$ **10.** $6p^2 - 3p - 2$ **11.** $a^3 - 8$ **12.** $2x^2 - 32$

13. $x^2y^2 - 2x^2 - y^2 + 2$ **14.** $a^2c + a^2d^2 + bc + bd^2$

15. $70p^4q^3 - 35p^4q^2 + 49p^5q^2$ **16.** $a^2 + 2ab + b^2 - x^2 - 2xy - y^2$

17. $2ab^2 + 8ab - 24a$ **18.** $t^4 - 16$ **19.** $-8p^3q^7 - 4p^2q^3$ **20.** $8m^2n^3 - 24mn^4$

21. $4a^2 - 4ab + b^2 - 9$ **22.** $3rs + 6r^2 - 18s^2$ **23.** $x^2 + 7x + 1$ **24.** $3a^3 + 24b^3$

25. $-2x^5 + 128x^2$ **26.** $16 - 40z + 25z^2$ **27.** $14t^3 - 40t^2 + 6t^4$ **28.** $6x^2 + 7x - 20$

29. $a^2(x - a) - b^2(x - a)$ **30.** $5x^3y^3z^4 + 25x^2y^3z^2 - 35x^3y^2z^5$

31. $8p^6 - 27q^6$ **32.** $2c^2 - 5cd - 3d^2$

33. $125p^3 - 64y^3$ **34.** $8a^2x^3y - 2b^2xy$

35. $-16x^4y^2z + 24x^5y^3z^4 - 15x^2y^3z^7$ **36.** $2ac + 4ad + bc + 2bd$

37. $81p^4 - 16q^4$ **38.** $6x^2 - x - 16$

39. $4x^2 + 9y^2$ **40.** $30a^4 + 5a^3 - 200a^2$

41. $54x^3 + 250y^6$ **42.** $6a^3 + 35a^2 - 6a$

43. $10r^2 - 13r - 4$ **44.** $4x^2 + 4x + 1 - y^2$

45. $21t^3 - 10t^2 + t$ **46.** $16x^2 - 40x^3 + 25x^4$

47. $x^5 - x^3y^2 + x^2y^3 - y^5$ **48.** $a^3x^3 - a^3y^3 + b^3x^3 - b^3y^3$

49. $2a^2c - 2b^2c + 4a^2d - 4b^2d$ **50.** $3a^2x^2 + 6a^2x + 3a^2 - 6b^2x^2 - 12b^2x - 6b^2$

Writing Exercises ■ *Write a paragraph using your own words.*

1. Explain how to identify the type of factoring required to factor a polynomial.

2. Which factoring technique do you find most difficult? Why?

Something to Think About ■ **1.** Write $x^6 - y^6$ as $(x^3)^2 - (y^3)^2$, factor it as the difference of two squares, and show that you get

$$(x + y)(x^2 - xy + y^2)(x - y)(x^2 + xy + y^2)$$

Write $x^6 - y^6$ as $(x^2)^3 - (y^2)^3$, factor it as the difference of two cubes, and show that you get

$$(x + y)(x - y)(x^4 + x^2y^2 + y^4)$$

2. Verify that the results of Question 1 agree by showing that the colored parts agree. Which do you think is completely factored?

Review Exercises ■ *Solve each equation, if possible.*

1. $2(t - 5) + t = 3(2 - t)$

2. $5 + 3(2x - 1) = 2(4 + 3x) - 24$

3. $5x^2 - 35x = 0$

4. $6x^2 - x = 35$

4.8 Problem Solving

GETTING READY

1. One side of a square is s inches long. Find an expression that represents its area.

2. The length of a rectangle is 4 centimeters more than twice the width. If w represents the width, find an expression that represents the length.

3. If x represents the smaller of two consecutive integers, find an expression that represents their product.

4. The length of a rectangle is 3 inches greater than the width. If w represents the width of the rectangle, find an expression that represents the area.

The solutions of many problems involve the use of quadratic equations.

EXAMPLE 1 **Integer problem**   One negative integer is 5 less than another, and their product is 84. Find the integers.

Solution   Let x represent the larger number. Then $x - 5$ represents the smaller number.

Because their product is 84, we form the equation $x(x - 5) = 84$ and solve it.

$$
\begin{aligned}
x(x - 5) &= 84 \\
x^2 - 5x &= 84 & \text{Remove parentheses.} \\
x^2 - 5x - 84 &= 0 & \text{Subtract 84 from both sides.} \\
(x - 12)(x + 7) &= 0 & \text{Factor.}
\end{aligned}
$$

$x - 12 = 0$ or $x + 7 = 0$ Set each factor equal to 0.

$x = 12$ | $x = -7$ Solve each linear equation.

Because we need two negative numbers, we discard the result $x = 12$. The two integers are

$$x = -7 \qquad \text{and} \qquad x - 5 = -7 - 5$$
$$= -12$$

Check: The number -12 is 5 less than -7, and $(-12)(-7) = 84$.   ∎

EXAMPLE 2   **Ballistics**   If an object is thrown straight up into the air with an initial velocity of 112 feet per second, its height after t seconds is given by the formula

$$h = 112t - 16t^2$$

where h represents the height of the object in feet. After this object has been thrown, in how many seconds will it hit the ground?

Solution   When the object hits the ground, its height will be 0. Thus, we set h equal to 0 and solve for t.

$$\mathbf{h} = 112t - 16t^2$$
$$\mathbf{0} = 112t - 16t^2$$
$$0 = 16t(7 - t) \qquad \text{Factor out } 16t.$$

$$16t = 0 \quad \text{or} \quad 7 - t = 0 \qquad \text{Set each factor equal to 0.}$$
$$t = 0 \quad | \quad \quad t = 7 \qquad \text{Solve each linear equation.}$$

When $t = 0$, the object's height above the ground is 0 feet, because it has just been released. When $t = 7$, the height is again 0 feet. The object has hit the ground. The solution is 7 seconds.   ∎

Recall that the area of a rectangle is given by the formula

$$A = lw$$

where A represents the area, l the length, and w the width of the rectangle. The perimeter of a rectangle is given by the formula

$$P = 2l + 2w$$

where P represents the perimeter, l the length, and w the width of the rectangle.

EXAMPLE 3   **Perimeter of a rectangle**   Assume that the rectangle in Figure 4-1 has an area of 52 square centimeters and that its length is 1 centimeter more than 3 times its width. Find the perimeter of the rectangle.

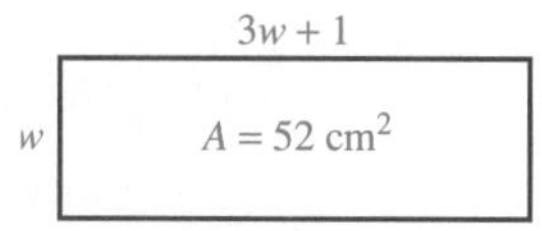

$3w + 1$

w $A = 52 \text{ cm}^2$

FIGURE 4-1

Solution   Let w represent the width of the rectangle. Then $3w + 1$ represents its length. Because the area is 52 square centimeters, we substitute 52 for A and $3w + 1$ for l in the formula $A = lw$ and solve for w.

$$A = lw$$
$$52 = (3w + 1)w$$
$$52 = 3w^2 + w \qquad \text{Remove parentheses.}$$
$$0 = 3w^2 + w - 52 \qquad \text{Subtract 52 from both sides.}$$
$$0 = (3w + 13)(w - 4) \qquad \text{Factor.}$$

$$3w + 13 = 0 \qquad \text{or} \qquad w - 4 = 0 \qquad \text{Set each factor equal to 0.}$$
$$3w = -13 \qquad\qquad\quad w = 4 \qquad \text{Solve each linear equation.}$$
$$w = -\frac{13}{3}$$

Because the width of a rectangle cannot be negative, we discard the result $w = -\frac{13}{3}$. Thus, the width of the rectangle is 4, and the length is given by

$$3w + 1 = 3(4) + 1$$
$$= 12 + 1$$
$$= 13$$

The dimensions of the rectangle are 4 centimeters by 13 centimeters. We find the perimeter by substituting 13 for l and 4 for w in the formula for the perimeter.

$$P = 2l + 2w$$
$$= 2(13) + 2(4)$$
$$= 26 + 8$$
$$= 34$$

The perimeter of the rectangle is 34 centimeters.

Check: A rectangle with dimensions of 13 centimeters by 4 centimeters does have an area of 52 square centimeters, and the length is 1 centimeter more than 3 times the width. A rectangle with these dimensions has a perimeter of 34 centimeters.   ∎

EXAMPLE 4   **Dimensions of a triangle**   The triangle in Figure 4-2 has an area of 10 square centimeters and a height that is 3 centimeters less than twice the length of its base. Find the length of the base and the height of the triangle.

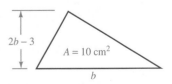

$2b - 3$

$A = 10 \text{ cm}^2$

b

FIGURE 4-2

Solution   Let b represent the length of the base of the triangle. Then $2b - 3$ represents the height. Because the area is 10 square centimeters, we substitute 10 for A and $2b - 3$ for h in the formula $A = \frac{1}{2}bh$ and solve for b.

$$A = \frac{1}{2}bh$$

$$10 = \frac{1}{2}b(2b - 3)$$

$$20 = b(2b - 3) \qquad \text{Multiply both sides by 2.}$$
$$20 = 2b^2 - 3b \qquad \text{Remove parentheses.}$$
$$0 = 2b^2 - 3b - 20 \qquad \text{Subtract 20 from both sides.}$$
$$0 = (2b + 5)(b - 4) \qquad \text{Factor.}$$

$$2b + 5 = 0 \qquad \text{or} \qquad b - 4 = 0 \qquad \text{Set both factors equal to 0.}$$
$$2b = -5 \qquad\qquad\qquad b = 4 \qquad \text{Solve each linear equation.}$$
$$b = -\frac{5}{2}$$

Because a triangle cannot have a negative number for the length of its base, we discard the result $b = -\frac{5}{2}$. The length of the base of the triangle is 4 centimeters. Its height is $2(4) - 3$, or 5 centimeters.

Check: If the base of a triangle has a length of 4 centimeters and the height of the triangle is 5 centimeters, its height is 3 centimeters less than twice the length of its base. Its area is 10 centimeters.

$$A = \frac{1}{2}bh$$

$$= \frac{1}{2}(4)(5)$$

$$= 2(5)$$

$$= 10 \qquad\qquad\qquad\qquad\qquad\qquad\qquad\qquad \blacksquare$$

ORALS   *Give the formula for* . . .

1. the area of a rectangle.

2. the area of a triangle.

3. the area of a square.

4. the volume of a rectangular solid.

5. the perimeter of a rectangle.

6. the perimeter of a square.

EXERCISE 4.8

1. Integer problem   One positive integer is 2 more than another. Their product is 35. Find the integers.

2. Integer problem   One positive integer is 5 less than 4 times another. Their product is 21. Find the integers.

3. Integer problem   If 4 is added to the square of a composite integer, the result is 5 less than 10 times that integer. Find the integer.

4. Number problem   If 3 times the square of a certain natural number is added to the number itself, the result is 14. Find the number.

In Exercises 5–8, an object has been thrown straight up into the air. The formula

$$h = vt - 16t^2$$

gives the height, h, of the object above the ground after t seconds, when it is thrown upward with an initial velocity v.

5. Time of flight   After how many seconds will an object hit the ground if it was thrown with a velocity of 144 feet per second?

6. Time of flight   After how many seconds will an object hit the ground if it was thrown with a velocity of 160 feet per second?

7. Ballistics   If a cannonball is fired with an upward velocity of 220 feet per second, at what times will it be at a height of 600 feet?

8. Ballistics   A cannonball's initial upward velocity is 128 feet per second. At what times will it be 192 feet above the ground?

9. Exhibition diving   A popular tourist attraction consists of swimmers diving from a cliff to the water 64 feet below. A diver's height, h, above the water t seconds after diving is given by $h = -16t^2 + 64$. How long does a dive last?

10. Forensic medicine   The kinetic energy, E, of a moving object is given by $E = \frac{1}{2}mv^2$, where m is the mass of the object (in kilograms) and v is the object's velocity (in meters per second). Kinetic energy is measured in joules. By measuring the damage done to a victim who has been struck by a 3-kilogram club, a police pathologist finds that the energy at impact was 54 joules. Find the velocity of the club at impact.

11. Insulation   The area of the rectangular slab of foam insulation in Illustration 1 is 36 square meters. Find the dimensions of the slab.

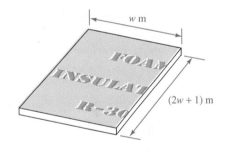

Illustration 1

12. Shipping pallet   The length of a rectangular shipping pallet is 2 feet less than 3 times its width. Its area is 21 square feet. Find the dimensions of the pallet.

13. Carpentry   A room containing 143 square feet is 2 feet longer than it is wide. How long a crown molding is needed to trim the perimeter of the ceiling?

14. Designing solar panels   The length of a standard rectangular solar heat exchange panel is 2 meters longer than the width. If the length remained the same but the width were doubled, the area would be 48 square meters. Find the perimeter of a standard panel.

15. **Designing a tent** The length of the base of the triangular sheet of canvas above the door of the tent shown in Illustration 2 is 2 feet more than twice its height. The area is 30 square feet. Find the height and the length of the base of the triangle.

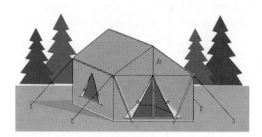

Illustration 2

16. **Dimensions of a triangle** The height of a triangle is 2 inches less than 5 times the length of its base. The area is 36 square inches. Find the length of the base and the height of the triangle.

17. **Area of a triangle** The base of a triangle is numerically 3 less than its area, and the height is numerically 6 less than its area. Find the area of the triangle.

18. **Area of a triangle** The length of the base and the height of a triangle are numerically equal. Their sum is 6 less than the number of units in the area of the triangle. Find the area of the triangle.

19. **Dimensions of a parallelogram** The formula for the area of a parallelogram is $A = bh$. The area of the parallelogram in Illustration 3 is 200 square centimeters. If its base is twice its height, how long is the base?

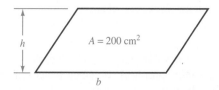

Illustration 3

20. **Swimming pool border** The owners of the rectangular swimming pool in Illustration 4 want to surround the pool with a crushed-stone border of uniform width. They have enough stone to cover 74 square meters. How wide should they make the border? (*Hint:* The area of the larger rectangle minus the area of the smaller is the area of the border.)

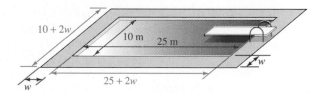

Illustration 4

21. **House construction** The formula for the area of a trapezoid is $A = \frac{h(B + b)}{2}$. The area of the trapezoidal truss in Illustration 5 is 24 square meters. Find the height of the trapezoid if one base is 8 meters and the other base is the same as the height.

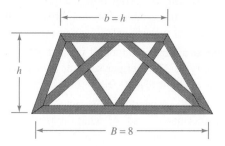

Illustration 5

22. **Volume** The volume of a rectangular solid is given by the formula $V = lwh$, where l is the length, w is the width, and h is the height. The volume of the rectangular solid in Illustration 6 is 210 cubic centimeters. Find the width of the rectangular solid if its length is 10 centimeters and its height is 1 centimeter longer than twice its width.

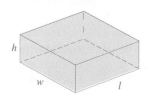

Illustration 6

23. Volume of a pyramid The volume of a pyramid is given by the formula $V = \frac{Bh}{3}$, where B is the area of its base and h is its height. The volume of the pyramid in Illustration 7 is 192 cubic centimeters. Find the dimensions of its rectangular base if one edge of the base is 2 centimeters longer than the other and the height of the pyramid is 12 centimeters.

24. Volume of a pyramid The volume of a pyramid is 84 cubic centimeters. Its height is 9 centimeters, and one side of its rectangular base is 3 centimeters shorter than the other. Find the dimensions of its base. (See Exercise 23.)

25. Volume of a solid The volume of a rectangular solid is 72 cubic centimeters. Its height is 4 centimeters, and its width is 3 centimeters shorter than its length. Find the sum of its length and width. (See Exercise 22.)

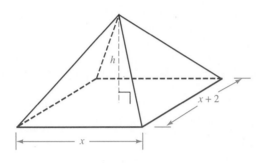

Illustration 7

Writing Exercises ■ *Write a paragraph using your own words.*

1. Explain the steps you would use to set up and solve a word problem.

2. Explain how you should check the solution to a word problem.

Something to Think About ■

1. Here is an easy-sounding word problem:
The length of a rectangle is 2 feet greater than the width, and the area is 18 square feet. Find the width of the rectangle.
Set up the equation. Can you solve it? Why not?

2. Does the equation in Question 1 have a solution, even if you can't find it? Find an estimate of the solution.

Review Exercises ■ *Solve each equation.*

1. $-2(5z + 2) = 3(2 - 3z)$

2. $3(2a - 1) - 9 = 2a$

3. A rectangle is 3 times as long as it is wide, and its perimeter is 120 centimeters. Find its area.

4. A woman invests $15,000, part at 7% annual interest and the rest at 8% annual interest. If she receives $1100 in interest per year, how much has she invested at 7%?

MATHEMATICS IN ECOLOGY

To double the existing capacity of the sewage treatment plant discussed at the beginning of the chapter, the surface area of the new pool must equal the total of the surface areas of the two existing pools. The area of a circle is given by $A = \pi r^2$. Because the radius of the smaller pool is 20 meters (one-half of the diameter), its area is $\pi 20^2$, or 400π square meters. Similarly, the area of the larger pool is $\pi 21^2$, or 441π square meters. The total of these areas is 841π square meters. The engineers must design a third pool with an area of 841π square meters.

Let r represent the radius of the new pool.

The area of the new pool	equals	the total existing area.

$\pi r^2 = 841\pi$	The area of a circle is πr^2.
$r^2 = 841$	Divide both sides by π.
$r^2 - 841 = 0$	Subtract 841 from both sides.
$(r + 29)(r - 29) = 0$	Factor the difference of two squares. $(841 = 29^2)$

$r + 29 = 0$	$r - 29 = 0$	Set each factor equal to 0.
$r = -29$	$r = 29$	

Because the radius of a pool cannot be negative, we discard the negative solution. The design engineers should specify a pool with a radius of 29 meters.

PROJECT ■ Pictures of Polynomial Products

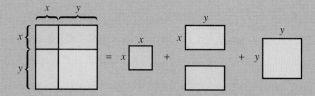

Figure 4-3

Because the length of each side of the largest square in Figure 4-3 is $x + y$, its area is $(x + y)^2$. This area is also the sum of four smaller areas, which illustrates the factorization

$$x^2 + 2xy + y^2 = (x + y)^2$$

What factorization is illustrated by each of the following figures?

1. 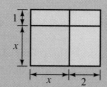   **3.**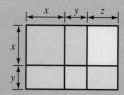

2. 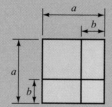   **4.**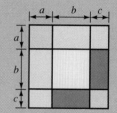

5. Factor the expression

$$a^2 + ac + 2a + ab + bc + 2b$$

and draw a figure that illustrates the factorization.

6. Verify the factorization

$$x^3 + 3x^2y + 3xy^2 + y^3 = (x + y)^3$$

Hint: Expand the right-hand side: $(x + y)^3 = (x + y)(x + y)(x + y)$

Then draw a figure that illustrates the factorization.

CHAPTER SUMMARY

KEY WORDS

difference of two cubes (4.6)

difference of two squares (4.3)

factor (4.1)

factoring by grouping (4.2)

factoring out the greatest common factor (4.1)

fundamental theorem of arithmetic (4.1)

greatest common factor (or divisor) (4.1)

irreducible polynomial (4.3)

key number (4.5)

key number method (4.5)

linear equation (4.1)

lowest common multiple (4.1)

multiple (4.1)

perfect square trinomials (4.4)

prime factor (4.1)

prime-factored form (4.1)

prime factorizations (4.1)

prime number (4.1)

prime polynomial (4.3)

relatively prime (4.1)

quadratic equation (4.1)

sum of two cubes (4.6)

sum of two squares (4.3)

zero-factor property (4.1)

(4.1) A natural number is in prime-factored form if it is written as the product of prime-number factors.

The greatest common factor of several monomials is found by taking each common prime factor and each common variable factor the fewest number of times that it appears in any one monomial.

To factor a polynomial, first factor out all common factors.

Zero-factor property. If a and b represent two real numbers and if $ab = 0$, then

$$a = 0 \quad \text{or} \quad b = 0$$

(4.2) If a polynomial has four or more terms, consider factoring it by grouping.

(4.3) To factor the difference of two squares, use the pattern

$$x^2 - y^2 = (x + y)(x - y)$$

(4.4–4.5) Factor trinomials by trying these steps:

1. Write the trinomial with the exponents of one variable in descending order.

2. Factor out any greatest common factor (including -1 if that is necessary to make the coefficient of the first term positive).

3. If the sign of the third term is plus (+), the signs between the terms of each binomial factor are the same as the sign of the trinomial's second term. If the sign of the third term is minus (−), the signs between the terms of the binomials are opposite.

4. Mentally try various combinations of first terms and last terms until you find the one that works or you exhaust all the possibilities. In that case, the trinomial does not factor using only integer coefficients.

5. Check the factorization by multiplication.

(4.6) The sum and the difference of two cubes factors according to the patterns

$$x^3 + y^3 = (x + y)(x^2 - xy + y^2)$$
$$x^3 - y^3 = (x - y)(x^2 + xy + y^2)$$

(4.7) Steps for factoring a polynomial.

1. Factor out all common factors.

2. If an expression has two terms, check to see if the problem type is

   a. the **difference of two squares:**
      $a^2 - b^2 = (a + b)(a - b),$

   b. the **sum of two cubes:**
      $a^3 + b^3 = (a + b)(a^2 - ab + b^2),$ or

   c. the **difference of two cubes:**
      $a^3 - b^3 = (a - b)(a^2 + ab + b^2).$

3. If an expression has three terms, check to see if the problem type is a **perfect trinomial square:**

   $$a^2 + 2ab + b^2 = (a + b)(a + b)$$
   $$a^2 - 2ab + b^2 = (a - b)(a - b)$$

   If the trinomial is not a trinomial square, attempt to factor the trinomial as a **general trinomial.**

4. If an expression has four or more terms, try to factor the expression by **grouping.**

5. Continue until all factors are prime.

■ Chapter 4 Review Exercises

(4.1) *In Review Exercises 1–8, find the prime factorization of each number.*

1. 35   **2.** 45   **3.** 96   **4.** 102

5. 87   **6.** 99   **7.** 2050   **8.** 4096

(4.1–4.7) *In Review Exercises 9–52, factor each expression completely.*

9. $3x + 9y$ **10.** $5ax^2 + 15a$ **11.** $7x^2 + 14x$ **12.** $3x^2 - 3x$

13. $2x^3 + 4x^2 - 8x$ **14.** $ax + ay - az$ **15.** $ax + ay - a$ **16.** $x^2yz + xy^2z$

17. $5a^2 + 5ab^2 + 10acd - 15a$ **18.** $7axy + 21x^2y - 35x^3y + 7xy^2$

19. $(x + y)a + (x + y)b$ **20.** $(x + y)^2 + (x + y)$

21. $2x^2(x + 2) + 6x(x + 2)$ **22.** $3x(y + z) - 9x(y + z)^2$

23. $3p + 9q + ap + 3aq$ **24.** $ar - 2as + 7r - 14s$

25. $x^2 + ax + bx + ab$ **26.** $xy + 2x - 2y - 4$

27. $3x^2y - xy^2 - 6xy + 2y^2$ **28.** $5x^2 + 10x - 15xy - 30y$

29. $x^2 - 9$ **30.** $x^2y^2 - 16$ **31.** $(x + 2)^2 - y^2$ **32.** $z^2 - (x + y)^2$

33. $6x^2y - 24y^3$ **34.** $(x + y)^2 - z^2$ **35.** $x^2 + 10x + 21$ **36.** $x^2 + 4x - 21$

37. $x^2 + 2x - 24$ **38.** $x^2 - 4x - 12$ **39.** $2x^2 - 5x - 3$ **40.** $3x^2 - 14x - 5$

41. $6x^2 + 7x - 3$ **42.** $6x^2 + 3x - 3$ **43.** $6x^3 + 17x^2 - 3x$ **44.** $4x^3 - 5x^2 - 6x$

45. $x^2 + 2ax + a^2 - y^2$ **46.** $ax^2 + 4ax + 3a - bx - b$

47. $xa + yb + ya + xb$ **48.** $2a^2x + 2abx + a^3 + a^2b$

49. $c^3 - 27$ **50.** $d^3 + 8$ **51.** $2x^3 + 54$ **52.** $2ab^4 - 2ab$

(4.1, 4.3–4.5) *In Review Exercises 53–68, solve each equation.*

53. $x^2 + 2x = 0$ **54.** $2x^2 - 6x = 0$ **55.** $x^2 - 9 = 0$ **56.** $x^2 - 25 = 0$

57. $a^2 - 7a + 12 = 0$ **58.** $x^2 - 2x - 15 = 0$ **59.** $2x - x^2 + 24 = 0$ **60.** $16 + x^2 - 10x = 0$

61. $2x^2 - 5x - 3 = 0$ **62.** $2x^2 + x - 3 = 0$ **63.** $4x^2 = 1$ **64.** $9x^2 = 4$

65. $x^3 - 7x^2 + 12x = 0$ **66.** $x^3 + 5x^2 + 6x = 0$ **67.** $2x^3 + 5x^2 = 3x$ **68.** $3x^3 - 2x = x^2$

(4.8) *In Review Exercises 69–74, solve each problem.*

69. Integer problem   The sum of two integers is 12, and their product is 35. Find the integers.

70. Integer problem   If 3 times the square of a positive integer is added to 5 times the integer, the result is 2. Find the integer.

71. Construction The base of the triangular preformed concrete panel in Illustration 1 is 3 feet longer than twice its height, and its area is 45 square feet. How long is the base?

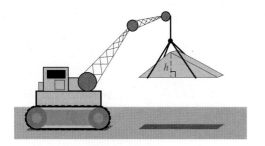

Illustration 1

72. Bombing run A pilot releases a bomb from an altitude of 3000 feet. The bomb's height, h, above the target t seconds after its release is given by the formula

$$h = 3000 + 40t - 16t^2$$

How long will it be before the bomb hits the target?

73. Gardening A rectangular flower bed is 3 feet longer than twice its width, and its area is 27 square feet. Find its dimensions.

74. Geometry A rectangle is 3 feet longer than it is wide. Its area is numerically equal to its perimeter. Find its dimensions.

Chapter 4 Test

1. Find the prime factorization of 196.

2. Find the prime factorization of 111.

In Problems 3–4, factor out the greatest common factor.

3. $60ab^2c^3 + 30a^3b^2c - 25a$

4. $3x^2(a + b) - 6xy(a + b)$

In Problems 5–20, factor each expression.

5. $ax + ay + bx + by$

6. $x^2 - 25$

7. $3a^2 - 27b^2$

8. $16x^4 - 81y^4$

9. $x^2 + 4x + 3$

10. $x^2 - 9x - 22$

11. $x^2 + 10xy + 9y^2$

12. $6x^2 - 30xy + 24y^2$

13. $3x^2 + 13x + 4$

14. $2a^2 + 5a - 12$

15. $2x^2 + 3xy - 2y^2$

16. $12 - 25x + 12x^2$

17. $12a^2 + 6ab - 36b^2$

18. $x^3 - 64$

19. $216 + 8a^3$

20. $x^9z^3 - y^3z^6$

In Problems 21–28, solve each equation.

21. $x^2 + 3x = 0$

22. $2x^2 + 5x + 3 = 0$

23. $9y^2 - 81 = 0$

24. $-3(y - 6) + 2 = y^2 + 2$

25. $10x^2 - 13x = 9$

26. $10x^2 - x = 9$

27. $10x^2 + 43x = 9$

28. $10x^2 - 89x = 9$

29. Cannon fire A cannonball is fired into the air with a velocity of 192 feet per second. In how many seconds will it hit the ground? (Its height above the ground is given by the formula $h = vt - 16t^2$, where v is the velocity and t is time.)

30. Base of a triangle The base of a triangle with an area of 40 square meters is 2 meters longer than its height. Find the base of the triangle.

5

Rational Expressions

An architect is designing a home with the combined living room-dining room shown in the sketch. The total length available is 32 feet, and the shortest wall in the living room must be 4 feet long to accommodate a custom bookshelf. The area of the living room is to be 288 square feet, and the area of the dining room is 168 square feet.

What must be the design width, *w*, of the dining room?

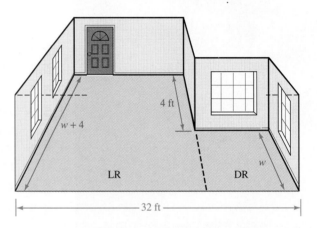

■ ■ ■ ■ ■ ■ ■ ■   After you have read this chapter, you will be able to answer this question.

We have seen that expressions such as $\frac{1}{2}$ and $\frac{-3}{4}$ that indicate the quotient of two integers are called **arithmetic fractions** or **rational numbers.** Expressions such as

$$\frac{x}{x+2} \qquad \text{and} \qquad \frac{5a^2 + b^2}{3a - b}$$

consisting of the quotient of two polynomials, are called **rational expressions.** We begin the study of these algebraic fractions by reviewing the basic properties of fractions.

5.1 The Basic Properties of Fractions

■ Simplifying Fractions

GETTING READY

Simplify.

1. $\dfrac{12}{16}$ **2.** $\dfrac{16}{8}$ **3.** $\dfrac{25}{55}$ **4.** $\dfrac{36}{72}$

We have seen that any number that can be written in the form $\frac{a}{b}$, where a and b are integers and $b \neq 0$, is a rational number. The number 0.5, for example, is a rational number, because it can be written in the form $\frac{5}{10}$ or $\frac{1}{2}$.

The fraction bar in the symbol $\frac{a}{b}$ indicates that a is to be divided by b. The number above the fraction bar is called the **numerator,** and the number below is called the **denominator.**

 WARNING! Remember that the denominator of a fraction can never be 0.

There are three signs associated with every fraction: the sign of the fraction, the sign of the numerator, and the sign of the denominator.

$$\text{Sign of the fraction} \longrightarrow -\frac{+12}{-4} \quad \begin{matrix} \text{Sign of the numerator} \\ \\ \text{Sign of the denominator} \end{matrix}$$

Any two of these signs can be changed without altering the value of the fraction. (If no sign is indicated, a plus (+) sign is understood.) For example,

$$-\frac{+12}{-4} = -\frac{-12}{+4} = +\frac{-12}{-4} = +\frac{+12}{+4} = +3$$

In general, we have

Properties of Fractions

$$\frac{a}{b} = \frac{-a}{-b} = -\frac{a}{-b} = -\frac{-a}{b} \qquad (b \neq 0)$$

and

$$-\frac{a}{b} = \frac{-a}{b} = \frac{a}{-b} = -\frac{-a}{-b} \qquad (b \neq 0)$$

■ Simplifying Fractions

We have seen that a fraction can be simplified by dividing out any common factors shared by its numerator and denominator. For example,

$$\frac{18}{30} = \frac{3 \cdot 6}{5 \cdot 6} = \frac{3 \cdot \overset{1}{\cancel{6}}}{5 \cdot \cancel{6}} = \frac{3}{5}$$

and

$$-\frac{6}{15} = -\frac{3 \cdot 2}{3 \cdot 5} = -\frac{\overset{1}{\cancel{3}} \cdot 2}{\cancel{3} \cdot 5} = -\frac{2}{5}$$

To simplify the fraction $\frac{ac}{bc}$, we can divide out the common factor of c to obtain

$$\frac{ac}{bc} = \frac{a\overset{1}{\cancel{c}}}{b\cancel{c}} = \frac{a}{b}$$

This fact establishes the fundamental property of fractions, previously stated in Section 1.2.

Fundamental Property of Fractions   If a is a real number and b and c are nonzero real numbers, then

$$\frac{ac}{bc} = \frac{a}{b}$$

The fundamental property of fractions implies that factors common to both the numerator and denominator of a fraction can be divided out. When all common factors have been divided out, we say that the fraction has been **expressed in lowest terms**. To **simplify a fraction** means to write it in lowest terms.

EXAMPLE 1   Simplify $\dfrac{21x^2y}{14xy^2}$.

Solution   To simplify the fraction means to write it in lowest terms.

$$\frac{21x^2y}{14xy^2} = \frac{3 \cdot 7 \cdot x \cdot x \cdot y}{2 \cdot 7 \cdot x \cdot y \cdot y} \qquad \text{Factor the numerator and denominator.}$$

$$= \frac{3 \cdot 7 \cdot \overset{1}{\cancel{x}} \cdot x \cdot \overset{1}{\cancel{y}}}{2 \cdot 7 \cdot \cancel{x} \cdot y \cdot \cancel{y}} \qquad \text{Divide out the common factors of 7, } x \text{, and } y.$$

$$= \frac{3x}{2y}$$

This fraction can also be simplified by using the rules of exponents:

$$\frac{21x^2y}{14xy^2} = \frac{3 \cdot 7}{2 \cdot 7}x^{2-1}y^{1-2}$$

$$= \frac{3}{2}xy^{-1} \qquad\qquad \text{Divide out the 7's and simplify.}$$

$$= \frac{3}{2} \cdot \frac{x}{y} \qquad\qquad y^{-1} = \frac{1}{y}.$$

$$= \frac{3x}{2y}$$

■

EXAMPLE 2   Write the fraction $\dfrac{x^2 + 3x}{3x + 9}$ in lowest terms.

Solution $$\frac{x^2 + 3x}{3x + 9} = \frac{x(x + 3)}{3(x + 3)} \qquad \text{Factor the numerator and the denominator.}$$

$$= \frac{\overset{1}{x(x + 3)}}{\underset{1}{3(x + 3)}} \qquad \text{Divide out the common factor of } x + 3.$$

$$= \frac{x}{3}$$

■

Any number divided by 1 remains unchanged. For example,

$$\frac{37}{1} = 37 \qquad \text{and} \qquad \frac{5x}{1} = 5x$$

In general, for any real number a, the equality at the top of the next page is true.

■ PERSPECTIVE

A Fraction of Calculus

The fraction $\frac{8}{4}$ is equal to 2, because $4 \cdot 2 = 8$. The expression $\frac{8}{0}$ is undefined, because there is *no* number x for which $0 \cdot x = 8$. The expression $\frac{0}{0}$ presents a different problem, however, because $\frac{0}{0}$ seems to be equal to *any* number. For example, $\frac{0}{0} = 17$, because $0 \cdot 17 = 0$. Similarly, $\frac{0}{0} = \pi$, because $0 \cdot \pi = 0$. Because "no answer" and "any answer" are both unacceptable, division by 0 is not allowed.

Although $\frac{0}{0}$ represents many numbers, there is often one best answer. In the 17th century,

mathematicians such as Sir Isaac Newton (1642–1727) and Gottfried Wilhelm von Leibniz (1646–1716) began to look more closely at expressions related to the fraction $\frac{0}{0}$, and discovered that under certain conditions, there was one best answer. Expressions related to $\frac{0}{0}$ are called **indeterminant forms.** One of these expressions, called a **derivative,** is the foundation of **calculus,** an important area of mathematics discovered independently by both Newton and Leibniz.

Division by 1 $\dfrac{a}{1} = a$

EXAMPLE 3   Simplify $\dfrac{x^3 + x^2}{x + 1}$.

Solution $\dfrac{x^3 + x^2}{x + 1} = \dfrac{x^2(x + 1)}{x + 1}$ Factor the numerator.

$$= \dfrac{x^2 \overset{1}{\cancel{(x + 1)}}}{\underset{1}{\cancel{x + 1}}}$$ Divide out the common factor of $x + 1$.

$$= \dfrac{x^2}{1}$$

$$= x^2$$ Denominators of 1 need not be written.   ∎

(b)   **EXAMPLE 4**   Simplify   **a.** $\dfrac{x - y}{y - x}$ and   **b.** $\dfrac{2a - 1}{1 - 2a}$.

Solution   We can rearrange terms in each numerator, factor out -1, and proceed as follows:

a. $\dfrac{x - y}{y - x} = \dfrac{-y + x}{y - x}$ **b.** $\dfrac{2a - 1}{1 - 2a} = \dfrac{-1 + 2a}{1 - 2a}$

$$= \dfrac{-(y - x)}{y - x}$$ $$= \dfrac{-(1 - 2a)}{1 - 2a}$$

$$= \dfrac{-\overset{1}{\cancel{(y - x)}}}{\underset{1}{\cancel{y - x}}}$$ $$= \dfrac{-\overset{1}{\cancel{(1 - 2a)}}}{\underset{1}{\cancel{1 - 2a}}}$$

$$= -1$$ $$= -1$$ ∎

The results of Example 4 illustrate this important fact.

Division of Negatives   The quotient of any nonzero expression and its negative is -1.

EXAMPLE 5   Simplify $\dfrac{x^2 + 13x + 12}{x^2 - 144}$.

Solution
$$\frac{x^2 + 13x + 12}{x^2 - 144} = \frac{(x + 1)(x + 12)}{(x + 12)(x - 12)}$$ Factor the numerator and denominator.

$$= \frac{(x + 1)(\cancel{x + 12})^{1}}{(\cancel{x + 12})_{1}(x - 12)}$$ Divide out the common factors of $x + 12$.

$$= \frac{x + 1}{x - 12}$$ ■

WARNING!   It is important to remember that only *factors* that are common to the *entire numerator* and the *entire denominator* can be divided out. *Terms* that are common to both the numerator and denominator *cannot* be divided out. For example, consider the correct simplification

$$\frac{5 + 8}{5} = \frac{13}{5}$$

It would be incorrect to divide out the common *term* of 5 in the above simplification. Note that doing so gives an incorrect answer.

$$\cancel{\frac{5 + 8}{5}} = \frac{\cancel{5}^{1} + 8}{\cancel{5}_{1}} = \frac{1 + 8}{1} = 9$$

EXAMPLE 6   Express the fraction $\dfrac{5(x + 3) - 5}{7(x + 3) - 7}$ in lowest terms.

Solution   We must not divide out the binomials $x + 3$, because $x + 3$ is not a *factor* of the entire numerator, nor is it a *factor* of the entire denominator. Instead, we simplify, factor the numerator and the denominator separately, and then divide out any common factors.

$$\frac{5(x + 3) - 5}{7(x + 3) - 7} = \frac{5x + 15 - 5}{7x + 21 - 7}$$ Remove parentheses.

$$= \frac{5x + 10}{7x + 14}$$ Combine like terms.

$$= \frac{5(x + 2)}{7(x + 2)}$$ Factor the numerator and denominator.

$$= \frac{5(\cancel{x + 2})^{1}}{7(\cancel{x + 2})_{1}}$$ Divide out the common factor of $x + 2$.

$$= \frac{5}{7}$$ ■

EXAMPLE 7 Simplify $\dfrac{x(x+3) - 3(x-1)}{x^2 + 3}$.

Solution

$$\dfrac{x(x+3) - 3(x-1)}{x^2 + 3} = \dfrac{x^2 + 3x - 3x + 3}{x^2 + 3}$$ Remove parentheses in the numerator.

$$= \dfrac{x^2 + 3}{x^2 + 3}$$ Combine like terms in the numerator.

$$= \dfrac{\overset{1}{\cancel{x^2 + 3}}}{\underset{1}{\cancel{x^2 + 3}}}$$ Divide out the common factor of $x^2 + 3$.

$$= 1$$ ∎

EXAMPLE 8 Simplify $\dfrac{xy + 2x + 3y + 6}{x^2 + x - 6}$.

Solution We can factor the numerator by grouping and factor the denominator as a general trinomial. Then we proceed as follows.

$$\dfrac{xy + 2x + 3y + 6}{x^2 + x - 6} = \dfrac{x(y+2) + 3(y+2)}{(x-2)(x+3)}$$ Begin to factor the numerator and factor the denominator.

$$= \dfrac{(y+2)(x+3)}{(x-2)(x+3)}$$ Factor the numerator.

$$= \dfrac{(y+2)\overset{1}{\cancel{(x+3)}}}{(x-2)\underset{1}{\cancel{(x+3)}}}$$ Divide out the common factor of $x + 3$.

$$= \dfrac{y+2}{x-2}$$ ∎

Sometimes a fraction does not simplify. Such a fraction is already in lowest terms.

EXAMPLE 9 Simplify $\dfrac{x^2 + x - 2}{x^2 + x}$.

Solution

$$\dfrac{x^2 + x - 2}{x^2 + x} = \dfrac{(x+2)(x-1)}{x(x+1)}$$ Factor the numerator and the denominator.

Because there are no factors common to the numerator and denominator, this fraction is already in lowest terms. ∎

ORALS   *Simplify each fraction.*

1. $\dfrac{14}{21}$ 2. $\dfrac{34}{17}$ 3. $\dfrac{xyz}{wxy}$ 4. $\dfrac{8x^2}{4x}$

5. $\dfrac{6x^2y}{6xy^2}$ 6. $\dfrac{x^2y^3}{x^2y^4}$ 7. $\dfrac{x+y}{y+x}$ 8. $\dfrac{x-y}{y-x}$

EXERCISE 5.1

In Exercises 1–68, express each fraction in lowest terms. If a fraction is already in lowest terms, so indicate. Assume that no variable has a value that would make a denominator equal to zero.

1. $\dfrac{8}{10}$ 2. $\dfrac{16}{28}$ 3. $\dfrac{28}{35}$ 4. $\dfrac{14}{20}$

5. $\dfrac{8}{52}$ 6. $\dfrac{15}{21}$ 7. $\dfrac{10}{45}$ 8. $\dfrac{21}{35}$

9. $\dfrac{-18}{54}$ 10. $\dfrac{16}{40}$ 11. $\dfrac{4x}{2}$ 12. $\dfrac{2x}{4}$

13. $\dfrac{-6x}{18}$ 14. $\dfrac{-25y}{5}$ 15. $\dfrac{45}{9a}$ 16. $\dfrac{48}{16y}$

17. $\dfrac{7+3}{5z}$ 18. $\dfrac{(3-18)k}{25}$ 19. $\dfrac{(3+4)a}{24-3}$ 20. $\dfrac{x+x}{2}$

21. $\dfrac{2x}{3x}$ 22. $\dfrac{5y}{7y}$ 23. $\dfrac{6x^2}{4x^2}$ 24. $\dfrac{9xy}{6xy}$

25. $\dfrac{2x^2}{3y}$ 26. $\dfrac{5y^2}{2y^2}$ 27. $\dfrac{15x^2y}{5xy^2}$ 28. $\dfrac{12xz}{4xz^2}$

29. $\dfrac{28x}{32y}$ 30. $\dfrac{14xz^2}{7x^2z^2}$ 31. $\dfrac{x+3}{3(x+3)}$ 32. $\dfrac{2(x+7)}{x+7}$

33. $\dfrac{5x+35}{x+7}$ 34. $\dfrac{x-9}{3x-27}$ 35. $\dfrac{x^2+3x}{2x+6}$ 36. $\dfrac{xz-2x}{yz-2y}$

37. $\dfrac{15x-3x^2}{25y-5xy}$ 38. $\dfrac{3y+xy}{3x+xy}$ 39. $\dfrac{6a-6b+6c}{9a-9b+9c}$ 40. $\dfrac{3a-3b-6}{2a-2b-4}$

41. $\dfrac{x-7}{7-x}$ 42. $\dfrac{d-c}{c-d}$ 43. $\dfrac{6x-3y}{3y-6x}$ 44. $\dfrac{3c-4d}{4c-3d}$

45. $\dfrac{a+b-c}{c-a-b}$ 46. $\dfrac{x-y-z}{z+y-x}$ 47. $\dfrac{x^2+3x+2}{x^2+x-2}$ 48. $\dfrac{x^2+x-6}{x^2-x-2}$

49. $\dfrac{x^2-8x+15}{x^2-x-6}$ 50. $\dfrac{x^2-6x-7}{x^2+8x+7}$ 51. $\dfrac{2x^2-8x}{x^2-6x+8}$ 52. $\dfrac{3y^2-15y}{y^2-3y-10}$

53. $\dfrac{xy + 2x^2}{2xy + y^2}$

54. $\dfrac{3x + 3y}{x^2 + xy}$

55. $\dfrac{x^2 + 3x + 2}{x^3 + x^2}$

56. $\dfrac{6x^2 - 13x + 6}{3x^2 + x - 2}$

57. $\dfrac{x^2 - 8x + 16}{x^2 - 16}$

58. $\dfrac{3x + 15}{x^2 - 25}$

59. $\dfrac{2x^2 - 8}{x^2 - 3x + 2}$

60. $\dfrac{3x^2 - 27}{x^2 + 3x - 18}$

61. $\dfrac{x^2 - 2x - 15}{x^2 + 2x - 15}$

62. $\dfrac{x^2 + 4x - 77}{x^2 - 4x - 21}$

63. $\dfrac{x^2 - 3(2x - 3)}{9 - x^2}$

64. $\dfrac{x(x - 8) + 16}{16 - x^2}$

65. $\dfrac{4(x + 3) + 4}{3(x + 2) + 6}$

66. $\dfrac{4 + 2(x - 5)}{3x - 5(x - 2)}$

67. $\dfrac{x^2 - 9}{(2x + 3) - (x + 6)}$

68. $\dfrac{x^2 + 5x + 4}{2(x + 3) - (x + 2)}$

In Exercises 69–76, simplify each fraction. Assume that no variable has a value that would make a denominator equal to 0. In each exercise, you will have to factor a sum or difference of two cubes or factor by grouping.

69. $\dfrac{x^3 + 1}{x^2 - x + 1}$

70. $\dfrac{x^3 - 1}{x^2 + x + 1}$

71. $\dfrac{2a^3 - 16}{2a^2 + 4a + 8}$

72. $\dfrac{3y^3 + 81}{y^2 - 3y + 9}$

73. $\dfrac{ab + b + 2a + 2}{ab + a + b + 1}$

74. $\dfrac{xy + 2y + 3x + 6}{x^2 + 5x + 6}$

75. $\dfrac{xy + 3y + 3x + 9}{x^2 - 9}$

76. $\dfrac{ab + b^2 + 2a + 2b}{a^2 + 2a + ab + 2b}$

Writing Exercises ■ *Write a paragraph using your own words.*

1. Explain why the fraction $\dfrac{x - 7}{7 - x}$ simplifies to -1.

2. Explain why the fraction $\dfrac{x + 7}{7 + x}$ simplifies to 1.

Something to Think About ■

1. Exercise 63, $\dfrac{x^2 - 3(2x - 3)}{9 - x^2}$, has two possible answers, $\dfrac{3 - x}{3 + x}$ and $-\dfrac{x - 3}{x + 3}$. Why is either answer correct?

2. Find two different-looking, but correct simplifications of the fraction $\dfrac{y^2 + 5(2y + 5)}{25 - y^2}$.

Review Exercises ■

1. State the associative property of addition.

2. State the distributive property.

3. What is the additive identity?

4. What is the multiplicative identity?

5. Find the additive inverse of $-\dfrac{5}{3}$.

6. Find the multiplicative inverse of $-\dfrac{5}{3}$.

5.2 Multiplying Fractions

■ Multiplying a Fraction by a Polynomial

GETTING READY *Multiply the fractions and simplify.*

1. $\dfrac{3}{7} \cdot \dfrac{14}{9}$
2. $\dfrac{21}{15} \cdot \dfrac{10}{3}$
3. $\dfrac{19}{38} \cdot 6$
4. $42 \cdot \dfrac{3}{21}$

5. $\dfrac{4}{9} \cdot \dfrac{45}{8}$
6. $\dfrac{11}{7} \cdot \dfrac{14}{22}$
7. $\dfrac{75}{12} \cdot \dfrac{6}{50}$
8. $\dfrac{13}{5} \cdot \dfrac{20}{26}$

Recall that to multiply fractions, we multiply their numerators and multiply their denominators. For example, to find the product of $\frac{4}{7}$ and $\frac{3}{5}$, we proceed as follows:

$$\frac{4}{7} \cdot \frac{3}{5} = \frac{4 \cdot 3}{7 \cdot 5}$$

$$= \frac{12}{35}$$

In general, we have the following rule.

Rule for Multiplying Fractions	If a, b, c, and d are real numbers and $b \neq 0$ and $d \neq 0$, then $$\frac{a}{b} \cdot \frac{c}{d} = \frac{ac}{bd}$$

(b, d) **EXAMPLE 1** Multiply **a.** $\dfrac{1}{3} \cdot \dfrac{2}{5}$, **b.** $\dfrac{7}{9} \cdot \dfrac{-5}{3x}$, **c.** $\dfrac{x^2}{2} \cdot \dfrac{3}{y^2}$, and **d.** $\dfrac{t+1}{t} \cdot \dfrac{t-1}{t-2}$.

Solution **a.** $\dfrac{1}{3} \cdot \dfrac{2}{5} = \dfrac{1 \cdot 2}{3 \cdot 5}$ **b.** $\dfrac{7}{9} \cdot \dfrac{-5}{3x} = \dfrac{7(-5)}{9 \cdot 3x}$

$= \dfrac{2}{15}$ $= \dfrac{-35}{27x}$

c. $\dfrac{x^2}{2} \cdot \dfrac{3}{y^2} = \dfrac{x^2 \cdot 3}{2 \cdot y^2}$ **d.** $\dfrac{t+1}{t} \cdot \dfrac{t-1}{t-2} = \dfrac{(t+1)(t-1)}{t(t-2)}$

$= \dfrac{3x^2}{2y^2}$

■

EXAMPLE 2   Multiply: $\dfrac{35x^2y}{7y^2z} \cdot \dfrac{z}{5xy}$.

Solution

$$\dfrac{35x^2y}{7y^2z} \cdot \dfrac{z}{5xy} = \dfrac{5 \cdot 7 \cdot x \cdot x \cdot y \cdot z}{7 \cdot y \cdot y \cdot z \cdot 5 \cdot x \cdot y}$$

Multiply the fractions and factor where possible.

$$= \dfrac{\overset{1}{\cancel{5}} \cdot \overset{1}{\cancel{7}} \cdot \overset{1}{\cancel{x}} \cdot x \cdot \cancel{y} \cdot \cancel{z}}{\underset{1}{\cancel{7}} \cdot \underset{1}{\cancel{y}} \cdot y \cdot \underset{1}{\cancel{z}} \cdot \underset{1}{\cancel{5}} \cdot \underset{1}{\cancel{x}} \cdot y}$$

Divide out all common factors.

$$= \dfrac{x}{y^2}$$

∎

EXAMPLE 3   Find the product of $\dfrac{x^2 - x}{2x + 4}$ and $\dfrac{x + 2}{x}$.

Solution

$$\dfrac{x^2 - x}{2x + 4} \cdot \dfrac{x + 2}{x} = \dfrac{x(x - 1)(x + 2)}{2(x + 2)x}$$

Multiply the fractions and factor where possible.

$$= \dfrac{\overset{1}{\cancel{x}}(x - 1)\overset{1}{(x + 2)}}{2(x + 2)\underset{1}{\cancel{x}}}$$

Divide out all common factors.

$$= \dfrac{x - 1}{2}$$

∎

EXAMPLE 4   Find the product of $\dfrac{x^2 - 3x}{x^2 - x - 6}$ and $\dfrac{x^2 + x - 2}{x^2 - x}$.

Solution

$$\dfrac{x^2 - 3x}{x^2 - x - 6} \cdot \dfrac{x^2 + x - 2}{x^2 - x} = \dfrac{x(x - 3)(x + 2)(x - 1)}{(x + 2)(x - 3)x(x - 1)}$$

Multiply the fractions and factor where possible.

$$= \dfrac{\overset{1}{\cancel{x}}\overset{1}{\cancel{(x - 3)}}\overset{1}{\cancel{(x + 2)}}\overset{1}{\cancel{(x - 1)}}}{\underset{1}{\cancel{(x + 2)}}\underset{1}{\cancel{(x - 3)}}\underset{1}{\cancel{x}}\underset{1}{\cancel{(x - 1)}}}$$

Divide out all common factors.

$$= 1$$

∎

■ **Multiplying a Fraction by a Polynomial**

EXAMPLE 5   Multiply the fraction $\dfrac{x^2 + x}{x^2 + 8x + 7}$ by $x + 7$.

Solution
$$\frac{x^2 + x}{x^2 + 8x + 7} \cdot (x + 7) = \frac{x^2 + x}{x^2 + 8x + 7} \cdot \frac{x + 7}{1}$$

Write $x + 7$ as a fraction with a denominator of 1.

$$= \frac{x(x + 1)(x + 7)}{(x + 1)(x + 7)1}$$

Multiply the fractions and factor where possible.

$$= \frac{x\cancel{(x+1)}\cancel{(x+7)}}{1\cancel{(x+1)}\cancel{(x+7)}}$$

Divide out all common factors.

$$= x$$ ∎

EXAMPLE 6 Multiply: $\dfrac{x^2 + 2x}{xy - 2y} \cdot \dfrac{x + 1}{x^2 - 4} \cdot \dfrac{x - 2}{x^2 + x}$.

Solution
$$\frac{x^2 + 2x}{xy - 2y} \cdot \frac{x + 1}{x^2 - 4} \cdot \frac{x - 2}{x^2 + x} = \frac{x(x + 2)(x + 1)(x - 2)}{y(x - 2)(x + 2)(x - 2)x(x + 1)}$$

Multiply the fractions and factor where possible.

$$= \frac{\cancel{x}\cancel{(x+2)}\cancel{(x+1)}\cancel{(x-2)}}{y\cancel{(x-2)}\cancel{(x+2)}(x - 2)\cancel{x}\cancel{(x+1)}}$$

Divide out all common factors.

$$= \frac{1}{y(x - 2)}$$ ∎

ORALS *Multiply, and simplify where possible.*

1. $\dfrac{2}{3} \cdot \dfrac{5}{7}$

2. $\dfrac{x}{2} \cdot \dfrac{3}{x}$

3. $\dfrac{x + 1}{5} \cdot \dfrac{7}{x + 1}$

4. $\dfrac{x^2 y}{y} \cdot \dfrac{1}{x}$

5. $\dfrac{5}{x + 7} \cdot (x + 7)$

6. $\dfrac{7}{7(x - 9)} \cdot 2(x - 9)$

E X E R C I S E   5 . 2

In Exercises 1–68, do the multiplications. Simplify answers if possible.

1. $\dfrac{2}{3} \cdot \dfrac{4}{5}$

2. $\dfrac{1}{2} \cdot \dfrac{3}{5}$

3. $\dfrac{5}{7} \cdot \dfrac{9}{13}$

4. $\dfrac{2}{7} \cdot \dfrac{5}{11}$

5. $\dfrac{2}{3} \cdot \dfrac{3}{5}$

6. $\dfrac{3}{7} \cdot \dfrac{7}{5}$

7. $-\dfrac{3}{7} \cdot \dfrac{14}{9}$

8. $-\dfrac{6}{9} \cdot \dfrac{15}{35}$

9. $\dfrac{25}{35} \cdot \dfrac{21}{55}$

10. $\dfrac{27}{24} \cdot \dfrac{56}{35}$

11. $\dfrac{-21}{18} \cdot \dfrac{-45}{14}$

12. $\dfrac{-33}{7} \cdot \dfrac{-5}{55}$

13. $\dfrac{2}{3} \cdot \dfrac{15}{2} \cdot \dfrac{1}{7}$

14. $\dfrac{2}{5} \cdot \dfrac{10}{9} \cdot \dfrac{3}{2}$

15. $\dfrac{3x}{y} \cdot \dfrac{y}{2}$

16. $\dfrac{2y}{z} \cdot \dfrac{z}{3}$

17. $\dfrac{5y}{7} \cdot \dfrac{7x}{5z}$

18. $\dfrac{4x}{3y} \cdot \dfrac{3y}{7x}$

19. $\dfrac{3y}{4x} \cdot \dfrac{2x}{5}$

20. $\dfrac{5x}{14y} \cdot \dfrac{7y}{x}$

21. $\dfrac{7z}{9z} \cdot \dfrac{4z}{2z}$

22. $\dfrac{8z}{2x} \cdot \dfrac{16x}{3x}$

23. $\dfrac{13x^2}{7x} \cdot \dfrac{28}{2x}$

24. $\dfrac{z^2}{z} \cdot \dfrac{5x}{5}$

25. $\dfrac{2x^2y}{3xy} \cdot \dfrac{3xy^2}{2}$

26. $\dfrac{2x^2z}{z} \cdot \dfrac{5x}{z}$

27. $\dfrac{8x^2y^2}{4x^2} \cdot \dfrac{2xy}{2y}$

28. $\dfrac{9x^2y}{3x} \cdot \dfrac{3xy}{3y}$

29. $\dfrac{-2xy}{x^2} \cdot \dfrac{3xy}{2}$

30. $\dfrac{-3x}{x^2} \cdot \dfrac{2xz}{3}$

31. $\dfrac{ab^2}{a^2b} \cdot \dfrac{b^2c^2}{abc} \cdot \dfrac{abc^2}{a^3c^2}$

32. $\dfrac{x^3y}{z} \cdot \dfrac{xz^3}{x^2y^2} \cdot \dfrac{yz}{xyz}$

33. $\dfrac{10r^2st^3}{6rs^2} \cdot \dfrac{3r^3t}{2rst} \cdot \dfrac{2s^3t^4}{5s^2t^3}$

34. $\dfrac{3a^3b}{25cd^3} \cdot \dfrac{-5cd^2}{6ab} \cdot \dfrac{10abc^2}{2bc^2d}$

35. $\dfrac{z+7}{7} \cdot \dfrac{z+2}{z}$

36. $\dfrac{a-3}{a} \cdot \dfrac{a+3}{5}$

37. $\dfrac{x-2}{2} \cdot \dfrac{2x}{x-2}$

38. $\dfrac{y+3}{y} \cdot \dfrac{3y}{y+3}$

39. $\dfrac{x+5}{5} \cdot \dfrac{x}{x+5}$

40. $\dfrac{y-9}{y+9} \cdot \dfrac{y}{9}$

41. $\dfrac{(x+1)^2}{x+1} \cdot \dfrac{x+2}{x+1}$

42. $\dfrac{(y-3)^2}{y-3} \cdot \dfrac{y-3}{y-3}$

43. $\dfrac{2x+6}{x+3} \cdot \dfrac{3}{4x}$

44. $\dfrac{3y-9}{y-3} \cdot \dfrac{y}{3y^2}$

45. $\dfrac{x^2-x}{x} \cdot \dfrac{3x-6}{3x-3}$

46. $\dfrac{5z-10}{z+2} \cdot \dfrac{3}{3z-6}$

47. $\dfrac{7y-14}{y-2} \cdot \dfrac{x^2}{7x}$

48. $\dfrac{y^2+3y}{9} \cdot \dfrac{3x}{y+3}$

49. $\dfrac{x^2+x-6}{5x} \cdot \dfrac{5x-10}{x+3}$

50. $\dfrac{z^2+4z-5}{5z-5} \cdot \dfrac{5z}{z+5}$

51. $\dfrac{m^2-2m-3}{2m+4} \cdot \dfrac{m^2-4}{m^2+3m+2}$

52. $\dfrac{p^2-p-6}{3p-9} \cdot \dfrac{p^2-9}{p^2+6p+9}$

53. $\dfrac{x^2+7xy+12y^2}{x^2+2xy-8y^2} \cdot \dfrac{x^2-xy-2y^2}{x^2+4xy+3y^2}$

54. $\dfrac{m^2+9mn+20n^2}{m^2-25n^2} \cdot \dfrac{m^2-9mn+20n^2}{m^2-16n^2}$

55. $\dfrac{3r^2+15rs+18s^2}{6r^2-24s^2} \cdot \dfrac{2r-4s}{3r+9s}$

56. $\dfrac{2u^2+8u}{2u+8} \cdot \dfrac{4u^2+8uv+4v^2}{u^2+5uv+4v^2}$

57. $\dfrac{abc^2}{a+1} \cdot \dfrac{c}{a^2b^2} \cdot \dfrac{a^2+a}{ac}$

58. $\dfrac{x^3yz^2}{4x+8} \cdot \dfrac{x^2-4}{2x^2y^2z^2} \cdot \dfrac{8yz}{x-2}$

59. $\dfrac{3x^2+5x+2}{x^2-9} \cdot \dfrac{x-3}{x^2-4} \cdot \dfrac{x^2+5x+6}{6x+4}$

60. $\dfrac{x^2-25}{3x+6} \cdot \dfrac{x^2+x-2}{2x+10} \cdot \dfrac{6x}{3x^2-18x+15}$

61. $\dfrac{x^2+5x+6}{x^2} \cdot \dfrac{x^2-2x}{x^2-9} \cdot \dfrac{x^2-3x}{x^2-4}$

62. $\dfrac{x^2-1}{1-x} \cdot \dfrac{4x}{x+x^2} \cdot \dfrac{x^2+2x+1}{2x+2}$

63. $\dfrac{x^2 + 4x}{xz} \cdot \dfrac{z^2 + z}{x^2 - 16} \cdot \dfrac{z + 3}{z^2 + 4z + 3}$

64. $(x + 1) \cdot \dfrac{x^3 - 1}{x^3 + 1} \cdot \dfrac{x^2 - x + 1}{x^2 - 1}$

65. $\dfrac{x^3 + 8}{x^3 - 8} \cdot \dfrac{x - 2}{x^2 - 4} \cdot (x^2 + 2x + 4)$

66. $(4x^3 - 16x) \cdot \dfrac{1}{12x^2 + 24x} \cdot (3x + 12)$

67. $\dfrac{x^2 + x - 6}{5x^2 + 7x + 2} \cdot \dfrac{5x + 2}{-x^2 - x + 6} \cdot \dfrac{x + 1}{x^2 + 4x + 3}$

68. $\dfrac{x^2 - 3x - 4}{3x^2 - 2x - 1} \cdot \dfrac{3x + 1}{x^2 - 6x + 8} \cdot \dfrac{x^2 - 3x + 2}{x^2 + x}$

In Exercises 69–72, do the multiplications. You will need to factor by grouping and factor a sum or difference of two cubes to simplify the answers.

69. $\dfrac{ax + bx + ay + by}{x^3 - y^3} \cdot \dfrac{x^2 + xy + y^2}{ax + bx}$

70. $\dfrac{a^2 - ab + b^2}{a^3 + b^3} \cdot \dfrac{ac + ad + bc + bd}{c^2 - d^2}$

71. $\dfrac{x^2 - y^2}{y^2 - xy} \cdot \dfrac{yx^3 - y^4}{ax + ay + bx + by}$

72. $\dfrac{xw - xz + wy - yz}{x^2 + 2xy + y^2} \cdot \dfrac{x^3 - y^3}{z^2 - w^2}$

Writing Exercises ■ *Write a paragraph using your own words.*

1. Explain how to multiply two fractions and how to simplify the result.

2. Explain why any mathematical expression can be written as a fraction.

Something to Think About ■ **1.** ▦ Let x equal a number of your choosing. Without simplifying first, use a calculator to evaluate

$$\dfrac{x^2 + x - 6}{x^2 + 3x} \cdot \dfrac{x^2}{x - 2}$$

Try again, with a different value of x. If you were to simplify the expression, what do you think you would get?

2. Simplify the expression in Problem 1 to see if your guess was correct.

Review Exercises ■ *Simplify each expression. Write all answers without using negative exponents.*

1. $2x^3y^2(-3x^2y^4z)$

2. $\dfrac{8x^4y^5}{-2x^3y^2}$

3. $(3y)^{-4}$

4. $(a^{-2}a)^{-3}$

5. $\dfrac{x^{3m}}{x^{4m}}$

6. $(3x^2y^3)^0$

5.3 Dividing Fractions

■ Combined Operations

GETTING READY *Divide the fractions and simplify.*

1. $\dfrac{1}{2} \div \dfrac{3}{2}$ 2. $\dfrac{3}{5} \div \dfrac{5}{3}$ 3. $\dfrac{12}{17} \div \dfrac{6}{34}$ 4. $\dfrac{27}{25} \div \dfrac{9}{125}$

5. $\dfrac{7}{8} \div 8$ 6. $\dfrac{7}{8} \div 7$ 7. $15 \div \dfrac{20}{3}$ 8. $\dfrac{37}{17} \div \dfrac{37}{17}$

Recall that division by a nonzero number is equivalent to multiplying by the reciprocal of that number. Thus, to divide two fractions, we must invert the **divisor** (the fraction following the $\div$ sign) and multiply. For example, to divide $\frac{4}{7}$ by $\frac{3}{5}$, we proceed as follows:

$$\frac{4}{7} \div \frac{3}{5} = \frac{4}{7} \cdot \frac{5}{3}$$

$$= \frac{20}{21}$$

In general, the following is true.

Division of Fractions If a is a real number and b, c, and d are nonzero real numbers, then

$$\frac{a}{b} \div \frac{c}{d} = \frac{a}{b} \cdot \frac{d}{c}$$

EXAMPLE 1 Do the divisions **a.** $\dfrac{7}{13} \div \dfrac{21}{26}$ and **b.** $\dfrac{-9x}{35y} \div \dfrac{15x^2}{14}$.

Solution **a.** $\dfrac{7}{13} \div \dfrac{21}{26} = \dfrac{7}{13} \cdot \dfrac{26}{21}$ Invert the divisor and multiply.

$$= \frac{7 \cdot 2 \cdot 13}{13 \cdot 3 \cdot 7}$$ Multiply the fractions and factor where possible.

$$= \frac{\overset{1}{\cancel{7}} \cdot 2 \cdot \overset{1}{\cancel{13}}}{\underset{1}{\cancel{13}} \cdot 3 \cdot \underset{1}{\cancel{7}}}$$ Divide out all common factors.

$$= \frac{2}{3}$$

b. $\dfrac{-9x}{35y} \div \dfrac{15x^2}{14} = \dfrac{-9x}{35y} \cdot \dfrac{14}{15x^2}$ Invert the divisor and multiply.

$= \dfrac{-3 \cdot 3 \cdot x \cdot 2 \cdot 7}{5 \cdot 7 \cdot y \cdot 3 \cdot 5 \cdot x \cdot x}$ Multiply the fractions and factor where possible.

$= \dfrac{-3 \cdot \overset{1}{\cancel{3}} \cdot \overset{1}{\cancel{x}} \cdot 2 \cdot \overset{1}{\cancel{7}}}{5 \cdot \underset{1}{\cancel{7}} \cdot y \cdot \underset{1}{\cancel{3}} \cdot 5 \cdot \underset{1}{\cancel{x}} \cdot x}$ Divide out all common factors.

$= -\dfrac{6}{25xy}$ Multiply the remaining factors. ∎

EXAMPLE 2   Do the division: $\dfrac{x^2 + x}{3x - 15} \div \dfrac{x^2 + 2x + 1}{6x - 30}$.

Solution $\dfrac{x^2 + x}{3x - 15} \div \dfrac{x^2 + 2x + 1}{6x - 30} = \dfrac{x^2 + x}{3x - 15} \cdot \dfrac{6x - 30}{x^2 + 2x + 1}$ Invert the divisor and multiply.

$= \dfrac{x(x + 1) \cdot 2 \cdot 3(x - 5)}{3(x - 5)(x + 1)(x + 1)}$ Multiply the fractions and factor.

$= \dfrac{x\overset{1}{\cancel{(x + 1)}} \cdot 2 \cdot \overset{1}{\cancel{3}}\overset{1}{\cancel{(x - 5)}}}{\underset{1}{\cancel{3}}\underset{1}{\cancel{(x - 5)}}\underset{1}{\cancel{(x + 1)}}(x + 1)}$ Divide out all common factors.

$= \dfrac{2x}{x + 1}$ ∎

EXAMPLE 3   Do the division: $\dfrac{2x^2 - 3x - 2}{2x + 1} \div (4 - x^2)$.

Solution $\dfrac{2x^2 - 3x - 2}{2x + 1} \div (4 - x^2) = \dfrac{2x^2 - 3x - 2}{2x + 1} \div \dfrac{4 - x^2}{1}$ Write $4 - x^2$ as a fraction with a denominator of 1.

$= \dfrac{2x^2 - 3x - 2}{2x + 1} \cdot \dfrac{1}{4 - x^2}$ Invert the divisor and multiply.

$= \dfrac{(2x + 1)(x - 2) \cdot 1}{(2x + 1)(2 + x)(2 - x)}$ Multiply the fractions and factor where possible.

$= \dfrac{\overset{1}{\cancel{(2x + 1)}}\overset{-1}{\cancel{(x - 2)}} \cdot 1}{\underset{1}{\cancel{(2x + 1)}}(2 + x)\underset{1}{\cancel{(2 - x)}}}$ Divide out all common factors.

$= \dfrac{-1}{2 + x}$

$= -\dfrac{1}{2 + x}$ ∎

■ Combined Operations

Unless parentheses indicate otherwise, we do multiplications and divisions in order from left to right.

EXAMPLE 4   Simplify the expression $\dfrac{x^2 - x - 6}{x - 2} \div \dfrac{x^2 - 4x}{x^2 - x - 2} \cdot \dfrac{x - 4}{x^2 + x}$.

Solution   There are no parentheses to indicate otherwise, so the operation of division is done first.

$$\dfrac{x^2 - x - 6}{x - 2} \div \dfrac{x^2 - 4x}{x^2 - x - 2} \cdot \dfrac{x - 4}{x^2 + x}$$

$$= \dfrac{x^2 - x - 6}{x - 2} \cdot \dfrac{x^2 - x - 2}{x^2 - 4x} \cdot \dfrac{x - 4}{x^2 + x}$$ Invert the divisor and multiply.

$$= \dfrac{(x + 2)(x - 3)(x + 1)(x - 2)(x - 4)}{(x - 2)x(x - 4)x(x + 1)}$$ Multiply the fractions and factor.

$$= \dfrac{(x + 2)(x - 3)\overset{1}{\cancel{(x + 1)}}\overset{1}{\cancel{(x - 2)}}\overset{1}{\cancel{(x - 4)}}}{\underset{1}{\cancel{(x - 2)}}x\underset{1}{\cancel{(x - 4)}}x\underset{1}{\cancel{(x + 1)}}}$$ Divide out all common factors.

$$= \dfrac{(x + 2)(x - 3)}{x^2}$$ ■

EXAMPLE 5   Simplify the expression $\dfrac{x^2 + 6x + 9}{x^2 - 2x}\left(\dfrac{x^2 - 4}{x^2 + 3x} \div \dfrac{x + 2}{x}\right)$.

Solution   We do the division within the parentheses first.

$$\dfrac{x^2 + 6x + 9}{x^2 - 2x}\left(\dfrac{x^2 - 4}{x^2 + 3x} \div \dfrac{x + 2}{x}\right) = \dfrac{x^2 + 6x + 9}{x^2 - 2x}\left(\dfrac{x^2 - 4}{x^2 + 3x} \cdot \dfrac{x}{x + 2}\right)$$ Invert the divisor and multiply.

$$= \dfrac{(x + 3)(x + 3)(x - 2)(x + 2)x}{x(x - 2)x(x + 3)(x + 2)}$$ Multiply the fractions and factor where possible.

$$= \dfrac{\overset{1}{\cancel{(x + 3)}}(x + 3)\overset{1}{\cancel{(x - 2)}}\overset{1}{\cancel{(x + 2)}}\overset{1}{\cancel{x}}}{\underset{1}{\cancel{x}}\underset{1}{\cancel{(x - 2)}}x\underset{1}{\cancel{(x + 3)}}\underset{1}{\cancel{(x + 2)}}}$$ Divide out all common factors.

$$= \dfrac{x + 3}{x}$$ ■

ORALS   *Do each division and simplify.*

1. $\dfrac{3}{7} \div \dfrac{7}{3}$ 2. $\dfrac{3}{7} \div \dfrac{3}{7}$

3. $\dfrac{3}{4} \div 3$ 4. $\dfrac{3}{4} \div 4$

5. $3 \div \dfrac{3}{4}$ 6. $4 \div \dfrac{3}{4}$

E X E R C I S E   5.3

In Exercises 1–44, do each division. Simplify answers when possible.

1. $\dfrac{1}{3} \div \dfrac{1}{2}$ 2. $\dfrac{3}{4} \div \dfrac{1}{3}$ 3. $\dfrac{1}{5} \div \dfrac{2}{3}$ 4. $\dfrac{1}{7} \div \dfrac{2}{5}$

5. $\dfrac{2}{5} \div \dfrac{1}{3}$ 6. $\dfrac{3}{7} \div \dfrac{8}{11}$ 7. $\dfrac{8}{5} \div \dfrac{7}{2}$ 8. $\dfrac{9}{19} \div \dfrac{4}{7}$

9. $\dfrac{21}{14} \div \dfrac{5}{2}$ 10. $\dfrac{14}{3} \div \dfrac{10}{3}$ 11. $\dfrac{6}{5} \div \dfrac{6}{7}$ 12. $\dfrac{6}{5} \div \dfrac{14}{5}$

13. $\dfrac{35}{2} \div \dfrac{15}{2}$ 14. $\dfrac{6}{14} \div \dfrac{10}{35}$ 15. $\dfrac{x}{2} \div \dfrac{1}{3}$ 16. $\dfrac{y}{3} \div \dfrac{1}{2}$

17. $\dfrac{2}{y} \div \dfrac{4}{3}$ 18. $\dfrac{3}{a} \div \dfrac{a}{9}$ 19. $\dfrac{3x}{2} \div \dfrac{x}{2}$ 20. $\dfrac{y}{6} \div \dfrac{2}{3y}$

21. $\dfrac{3x}{y} \div \dfrac{2x}{4}$ 22. $\dfrac{3y}{8} \div \dfrac{2y}{4y}$ 23. $\dfrac{4x}{3x} \div \dfrac{2y}{9y}$ 24. $\dfrac{14}{7y} \div \dfrac{10}{5z}$

25. $\dfrac{x^2}{3} \div \dfrac{2x}{4}$ 26. $\dfrac{z^2}{z} \div \dfrac{z}{3z}$ 27. $\dfrac{y^2}{5z} \div \dfrac{3z}{2z}$ 28. $\dfrac{xy}{x^2} \div \dfrac{y^2}{5}$

29. $\dfrac{x^2y}{3xy} \div \dfrac{xy^2}{6y}$ 30. $\dfrac{2xz}{z} \div \dfrac{4x^2}{z^2}$ 31. $\dfrac{x+2}{3x} \div \dfrac{x+2}{2}$ 32. $\dfrac{z-3}{3z} \div \dfrac{z+3}{z}$

33. $\dfrac{(z-2)^2}{3z^2} \div \dfrac{z-2}{6z}$ 34. $\dfrac{(x+7)^2}{x+7} \div \dfrac{(x-3)^2}{x+7}$

35. $\dfrac{(z-7)^2}{z+2} \div \dfrac{z(z-7)}{5z^2}$ 36. $\dfrac{y(y+2)}{y^2(y-3)} \div \dfrac{y^2(y+2)}{(y-3)^2}$

37. $\dfrac{x^2-4}{3x+6} \div \dfrac{x-2}{x+2}$ 38. $\dfrac{x^2-9}{5x+15} \div \dfrac{x-3}{x+3}$

39. $\dfrac{x^2-1}{3x-3} \div \dfrac{x+1}{3}$ 40. $\dfrac{x^2-16}{x-4} \div \dfrac{3x+12}{x}$

41. $\dfrac{5x^2+13x-6}{x+3} \div \dfrac{5x^2-17x+6}{x-2}$ 42. $\dfrac{x^2-x-6}{2x^2+9x+10} \div \dfrac{x^2-25}{2x^2+15x+25}$

43. $\dfrac{2x^2 + 8x - 42}{x - 3} \div \dfrac{2x^2 + 14x}{x^2 + 5x}$

44. $\dfrac{x^2 - 2x - 35}{3x^2 + 27x} \div \dfrac{x^2 + 7x + 10}{6x^2 + 12x}$

In Exercises 45–68, do the indicated operations. In the absence of grouping symbols, multiplications and divisions are performed as they are encountered from left to right.

45. $\dfrac{2}{3} \cdot \dfrac{15}{5} \div \dfrac{10}{5}$

46. $\dfrac{6}{5} \div \dfrac{3}{5} \cdot \dfrac{5}{15}$

47. $\dfrac{6}{7} \div \dfrac{5}{2} \cdot \dfrac{5}{4}$

48. $\dfrac{15}{7} \div \dfrac{5}{2} \div \dfrac{4}{2}$

49. $\dfrac{x}{3} \cdot \dfrac{9}{4} \div \dfrac{x^2}{6}$

50. $\dfrac{y^2}{2} \div \dfrac{4}{y} \cdot \dfrac{y^2}{8}$

51. $\dfrac{x^2}{18} \div \dfrac{x^3}{6} \div \dfrac{12}{x^2}$

52. $\dfrac{y^3}{3y} \cdot \dfrac{3y^2}{4} \div \dfrac{15}{20}$

53. $\dfrac{x^2 - 1}{x^2 - 9} \cdot \dfrac{x + 3}{x + 2} \div \dfrac{5}{x + 2}$

54. $\dfrac{2}{3x - 3} \div \dfrac{2x + 2}{x - 1} \cdot \dfrac{5}{x + 1}$

55. $\dfrac{x^2 - 4}{2x + 6} \div \dfrac{x + 2}{4} \cdot \dfrac{x + 3}{x - 2}$

56. $\dfrac{x^2 - 5x}{x + 1} \cdot \dfrac{x + 1}{x^2 + 3x} \div \dfrac{x - 5}{x - 3}$

57. $\dfrac{x - x^2}{x^2 - 4} \left(\dfrac{2x + 4}{x + 2} \div \dfrac{5}{x + 2} \right)$

58. $\dfrac{2}{3x - 3} \div \left(\dfrac{2x + 2}{x - 1} \cdot \dfrac{5}{x + 1} \right)$

59. $\dfrac{y^2}{x + 1} \cdot \dfrac{x^2 + 2x + 1}{x^2 - 1} \div \dfrac{3y}{xy - y}$

60. $\dfrac{x^2 - y^2}{x^4 - x^3} \div \dfrac{x - y}{x^2} \div \dfrac{x^2 + 2xy + y^2}{x + y}$

61. $\dfrac{x^2 + x - 6}{x^2 - 4} \cdot \dfrac{x^2 + 2x}{x - 2} \div \dfrac{x^2 + 3x}{x + 2}$

62. $\dfrac{x^2 - x - 6}{x^2 + 6x - 7} \cdot \dfrac{x^2 + x - 2}{x^2 + 2x} \div \dfrac{x^2 + 7x}{x^2 - 3x}$

63. $(a + 2b) \div \left(\dfrac{a^2 + 4ab + 4b^2}{a + b} \div \dfrac{a^2 + 7ab + 10b^2}{a^2 + 6ab + 5b^2} \right)$

64. $(ab - 2b^2) \div \left(\dfrac{a^2 - ab}{b - a} \cdot \dfrac{a^2 - b^2}{a^3 - 3a^2b + 2ab^2} \right)$

65. $\dfrac{x^2 + 2x - 3}{x^2 + x} \cdot \dfrac{x^2}{x^2 - 1} \div (x^2 + 3x)$

66. $\dfrac{x^2 - 6x + 5}{x + 2} \div (x^2 + 3x - 4) \cdot \dfrac{x^2 + 5x + 6}{x^2 - 2x - 15}$

67. $\dfrac{x^2 + 4x + 3}{x^2 - y^2} \div \dfrac{xy + y}{xy - x^2} \cdot \dfrac{x^2y + 2xy^2 + y^3}{x^2 + 3x}$

68. $\dfrac{a^2 - b^2}{a^2 - a - 2} \cdot \dfrac{a^2 - 2a - 3}{b - a} \div \dfrac{a^2 + ab}{a^2 - 2a}$

In Exercises 69–72, do the divisions. You will need to factor by grouping and factor a sum or difference of two cubes to simplify the answers.

69. $\dfrac{ab + 4a + 2b + 8}{b^2 + 4b + 16} \div \dfrac{b^2 - 16}{b^3 - 64}$

70. $\dfrac{r^3 - s^3}{r^2 - s^2} \div \dfrac{r^2 + rs + s^2}{mr + ms + nr + ns}$

71. $\dfrac{p^3 - p^2q + pq^2}{mp - mq + np - nq} \div \dfrac{q^3 + p^3}{q^2 - p^2}$

72. $\dfrac{s^3 - r^3}{r^2 + rs + s^2} \div \dfrac{pr - ps - qr + qs}{q^2 - p^2}$

Writing Exercises ■ *Write a paragraph using your own words.*

1. To divide fractions, you must first know how to multiply fractions. Explain.

2. Explain how to do the division $\dfrac{a}{b} \div \dfrac{c}{d} \div \dfrac{e}{f}$.

Something to Think About ■

1. When the division problem

$$\frac{a}{b} \div \left[\frac{c}{d} \div \left(\frac{e}{f} \div \frac{g}{h} \right) \right]$$

is written to involve only multiplications, which fractions will have been inverted?

2. Answer Question 1 again for the division problem

$$\left[\frac{a}{b} \div \left(\frac{c}{d} \div \frac{e}{f} \right) \right] \div \frac{g}{h}$$

Review Exercises ■ *Do the indicated operations and simplify.*

1. $-4(y^3 - 4y^2 + 3y - 2) + 6(-2y^2 + 4) - 4(-2y^3 - y)$

2. $6(3a^3 + 2a^2 + 3) - (-2a^2 + 4a - 2) + 5(-2a^3 - a^2 + 2a - 3)$

3. $(3m + 2)(-2m + 1)(m - 1)$ **4.** $(2p - q)(p + q)^2$

5. $y - 5 \overline{)5y^3 - 3y^2 + 4y - 1}$ **6.** $x + 4 \overline{)6x^3 + 5 - 4x}$

5.4   Adding and Subtracting Fractions with Like Denominators

■ Adding Fractions with Like Denominators ■ Subtracting Fractions with Like Denominators

GETTING READY   *Add the fractions and simplify.*

1. $\dfrac{1}{5} + \dfrac{3}{5}$ **2.** $\dfrac{3}{7} + \dfrac{4}{7}$ **3.** $\dfrac{3}{8} + \dfrac{4}{8}$ **4.** $\dfrac{18}{19} + \dfrac{20}{19}$

Subtract the fractions and simplify.

5. $\dfrac{5}{9} - \dfrac{4}{9}$ **6.** $\dfrac{7}{12} - \dfrac{1}{12}$ **7.** $\dfrac{7}{13} - \dfrac{9}{13}$ **8.** $\dfrac{20}{10} - \dfrac{7}{10}$

■ Adding Fractions with Like Denominators

Recall that to add or subtract fractions with the same denominator, called a **common denominator,** we add or subtract their numerators and keep the same denominator. For example, we add $\frac{2}{7}$ and $\frac{3}{7}$ as follows:

$$\frac{2}{7} + \frac{3}{7} = \frac{2 + 3}{7}$$

$$= \frac{5}{7}$$

In general, the following is true.

Adding Fractions with Like Denominators

If a, b, and d represent real numbers, and $d \neq 0$, then

$$\frac{a}{d} + \frac{b}{d} = \frac{a+b}{d}$$

EXAMPLE 1 Do the following additions. Simplify each result if possible.

a. $\dfrac{5}{9} + \dfrac{2}{9} = \dfrac{5+2}{9}$

$= \dfrac{7}{9}$

b. $\dfrac{8}{41} + \dfrac{21}{41} = \dfrac{8+21}{41}$

$= \dfrac{29}{41}$

c. $\dfrac{x}{7} + \dfrac{y}{7} = \dfrac{x+y}{7}$

d. $\dfrac{x}{7} + \dfrac{3x}{7} = \dfrac{x+3x}{7}$

$= \dfrac{4x}{7}$

e. $\dfrac{3x+y}{5x} + \dfrac{x+y}{5x} = \dfrac{3x+y+x+y}{5x}$

$= \dfrac{4x+2y}{5x}$

f. $\dfrac{3x}{7y} + \dfrac{4x}{7y} = \dfrac{3x+4x}{7y}$

$= \dfrac{7x}{7y}$

$= \dfrac{x}{y}$ ∎

EXAMPLE 2 Add the fractions $\dfrac{3x+21}{5x+10}$ and $\dfrac{8x+1}{5x+10}$.

Solution Because the fractions have the same denominator, we add their numerators and keep the common denominator.

$$\frac{3x+21}{5x+10} + \frac{8x+1}{5x+10} = \frac{3x+21+8x+1}{5x+10} \qquad \text{Add the fractions.}$$

$$= \frac{11x+22}{5x+10} \qquad \text{Combine like terms.}$$

$$= \frac{11(\overset{1}{\cancel{x+2}})}{5(\underset{1}{\cancel{x+2}})} \qquad \begin{array}{l}\text{Factor and divide out the}\\ \text{common factor of } x+2.\end{array}$$

$$= \frac{11}{5} \qquad \blacksquare$$

■ Subtracting Fractions with Like Denominators

Recall that to subtract fractions with the same denominator, we subtract their numerators and keep the same denominator.

Subtracting Fractions with Like Denominators	If a, b, and d represent real numbers, and $d \neq 0$, then $$\frac{a}{d} - \frac{b}{d} = \frac{a-b}{d}$$

 (b)

EXAMPLE 3   Do each subtraction and simplify: **a.** $\dfrac{5x}{3} - \dfrac{2x}{3}$ and   **b.** $\dfrac{5x+1}{x-3} - \dfrac{4x-2}{x-3}$.

Solution   In each part, both fractions have the same denominator. To subtract them, we subtract their numerators and keep the common denominator.

a. $\dfrac{5x}{3} - \dfrac{2x}{3} = \dfrac{5x - 2x}{3}$ Subtract the fractions.

$\qquad\qquad = \dfrac{3x}{3}$ Combine like terms.

$\qquad\qquad = \dfrac{x}{1}$ Simplify the fraction.

$\qquad\qquad = x$ Denominators of 1 need not be written.

b. $\dfrac{5x+1}{x-3} - \dfrac{4x-2}{x-3} = \dfrac{(5x+1) - (4x-2)}{x-3}$ Subtract the fractions.

$\qquad\qquad\qquad = \dfrac{5x + 1 - 4x + 2}{x-3}$ Remove parentheses.

$\qquad\qquad\qquad = \dfrac{x+3}{x-3}$ Combine like terms.   ■

The denominator of a fraction can never be zero. However, if the numerator of a fraction is 0, then the value of the fraction is 0. This fact is used in the next example.

EXAMPLE 4   Do the operations: $\dfrac{3x+1}{x-7} - \dfrac{5x+2}{x-7} + \dfrac{2x+1}{x-7}$.

Solution   This example combines both addition and subtraction of fractions with like denominators. Unless parentheses indicate otherwise, additions and subtractions are done in order from left to right.

$$\frac{3x+1}{x-7} - \frac{5x+2}{x-7} + \frac{2x+1}{x-7}$$

$$= \frac{(3x+1)-(5x+2)+(2x+1)}{x-7}$$ Combine the numerators and keep the common denominator.

$$= \frac{3x+1-5x-2+2x+1}{x-7}$$ Remove parentheses.

$$= \frac{0}{x-7}$$ Combine like terms.

$$= 0$$ Simplify.   ■

ORALS *Find each sum.*

1. $\dfrac{5}{7} + \dfrac{9}{7}$ **2.** $\dfrac{3}{11} + \dfrac{8}{11}$ **3.** $\dfrac{5}{x} + \dfrac{8}{x}$ **4.** $\dfrac{y}{5} + \dfrac{1}{5}$

Find each difference.

5. $\dfrac{9}{4} - \dfrac{5}{4}$ **6.** $\dfrac{15}{7} - \dfrac{1}{7}$ **7.** $\dfrac{7}{z} - \dfrac{1}{z}$ **8.** $\dfrac{r}{t} - \dfrac{1}{t}$

EXERCISE 5.4

In Exercises 1–30, do each addition. Write all answers in lowest terms.

1. $\dfrac{1}{3} + \dfrac{1}{3}$ **2.** $\dfrac{3}{4} + \dfrac{3}{4}$ **3.** $\dfrac{1}{5} + \dfrac{2}{5}$ **4.** $\dfrac{3}{7} + \dfrac{2}{7}$

5. $\dfrac{2}{9} + \dfrac{1}{9}$ **6.** $\dfrac{5}{7} + \dfrac{9}{7}$ **7.** $\dfrac{8}{7} + \dfrac{6}{7}$ **8.** $\dfrac{9}{11} + \dfrac{2}{11}$

9. $\dfrac{21}{14} + \dfrac{7}{14}$ **10.** $\dfrac{14}{3} + \dfrac{10}{3}$ **11.** $\dfrac{6}{7} + \dfrac{6}{7}$ **12.** $\dfrac{6}{5} + \dfrac{14}{5}$

13. $\dfrac{35}{8} + \dfrac{15}{8}$ **14.** $\dfrac{6}{14} + \dfrac{10}{14}$ **15.** $\dfrac{14x}{11} + \dfrac{30x}{11}$ **16.** $\dfrac{6a}{10} + \dfrac{28a}{10}$

17. $\dfrac{-77y}{126} + \dfrac{-7y}{126}$ **18.** $\dfrac{-39a}{15} + \dfrac{-21a}{15}$ **19.** $\dfrac{15z}{22} + \dfrac{-15z}{22}$ **20.** $\dfrac{-30rs}{21} + \dfrac{30rs}{21}$

21. $\dfrac{2x}{y} + \dfrac{2x}{y}$ **22.** $\dfrac{3y}{5} + \dfrac{2y}{5}$ **23.** $\dfrac{4y}{3x} + \dfrac{2y}{3x}$ **24.** $\dfrac{4}{7y} + \dfrac{10}{7y}$

25. $\dfrac{x^2}{4y} + \dfrac{x^2}{4y}$ **26.** $\dfrac{r^2}{r} + \dfrac{r^2}{r}$ **27.** $\dfrac{y+2}{5z} + \dfrac{y+4}{5z}$ **28.** $\dfrac{x+3}{x^2} + \dfrac{x+5}{x^2}$

29. $\dfrac{3x-5}{x-2} + \dfrac{6x-13}{x-2}$

30. $\dfrac{8x-7}{x+3} + \dfrac{2x+37}{x+3}$

In Exercises 31–60, do each subtraction. Simplify answers if possible.

31. $\dfrac{5}{7} - \dfrac{4}{7}$

32. $\dfrac{5}{9} - \dfrac{3}{9}$

33. $\dfrac{4}{3} - \dfrac{8}{3}$

34. $\dfrac{7}{11} - \dfrac{4}{11}$

35. $\dfrac{17}{13} - \dfrac{15}{13}$

36. $\dfrac{18}{31} - \dfrac{18}{31}$

37. $\dfrac{21}{23} - \dfrac{45}{23}$

38. $\dfrac{35}{72} - \dfrac{44}{72}$

39. $\dfrac{39}{37} - \dfrac{2}{37}$

40. $\dfrac{35}{99} - \dfrac{13}{99}$

41. $\dfrac{-47}{123} - \dfrac{4}{123}$

42. $\dfrac{-23}{17} - \dfrac{11}{17}$

43. $\dfrac{15}{21} - \left(\dfrac{-15}{21}\right)$

44. $\dfrac{-37}{25} - \left(\dfrac{-22}{25}\right)$

45. $\dfrac{2x}{y} - \dfrac{x}{y}$

46. $\dfrac{7y}{5} - \dfrac{4y}{5}$

47. $\dfrac{9y}{3x} - \dfrac{6y}{3x}$

48. $\dfrac{24}{7y} - \dfrac{10}{7y}$

49. $\dfrac{3x^2}{4x} - \dfrac{x^2}{4x}$

50. $\dfrac{5r^2}{2r} - \dfrac{r^2}{2r}$

51. $\dfrac{y+2}{5z} - \dfrac{y+4}{5z}$

52. $\dfrac{x+3}{x^2} - \dfrac{x+5}{x^2}$

53. $\dfrac{6x-5}{3xy} - \dfrac{3x-5}{3xy}$

54. $\dfrac{7x+7}{5y} - \dfrac{2x+7}{5y}$

55. $\dfrac{y+2}{2z} - \dfrac{y+4}{2z}$

56. $\dfrac{2x-3}{x^2} - \dfrac{x-3}{x^2}$

57. $\dfrac{5x+5}{3xy} - \dfrac{2x-4}{3xy}$

58. $\dfrac{8x-7}{2y} - \dfrac{2x+7}{2y}$

59. $\dfrac{3y-2}{y+3} - \dfrac{2y-5}{y+3}$

60. $\dfrac{5x+8}{x+5} - \dfrac{3x-2}{x+5}$

In Exercises 61–76, do the operations. Simplify answers if possible.

61. $\dfrac{3}{7} - \dfrac{5}{7} + \dfrac{2}{7}$

62. $\dfrac{3}{4} - \dfrac{5}{4} + \dfrac{8}{4}$

63. $\dfrac{3}{5} - \dfrac{2}{5} + \dfrac{7}{5}$

64. $\dfrac{5}{11} - \dfrac{8}{11} + \dfrac{14}{11}$

65. $\dfrac{13x}{15} + \dfrac{12x}{15} - \dfrac{5x}{15}$

66. $\dfrac{13y}{32} + \dfrac{13y}{32} - \dfrac{10y}{32}$

67. $\dfrac{x}{3y} + \dfrac{2x}{3y} - \dfrac{x}{3y}$

68. $\dfrac{5y}{8x} + \dfrac{4y}{8x} - \dfrac{y}{8x}$

69. $\dfrac{3x}{y+2} - \dfrac{3y}{y+2} + \dfrac{x+y}{y+2}$

70. $\dfrac{3y}{x-5} + \dfrac{x}{x-5} - \dfrac{y-x}{x-5}$

71. $\dfrac{x+1}{x-2} - \dfrac{2(x-3)}{x-2} + \dfrac{3(x+1)}{x-2}$

72. $\dfrac{x^2-4}{x+2} + \dfrac{2(x^2-9)}{x+2} - \dfrac{3(x^2-5)}{x+2}$

73. $\dfrac{3xy}{x-y} - \dfrac{x(3y-x)}{x-y} - \dfrac{x(x-y)}{x-y}$

74. $\dfrac{x^2+4x+1}{(x-1)^2} - \dfrac{x(x+1)}{(x-1)^2} - \dfrac{x}{(x-1)^2}$

75. $\dfrac{2(2a+b)}{(a-b)^2} - \dfrac{2(2b+a)}{(a-b)^2} + \dfrac{3(b-a)}{(a-b)^2}$

76. $\dfrac{2(x-2)}{2x-3} + \dfrac{2(2x+1)}{2x-3} - \dfrac{2(5x-4)}{2x-3}$

Writing Exercises ■ *Write a paragraph using your own words.*

1. Explain how to add fractions with the same denominator.

2. Explain how to subtract fractions with the same denominator.

Something to Think About ■   **1.** Find the mistake:

$$\frac{2x+3}{x+5} - \frac{x+2}{x+5} = \frac{2x+3-x+2}{x+5}$$
$$= \frac{x+5}{x+5}$$
$$= 1$$

2. Find the mistake:

$$\frac{5x-4}{y} + \frac{x}{y} = \frac{5x-4+x}{y+y}$$
$$= \frac{6x-4}{2y}$$
$$= \frac{3x-2}{y}$$

Review Exercises ■   *Write each number in prime-factored form.*

1. 49   **2.** 64   **3.** 136   **4.** 242

5. 102   **6.** 315   **7.** 144   **8.** 145

5.5 Adding and Subtracting Fractions with Unlike Denominators

■ Least Common Denominator

GETTING READY   *Decide if the fractions in each pair are equal.*

1. $\frac{1}{2}, \frac{7}{14}$ **2.** $\frac{2}{3}, \frac{10}{15}$ **3.** $\frac{3}{7}, \frac{14}{18}$ **4.** $\frac{6}{13}, \frac{18}{39}$

5. $\frac{4}{3}, \frac{12}{9}$ **6.** $\frac{5}{2}, \frac{12}{9}$ **7.** $\frac{7}{2}, \frac{21}{6}$ **8.** $\frac{5}{23}, \frac{10}{46}$

Recall that to add fractions with unlike denominators, we must first convert them to fractions with the same denominator. To do so, we can use the fundamental property of fractions to multiply both the numerator and denominator of each fraction by some appropriate nonzero number. The fractions can then be added. For example, we add the fractions $\frac{4}{7}$ and $\frac{3}{5}$ as follows:

$$\frac{4}{7} + \frac{3}{5} = \frac{4 \cdot 5}{7 \cdot 5} + \frac{3 \cdot 7}{5 \cdot 7} \qquad \text{Multiply both the numerator and the denominator of the first fraction by 5 and those of the second fraction by 7.}$$

$$= \frac{20}{35} + \frac{21}{35} \qquad \text{Do the multiplications.}$$

$$= \frac{41}{35} \qquad \text{Add the fractions.}$$

The process of multiplying both the numerator and denominator of a fraction by the same nonzero number is often called **building a fraction.**

EXAMPLE 1 Change **a.** $\dfrac{1}{2}$, **b.** $\dfrac{3}{5}$, and **c.** $\dfrac{7}{10}$ into fractions that have a common denominator of 30.

Solution Build each fraction as follows.

$$\textbf{a.}\ \frac{1}{2} = \frac{1 \cdot 15}{2 \cdot 15} \qquad\qquad \textbf{b.}\ \frac{3}{5} = \frac{3 \cdot 6}{5 \cdot 6} \qquad\qquad \textbf{c.}\ \frac{7}{10} = \frac{7 \cdot 3}{10 \cdot 3}$$

$$= \frac{15}{30} \qquad\qquad\qquad\quad = \frac{18}{30} \qquad\qquad\qquad\quad = \frac{21}{30} \qquad\blacksquare$$

To add fractions with unlike denominators, we must express those fractions in equivalent forms with the same denominator.

EXAMPLE 2 Add the fractions $\dfrac{5}{14}$ and $\dfrac{2}{21}$.

Solution Build the fractions so that each has a denominator of 42.

$$\frac{5}{14} + \frac{2}{21} = \frac{5 \cdot 3}{14 \cdot 3} + \frac{2 \cdot 2}{21 \cdot 2} \qquad \begin{array}{l}\text{Multiply both the numerator and denominator}\\ \text{of the first fraction by 3 and those of the}\\ \text{second fraction by 2.}\end{array}$$

$$= \frac{15}{42} + \frac{4}{42} \qquad\qquad \text{Do the multiplications.}$$

$$= \frac{19}{42} \qquad\qquad\qquad \text{Add the fractions.} \qquad\blacksquare$$

■ Least Common Denominator

In Example 2, the number 42 was used as the common denominator because it is divisible by both 14 and 21. The number 42 is also the *smallest* number that is divisible by both 14 and 21. The smallest number that can be divided by each of the denominators of a set of fractions is called the **least** (or **lowest**) **common denominator (LCD)** of those fractions.

We have seen that there is a process to find the least common denominator of a set of fractions.

Finding the Least Common Denominator (LCD)

1. Write down each of the different denominators that appear in the given fractions.

2. Factor each of these denominators completely.

3. Form a product using each of the different factors obtained in Step 2. Use each different factor the *greatest* number of times it appears in any *single* factorization. This product is the least common denominator.

EXAMPLE 3 Several fractions have denominators of 24, 18, and 36. Find the least common denominator.

Solution First, we write down and factor each denominator into products of prime numbers.

$$24 = 2 \cdot 2 \cdot 2 \cdot 3 = 2^3 \cdot 3$$
$$18 = 2 \cdot 3 \cdot 3 = 2 \cdot 3^2$$
$$36 = 2 \cdot 2 \cdot 3 \cdot 3 = 2^2 \cdot 3^2$$

We then form a product with the factors of 2 and 3, using each of these factors the greatest number of times it appears in any single factorization. That is, we use the factor 2 three times because 2 appears three times as a factor of 24. We use the factor of 3 twice because it occurs twice as a factor of 18 and 36. Thus, the least common denominator of 24, 18 and 36 is

$$\text{LCD} = 2 \cdot 2 \cdot 2 \cdot 3 \cdot 3$$
$$= 8 \cdot 9$$
$$= 72 \qquad \blacksquare$$

EXAMPLE 4 Add the fractions $\dfrac{1}{24}, \dfrac{5}{18},$ and $\dfrac{7}{36}$.

Solution The fractions $\frac{1}{24}, \frac{5}{18},$ and $\frac{7}{36}$ have different denominators, but each one can be written as a fraction with the least common denominator of $2 \cdot 2 \cdot 2 \cdot 3 \cdot 3$, or 72. (See Example 3.) We proceed as follows:

$$\frac{1}{24} + \frac{5}{18} + \frac{7}{36} = \frac{1}{2 \cdot 2 \cdot 2 \cdot 3} + \frac{5}{2 \cdot 3 \cdot 3} + \frac{7}{2 \cdot 2 \cdot 3 \cdot 3} \qquad \text{Factor each denominator.}$$

In each of these fractions, we multiply both the numerator and the denominator by whatever is necessary to build the denominator to the required $2 \cdot 2 \cdot 2 \cdot 3 \cdot 3$.

$$\frac{1}{24} + \frac{5}{18} + \frac{7}{36} = \frac{1 \cdot 3}{2 \cdot 2 \cdot 2 \cdot 3 \cdot 3} + \frac{5 \cdot 2 \cdot 2}{2 \cdot 3 \cdot 3 \cdot 2 \cdot 2} + \frac{7 \cdot 2}{2 \cdot 2 \cdot 3 \cdot 3 \cdot 2} \qquad \text{Build each fraction.}$$

$$= \frac{3 + 20 + 14}{72} \qquad \text{Do the multiplications and add the fractions.}$$

$$= \frac{37}{72} \qquad \text{Simplify.} \qquad \blacksquare$$

EXAMPLE 5 Add $\dfrac{1}{x} + \dfrac{x}{y}$.

Solution The least common denominator is xy.

$$\frac{1}{x} + \frac{x}{y} = \frac{(1)y}{(x)y} + \frac{x(x)}{x(y)} \qquad \text{Build the fractions to get the common denominator of } xy.$$

$$= \frac{y}{xy} + \frac{x^2}{xy} \qquad \text{Do the multiplications.}$$

$$= \frac{y + x^2}{xy} \qquad \text{Add the fractions.} \qquad \blacksquare$$

EXAMPLE 6 Do the operations: $\dfrac{3}{x^2y} + \dfrac{2}{xy} - \dfrac{1}{xy^2}$.

Solution Find the least common denominator.

$$\left.\begin{array}{l} x^2y = x \cdot x \cdot y \\[4pt] xy = x \cdot y \\[4pt] xy^2 = x \cdot y \cdot y \end{array}\right\} \quad \text{Factor each denominator.}$$

In any one of these denominators, the factor x occurs at most twice, and the factor y occurs at most twice. Thus, the LCD is

$$\text{LCD} = x \cdot x \cdot y \cdot y$$
$$= x^2y^2$$

We build each given fraction into a new fraction with a denominator of x^2y^2.

$$\frac{3}{x^2y} + \frac{2}{xy} - \frac{1}{xy^2}$$

$$= \frac{3 \cdot y}{x \cdot x \cdot y \cdot y} + \frac{2 \cdot x \cdot y}{x \cdot y \cdot x \cdot y} - \frac{1 \cdot x}{x \cdot y \cdot y \cdot x} \qquad \begin{array}{l}\text{Factor each denominator}\\\text{and build each fraction.}\end{array}$$

$$= \frac{3y + 2xy - x}{x^2y^2} \qquad \begin{array}{l}\text{Do the multiplications}\\\text{and add the fractions.}\end{array} \qquad \blacksquare$$

EXAMPLE 7 Do the subtraction: $\dfrac{x}{x + 1} - \dfrac{3}{x}$.

Solution The least common denominator is $(x + 1)x$.

$$\frac{x}{x + 1} - \frac{3}{x} = \frac{x(x)}{(x + 1)x} - \frac{3(x + 1)}{x(x + 1)} \qquad \begin{array}{l}\text{Build the fractions to get the common}\\\text{denominator.}\end{array}$$

$$= \frac{x^2}{(x + 1)x} - \frac{3x + 3}{x(x + 1)} \qquad \text{Do the multiplication in the numerator.}$$

$$= \frac{x^2 - 3x - 3}{x(x + 1)} \qquad \begin{array}{l}\text{Subtract the numerators and keep}\\\text{the common denominator.}\end{array} \qquad \blacksquare$$

EXAMPLE 8   Do the indicated operations: $\dfrac{3}{x^2 - y^2} + \dfrac{2}{x - y} - \dfrac{1}{x + y}$.

Solution   Find the least common denominator.

$$\left. \begin{array}{l} x^2 - y^2 = (x - y)(x + y) \\ x - y = x - y \\ x + y = x + y \end{array} \right\} \text{ Factor each denominator where possible.}$$

Since the least common denominator is $(x - y)(x + y)$, we build each given fraction into a new fraction with that common denominator.

$$\dfrac{3}{x^2 - y^2} + \dfrac{2}{x - y} - \dfrac{1}{x + y} = \dfrac{3}{(x - y)(x + y)} + \dfrac{2}{x - y} - \dfrac{1}{x + y} \qquad \text{Factor } x^2 - y^2.$$

$$= \dfrac{3}{(x - y)(x + y)} + \dfrac{2(x + y)}{(x - y)(x + y)} - \dfrac{1(x - y)}{(x + y)(x - y)} \qquad \begin{array}{l}\text{Build the fractions}\\\text{to get a common}\\\text{denominator.}\end{array}$$

$$= \dfrac{3}{(x - y)(x + y)} + \dfrac{2x + 2y}{(x - y)(x + y)} - \dfrac{x - y}{(x + y)(x - y)} \qquad \text{Remove parentheses.}$$

$$= \dfrac{3 + 2x + 2y - x + y}{(x - y)(x + y)} \qquad \text{Add the fractions.}$$

$$= \dfrac{3 + x + 3y}{(x - y)(x + y)} \qquad \begin{array}{l}\text{Combine like}\\\text{terms.}\end{array} \blacksquare$$

EXAMPLE 9   Do the subtraction $\dfrac{a}{a - 1} - \dfrac{2}{a^2 - 1}$ and simplify.

Solution   Factor $a^2 - 1$ and write each fraction as a fraction with a denominator of $(a + 1)(a - 1)$.

$$\dfrac{a}{a - 1} - \dfrac{2}{a^2 - 1} = \dfrac{a(a + 1)}{(a - 1)(a + 1)} - \dfrac{2}{(a + 1)(a - 1)} \qquad \begin{array}{l}\text{Factor } a^2 - 1 \text{ and build}\\\text{the first fraction.}\end{array}$$

$$= \dfrac{a^2 + a}{(a - 1)(a + 1)} - \dfrac{2}{(a + 1)(a - 1)} \qquad \begin{array}{l}\text{Remove parentheses in}\\\text{the numerator.}\end{array}$$

$$= \dfrac{a^2 + a - 2}{(a - 1)(a + 1)} \qquad \text{Subtract the fractions.}$$

$$= \dfrac{(a + 2)\overset{1}{\cancel{(a - 1)}}}{\underset{1}{\cancel{(a - 1)}}(a + 1)} \qquad \text{Factor.}$$

$$= \dfrac{a + 2}{a + 1} \qquad \begin{array}{l}\text{Divide out the common}\\\text{factor of } a - 1.\end{array} \blacksquare$$

EXAMPLE 10   Do the addition: $\dfrac{m+1}{2m+6} + \dfrac{4-m^2}{2m^2+2m-12}$.

Solution   Factor $2m+6$ and $2m^2+2m-12$ and write each fraction with a denominator of $2(m+3)(m-2)$:

$$\frac{m+1}{2m+6} + \frac{4-m^2}{2m^2+2m-12} = \frac{m+1}{2(m+3)} + \frac{4-m^2}{2(m+3)(m-2)}$$

$$= \frac{(m+1)(m-2)}{2(m+3)(m-2)} + \frac{4-m^2}{2(m+3)(m-2)}$$

Now add the fractions by adding the numerators and keeping the common denominator, and simplify.

$$\frac{m+1}{2m+6} + \frac{4-m^2}{2m^2+2m-12} = \frac{(m+1)(m-2)+4-m^2}{2(m+3)(m-2)}$$

$$= \frac{m^2-m-2+4-m^2}{2(m+3)(m-2)} \qquad \text{Multiply } (m+1)(m-2).$$

$$= \frac{-m+2}{2(m+3)(m-2)} \qquad \text{Combine like terms.}$$

$$= \frac{\overset{1}{-\cancel{(m-2)}}}{2(m+3)\underset{1}{\cancel{(m-2)}}} \qquad \text{Factor } -1 \text{ from } -m+2.$$

$$= \frac{-1}{2(m+3)} \qquad \begin{array}{l}\text{Divide out the} \\ \text{common factor} \\ \text{of } m-2. \quad \blacksquare\end{array}$$

ORALS   *Tell whether the fractions are equal.*

1. $\dfrac{1}{2}, \dfrac{6}{12}$ **2.** $\dfrac{3}{8}, \dfrac{15}{40}$ **3.** $\dfrac{7}{9}, \dfrac{14}{27}$ **4.** $\dfrac{5}{10}, \dfrac{15}{30}$

5. $\dfrac{x}{3}, \dfrac{3x}{9}$ **6.** $\dfrac{5}{3}, \dfrac{5x}{3y}$ **7.** $\dfrac{5}{3}, \dfrac{5x}{3x}$ **8.** $\dfrac{5y}{10}, \dfrac{y}{2}$

E X E R C I S E   5.5

In Exercises 1–20, build each fraction into an equivalent fraction with the indicated denominator.

1. $\dfrac{2}{3}$; 6   **2.** $\dfrac{3}{4}$; 12   **3.** $\dfrac{25}{4}$; 20   **4.** $\dfrac{19}{21}$; 42

5. $\dfrac{2}{x}$; x^2

6. $\dfrac{3}{y}$; y^2

7. $\dfrac{5}{y}$; xy

8. $\dfrac{3}{x}$; xy

9. $\dfrac{8}{x}$; x^2y

10. $\dfrac{7}{y}$; xy^2

11. $\dfrac{3x}{x+1}$; $(x+1)^2$

12. $\dfrac{5y}{y-2}$; $(y-2)^2$

13. $\dfrac{2y}{x}$; x^2+x

14. $\dfrac{3x}{y}$; y^2-y

15. $\dfrac{z}{z-1}$; z^2-1

16. $\dfrac{y}{y+2}$; y^2-4

17. $\dfrac{x+2}{x-2}$; x^2-4

18. $\dfrac{x-3}{x+3}$; x^2-9

19. $\dfrac{2}{x+1}$; x^2+3x+2

20. $\dfrac{3}{x-1}$; x^2+x-2

In Exercises 21–38, several denominators are given. Find the least common denominator.

21. 15, 12

22. 18, 24

23. 14, 21, 42

24. 12, 15, 10

25. $2x$, $6x$

26. $3y$, $9y$

27. x^2y, x^2y^2, xy^2

28. xy, x^2, y^2

29. $3x$, $6y$, $9xy$

30. $2x^2$, $6y$, $3xy$

31. x^2-1, $x+1$

32. y^2-9, $y-3$

33. x^2+6x, $x+6$, x

34. xy^2-xy, xy, $y-1$

35. x^2-x-2, $(x-2)^2$

36. x^2+2x-3, $(x+3)^2$

37. x^2-4x-5, x^2-25

38. x^2-x-6, x^2-9

In Exercises 39–82, do the operations. Simplify answers if possible.

39. $\dfrac{1}{2}+\dfrac{2}{3}$

40. $\dfrac{3}{4}+\dfrac{1}{2}$

41. $\dfrac{2}{3}-\dfrac{5}{6}$

42. $\dfrac{4}{9}-\dfrac{2}{3}$

43. $\dfrac{2y}{9}+\dfrac{y}{3}$

44. $\dfrac{3x}{8}+\dfrac{3x}{4}$

45. $\dfrac{8a}{15}-\dfrac{5a}{12}$

46. $\dfrac{2b}{15}-\dfrac{2b}{5}$

47. $\dfrac{21x}{14}-\dfrac{5x}{21}$

48. $\dfrac{7y}{6}+\dfrac{10y}{9}$

49. $\dfrac{4x}{3}+\dfrac{2x}{y}$

50. $\dfrac{2y}{5x}-\dfrac{y}{2}$

51. $\dfrac{2}{x}-3x$

52. $14+\dfrac{10}{y^2}$

53. $\dfrac{x^2}{2y^2}+\dfrac{x^2}{3xy}$

54. $\dfrac{r^2}{2rs}+\dfrac{r^2}{6s^2}$

55. $\dfrac{y+2}{5y}+\dfrac{y+4}{15y}$

56. $\dfrac{x+3}{x^2}+\dfrac{x+5}{2x}$

57. $\dfrac{x+5}{xy}-\dfrac{x-1}{x^2y}$

58. $\dfrac{y-7}{y^2}-\dfrac{y+7}{2y}$

59. $\dfrac{x}{x+1}+\dfrac{x-1}{x}$

60. $\dfrac{3x}{xy}+\dfrac{x+1}{y-1}$

61. $\dfrac{3}{x-2}-(x-1)$

62. $a+1-\dfrac{3}{a+3}$

63. $\dfrac{x-1}{x}+\dfrac{y+1}{y}$

64. $\dfrac{a+2}{b}+\dfrac{b-2}{a}$

65. $\dfrac{x}{x-2} + \dfrac{4+2x}{x^2-4}$

66. $\dfrac{y}{y+3} - \dfrac{2y-6}{y^2-9}$

67. $\dfrac{x+1}{x-1} + \dfrac{x-1}{x+1}$

68. $\dfrac{2x}{x+2} + \dfrac{x+1}{x-3}$

69. $\dfrac{2x+2}{x-2} - \dfrac{2x}{x+2}$

70. $\dfrac{y+3}{y-1} - \dfrac{y+4}{y+1}$

71. $\dfrac{x}{(x-2)^2} + \dfrac{x-4}{(x+2)(x-2)}$

72. $\dfrac{a-2}{(a+3)^2} - \dfrac{a}{a-3}$

73. $\dfrac{2x}{x^2-3x+2} + \dfrac{2x}{x-1} - \dfrac{x}{x-2}$

74. $\dfrac{4a}{a-2} - \dfrac{3a}{a-3} + \dfrac{4a}{a^2-5a+6}$

75. $\dfrac{2x}{x-1} + \dfrac{3x}{x+1} - \dfrac{x+3}{x^2-1}$

76. $\dfrac{a}{a-1} - \dfrac{2}{a+2} + \dfrac{3(a-2)}{a^2+a-2}$

77. $-2 - \dfrac{y+1}{y-3} + \dfrac{3(y-2)}{y}$

78. $\dfrac{3(a-b)}{a+b} + \dfrac{2(a+b)}{a-b} - 1$

79. $\dfrac{x+1}{2x+4} - \dfrac{x^2}{2x^2-8}$

80. $\dfrac{x+1}{x+2} - \dfrac{x^2+1}{x^2-x-6}$

81. $\dfrac{x-1}{x+2} + \dfrac{x}{3-x} + \dfrac{9x+3}{x^2-x-6}$

82. $\dfrac{x+1}{x-3} - \dfrac{x^2+9x}{x^2-2x-3} - \dfrac{5}{3-x}$

In Exercises 83–84, show that each formula is true.

83. $\dfrac{a}{b} + \dfrac{c}{d} = \dfrac{ad+bc}{bd}$

84. $\dfrac{a}{b} - \dfrac{c}{d} = \dfrac{ad-bc}{bd}$

Writing Exercises ■ *Write a paragraph using your own words.*

1. Explain how to find a lowest common denominator.

2. Explain how to add two fractions with different denominators.

Something to Think About ■ **1.** Add the fractions $\dfrac{1}{2x^2} + \dfrac{3}{8x^2}$, but use the common denominator $16x^4$. Then add them again, using the lowest common denominator $8x^2$. Do the results agree? What are the advantages of using the *lowest* common denominator?

2. Find the mistake:

$$\frac{3(x-7)}{y} + \frac{x}{x-7} = \frac{3(\cancel{x-7})^{1}}{y} + \frac{x}{\cancel{x-7}_{1}} = \frac{3+x}{y}$$

Review Exercises ■ **1.** Define a prime number.

2. Define a natural number.

3. Define a composite number.

4. Define an integer.

5.6 Complex Fractions

Use the distributive property to remove parentheses, and simplify.

1. $3\left(1 + \dfrac{1}{3}\right)$ **2.** $10\left(\dfrac{1}{5} - 2\right)$ **3.** $4\left(\dfrac{3}{2} + \dfrac{1}{4}\right)$ **4.** $14\left(\dfrac{3}{7} - 1\right)$

5. $x\left(\dfrac{3}{x} + 3\right)$ **6.** $y\left(\dfrac{2}{y} - 1\right)$ **7.** $4x\left(3 - \dfrac{1}{2x}\right)$ **8.** $6xy\left(\dfrac{1}{2x} + \dfrac{1}{3y}\right)$

Fractions such as

$$\frac{\dfrac{1}{3}}{4}, \qquad \frac{\dfrac{5}{3}}{\dfrac{2}{9}}, \qquad \frac{x + \dfrac{1}{2}}{3 - x}, \qquad \text{and} \qquad \frac{\dfrac{x + 1}{2}}{x + \dfrac{1}{x}}$$

which contain fractions in their numerators or denominators, are called **complex fractions.** Complex fractions can often be simplified. For example, we can simplify the complex fraction

$$\frac{\dfrac{5}{3}}{\dfrac{2}{9}}$$

by doing the division:

$$\frac{\dfrac{5}{3}}{\dfrac{2}{9}} = \frac{5}{3} \div \frac{2}{9} = \frac{5}{3} \cdot \frac{9}{2} = \frac{15}{2}$$

We will use the following two methods for simplifying complex fractions.

Methods for Simplifying a Complex Fraction

Method 1
Write the numerator and denominator of the complex fraction as single fractions. Then do the indicated division of the two fractions and simplify.

Method 2
Multiply both the numerator and denominator of the complex fraction by the LCD of all of the fractions that appear in that numerator and denominator. Then simplify.

Hypatia (370 A.D.–415 A.D.)
Hypatia is the earliest known woman in the history of mathematics. She was a professor at the University of Alexandria. Because of her scientific beliefs, she was considered to be a heretic. At the age of 45, she was attacked by a mob and murdered for her beliefs.

To simplify the complex fraction $\dfrac{\frac{3}{5}+1}{2-\frac{1}{5}}$, for example, we can use Method 1

and proceed as follows:

$$\frac{\dfrac{3}{5}+1}{2-\dfrac{1}{5}} = \frac{\dfrac{3}{5}+\dfrac{5}{5}}{\dfrac{10}{5}-\dfrac{1}{5}}$$ Change 1 to $\frac{5}{5}$ and 2 to $\frac{10}{5}$.

$$= \frac{\dfrac{8}{5}}{\dfrac{9}{5}}$$ Add the fractions in the numerator and subtract the fractions in the denominator.

$$= \frac{8}{5} \div \frac{9}{5}$$ Express the complex fraction as an equivalent division problem.

$$= \frac{8}{5} \cdot \frac{5}{9}$$ Invert the divisor and multiply.

$$= \frac{8 \cdot 5}{5 \cdot 9}$$ Multiply the fractions.

$$= \frac{8}{9}$$ Simplify.

To use Method 2, we proceed as follows:

$$\frac{\dfrac{3}{5}+1}{2-\dfrac{1}{5}} = \frac{5\left(\dfrac{3}{5}+1\right)}{5\left(2-\dfrac{1}{5}\right)}$$ Multiply both the numerator and denominator by 5, the LCD of $\frac{3}{5}$ and $\frac{1}{5}$.

$$= \frac{5 \cdot \dfrac{3}{5}+5 \cdot 1}{5 \cdot 2-5 \cdot \dfrac{1}{5}}$$ Remove parentheses.

$$= \frac{3+5}{10-1}$$ Simplify.

$$= \frac{8}{9}$$ Simplify.

In this case, Method 2 is easier than Method 1. Any complex fraction can be simplified by using either method. With practice we will be able to see which method is best to use in any given situation.

EXAMPLE 1 Simplify $\dfrac{\dfrac{x}{3}}{\dfrac{y}{3}}$.

Solution

Method 1

$$\dfrac{\dfrac{x}{3}}{\dfrac{y}{3}} = \dfrac{x}{3} \div \dfrac{y}{3}$$

$$= \dfrac{x}{3} \cdot \dfrac{3}{y}$$

$$= \dfrac{3x}{3y}$$

$$= \dfrac{x}{y}$$

Method 2

$$\dfrac{\dfrac{x}{3}}{\dfrac{y}{3}} = \dfrac{\left(\dfrac{x}{3}\right)3}{\left(\dfrac{y}{3}\right)3}$$

$$= \dfrac{\dfrac{x}{1}}{\dfrac{y}{1}}$$

$$= \dfrac{x}{y}$$ ∎

EXAMPLE 2 Simplify $\dfrac{\dfrac{x}{x+1}}{\dfrac{y}{x}}$.

Solution

Method 1

$$\dfrac{\dfrac{x}{x+1}}{\dfrac{y}{x}} = \dfrac{x}{x+1} \div \dfrac{y}{x}$$

$$= \dfrac{x}{x+1} \cdot \dfrac{x}{y}$$

$$= \dfrac{x^2}{y(x+1)}$$

Method 2

$$\dfrac{\dfrac{x}{x+1}}{\dfrac{y}{x}} = \dfrac{\left(\dfrac{x}{x+1}\right)x(x+1)}{\left(\dfrac{y}{x}\right)x(x+1)}$$

$$= \dfrac{\dfrac{x^2}{1}}{\dfrac{y(x+1)}{1}}$$

$$= \dfrac{x^2}{y(x+1)}$$ ∎

EXAMPLE 3 Simplify $\dfrac{1+\dfrac{1}{x}}{1-\dfrac{1}{x}}$.

Solution

Method 1	*Method 2*

Method 1

$$\dfrac{1 + \dfrac{1}{x}}{1 - \dfrac{1}{x}} = \dfrac{\dfrac{x}{x} + \dfrac{1}{x}}{\dfrac{x}{x} - \dfrac{1}{x}}$$

$$= \dfrac{\dfrac{x + 1}{x}}{\dfrac{x - 1}{x}}$$

$$= \dfrac{x + 1}{x} \div \dfrac{x - 1}{x}$$

$$= \dfrac{x + 1}{x} \cdot \dfrac{x}{x - 1}$$

$$= \dfrac{x + 1}{x - 1}$$

Method 2

$$\dfrac{1 + \dfrac{1}{x}}{1 - \dfrac{1}{x}} = \dfrac{\left(1 + \dfrac{1}{x}\right)x}{\left(1 - \dfrac{1}{x}\right)x}$$

$$= \dfrac{x + 1}{x - 1}$$

■

EXAMPLE 4   Simplify the complex fraction $\dfrac{1}{1 + \dfrac{1}{x + 1}}$.

Solution   We use Method 2.

$$\dfrac{1}{1 + \dfrac{1}{x + 1}} = \dfrac{1(x + 1)}{\left(1 + \dfrac{1}{x + 1}\right)(x + 1)}$$ Multiply numerator and denominator by $x + 1$.

$$= \dfrac{x + 1}{1(x + 1) + 1}$$ Remove parentheses and simplify.

$$= \dfrac{x + 1}{x + 2}$$ Simplify.   ■

EXAMPLE 5   Simplify the fraction $\dfrac{x^{-1} + y^{-2}}{x^{-2} - y^{-1}}$.

Solution   Write the fraction in complex fraction form and simplify:

$$\dfrac{x^{-1} + y^{-2}}{x^{-2} - y^{-1}} = \dfrac{\dfrac{1}{x} + \dfrac{1}{y^2}}{\dfrac{1}{x^2} - \dfrac{1}{y}}$$

$$= \dfrac{x^2y^2\left(\dfrac{1}{x} + \dfrac{1}{y^2}\right)}{x^2y^2\left(\dfrac{1}{x^2} - \dfrac{1}{y}\right)}$$

Multiply numerator and denominator by x^2y^2.

$$= \dfrac{xy^2 + x^2}{y^2 - x^2y}$$

Remove parentheses.

$$= \dfrac{x(y^2 + x)}{y(y - x^2)}$$

Attempt to simplify the fraction by factoring the numerator and denominator.

The result cannot be simplified. ■

ORALS *Simplify each complex fraction.*

1. $\dfrac{\dfrac{2}{3}}{\dfrac{1}{2}}$

2. $\dfrac{\dfrac{2}{1}}{\dfrac{1}{2}}$

3. $\dfrac{\dfrac{1}{2}}{2}$

4. $\dfrac{1 + \dfrac{1}{2}}{\dfrac{1}{2}}$

E X E R C I S E   5.6

In Exercises 1–34, simplify each complex fraction.

1. $\dfrac{\dfrac{2}{3}}{\dfrac{3}{4}}$

2. $\dfrac{\dfrac{3}{5}}{\dfrac{2}{7}}$

3. $\dfrac{\dfrac{4}{5}}{\dfrac{32}{15}}$

4. $\dfrac{\dfrac{7}{8}}{\dfrac{49}{4}}$

5. $\dfrac{\dfrac{2}{3} + 1}{\dfrac{1}{3} + 1}$

6. $\dfrac{\dfrac{3}{5} - 2}{\dfrac{2}{5} - 2}$

7. $\dfrac{\dfrac{1}{2} + \dfrac{3}{4}}{\dfrac{3}{2} + \dfrac{1}{4}}$

8. $\dfrac{\dfrac{2}{3} - \dfrac{5}{2}}{\dfrac{2}{3} - \dfrac{3}{2}}$

9. $\dfrac{\dfrac{x}{y}}{\dfrac{1}{x}}$

10. $\dfrac{\dfrac{y}{x}}{\dfrac{x}{xy}}$

11. $\dfrac{\dfrac{5t^2}{9x^2}}{\dfrac{3t}{x^2t}}$

12. $\dfrac{\dfrac{5w^2}{4tz}}{\dfrac{15wt}{z^2}}$

13. $\dfrac{\dfrac{1}{x} - 3}{\dfrac{5}{x} + 2}$

14. $\dfrac{\dfrac{1}{y} + 3}{\dfrac{3}{y} - 2}$

15. $\dfrac{\dfrac{2}{x} + 2}{\dfrac{4}{x} + 2}$

16. $\dfrac{\dfrac{3}{x} - 3}{\dfrac{9}{x} - 3}$

17. $\dfrac{\dfrac{3y}{x} - y}{y - \dfrac{y}{x}}$

18. $\dfrac{\dfrac{y}{x} + 3y}{y + \dfrac{2y}{x}}$

19. $\dfrac{\dfrac{1}{x+1}}{1 + \dfrac{1}{x+1}}$

20. $\dfrac{\dfrac{1}{x-1}}{1 - \dfrac{1}{x-1}}$

21. $\dfrac{\dfrac{x}{x+2}}{\dfrac{x}{x+2} + x}$

22. $\dfrac{\dfrac{2}{x-2}}{\dfrac{2}{x-2} - 1}$

23. $\dfrac{1}{\dfrac{1}{x} + \dfrac{1}{y}}$

24. $\dfrac{1}{\dfrac{b}{a} - \dfrac{a}{b}}$

25. $\dfrac{\dfrac{2}{x}}{\dfrac{2}{y} - \dfrac{4}{x}}$

26. $\dfrac{\dfrac{2y}{3}}{\dfrac{2y}{3} - \dfrac{8}{y}}$

27. $\dfrac{3 + \dfrac{3}{x-1}}{3 - \dfrac{3}{x}}$

28. $\dfrac{2 - \dfrac{2}{x+1}}{2 + \dfrac{2}{x}}$

29. $\dfrac{\dfrac{3}{x} + \dfrac{4}{x+1}}{\dfrac{2}{x+1} - \dfrac{3}{x}}$

30. $\dfrac{\dfrac{5}{y-3} - \dfrac{2}{y}}{\dfrac{1}{y} + \dfrac{2}{y-3}}$

31. $\dfrac{\dfrac{2}{x} - \dfrac{3}{x+1}}{\dfrac{2}{x+1} - \dfrac{3}{x}}$

32. $\dfrac{\dfrac{5}{y} + \dfrac{4}{y+1}}{\dfrac{4}{y} - \dfrac{5}{y+1}}$

33. $\dfrac{\dfrac{1}{y^2+y} - \dfrac{1}{xy+x}}{\dfrac{1}{xy+x} - \dfrac{1}{y^2+y}}$

34. $\dfrac{\dfrac{2}{b^2-1} - \dfrac{3}{ab-a}}{\dfrac{3}{ab-a} - \dfrac{2}{b^2-1}}$

In Exercises 35–44, simplify each fraction.

35. $\dfrac{x^{-2}}{y^{-1}}$

36. $\dfrac{a^{-4}}{b^{-2}}$

37. $\dfrac{1 + x^{-1}}{x^{-1} - 1}$

38. $\dfrac{y^{-2} + 1}{y^{-2} - 1}$

39. $\dfrac{a^{-2} + a}{a + 1}$

40. $\dfrac{t - t^{-2}}{1 - t^{-1}}$

41. $\dfrac{2x^{-1} + 4x^{-2}}{2x^{-2} + x^{-1}}$

42. $\dfrac{x^{-2} - 3x^{-3}}{3x^{-2} - 9x^{-3}}$

43. $\dfrac{1 - 25y^{-2}}{1 + 10y^{-1} + 25y^{-2}}$

44. $\dfrac{1 - 9x^{-2}}{1 - 6x^{-1} + 9x^{-2}}$

Writing Exercises ■ *Write a paragraph using your own words.*

1. Explain how to use Method 1 to simplify

$$\dfrac{1 + \dfrac{1}{x}}{3 - \dfrac{1}{x}}$$

2. Explain how to use Method 2 to simplify the fraction in Problem 1.

Something to Think About ■ **1.** Simplify these four complex fractions:

$$\frac{1}{1+1}, \quad \frac{1}{1+\dfrac{1}{2}}, \quad \frac{1}{1+\dfrac{1}{1+\dfrac{1}{2}}}, \quad \text{and} \quad \frac{1}{1+\dfrac{1}{1+\dfrac{1}{1+\dfrac{1}{2}}}}$$

(*Hint:* Use Method 2, and work from the bottom up.)

2. What is the pattern in the numerators of the answers in Problem 1? What would be the next answer?

Review Exercises ■ *Write each expression as an expression involving one exponent.*

1. $t^3 t^4 t^2$ **2.** $(a^0 a^2)^3$ **3.** $-2r(r^3)^2$ **4.** $(s^3)^2(s^4)^0$

Write each expression without using parentheses or negative exponents.

5. $\left(\dfrac{3r}{4r^3}\right)^4$ **6.** $\left(\dfrac{12y^{-3}}{3y^2}\right)^{-2}$ **7.** $\left(\dfrac{6r^{-2}}{2r^3}\right)^{-2}$ **8.** $\left(\dfrac{4x^3}{5x^{-3}}\right)^{-2}$

5.7 Solving Equations That Contain Fractions

■ **Extraneous Solutions** ■ **Literal Equations**

GETTING READY   *Simplify.*

1. $3\left(x + \dfrac{1}{3}\right)$ **2.** $8\left(x - \dfrac{1}{8}\right)$ **3.** $x\left(\dfrac{3}{x} + 2\right)$

4. $3y\left(\dfrac{1}{3} - \dfrac{2}{y}\right)$ **5.** $6x\left(\dfrac{5}{2x} + \dfrac{2}{3x}\right)$ **6.** $9x\left(\dfrac{7}{9} + \dfrac{2}{3x}\right)$

7. $(y - 1)\left(\dfrac{1}{y-1} + 1\right)$ **8.** $(x + 2)\left(3 - \dfrac{1}{x+2}\right)$

To solve equations containing fractions, it is usually best to eliminate those fractions. To do so, we multiply both sides of the equation by the least common denominator of the fractions that appear in the equation. For example, to solve the equation $\frac{x}{3} + 1 = \frac{x}{6}$, we multiply both sides of the equation by 6:

$$\frac{x}{3} + 1 = \frac{x}{6}$$

$$6\left(\frac{x}{3} + 1\right) = 6\left(\frac{x}{6}\right)$$

We then use the distributive law to remove parentheses, simplify, and solve the resulting equation for x.

$$6 \cdot \frac{x}{3} + 6 \cdot 1 = 6 \cdot \frac{x}{6}$$

$2x + 6 = x$	Simplify.
$x + 6 = 0$	Subtract x from both sides.
$x = -6$	Subtract 6 from both sides.

Check: $\dfrac{x}{3} + 1 = \dfrac{x}{6}$

$\dfrac{-6}{3} + 1 \stackrel{?}{=} \dfrac{-6}{6}$	Replace x with -6.
$-2 + 1 \stackrel{?}{=} -1$	Simplify.
$-1 = -1$	

EXAMPLE 1 Solve the equation $\dfrac{4}{x} + 1 = \dfrac{6}{x}$.

Solution

$$\frac{4}{x} + 1 = \frac{6}{x}$$

$\left(\dfrac{4}{x} + 1\right)x = \left(\dfrac{6}{x}\right)x$	Multiply both sides by x.
$\dfrac{4}{x} \cdot x + 1 \cdot x = \dfrac{6}{x} \cdot x$	Remove parentheses.
$4 + x = 6$	Simplify.
$x = 2$	Subtract 4 from both sides.

Check: $\dfrac{4}{x} + 1 = \dfrac{6}{x}$

$\dfrac{4}{2} + 1 \stackrel{?}{=} \dfrac{6}{2}$	Replace x with 2.
$2 + 1 \stackrel{?}{=} 3$	Simplify.
$3 = 3$	

■

◼ Extraneous Solutions

If we multiply both sides of an equation by an expression that involves a variable, as we did in Example 1, we *must* check the apparent solutions. The next example shows why.

EXAMPLE 2   Solve the equation $\dfrac{x+3}{x-1} = \dfrac{4}{x-1}$.

Solution   Multiply both sides of the equation by $x - 1$, the least common denominator of the fractions contained in the equation.

$$\frac{x+3}{x-1} = \frac{4}{x-1}$$

$$\frac{x+3}{x-1}(x-1) = \frac{4}{x-1}(x-1) \qquad \text{Multiply both sides by } x-1.$$

$$x + 3 = 4 \qquad\qquad \text{Simplify.}$$

$$x = 1 \qquad\qquad \text{Subtract 3 from both sides.}$$

Because both sides of the equation were multiplied by an expression containing a variable, we must check the apparent solution.

$$\frac{x+3}{x-1} = \frac{4}{x-1}$$

$$\frac{1+3}{1-1} \overset{?}{=} \frac{4}{1-1} \qquad\qquad \text{Replace } x \text{ with 1.}$$

$$\frac{4}{0} = \frac{4}{0} \qquad\qquad \text{Simplify.}$$

Because zeros appear in the denominators of fractions, the fractions are undefined. Thus, 1 is a false solution, and the given equation has no solutions. Such false solutions are called **extraneous solutions.**   ◼

EXAMPLE 3   Solve the equation $\dfrac{3x+1}{x+1} - 2 = \dfrac{3(x-3)}{x+1}$.

Solution   We multiply both sides of the equation by $x + 1$, the least common denominator of the fractions contained in the equation.

$$\frac{3x+1}{x+1} - 2 = \frac{3(x-3)}{x+1}$$

$$(x+1)\left(\frac{3x+1}{x+1} - 2\right) = (x+1)\left[\frac{3(x-3)}{x+1}\right]$$

$$3x + 1 - 2(x+1) = 3(x-3) \qquad \text{Use the distributive property to remove parentheses.}$$

$$3x + 1 - 2x - 2 = 3x - 9 \qquad \text{Remove parentheses.}$$
$$x - 1 = 3x - 9 \qquad \text{Combine like terms.}$$
$$-2x = -8 \qquad \text{Subtract } 3x \text{ and add 1 to both sides.}$$
$$x = 4 \qquad \text{Divide both sides by } -2.$$

Check:
$$\frac{3x + 1}{x + 1} - 2 = \frac{3(x - 3)}{x + 1}$$
$$\frac{3(4) + 1}{4 + 1} - 2 \stackrel{?}{=} \frac{3(4 - 3)}{4 + 1}$$
$$\frac{13}{5} - \frac{10}{5} \stackrel{?}{=} \frac{3(1)}{5}$$
$$\frac{3}{5} = \frac{3}{5} \qquad \blacksquare$$

EXAMPLE 4   Solve the equation $\dfrac{x + 2}{x + 3} + \dfrac{1}{x^2 + 2x - 3} = 1$.

Solution
$$\frac{x + 2}{x + 3} + \frac{1}{x^2 + 2x - 3} = 1$$

$$\frac{x + 2}{x + 3} + \frac{1}{(x + 3)(x - 1)} = 1 \qquad \text{Factor } x^2 + 2x - 3.$$

$$(x + 3)(x - 1)\left[\frac{x + 2}{x + 3} + \frac{1}{(x + 3)(x - 1)}\right] = 1(x + 3)(x - 1) \qquad \text{Multiply both sides by } (x + 3)(x - 1).$$

$$(x + 3)(x - 1)\frac{x + 2}{x + 3} + (x + 3)(x - 1)\frac{1}{(x + 3)(x - 1)} = 1(x + 3)(x - 1) \qquad \text{Remove brackets.}$$

$$(x - 1)(x + 2) + 1 = (x + 3)(x - 1) \qquad \text{Simplify.}$$

$$x^2 + x - 2 + 1 = x^2 + 2x - 3 \qquad \text{Remove parentheses.}$$

$$x - 2 + 1 = 2x - 3 \qquad \text{Subtract } x^2 \text{ from both sides.}$$

$$x - 1 = 2x - 3 \qquad \text{Combine like terms.}$$

$$-x - 1 = -3 \qquad \text{Subtract } -2x \text{ from both sides.}$$

$$-x = -2 \qquad \text{Add 1 to both sides.}$$

$$x = 2 \qquad \text{Divide both sides by } -1.$$

Verify that 2 is a solution of the given equation. $\blacksquare$

EXAMPLE 5   Solve the equation $\dfrac{4}{5} + y = \dfrac{4y - 50}{5y - 25}$.

Solution

$$\frac{4}{5} + y = \frac{4y - 50}{5y - 25}$$

$$\frac{4}{5} + y = \frac{4y - 50}{5(y - 5)} \qquad \text{Factor } 5y - 25.$$

$$5(y - 5)\left[\frac{4}{5} + y\right] = 5(y - 5)\left[\frac{4y - 50}{5(y - 5)}\right] \qquad \begin{array}{l}\text{Multiply both sides by} \\ 5(y - 5).\end{array}$$

$$4(y - 5) + 5y(y - 5) = 4y - 50 \qquad \text{Remove brackets.}$$

$$4y - 20 + 5y^2 - 25y = 4y - 50 \qquad \text{Remove parentheses.}$$

$$5y^2 - 25y - 20 = -50 \qquad \begin{array}{l}\text{Add } -4y \text{ to both sides} \\ \text{and rearrange terms.}\end{array}$$

$$5y^2 - 25y + 30 = 0 \qquad \text{Add 50 to both sides.}$$

$$y^2 - 5y + 6 = 0 \qquad \text{Divide both sides by 5.}$$

$$(y - 3)(y - 2) = 0 \qquad \text{Factor } y^2 - 5y + 6.$$

$$y - 3 = 0 \quad \text{or} \quad y - 2 = 0 \qquad \text{Set each factor equal to 0.}$$

$$y = 3 \quad | \quad y = 2$$

Verify that 3 and 2 both satisfy the original equation. ∎

■ Literal Equations

EXAMPLE 6 The formula $\dfrac{1}{r} = \dfrac{1}{r_1} + \dfrac{1}{r_2}$ is used in electronics to calculate parallel resistances. Solve the equation for r.

Solution We clear the equation of fractions by multiplying both sides by the LCD, which is rr_1r_2.

$$\frac{1}{r} = \frac{1}{r_1} + \frac{1}{r_2}$$

$$rr_1r_2\left(\frac{1}{r}\right) = rr_1r_2\left(\frac{1}{r_1} + \frac{1}{r_2}\right) \qquad \text{Multiply both sides by } rr_1r_2.$$

$$\frac{rr_1r_2}{r} = \frac{rr_1r_2}{r_1} + \frac{rr_1r_2}{r_2} \qquad \text{Remove parentheses.}$$

$$r_1r_2 = rr_2 + rr_1 \qquad \text{Simplify.}$$

$$r_1r_2 = r(r_2 + r_1) \qquad \text{Factor out an } r.$$

$$\frac{r_1r_2}{r_2 + r_1} = r \qquad \text{Divide both sides by } r_2 + r_1.$$

or

$$r = \frac{r_1r_2}{r_1 + r_2}$$

∎

Indicate your first step in solving each equation.

1. $\dfrac{x-3}{5} = \dfrac{x}{2}$

2. $\dfrac{1}{x-1} = \dfrac{8}{x}$

3. $\dfrac{y}{9} + 5 = \dfrac{y+1}{3}$

4. $\dfrac{5x-8}{3} + 3x = \dfrac{x}{5}$

E X E R C I S E 5.7

In Exercises 1–58, solve each equation and check the solution. If an equation has no solution, so indicate.

1. $\dfrac{x}{2} + 4 = \dfrac{3x}{2}$

2. $\dfrac{y}{3} + 6 = \dfrac{4y}{3}$

3. $\dfrac{2y}{5} - 8 = \dfrac{4y}{5}$

4. $\dfrac{3x}{4} - 6 = \dfrac{x}{4}$

5. $\dfrac{x}{3} + 1 = \dfrac{x}{2}$

6. $\dfrac{x}{2} - 3 = \dfrac{x}{5}$

7. $\dfrac{x}{5} - \dfrac{x}{3} = -8$

8. $\dfrac{2}{3} + \dfrac{x}{4} = 7$

9. $\dfrac{3a}{2} + \dfrac{a}{3} = -22$

10. $\dfrac{x}{2} + x = \dfrac{9}{2}$

11. $\dfrac{x-3}{3} + 2x = -1$

12. $\dfrac{x+2}{2} - 3x = x + 8$

13. $\dfrac{z-3}{2} = z + 2$

14. $\dfrac{b+2}{3} = b - 2$

15. $\dfrac{5(x+1)}{8} = x + 1$

16. $\dfrac{3(x-1)}{2} + 2 = x$

17. $\dfrac{c-4}{4} = \dfrac{c+4}{8}$

18. $\dfrac{t+3}{2} = \dfrac{t-3}{3}$

19. $\dfrac{x+1}{3} + \dfrac{x-1}{5} = \dfrac{2}{15}$

20. $\dfrac{y-5}{7} + \dfrac{y-7}{5} = \dfrac{-2}{5}$

21. $\dfrac{3x-1}{6} - \dfrac{x+3}{2} = \dfrac{3x+4}{3}$

22. $\dfrac{2x+3}{3} + \dfrac{3x-4}{6} = \dfrac{x-2}{2}$

23. $\dfrac{3}{x} + 2 = 3$

24. $\dfrac{2}{x} + 9 = 11$

25. $\dfrac{5}{a} - \dfrac{4}{a} = 8 + \dfrac{1}{a}$

26. $\dfrac{11}{b} + \dfrac{13}{b} = 12$

27. $\dfrac{2}{y+1} + 5 = \dfrac{12}{y+1}$

28. $\dfrac{1}{t-3} = \dfrac{-2}{t-3} + 1$

29. $\dfrac{1}{x-1} + \dfrac{3}{x-1} = 1$

30. $\dfrac{3}{p+6} - 2 = \dfrac{7}{p+6}$

31. $\dfrac{a^2}{a+2} - \dfrac{4}{a+2} = a$

32. $\dfrac{z^2}{z+1} + 2 = \dfrac{1}{z+1}$

33. $\dfrac{x}{x-5} - \dfrac{5}{x-5} = 3$

34. $\dfrac{3}{y-2} + 1 = \dfrac{3}{y-2}$

35. $\dfrac{3r}{2} - \dfrac{3}{r} = \dfrac{3r}{2} + 3$

36. $\dfrac{2p}{3} - \dfrac{1}{p} = \dfrac{2p-1}{3}$

37. $\dfrac{1}{3} + \dfrac{2}{x-3} = 1$

38. $\dfrac{3}{5} + \dfrac{7}{x+2} = 2$

39. $\dfrac{u}{u-1} + \dfrac{1}{u} = \dfrac{u^2+1}{u^2-u}$

40. $\dfrac{v}{v+2} + \dfrac{1}{v-1} = 1$

41. $\dfrac{3}{x-2} + \dfrac{1}{x} = \dfrac{2(3x+2)}{x^2-2x}$

42. $\dfrac{5}{x} + \dfrac{3}{x+2} = \dfrac{-6}{x(x+2)}$

43. $\dfrac{7}{q^2 - q - 2} + \dfrac{1}{q + 1} = \dfrac{3}{q - 2}$

44. $\dfrac{-5}{s^2 + s - 2} + \dfrac{3}{s + 2} = \dfrac{1}{s - 1}$

45. $\dfrac{3y}{3y - 6} + \dfrac{8}{y^2 - 4} = \dfrac{2y}{2y + 4}$

46. $\dfrac{x - 3}{4x - 4} + \dfrac{1}{9} = \dfrac{x - 5}{6x - 6}$

47. $y + \dfrac{2}{3} = \dfrac{2y - 12}{3y - 9}$

48. $y + \dfrac{3}{4} = \dfrac{3y - 50}{4y - 24}$

49. $\dfrac{5}{4y + 12} - \dfrac{3}{4} = \dfrac{5}{4y + 12} - \dfrac{y}{4}$

50. $\dfrac{3}{5x - 20} + \dfrac{4}{5} = \dfrac{3}{5x - 20} - \dfrac{x}{5}$

51. $\dfrac{x}{x - 1} - \dfrac{12}{x^2 - x} = \dfrac{-1}{x - 1}$

52. $1 - \dfrac{3}{b} = \dfrac{-8b}{b^2 + 3b}$

53. $\dfrac{z - 4}{z - 3} = \dfrac{z + 2}{z + 1}$

54. $\dfrac{a + 2}{a + 8} = \dfrac{a - 3}{a - 2}$

55. $\dfrac{n}{n^2 - 9} + \dfrac{n + 8}{n + 3} = \dfrac{n - 8}{n - 3}$

56. $\dfrac{x - 3}{x - 2} = \dfrac{1}{x} + \dfrac{x - 3}{x}$

57. $\dfrac{b + 2}{b + 3} + 1 = \dfrac{-7}{b - 5}$

58. $\dfrac{x - 4}{x - 3} + \dfrac{x - 2}{x - 3} = x - 3$

59. Solve the formula $\dfrac{1}{a} + \dfrac{1}{b} = 1$ for a.

60. Solve the formula $\dfrac{1}{a} - \dfrac{1}{b} = 1$ for b.

61. Optics The focal length f of a lens is given by the formula

$$\frac{1}{f} = \frac{1}{d_1} + \frac{1}{d_2}$$

where d_1 is the distance from the object to the lens and d_2 is the distance from the lens to the image. Solve the formula for f.

62. Solve the formula in Exercise 61 for d_1.

Writing Exercises ■ *Write a paragraph using your own words.*

1. Explain how you would decide what to do first when you solve an equation that involves fractions.

2. Explain why it is important to check your solutions to an equation that contains fractions with variables in the denominator.

Something to Think About ■

1. What number is equal to its own reciprocal?

2. Solve $x^{-2} + x^{-1} = 0$

Review Exercises ■ *Factor each expression by grouping.*

1. $ab + 3a + 2b + 6$

2. $yz + z^2 + y + z$

3. $mr + ms + nr + ns$

4. $ac + bc - ad - bd$

5. $2a + 2b - a^2 - ab$

6. $xa - x + ya - y$

5.8 Applications of Equations That Contain Fractions

GETTING READY

1. If it takes 5 hours to fill a pool, what part could be filled in 1 hour?

2. If \$x is invested at 5% annual interest, how much is earned in one year?

3. What is the amount of an investment that earns \$y interest in one year at 5%?

4. How long does it take to travel y miles at 52 miles per hour?

Many applications involve equations that contain fractions. In this section, we will consider several of these applications.

EXAMPLE 1 **Number problem** If the same number is added to both the numerator and denominator of the fraction $\frac{3}{5}$, the result is $\frac{4}{5}$. Find the number.

Solution We can let n represent the number, add n to both the numerator and denominator of $\frac{3}{5}$, and set the result equal to $\frac{4}{5}$. Then we solve the equation for n.

$$\frac{3 + n}{5 + n} = \frac{4}{5}$$

$$5(5 + n)\frac{3 + n}{5 + n} = 5(5 + n)\frac{4}{5} \qquad \text{Multiply both sides by } 5(5 + n).$$

$$5(3 + n) = (5 + n)4 \qquad \text{Simplify.}$$

$$15 + 5n = 20 + 4n \qquad \text{Remove parentheses.}$$

$$5n = 5 + 4n \qquad \text{Subtract 15 from both sides.}$$

$$n = 5 \qquad \text{Subtract } 4n \text{ from both sides.}$$

The number is 5.

Check: We add 5 to both the numerator and denominator of $\frac{3}{5}$ and get

$$\frac{3 + 5}{5 + 5} = \frac{8}{10} = \frac{4}{5}$$

■

EXAMPLE 2 **Draining an oil tank** An inlet pipe can fill an oil tank in 7 days, and a second inlet pipe can fill that tank in 9 days. If both pipes are used, how long will it take to fill the tank?

Analysis The key in this shared work problem is to note what each inlet pipe can do in 1 day. If you add what the first inlet pipe can do in 1 day to what the second inlet pipe can do in 1 day, the sum is what they can do together in 1 day. Since the first inlet pipe can fill the tank in 7 days, it can do $\frac{1}{7}$ of the job in 1 day. Since it takes the second inlet pipe 9 days, it can do $\frac{1}{9}$ of the job in 1 day. Since it takes x days for both inlet pipes to fill the tank, together they can do $\frac{1}{x}$ of the job in 1 day.

Solution We let x represent the number of days it will take to fill the tank if both inlet pipes are used. Then we form the equation

What the first inlet pipe can do in 1 day	+	what the second inlet pipe can do in 1 day	=	what they can do together in 1 day.

$$\frac{1}{7} \quad + \quad \frac{1}{9} \quad = \quad \frac{1}{x}$$

$$63x\left(\frac{1}{7} + \frac{1}{9}\right) = 63x\left(\frac{1}{x}\right) \qquad \text{Multiply both sides by } 63x.$$

$$9x + 7x = 63 \qquad \text{Remove parentheses and simplify.}$$

$$16x = 63 \qquad \text{Combine like terms.}$$

$$x = \frac{63}{16} \qquad \text{Divide both sides by 16.}$$

It will take $\frac{63}{16}$ or $3\frac{15}{16}$ days for both inlet pipes to fill the tank.

Check: In $\frac{63}{16}$ days, the first inlet pipe fills $\frac{1}{7}\left(\frac{63}{16}\right)$ of the tank and the second inlet pipe fills $\frac{1}{9}\left(\frac{63}{16}\right)$ of the tank. The sum of these efforts, $\frac{9}{16} + \frac{7}{16}$, is equal to one full tank. ∎

EXAMPLE 3 **Track team** The coach can run 10 miles in the same amount of time that his best student-athlete can run 12 miles. If the student can run 1 mile per hour faster than the coach, how fast can the student run?

Analysis This is a uniform motion problem based on the formula $d = rt$, where d is the distance traveled, r is the rate, and t is the time. If we solve the formula for t, we obtain

$$t = \frac{d}{r}$$

The coach, running 10 miles at some unknown rate r, will take $\frac{10}{r}$ hours. The student, running 12 miles at a rate of $r + 1$ miles per hour, will take $\frac{12}{r+1}$ hours. As before, we organize the information of the problem in chart form, as in Figure 5-1.

	Distance	=	Rate	·	Time
The student	12		$r + 1$		$\dfrac{12}{r + 1}$
The coach	10		r		$\dfrac{10}{r}$

FIGURE 5-1

Because the times are given to be equal, we know that $\frac{12}{r+1} = \frac{10}{r}$.

Solution   Let r be the rate that the coach can run.
Then $r + 1$ is the rate that the student can run.

We can form the equation

The time it takes the student to run 12 miles	=	the time it takes the coach to run 10 miles.

$$\frac{12}{r + 1} = \frac{10}{r}$$

$$r(r + 1)\frac{12}{r + 1} = r(r + 1)\frac{10}{r} \qquad \text{Multiply both sides by } r(r + 1).$$

$$12r = 10(r + 1) \qquad \text{Simplify.}$$

$$12r = 10r + 10 \qquad \text{Remove parentheses.}$$

$$2r = 10 \qquad \text{Subtract } 10r \text{ from both sides.}$$

$$r = 5 \qquad \text{Divide both sides by 2.}$$

Thus, the coach can run 5 miles per hour. The student, running 1 mile per hour faster, can run 6 miles per hour. Verify that these results check. ■

EXAMPLE 4   **Comparing investments**   At a bank, a sum of money invested for one year will earn $96 interest. If invested in bonds, that same money would earn $108, because the interest rate paid by the bonds is 1% greater than that paid by the bank. Find the bank's rate.

Analysis   The interest paid by either investment is the product of the principal (the amount invested) and the interest rate. If we let r represent the bank's rate of interest, then $r + .01$ represents the rate paid by the bonds. For each investment, the principal is the interest paid divided by the interest rate. See Figure 5-2.

Interest	=	Principal	·	Rate
Bank	96		$\dfrac{96}{r}$	r
Bonds	108		$\dfrac{108}{r + .01}$	$r + .01$

FIGURE 5-2

Because the same principal would be invested in either account, we can set up and solve this equation:

$$\frac{96}{r} = \frac{108}{r + .01}$$

Solution We can solve the equation as follows:

$$\frac{96}{r} = \frac{108}{r + .01}$$

$$r(r + .01)\frac{96}{r} = \frac{108}{r + .01}r(r + .01) \qquad \text{Multiply both sides by } r(r + .01).$$

$$96(r + .01) = 108r$$

$$96r + .96 = 108r \qquad \text{Remove parentheses.}$$

$$.96 = 12r \qquad \text{Subtract } 96r \text{ from both sides.}$$

$$.08 = r \qquad \text{Divide both sides by 12.}$$

The bank's interest rate is .08, or 8%. The bonds pay 9% interest, a rate that is 1% greater than that paid by the bank. Verify that these rates check. ■

ORALS
1. What is the formula that relates the principal P that is invested, the earned interest I, and the rate r?
2. What is the formula that relates the distance d traveled at a speed r, for a time t?
3. What is the formula that relates the cost C of purchasing q items that cost $d each?

EXERCISE 5.8

1. **Number problem** If the denominator of the fraction $\frac{3}{4}$ is increased by a number and the numerator of the fraction is doubled, the result is 1. Find the number.

2. **Number problem** If a number is added to the numerator of the fraction $\frac{7}{8}$ and the same number is subtracted from the denominator, the result is 2. Find the number.

3. **Number problem** If a number is added to the numerator of the fraction $\frac{3}{4}$ and twice as much is added to the denominator, the result is $\frac{4}{7}$. Find the number.

4. **Number problem** If a number is added to the numerator of the fraction $\frac{5}{7}$ and twice as much is subtracted from the denominator, the result is 8. Find the number.

5. **Number problem** The sum of a number and its reciprocal is $\frac{13}{6}$. Find the numbers.

6. **Number problem** The sum of the reciprocals of two consecutive even integers is $\frac{7}{24}$. Find the integers.

7. **Filling a pool** An inlet pipe can fill an empty swimming pool in 5 hours, and another inlet pipe can fill the pool in 4 hours. How long will it take both pipes to fill the pool?

8. **Filling a pool** One inlet pipe can fill an empty pool in 4 hours, and a drain can empty the pool in 8 hours. How long will it take the pipe to fill the pool if the drain is left open?

9. **Roofing a house** A homeowner estimates that it will take 7 days to roof his house. A professional roofer estimates that he could roof the house in 4 days. How long will it take if the homeowner helps the roofer?

10. **Sewage treatment** A sludge pool is filled by two inlet pipes. One pipe can fill the pool in 15 days, and the other pipe can fill it in 21 days. However, if no sewage is added, continuous waste removal will empty the pool in 36 days. How long will it take the two inlet pipes to fill an empty pool?

11. **Seeing the sights** A tourist can bicycle 28 miles in the same time he can walk 8 miles. If he can ride 10 miles per hour faster than he can walk, how much time should he allow to walk a 30-mile trail? (See Illustration 1.) (*Hint:* How fast can he walk?)

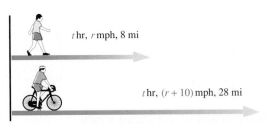

t hr, r mph, 8 mi

t hr, (r + 10) mph, 28 mi

ILLUSTRATION 1

12. **Comparing travel** A plane can fly 300 miles in the same time it takes a car to go 120 miles. If the car travels 90 miles per hour slower than the plane, how fast is the plane?

13. **Boating in a river** A boat that can travel 18 miles per hour in still water can travel 22 miles downstream in the same amount of time that it can travel 14 miles upstream. Find the speed of the current in the river. (See Illustration 2.)

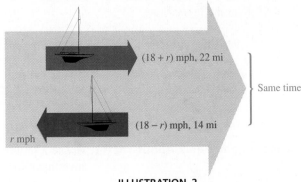

(18 + r) mph, 22 mi

(18 − r) mph, 14 mi

r mph

Same time

ILLUSTRATION 2

14. **Wind speed** A plane can fly 300 miles downwind in the same amount of time it can travel 210 miles upwind. Find the velocity of the wind if the plane can fly at 255 miles per hour in still air.

15. **Comparing investments** Two certificates of deposit pay interest at rates that differ by 1%. Money invested for one year in the first CD earns $175 interest. The same principal invested in the other CD earns $200. Find the two rates of interest.

16. **Comparing interest rates** Two bond funds pay interest at rates that differ by 2%. Money invested for one year in the first fund earns $315 interest. The same amount invested in the other fund earns $385. Find the lower rate of interest.

17. **Sharing costs** The office workers bought a $35 gift for their boss. If there had been two more employees to contribute, everyone's cost would have been $2 less. How many workers contributed to the gift?

18. **Merchandising** A dealer bought some radios for a total of $1200. She gave away 6 radios as gifts, sold each of the rest for $10 more than she paid for each radio, and broke even. How many radios did she buy?

19. Merchandising   A college bookstore can purchase several calculators for a total cost of $120. If each calculator cost $1 less, the bookstore could purchase 10 addition calculators at the same total cost. How many calculators can be purchased at the regular price?

20. Furnace repair   A repairman purchased several furnace-blower motors for a total cost of $210. If his cost per motor had been $5 less, he could have purchased 1 additional motor. How many motors did he buy at the regular rate?

21. River tours   A river boat tour begins by going 60 miles upstream against a 5 mph current. There, the boat turns around and returns with the current. What still-water speed should the captain use to complete the tour in 5 hours?

22. Travel time   The company president flew 680 miles in the corporate jet and returned in a smaller plane that could fly only half as fast. If the total travel time was 6 hours, find the speeds of the planes.

Writing Exercises ■ *Write a paragraph using your own words.*

1. The key to solving shared work problems is to ask, "How much of the job could be done in 1 hour?" Explain.

2. It is difficult to check the solution of a shared work problem. Explain how you could decide if the answer is at least reasonable.

Something to Think About ■

1. Create a problem, involving either investment income or shared work, that can be solved by an equation that contains fractions.

2. Solve the problem you created in Exercise 1.

Review Exercises ■ *Solve each equation.*

1. $x^2 - 5x - 6 = 0$

2. $x^2 - 25 = 0$

3. $(t + 2)(t^2 + 7t + 12) = 0$

4. $2(y - 4) = -y^2$

5. $y^3 - y^2 = 0$

6. $5a^3 - 125a = 0$

7. $(x^2 - 1)(x^2 - 4) = 0$

8. $6t^3 + 35t^2 = 6t$

5.9 Ratio and Proportion

■ Solving Proportions ■ Similar Triangles

GETTING READY   *Solve each equation.*

1. $\dfrac{5}{2} = \dfrac{x}{4}$

2. $\dfrac{7}{9} = \dfrac{y}{3}$

3. $\dfrac{y}{10} = \dfrac{2}{7}$

4. $\dfrac{1}{x} = \dfrac{8}{40}$

5. $\dfrac{w}{14} = \dfrac{7}{21}$

6. $\dfrac{c}{12} = \dfrac{5}{12}$

7. $\dfrac{3}{q} = \dfrac{1}{7}$

8. $\dfrac{16}{3} = \dfrac{8}{z}$

An indicated quotient of two numbers is often called a **ratio**.

| Ratio | A **ratio** is the comparison of two numbers by their indicated quotient. |

The previous definition implies that a ratio is a fraction. Some examples of ratios are

$$\frac{7}{8}, \quad \frac{21}{24}, \quad \text{and} \quad \frac{117}{223}$$

The fraction $\frac{7}{8}$ can be read as "the ratio of 7 to 8," the fraction $\frac{21}{24}$ can be read as "the ratio of 21 to 24," and the fraction $\frac{117}{223}$ can be read as "the ratio of 117 to 223." Because the fractions $\frac{7}{8}$ and $\frac{21}{24}$ represent equal numbers, they are called **equal ratios.**

EXAMPLE 1   Express each phrase as a ratio in lowest terms:

a. the ratio of 15 to 12

b. the ratio of 3 inches to 7 inches

c. the ratio of 2 feet to 1 yard

d. the ratio of 6 ounces to 1 pound

Solution   **a.** The ratio of 15 to 12 can be written as the fraction $\frac{15}{12}$. Expressed in lowest terms, it is $\frac{5}{4}$.

b. The ratio of 3 inches to 7 inches can be written as the fraction $\frac{3 \text{ inches}}{7 \text{ inches}}$, or just $\frac{3}{7}$.

c. The ratio of 2 feet to 1 yard should not be written as the fraction $\frac{2}{1}$. To express the ratio as a pure number, we must use the same units. Because there are 3 feet in 1 yard, the proper ratio is $\frac{2 \text{ feet}}{3 \text{ feet}}$, or just $\frac{2}{3}$.

d. The ratio 6 ounces to 1 pound should not be written as the fraction $\frac{6}{1}$. To express the ratio as a pure number, we must use the same units. Because there are 16 ounces in 1 pound, the proper ratio is $\frac{6 \text{ ounces}}{16 \text{ ounces}}$, which simplifies to $\frac{3}{8}$.   ∎

| Proportion | A **proportion** is a statement indicating that two ratios are equal. |

Some examples of proportions are

$$\frac{1}{2} = \frac{3}{6}, \quad \frac{3}{7} = \frac{9}{21}, \quad \text{and} \quad \frac{8}{1} = \frac{40}{5}$$

The proportion $\frac{1}{2} = \frac{3}{6}$ can be read as "1 is to 2 as 3 is to 6," the proportion $\frac{3}{7} = \frac{9}{21}$ can be read as "3 is to 7 as 9 is to 21," and the proportion $\frac{8}{1} = \frac{40}{5}$ can be read as "8 is to 1 as 40 is to 5."

In the proportion $\frac{1}{2} = \frac{3}{6}$, the numbers 1 and 6 are called the **extremes** of the proportion, and the numbers 2 and 3 are called the **means.** If we find the product of the extremes and the product of the means in this proportion, we see that the products are equal:

$$1 \cdot 6 = 6 \qquad \text{and} \qquad 3 \cdot 2 = 6$$

This is not a coincidence. To show that this is always true, we consider the proportion

$$\frac{a}{b} = \frac{c}{d}$$

where a and d are the extremes and b and c are the means. Then

$$\frac{a}{b} = \frac{c}{d}$$

$$bd\frac{a}{b} = \frac{c}{d}bd \qquad \text{Multiply both sides by } bd.$$

$$ad = bc \qquad \text{Simplify.}$$

and the product of the extremes, ad, is equal to the product of the means, bc. Thus, we have this important theorem.

Theorem   In any proportion, the product of the extremes is equal to the product of the means.

EXAMPLE 2   Determine if the equation $\dfrac{x}{3y} = \dfrac{xy + 3x}{3y^2 + 9y}$ is a proportion.

Solution   We check to see if the product of the means is equal to the product of the extremes.

$$3y(xy + 3x) = 3xy^2 + 9xy \qquad \text{The product of the means.}$$
$$x(3y^2 + 9y) = 3xy^2 + 9xy \qquad \text{The product of the extremes.}$$

Because the products are equal, the equation is a proportion.   ■

■ Solving Proportions

EXAMPLE 3   Solve the proportion $\dfrac{12}{18} = \dfrac{3}{x}$ for x.

Solution   We proceed as follows:

$$\frac{12}{18} = \frac{3}{x}$$

$12x = 3 \cdot 18$ The product of the extremes equals the product of the means.

$$x = \frac{54}{12}$$ Divide both sides by 12.

$$x = \frac{9}{2}$$ Simplify.

Thus, x represents the fraction $\frac{9}{2}$. ∎

EXAMPLE 4   Solve the proportion $\dfrac{y+1}{y} = \dfrac{y}{y+2}$ for y.

Solution $\dfrac{y+1}{y} = \dfrac{y}{y+2}$

$y^2 = (y+1)(y+2)$ The product of the means is equal to the product of the extremes.

$y^2 = y^2 + 3y + 2$ Remove parentheses.

$0 = 3y + 2$ Add $-y^2$ to both sides.

$-3y = 2$ Add $-3y$ to both sides.

$y = -\dfrac{2}{3}$ Divide both sides by -3. ∎

EXAMPLE 5   **Grocery shopping**   If 5 tomatoes cost \$1.15, how much will 16 tomatoes cost?

Solution   We can let c represent the cost of 16 tomatoes. Since the ratio of the numbers of tomatoes is the same as the ratio of their costs, we can express this relationship as a proportion and find c.

$$\frac{5}{16} = \frac{1.15}{c}$$

$5c = 1.15(16)$ The product of the extremes is equal to the product of the means.

$5c = 18.4$ Do the multiplication.

$c = \dfrac{18.4}{5}$ Divide both sides by 5.

$c = 3.68$ Simplify.

Sixteen tomatoes will cost \$3.68. ∎

■ Similar Triangles

If two triangles have the same shape, they are said to be **similar triangles.** The following theorem points out an important fact about similar triangles.

Theorem	If two triangles are similar, then all pairs of corresponding sides are in proportion.

This theorem often enables us to measure sides of triangles indirectly. For example, on a sunny day we can find the height of a tree and stay safely on the ground.

EXAMPLE 6 **Measuring a tree** On a sunny day, a large tree casts a shadow of 24 feet at the same time as a vertical yardstick casts a shadow of 2 feet. Find the height of the tree.

Solution Refer to Figure 5-3, which shows the triangles determined by the tree and its shadow and the yardstick and its shadow. Because the triangles have the same shape, they are similar, and the measures of their corresponding sides are in proportion. We can let h represent the height of the tree and find h by setting up and solving the following proportion.

$$\frac{h}{3} = \frac{24}{2}$$

$2h = 3(24)$ In a proportion, the product of the extremes is equal to the product of the means.

$2h = 72$ Simplify.

$h = 36$ Divide both sides by 2.

The tree is 36 feet tall.

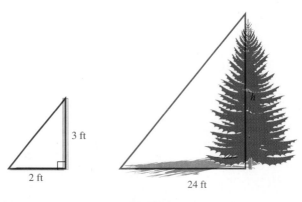

3 ft

2 ft

24 ft

FIGURE 5-3 ■

ORALS *Express as a ratio in lowest terms.*

1. 5 to 7   **2.** 50 to 1   **3.** 3 to 9   **4.** 7 to 10

Indicate which are proportions.

5. $\dfrac{3}{5} = \dfrac{6}{10}$ **6.** $\dfrac{1}{2} = \dfrac{1}{3}$ **7.** $\dfrac{1}{2} + \dfrac{2}{4}$ **8.** $\dfrac{1}{x} = \dfrac{2}{2x}$

E X E R C I S E   5 . 9

In Exercises 1–20, express each phrase as a ratio in lowest terms.

1. 5 to 7   **2.** 3 to 5   **3.** 17 to 34   **4.** 19 to 38

5. 22 to 33   **6.** 14 to 21

7. 4 ounces to 12 ounces   **8.** 3 inches to 15 inches

9. 12 minutes to 1 hour   **10.** 8 ounces to 1 pound

11. 3 days to 1 week   **12.** 2 quarts to 1 gallon

13. 4 inches to 2 yards   **14.** 1 mile to 5280 feet

15. 3 pints to 2 quarts   **16.** 4 dimes to 8 pennies

17. 6 nickels to 1 quarter   **18.** 3 people to 12 people

19. 3 meters to 12 centimeters   **20.** 3 dollars to 3 quarters

In Exercises 21–34, tell whether each statement is a proportion.

21. $\dfrac{9}{7} = \dfrac{81}{70}$ **22.** $\dfrac{5}{2} = \dfrac{20}{8}$ **23.** $\dfrac{-7}{3} = \dfrac{14}{-6}$ **24.** $\dfrac{13}{-19} = \dfrac{-65}{95}$

25. $\dfrac{9}{19} = \dfrac{38}{80}$ **26.** $\dfrac{40}{29} = \dfrac{29}{22}$ **27.** $\dfrac{x^2}{y} = \dfrac{x}{y^2}$ **28.** $\dfrac{x^2 y}{x^2 z} = \dfrac{y}{z}$

29. $\dfrac{3x^2 y}{3xy^2} = \dfrac{x}{y}$ **30.** $\dfrac{5y}{25x} = \dfrac{5x}{y}$ **31.** $\dfrac{x+2}{x(x+2)} = \dfrac{1}{x}$ **32.** $\dfrac{x+y}{x-2} = \dfrac{y}{2}$

33. $\dfrac{xy+x}{xy} = \dfrac{y+1}{y}$ **34.** $\dfrac{x^2+x}{x+1} = \dfrac{x}{1}$

In Exercises 35–60, solve for the variable in each proportion.

35. $\dfrac{2}{3} = \dfrac{x}{6}$ **36.** $\dfrac{3}{6} = \dfrac{x}{8}$ **37.** $\dfrac{5}{10} = \dfrac{3}{c}$ **38.** $\dfrac{7}{14} = \dfrac{2}{b}$

39. $\dfrac{-6}{x} = \dfrac{8}{4}$ **40.** $\dfrac{4}{x} = \dfrac{2}{8}$ **41.** $\dfrac{x}{3} = \dfrac{9}{3}$ **42.** $\dfrac{x}{2} = \dfrac{-18}{6}$

43. $\dfrac{x+1}{5} = \dfrac{3}{15}$ **44.** $\dfrac{x-1}{7} = \dfrac{2}{21}$ **45.** $\dfrac{x+3}{12} = \dfrac{-7}{6}$ **46.** $\dfrac{x+7}{-4} = \dfrac{3}{12}$

47. $\dfrac{4-x}{13} = \dfrac{11}{26}$ **48.** $\dfrac{5-x}{17} = \dfrac{13}{34}$ **49.** $\dfrac{2x+1}{9} = \dfrac{x}{27}$ **50.** $\dfrac{3x-2}{7} = \dfrac{x}{28}$

51. $\dfrac{3(x+5)}{2} = \dfrac{5(x-2)}{3}$ **52.** $\dfrac{7(x+6)}{6} = \dfrac{6(x+3)}{5}$

53. $\dfrac{2(x+3)}{3} = \dfrac{4(x-4)}{5}$ **54.** $\dfrac{x+4}{5} = \dfrac{3(x-2)}{3}$ **55.** $\dfrac{1}{x+3} = \dfrac{-2x}{x+5}$ **56.** $\dfrac{x-1}{x+1} = \dfrac{2}{3x}$

57. $\dfrac{2}{x+6} = \dfrac{-2x}{5}$ **58.** $\dfrac{x-3}{x-2} = \dfrac{x+3}{2x}$ **59.** $\dfrac{x+1}{x} = \dfrac{10}{2x}$ **60.** $\dfrac{2x}{x+4} = \dfrac{5x+2}{18}$

In Exercises 61–82, set up and solve the required proportion.

61. Buying yogurt   Three pints of yogurt cost $1. How much will 51 pints cost?

62. Selling shirts   Sport shirts are on sale. How much will 5 shirts cost? (See Illustration 1.)

Sale!

Buy 2!

Only $25!

ILLUSTRATION 1

63. Gardening   Garden seeds are on sale at three packets for 50 cents. How much will 39 packets cost?

64. Increasing a recipe   A recipe for spaghetti sauce requires four 16-ounce bottles of ketchup to make 2 gallons of sauce. How many bottles of ketchup are needed to make 10 gallons of sauce?

65. Gas consumption   A car gets 42 miles per gallon of gas. How much gas is needed to drive 315 miles?

66. Gas consumption   A truck gets 12 miles per gallon of gas. How far can the truck go on 17 gallons of gas?

67. Paychecks   Bill earns $412 for a 40-hour week. Last week he missed 10 hours of work. How much did he get paid?

68. Model railroading   An HO-scale model railroad engine is 9 inches long. The HO scale is 87 feet to 1 foot. How long is a real engine?

69. Model railroading   An N-scale model railroad caboose is 3.5 inches long. The N scale is 169 feet to 1 foot. How long is a real caboose?

70. Hobby   Standard dollhouse scale is 1 inch to 1 foot. Heidi's dollhouse is 32 inches wide. How wide would it be if it were a real house?

71. Staffing   A school board has determined that there should be 3 teachers for every 50 students. How many teachers are needed for an enrollment of 2700 students?

72. Drafting   In a scale drawing, a 280-foot antenna tower is drawn 7 inches high. The building next to it is drawn 2 inches high. How tall is the actual building?

73. Mixing fuel   The instructions on a can of oil intended to be added to lawn mower gasoline read as follows.

Recommended	Gasoline	Oil
50 to 1	6 gal	16 oz

Are these instructions correct? (*Hint:* There are 128 ounces in 1 gallon.)

74. **Height of a tree** A tree casts a shadow of 26 feet at the same time as a 6-foot man casts a shadow of 4 feet. (See Illustration 2.) Find the height of the tree.

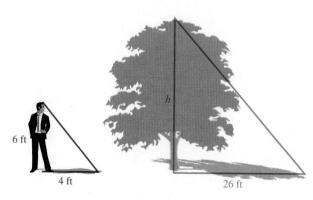

ILLUSTRATION 2

75. **Height of a flagpole** A man places a mirror on the ground and sees the reflection of the top of a flagpole, as in Illustration 3. The two triangles in the illustration are similar. Find the height, *h*, of the flagpole.

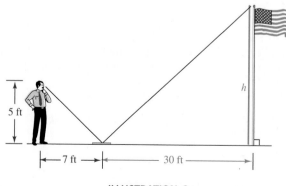

ILLUSTRATION 3

76. **Width of a river** Use the dimensions in Illustration 4 to find *w*, the width of the river. The two triangles in the illustration are similar.

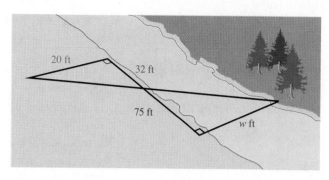

ILLUSTRATION 4

77. **Flight path** An airplane ascends 100 feet as it flies a horizontal distance of 1000 feet. How much altitude will it gain as it flies a horizontal distance of 1 mile? (See Illustration 5.) (*Hint:* 5280 feet = 1 mile.)

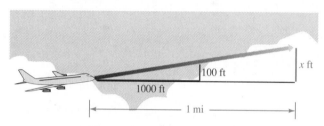

ILLUSTRATION 5

78. **Flight path** An airplane descends 1350 feet as it flies a horizontal distance of 1 mile. How much altitude is lost as it flies a horizontal distance of 5 miles?

79. **Ski runs** A $\frac{1}{2}$-mile long ski course falls 100 feet in every 300 feet of horizontal run. Find the height of the hill.

80. **Mountain travel** A mountain road ascends 375 feet in every 2500 feet of travel. By how much will the road rise in a trip of 10 miles?

81. **Recommended dosage** The recommended child's dose of the sedative hydroxine is 0.006 gram per kilogram of body mass. Find the dosage for a 30-kg child.

82. **Body mass** Proper dosage of the antibiotic cephalexin in children is 0.025 grams per kilogram of body mass. Find the dosage for a child who weighs 100 pounds. (*Hint:* 2.2 lbs = 1 kg)

Writing Exercises ■ *Write a paragraph using your own words.*

1. Explain the terms *means* and *extremes*.

2. Distinguish between a *ratio* and a *proportion*.

Something to Think About ■

1. Is a proportion useful for solving this problem?

A water bill for 1000 gallons was $15, and a bill for 2000 gallons was $25. Find the bill for 3000 gallons.

Explain.

2. How would you solve the problem in Exercise 1?

Review Exercises ■ *Find the perimeter or circumference of each figure.*

1. A square with sides measuring 6 inches.

2. A rectangle with sides measuring 2 by 7 centimeters.

3. A trapezoid with sides measuring 6, 5, 4, and 7 centimeters.

4. A circle with a diameter of 8 feet.

Find the area of each figure.

5. A square with sides measuring 6 inches.

6. A rectangle with sides measuring 2 by 7 centimeters.

7. A trapezoid with a height of 5 centimeters and bases measuring 6 and 7 centimeters.

8. A circle with a diameter of 8 feet.

MATHEMATICS IN ARCHITECTURE

The area A of either room in the home discussed at the beginning of the chapter is given by $A = lw$, where l and w are the length and the width. We solve this equation for l:

$$l = \frac{A}{w}$$

Because the area of the living room is 288 sq ft and its width is $(w + 4)$ ft, its length is $\frac{288}{w+4}$ ft. Similarly, the length of the dining room is $\frac{168}{w}$ ft. We summarize this information in the table.

Area	=	Length	·	Width
Living room	288	$\dfrac{288}{w + 4}$		$w + 4$
Dining room	168	$\dfrac{168}{w}$		w

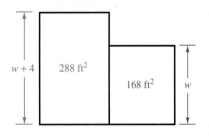

The length of the living room	+	the length of the dining room	=	the total length.

$$\frac{288}{w + 4} + \frac{168}{w} = 32$$

$$w(w + 4)\left(\frac{288}{w + 4} + \frac{168}{w}\right) = 32w(w + 4)$$ Clear the equation of fractions.

$$288w + 168w + 672 = 32w^2 + 128w$$ Remove parentheses.

$$0 = 32w^2 - 328w - 672$$ Subtract 456w and 672 from both sides.

$$0 = 4w^2 - 41w - 84$$ Divide both sides by 8.

$$0 = (w - 12)(4w + 7)$$ Factor.

$$w - 12 = 0 \quad \text{or} \quad 4w + 7 = 0$$ Set each factor equal to 0.

$$w = 12 \qquad\qquad w = -\frac{7}{4}$$ Solve each equation.

We discard the negative root. The architect must plan for a 12-foot-wide dining room.

P R O J E C T ■ The Solar System to Scale

Our solar system consists of nine planets and their moons, and some assorted asteroids, comets, and other debris, all orbiting the sun. If the sizes of the planets and their distances from the sun were reduced proportionally so that the sun was the size of an orange, earth would be a grain of sand, and the farthest planet, Pluto, would be half a mile away.

The diameters of the planets and their distances from the sun are given in Table 5-1.

	Diameter (km)	Distance from sun (AU)*
Sun	1.5×10^6	0
Mercury	4.9×10^3	0.39
Venus	1.2×10^4	0.72
Earth	1.3×10^4	1.0
Mars	6.8×10^3	1.5
Jupiter	1.4×10^5	5.2
Saturn	1.2×10^5	9.5
Uranus	5.1×10^5	19
Neptune	4.9×10^5	30
Pluto	2.3×10^3	39

*One AU (astronomical unit) is the distance from the earth to the sun, about 93 million miles.

TABLE 5-1

- Use the information in Table 5-1 to draw a scale diagram of the relative *positions* of the sun and the planets. You will need a large sheet of paper, or perhaps the classroom chalkboard. From your diagram, which planets do you think are called the *inner planets*, and which are the *outer planets*?
- Draw a scale diagram that shows the relative *sizes* of the sun and planets.
- What difficulty would you have drawing a scale diagram that shows both relative sizes and distances? Could you draw a scale diagram if you disregarded the enormous size of the sun? Write your observations in a brief paragraph.

Chapter Summary

KEY WORDS

arithmetic fraction (5.1)
building a fraction (5.5)
common denominator (5.4)
complex fractions (5.6)

denominator (5.1)
divisor (5.3)
extraneous solutions (5.7)
extremes of a proportion (5.9)

fraction (5.1)

least common denominator
(LCD) (5.5)

lowest terms (5.1)

means of a proportion (5.9)

numerator (5.1)

proportion (5.9)

ratio (5.9)

rational expression (5.1)

rational number (5.1)

similar triangles (5.9)

simplifying a fraction (5.1)

KEY IDEAS

(5.1) If b and c are not equal to zero, then $\dfrac{a}{b} = \dfrac{a \cdot c}{b \cdot c}$.

$\dfrac{a}{1} = a$ $\qquad$ $\dfrac{a}{0}$ is not defined.

(5.2) $\dfrac{a}{b} \cdot \dfrac{c}{d} = \dfrac{a \cdot c}{b \cdot d}$

(5.3) $\dfrac{a}{b} \div \dfrac{c}{d} = \dfrac{a}{b} \cdot \dfrac{d}{c} = \dfrac{a \cdot d}{b \cdot c}$

(5.4) $\dfrac{a}{d} + \dfrac{b}{d} = \dfrac{a + b}{d}$ $\qquad$ $\dfrac{a}{d} - \dfrac{b}{d} = \dfrac{a - b}{d}$

If $d \neq 0$, then $\dfrac{0}{d} = 0$.

(5.5) To add or subtract fractions with unlike denominators, first find the least common denominator of those fractions. Then express each of the fractions in equivalent form with the same common denominator. Finally, add or subtract the fractions.

(5.6) To simplify a complex fraction, use either of these methods:

1. Write the numerator and the denominator of the complex fraction as single fractions, do the division, and simplify.

2. Multiply both the numerator and the denominator of the complex fraction by the LCD of the fractions that appear in the numerator and the denominator; then simplify.

(5.7) To solve an equation that contains fractions, transform it into another equation without fractions. Do so by multiplying both sides by the LCD of the fractions. Check all solutions.

(5.9) In any proportion, the product of the means is equal to the product of the extremes.

The measures of corresponding sides of similar triangles are in proportion.

◼ Chapter 5 Review Exercises

(5.1) *In Review Exercises 1–14, write each fraction in lowest terms. If a fraction is already in lowest terms, so indicate.*

1. $\dfrac{10}{25}$

2. $\dfrac{-12}{18}$

3. $\dfrac{-51}{153}$

4. $\dfrac{105}{45}$

5. $\dfrac{3x^2}{6x^3}$

6. $\dfrac{5xy^2}{2x^2y^2}$

7. $\dfrac{x^2}{x^2 + x}$

8. $\dfrac{x + 2}{x^2 + 2x}$

9. $\dfrac{6xy}{3xy}$

10. $\dfrac{8x^2y}{2x(4xy)}$

11. $\dfrac{x^2 + 4x + 3}{x^2 - 4x - 5}$

12. $\dfrac{x^2 - x - 56}{x^2 - 5x - 24}$

13. $\dfrac{2x^2 - 16x}{2x^2 - 18x + 16}$

14. $\dfrac{x^2 + x - 2}{x^2 - x - 2}$

(5.2) *In Review Exercises 15–18, do each multiplication and simplify.*

15. $\dfrac{3xy}{2x} \cdot \dfrac{4x}{2y^2}$

16. $\dfrac{3x}{x^2 - x} \cdot \dfrac{2x - 2}{x^2}$

17. $\dfrac{x^2 + 3x + 2}{x^2 + 2x} \cdot \dfrac{x}{x + 1}$

18. $\dfrac{x^3 - y^3}{x^2 + xy + y^2} \cdot \dfrac{x}{x^2 - y^2}$

(5.3) *In Review Exercises 19–22, do the divisions and simplify.*

19. $\dfrac{3x^2}{5x^2y} \div \dfrac{6x}{15xy^2}$

20. $\dfrac{x^2 + 5x}{x^2 + 4x - 5} \div \dfrac{x^2}{x - 1}$

21. $\dfrac{x^2 - x - 6}{2x - 1} \div \dfrac{x^2 - 2x - 3}{2x^2 + x - 1}$

22. $\dfrac{x^3 + 125}{x + 5} \div \dfrac{x^2 - 5x + 25}{x^2 + 10x + 25}$

(5.4–5.5) *In Review Exercises 23–28, several denominators are given. Find the least common denominator.*

23. $4, 8$

24. $35, 14$

25. $3x^2y, xy^2$

26. $2x + 1, 2x^2 + x$

27. $x + 2, x - 3$

28. $x^2 + 4x + 3, x^2 + x$

In Review Exercises 29–36, do the operations. Simplify all answers.

29. $\dfrac{x}{x + y} + \dfrac{y}{x + y}$

30. $\dfrac{3x}{x - 7} - \dfrac{x - 2}{x - 7}$

31. $\dfrac{x}{x - 1} + \dfrac{1}{x}$

32. $\dfrac{1}{7} - \dfrac{1}{x}$

33. $\dfrac{3}{x + 1} - \dfrac{2}{x}$

34. $\dfrac{x + 2}{2x} - \dfrac{2 - x}{x^2}$

35. $\dfrac{x}{x + 2} + \dfrac{3}{x} - \dfrac{4}{x^2 + 2x}$

36. $\dfrac{2}{x - 1} - \dfrac{3}{x + 1} + \dfrac{x - 5}{x^2 - 1}$

(5.6) *In Review Exercises 37–42, simplify each complex fraction.*

37. $\dfrac{\dfrac{3}{2}}{\dfrac{2}{3}}$

38. $\dfrac{\dfrac{3}{2} + 1}{\dfrac{2}{3} + 1}$

39. $\dfrac{\dfrac{1}{x} + 1}{\dfrac{1}{x} - 1}$

40. $\dfrac{1 + \dfrac{3}{x}}{2 - \dfrac{1}{x^2}}$

41. $\dfrac{\dfrac{2}{x - 1} + \dfrac{x - 1}{x + 1}}{\dfrac{1}{x^2 - 1}}$

42. $\dfrac{\dfrac{a}{b} + c}{\dfrac{b}{a} + c}$

(5.7) *In Review Exercises 43–48, solve each equation. Check all answers.*

43. $\dfrac{3}{x} = \dfrac{2}{x-1}$

44. $\dfrac{5}{x+4} = \dfrac{3}{x+2}$

45. $\dfrac{2}{3x} + \dfrac{1}{x} = \dfrac{5}{9}$

46. $\dfrac{2x}{x+4} = \dfrac{3}{x-1}$

47. $\dfrac{2}{x-1} + \dfrac{3}{x+4} = \dfrac{-5}{x^2+3x-4}$

48. $\dfrac{4}{x+2} - \dfrac{3}{x+3} = \dfrac{6}{x^2+5x+6}$

49. The efficiency, E, of a Carnot engine is given by the equation

$$E = 1 - \dfrac{T_2}{T_1}$$

Solve the equation for T_1.

50. Radioactive tracers are used for diagnostic purposes in nuclear medicine. The **effective half-life, H,** of a radioactive material in a biological organism is given by the formula

$$H = \dfrac{RB}{R+B}$$

where R is the radioactive half-life, and B is the biological half-life of the tracer. Solve the equation for R.

(5.8) *In Review Exercises 51–54, solve each problem. Check all answers.*

51. Pumping a basement If one pump can empty a flooded basement in 18 hours and a second pump can empty the basement in 20 hours, how long will it take to empty the basement if both pumps are used?

52. Painting houses If a homeowner can paint a house in 14 days and a professional painter can paint it in 10 days, how long will it take if they work together?

53. Jogging A jogger can bicycle 30 miles in the same time as he can jog 10 miles. If he can ride 10 miles per hour faster than he can jog, how fast can he jog?

54. Wind speed A plane can fly 400 miles downwind in the same amount of time as it can travel 320 miles upwind. If the plane can fly at 360 miles per hour in still air, find the velocity of the wind.

(5.9) *In Review Exercises 55–58, write each ratio as a fraction in lowest terms.*

55. 3 to 6

56. $12x$ to $15x$

57. 2 feet to 1 yard

58. 5 pints to 3 quarts

In Review Exercises 59–62, write each ratio as a fraction in lowest terms.

59. $\dfrac{3}{x} = \dfrac{6}{9}$

60. $\dfrac{x}{3} = \dfrac{x}{5}$

61. $\dfrac{x-2}{5} = \dfrac{x}{7}$

62. $\dfrac{x+1}{4} = \dfrac{3}{x}$

In Review Exercises 63–68, set up and solve a proportion.

63. Making iron If 5 tons of iron ore yields 3 tons of pig iron, how much iron ore is needed to make 18 tons of pig iron?

64. Map reading On a certain map, 1 inch represents 60 miles. If two cities are 3.5 inches apart on the map, what is the distance between the cities?

65. Dispensing medicine A pharmacist mixes 3 grams of medicine with 300 milliliters of sugar syrup. Find how much medicine is in a single dose of 5 milliliters.

66. Length of a wire Ten feet of copper wire weighs 1.2 pounds. How long is a 564-pound roll of this wire?

67. Height of a building   Refer to Illustration 1. How tall is the building if a 5-foot-tall woman casts a shadow of 2 feet?

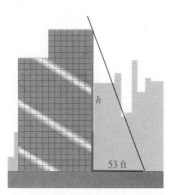

ILLUSTRATION 1

68. Flight path   A plane gains 1350 feet in altitude as it travels a horizontal distance of 10,000 feet. How much will it gain in altitude as it flies a horizontal distance of 5 miles?

Chapter 5 Test

1. Simplify: $\dfrac{48x^2y}{54xy^2}$.

2. Simplify: $\dfrac{2x^2 - x - 3}{4x^2 - 9}$

3. Simplify: $\dfrac{3(x + 2) - 3}{2x - 4 - (x - 5)}$.

4. Multiply and simplify: $\dfrac{12x^2y}{15xyz} \cdot \dfrac{25y^2z}{16xt}$.

5. Multiply and simplify: $\dfrac{x^2 + 3x + 2}{3x + 9} \cdot \dfrac{x + 3}{x^2 - 4}$.

6. Divide and simplify: $\dfrac{8x^2y}{25xt} \div \dfrac{16x^2y^3}{30xyt^3}$.

7. Divide and simplify: $\dfrac{x^2 - x}{3x^2 + 6x} \div \dfrac{3x - 3}{3x^3 + 6x^2}$.

8. Simplify: $\dfrac{x^2 + xy}{x - y} \cdot \dfrac{x^2 - y^2}{x^2 - 2x} \div \dfrac{x^2 + 2xy + y^2}{x^2 - 4}$.

9. Add: $\dfrac{5x - 4}{x - 1} + \dfrac{5x + 3}{x - 1}$.

10. Subtract: $\dfrac{3y + 7}{2y + 3} - \dfrac{3(y - 2)}{2y + 3}$.

11. Add: $\dfrac{x + 1}{x} + \dfrac{x - 1}{x + 1}$.

12. Subtract: $\dfrac{5x}{x - 2} - 3$.

13. Simplify: $\dfrac{\dfrac{8x^2}{xy^3}}{\dfrac{4y^3}{x^2y^3}}$.

14. Simplify: $\dfrac{1 + \dfrac{y}{x}}{\dfrac{y}{x} - 1}$.

15. Solve for x: $\dfrac{x}{10} - \dfrac{1}{2} = \dfrac{x}{5}$.

16. Solve for x: $3x - \dfrac{2(x + 3)}{3} = 16 - \dfrac{x + 2}{2}$.

17. Solve for x: $\dfrac{7}{x + 4} - \dfrac{1}{2} = \dfrac{3}{x + 4}$.

18. Solve $H = \dfrac{RB}{R + B}$ for B.

19. Express as a ratio in lowest terms: 6 feet to 3 yards.

20. Is the equation $\dfrac{3xy}{5xy} = \dfrac{3xt}{5xt}$ a proportion?

21. Solve the proportion for y: $\dfrac{y}{y-1} = \dfrac{y-2}{y}$.

22. Cleaning highways   One highway worker could pick up all the trash on a strip of highway in 7 hours, and his helper could pick up the trash in 9 hours. How long will it take them if they work together?

23. Boating   A boat can motor 28 miles downstream in the same amount of time as it can motor 18 miles upstream. Find the speed of the current if the boat can motor at 23 miles per hour in still water.

24. Flight path   A plane drops 575 feet as it flies a horizontal distance of $\frac{1}{2}$ mile. How much altitude will it lose as it flies a horizontal distance of 7 miles? (See Illustration 1.)

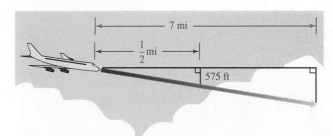

ILLUSTRATION 1

6

Graphing Linear Equations and Inequalities

An investor bought an office building for $465,000, excluding the value of the land. At that time, a real estate appraiser estimated that the building would maintain 80% of its value after 40 years. For tax purposes, the investor used linear depreciation to depreciate the building over a period of 40 years. If the investor sells the building for $400,000 after 35 years, find the taxable capital gain.

■ ■ ■ ■ ■ ■ ■   After you have read this chapter, you will be able to answer this question.

Most equations encountered so far have involved only one variable. However, many applications of mathematics involve equations with two or more variables. In this chapter, we will discuss equations and inequalities with two variables.

6.1   Graphing Linear Equations

■ The Rectangular Coordinate System ■ An Application of the Coordinate System ■ Graphing Linear Equations ■ The Intercept Method of Graphing a Line ■ Graphing Lines Parallel to the *x*- and *y*-Axes

GETTING READY   *In Problems 1–4, let* $y = 2x + 1$.

1. Find y when $x = 0$.   **2.** Find y when $x = 2$.

3. Find y when $x = -2$.   **4.** Find y when $x = \frac{1}{2}$.

5. Find five pairs of numbers with a sum of 8.   **6.** Find five pairs of numbers with a difference of 5.

The equation $3x + 2 = 5$ has the single variable x, and its only solution is 1. This solution can be graphed (or plotted) on the number line as in Figure 6-1.

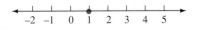

FIGURE 6-1

The equation $x + 2y = 5$ contains the two variables x and y. The solutions of such equations are pairs of numbers. For example, the pair $x = 1$ and $y = 2$ is a solution, because the equation is satisfied when $x = 1$ and $y = 2$.

$$x + 2y = 5$$
$$1 + 2(2) = 5 \qquad \text{Substitute 1 for } x \text{ and 2 for } y.$$
$$1 + 4 = 5$$
$$5 = 5$$

The pair $x = 5$ and $y = 0$ is also a solution, because the pair satisfies the equation:

$$x + 2y = 5$$
$$5 + 2(0) = 5 \qquad \text{Substitute 5 for } x \text{ and 0 for } y.$$
$$5 + 0 = 5$$
$$5 = 5$$

■ The Rectangular Coordinate System

Solutions of equations with two variables can be plotted on a **rectangular coordinate system,** sometimes called a **Cartesian coordinate system** after the 17th-century French mathematician René Descartes. The rectangular coordinate system consists of two number lines, called the **x-axis** and the **y-axis,** drawn at right angles to each other, as shown in Figure 6-2(a). The two axes intersect at a point called the **origin,** which is the 0 point on each axis. The positive direction on the x-axis is to the right, and the positive direction on the y-axis is upward. The two axes divide the coordinate system into four regions, called **quadrants,** which are numbered as shown in Figure 6-2(a).

René Descartes (1596–1650)
Descartes is famous for his work in philosophy as well as for his work in mathematics. His philosophy is expressed in the words "I think, therefore I am." He is best known in mathematics for his invention of a coordinate system and his work with conic sections.

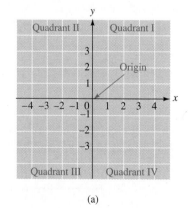

(a)

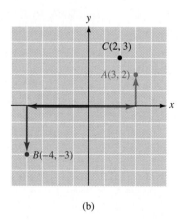

(b)

FIGURE 6-2

To plot the pair $x = 3$ and $y = 2$, we start at the origin and move 3 units to the right along the x-axis and then 2 units up in the positive y direction, as in Figure 6-2(b). This locates point A. Point A has an **x-coordinate** or **abscissa** of 3 and a **y-coordinate** or **ordinate** of 2. This information is denoted by the pair of numbers $(3, 2)$, called the **coordinates** of point A. Point A is the **graph** of the pair $(3, 2)$.

To plot the pair $(-4, -3)$, we start at the origin, move 4 units to the left along the x-axis, and then 3 units down to point B in the figure. Point C has coordinates of $(2, 3)$.

WARNING!   The order of the coordinates of a point is important. Point A with coordinates of $(3, 2)$ is not the same as point C with coordinates $(2, 3)$. For this reason, the pair of coordinates (x, y) of a point is called an **ordered pair.** The x-coordinate is the first number in the pair, and the y-coordinate is the second number.

EXAMPLE 1   On a rectangular coordinate system, plot the points   **a.** $A(-1, 2)$,   **b.** $B(0, 0)$,   **c.** $C(5, 0)$,   **d.** $D(-\frac{5}{2}, -3)$,   and   **e.** $E(3, -2)$.

Solution   **a.** The point $A(-1, 2)$ has an x-coordinate of -1 and a y-coordinate of 2. To plot point A, we start at the origin and move 1 unit to the left and then 2 units up. (See Figure 6-3.) Point A lies in quadrant II.

b. To plot point $B(0, 0)$, we start at the origin and move 0 units to the right and 0 units up. Since there was no movement, point B is the origin.

PERSPECTIVE

Stay in Bed!

As a child, René Descartes, the inventor of analytic geometry, was frail and often sick. To improve his health, eight-year-old René was sent to a Jesuit school. The headmaster encouraged him to sleep in the morning as long as he wished, often past noon. As a young man, Descartes spent several years as a soldier and world traveler, but his interests included mathematics and philosophy, as well as science, literature, and writing, and taking it easy: The habit of sleeping late continued throughout his life. Descartes claimed that his times of most productive thinking were spent lying in bed, long past breakfast. According to one story, Descartes first thought of analytic geometry as he watched the path of a fly walking on his bedroom ceiling.

Descartes might have lived longer if he had stayed in bed. In 1649, Sweden's Queen Christina decided that she needed a tutor in philosophy, and she requested the services of Descartes. Tutoring would not have been difficult, except that the stubborn queen scheduled her lessons before dawn in her library, with the windows open. The cold Stockholm weather and the early hour was too much for a man who was used to sleeping until noon. Within a few months, Descartes developed a fever and died, probably of pneumonia.

c. To plot point $C(5, 0)$, we start at the origin and move 5 units to the right and 0 units up. Point C lies on the x-axis, 5 units to the right of the origin.

d. To plot point $D(-\frac{5}{2}, -3)$, we start at the origin and move $\frac{5}{2}$ units $\left(\frac{5}{2}\text{ units} = 2\frac{1}{2}\text{ units}\right)$ to the left and then 3 units down. Point D lies in quadrant III.

e. To plot point $E(3, -2)$, we start at the origin and move 3 units to the right and then 2 units down. Point E lies in quadrant IV.

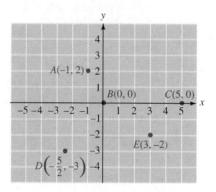

FIGURE 6-3

■ An Application of the Coordinate System

EXAMPLE 2

Braking distances   Approximate braking distances for a car in good condition on dry payment are given in Table 6-1.

Speed (in miles per hour)	20	40	55	65
Braking distance (in meters)	6	22	46	60

TABLE 6-1

Plot these points on a rectangular coordinate system.

Solution   If we plot the speeds along the x-axis and the braking distances along the y-axis, we obtain the graph shown in Figure 6-4.

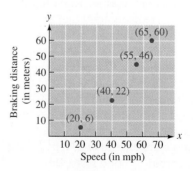

FIGURE 6-4

From the position of the points plotted in the figure, we can see that the braking distances are much longer at greater speeds.   ■

■ Graphing Linear Equations

We have seen that solutions to equations in x and y are ordered pairs of numbers. To find some ordered pairs (x, y) that satisfy $y = 5 - x$, we pick numbers at random, substitute them for x, and find the corresponding values of y. If we pick $x = 1$, we have

$$y = 5 - x$$
$$y = 5 - 1 \qquad \text{Substitute 1 for } x.$$
$$y = 4$$

The ordered pair $(1, 4)$ satisfies the equation. If we let $x = 2$, we have

$$y = 5 - x$$
$$y = 5 - 2 \qquad \text{Substitute 2 for } x.$$
$$y = 3$$

A second solution of the equation is $(2, 3)$. If we let $x = 5$, we have

$$y = 5 - x$$
$$y = 5 - 5 \qquad \text{Substitute 5 for } x.$$
$$y = 0$$

A third solution of the equation is $(5, 0)$. If we let $x = -1$, we have

$$y = 5 - x$$
$$y = 5 - (-1) \qquad \text{Substitute } -1 \text{ for } x.$$
$$y = 6$$

A fourth solution is $(-1, 6)$. If we let $x = 6$, we have

$$y = 5 - x$$
$$y = 5 - 6 \qquad \text{Substitute 6 for } x.$$
$$y = -1$$

A fifth solution is $(6, -1)$.

We list the ordered pairs $(1, 4)$, $(2, 3)$, $(5, 0)$, $(-1, 6)$, and $(6, -1)$ in the table shown in Figure 6-5. Their graphs lie on a line, called the **graph** of the equation.

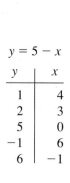

$y = 5 - x$

y	x
1	4
2	3
5	0
-1	6
6	-1

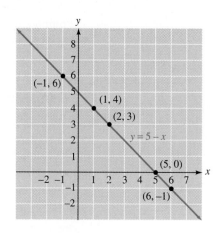

FIGURE 6-5

Any equation, such as $y = 5x$, whose graph is a line is called a **linear equation in two variables.** Any point on the line has coordinates that satisfy the equation, and the graph of any pair (x, y) that satisfies the equation is a point on the line.

Although only two points are needed to graph a linear equation, we often plot a third point as a check. If the three points do not lie on a line, at least one of them is in error.

Procedure for Graphing Linear Equations in *x* and *y*	**1.** Find two pairs (x, y) that satisfy the equation by picking arbitrary numbers for x and solving for the corresponding values of y. A third point provides a check.
	2. Plot each resulting pair (x, y) on a rectangular coordinate system. If they do not lie on a line, check your calculations.
	3. Draw the line passing through the points.

EXAMPLE 3   Graph the equation $y = 3x - 4$.

Solution   If we substitute 1 for x, we get $y = -1$, and the pair $(1, -1)$ is a solution. If we substitute 2 for x, we get $y = 2$, and the pair $(2, 2)$ is a solution. If we substitute 3 for x, we get $y = 5$, and the pair $(3, 5)$ is a solution. These ordered pairs appear in the table in Figure 6-6. We plot these points and join them with a line, which is the graph of $y = 3x - 4$.

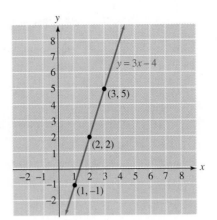

$y = 3x - 4$

x	y
1	-1
2	2
3	5

FIGURE 6-6   ■

EXAMPLE 4   Graph the equation $y - 4 = \frac{1}{2}(x - 8)$.

Solution   It is easier to find pairs (x, y) that satisfy the equation if we remove parentheses and solve for y.

$$y - 4 = \frac{1}{2}(x - 8)$$

$$y - 4 = \frac{1}{2}x - 4 \qquad \text{Use the distributive property to remove parentheses.}$$

$$y = \frac{1}{2}x \qquad \text{Add 4 to both sides.}$$

If we substitute 0 for x, we obtain $y = 0$, and the pair $(0, 0)$ is a solution. Since $y = 1$ when $x = 2$, the pair $(2, 1)$ is another solution. Finally, since $y = -2$ when $x = -4$, the pair $(-4, -2)$ is a third solution. The graph of the equation appears in Figure 6-7.

$y - 4 = \frac{1}{2}(x - 8)$

x	y
0	0
2	1
-4	-2

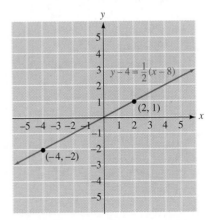

FIGURE 6-7   ■

◼ The Intercept Method of Graphing a Line

The points where a line intersects the y- and x-axes are called **intercepts** of the line.

y- and x-Intercepts

The **y-intercept** of a line is the point $(0, b)$ where the line intersects the y-axis. To find b, substitute 0 for x in the equation of the line and solve for y.

The **x-intercept** of a line is the point $(a, 0)$ where the line intersects the x-axis. To find a, substitute 0 for y in the equation of the line and solve for x.

Plotting the y- and x-intercepts and drawing a line through them is called the **intercept method of graphing,** a method useful for graphing equations written in **general form.**

General Form of the Equation of a Line

If A, B, and C are real numbers and A and B are not both 0, then the equation
$$Ax + By = C$$
is called the **general form** of the equation of a line.

Whenever possible, we will write the general form $Ax + By = C$ so that A, B, and C are integers and $A \geq 0$.

EXAMPLE 5 Graph the equation $3x + 2y = 6$.

Solution To find the y-intercept, we let $x = 0$ and solve for y.

$$3x + 2y = 6$$
$$3(0) + 2y = 6 \quad \text{Substitute 0 for } x.$$
$$2y = 6 \quad \text{Simplify.}$$
$$y = 3 \quad \text{Divide both sides by 2.}$$

The y-intercept is the pair $(0, 3)$.

To find the x-intercept, we let $y = 0$ and solve for x.

$$3x + 2y = 6$$
$$3x + 2(0) = 6 \quad \text{Substitute 0 for } y.$$
$$3x = 6 \quad \text{Simplify.}$$
$$x = 2 \quad \text{Divide both sides by 3.}$$

The *x*-intercept is the pair $(2, 0)$.

As a check, we plot one more point. If $x = 4$, then

$$3x + 2y = 6$$
$$3(4) + 2y = 6 \qquad \text{Substitute 4 for } x.$$
$$12 + 2y = 6 \qquad \text{Simplify.}$$
$$2y = -6 \qquad \text{Add } -12 \text{ to both sides.}$$
$$y = -3 \qquad \text{Divide both sides by 2.}$$

The point $(4, -3)$ lies on the graph. We plot these three points and join them with a line. The graph of $3x + 2y = 6$ is shown in Figure 6-8.

$3x + 2y = 6$

x	y
0	3
2	0
4	-3

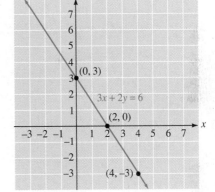

FIGURE 6-8 ■

■ Graphing Lines Parallel to the *x*- and *y*-Axes

EXAMPLE 6   Graph the equations   **a.** $y = 3$ and   **b.** $x = -2$.

Solution   **a.** We can write the equation $y = 3$ in general form as $0x + y = 3$. Since the coefficient of *x* is 0, the numbers assigned to *x* have no effect on *y*. The value of *y* is always 3. For example, if we replace *x* with -3, we get

$$0x + y = 3$$
$$0(-3) + y = 3$$
$$0 + y = 3$$
$$y = 3$$

The table in Figure 6-9(a) gives several ordered pairs that satisfy the equation $y = 3$. After plotting these pairs (x, y) and joining them with a line, we see that the graph of $y = 3$ is a horizontal line, parallel to the x-axis and intersecting the y-axis at 3. The y-intercept is $(0, 3)$, and there is no x-intercept.

b. We can write the equation $x = -2$ in general form as $x + 0y = -2$. Since the coefficient of y is 0, the values of y have no effect on x. The number x is always -2. A table of values and the graph are shown in Figure 6-9(b).

　　　The graph of $x = -2$ is a vertical line, parallel to the y-axis and intersecting the x-axis at -2. The x-intercept is $(-2, 0)$, and there is no y-intercept.

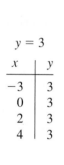

$y = 3$

x	y
-3	3
0	3
2	3
4	3

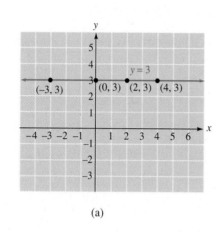

(a)

$x = -2$

x	y
-2	-2
-2	0
-2	2
-2	3

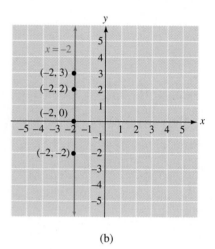

(b)

FIGURE 6-9　■

From the results of Example 6, we have the following facts.

Equations of Lines Parallel to the Coordinate Axes	The equation $y = b$ represents a horizontal line that intersects the y-axis at $(0, b)$. If $b = 0$, the line is the x-axis. The equation $x = a$ represents a vertical line that intersects the x-axis at $(a, 0)$. If $a = 0$, the line is the y-axis.

ORALS

1. How many points should be plotted to graph a line?
2. Define the intercepts of a line.
3. Find three pairs (x, y) that satisfy $x + y = 8$.
4. Find three pairs (x, y) that satisfy $x - y = 6$.
5. Which lines have no y-intercepts?
6. Which lines have no x-intercepts?

EXERCISE 6.1

In Exercises 1–8, plot each point on a rectangular coordinate system. Indicate in which quadrant each point lies.

1. $A(2, 5)$

2. $B(5, 2)$

3. $C(-3, 1)$

4. $D(1, -3)$

5. $E(-2, -3)$

6. $F(-3, -2)$

7. $G(3, -2)$

8. $H(-4, 5)$

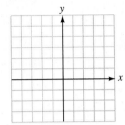

In Exercises 9–16, plot each point on a rectangular coordinate system. Indicate in which quadrant each point lies.

9. $A(-3, 5)$

10. $B(-5, 3)$

11. $C(3, -5)$

12. $D(5, -3)$

13. $E\left(-\dfrac{3}{2}, -4\right)$

14. $F\left(-5, \dfrac{9}{2}\right)$

15. $G\left(\dfrac{5}{2}, \dfrac{7}{2}\right)$

16. $H\left(\dfrac{7}{2}, -\dfrac{7}{2}\right)$

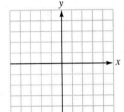

In Exercises 17–24, plot each point on a rectangular coordinate system. Indicate on which axis each point lies.

17. $A(0, 5)$

18. $B(0, -2)$

19. $C(2, 0)$

20. $D(-2, 0)$

21. $E(-4, 0)$

22. $F(0, -4)$

23. $G(0, 0)$

24. $H(-5, 0)$

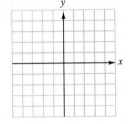

In Exercises 25–32, refer to Illustration 1 and find the coordinates of each point.

25. A

26. B

27. C

28. D

29. E

30. F

31. G

32. H

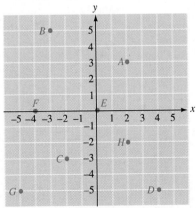

ILLUSTRATION 1

33. Stopping distances Approximate stopping distances for a car in good condition on dry pavement, including the normal $\frac{3}{4}$ second reaction time, are as shown in Table 1. Plot the points and draw any appropriate conclusions.

Speed (in miles per hour)	20	40	55	65
Stopping distance (in meters)	12	35	64	80

TABLE 1

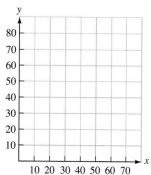

34. Stopping distances Consuming two or three drinks can easily increase the reaction time to put on the brakes by 1 second. Approximate stopping distances for a car in good condition on dry pavement, including $1\frac{3}{4}$ seconds for reaction time, are as shown in Table 2. Plot the points and draw any appropriate conclusions.

Speed (in miles per hour)	20	40	55	65
Stopping distance (in meters)	22	53	87	109

TABLE 2

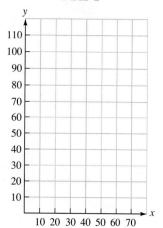

35. Tutoring mathematics A tutor charges $12 an hour for tutoring mathematics. Complete the following table showing the tutor's fee for working 1, 2, 3, and 5 hours.

Hours worked	1	2	3	5
Tutor's fee				

36. Let y represent the tutor's fee for working x hours. Plot the pairs found in Exercise 35. Do the points lie on a line?

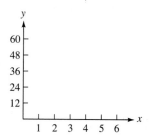

37. Installing computer systems A computer-system installer charges $10 for materials and $40 per hour. Complete the following table showing the charge for jobs taking 1, 2, 3, and $5\frac{1}{2}$ hours.

Hours worked	1	2	3	$5\frac{1}{2}$
Charge				

38. Let y represent the charge for working x hours. Plot the pairs found in Exercise 37. Do the points lie on a line?

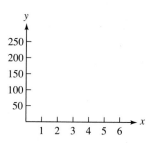

39. Maximum heart rate   A rule of thumb for a
person's maximum heart rate for safe aerobic
exercise is 220 minus the person's age. Complete the
following table showing the maximum heart rate for
persons 20, 30, 40, and 60 years old.

Age	20	30	40	60
Heart rate				

40. Let y represent the maximum heart rate for a person
x years old. (See Exercise 39.) Write an equation that
relates x and y and graph it.

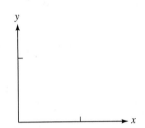

In Exercises 41–48, complete the table of solutions for each equation. Then graph the equation.

41. $y = x + 2$

x	y
3	
1	
-2	

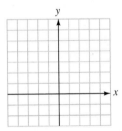

42. $y = x - 4$

x	y
5	
4	
-1	

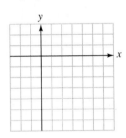

43. $y = -2x$

x	y
2	
1	
-3	

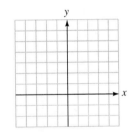

44. $y = \dfrac{x}{2}$

x	y
1	
-1	
-4	

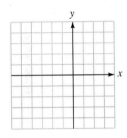

45. $y = 2x - 1$

x	y
3	
-1	
-2	

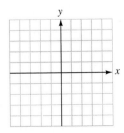

46. $y = 3x + 1$

x	y
-2	
0	
1	

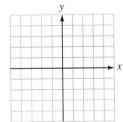

47. $y = \dfrac{x}{2} - 2$

x	y
8	
0	
-2	

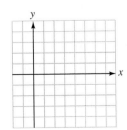

48. $y = \dfrac{x}{3} - 3$

x	y
6	
0	
-3	

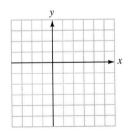

In Exercises 49–60, write each equation in general form, when necessary. Then graph it by using the intercept method.

49. $x + y = 7$

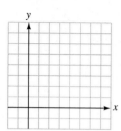

50. $x + y = -2$

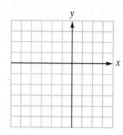

51. $x - y = 7$

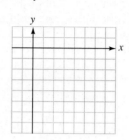

52. $x - y = -2$

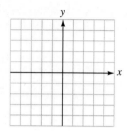

53. $2x + y = 5$

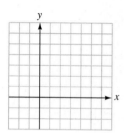

54. $3x + y = -1$

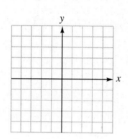

55. $2x + 3y = 12$

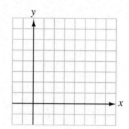

56. $3x - 2y = 6$

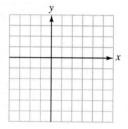

57. $3x + 12 = 4y$

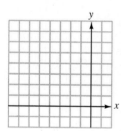

58. $2x + 12 = 9y$

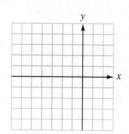

59. $2(x + 2) - y = 4$

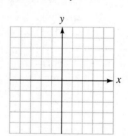

60. $3(y + 1) - x = 4$

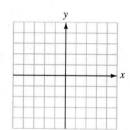

In Exercises 61–72, graph each equation.

61. $y = -5$

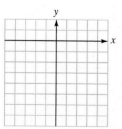

62. $x = 4$

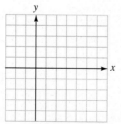

63. $x = 5$

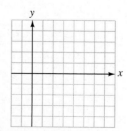

64. $y = 4$

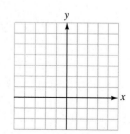

65. $y = 0$

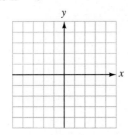

66. $x = 0$

67. $2x = 5$

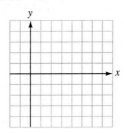

68. $3y = 7$

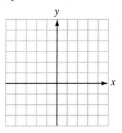

69. $3(x + 2) + x = 4$

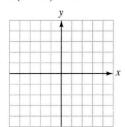

70. $2(y + 3) - y = 5$

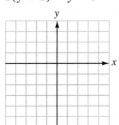

71. $3(y - 2) + 2 = y$

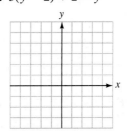

72. $4(x + 2) - 2(x - 1) = 6$

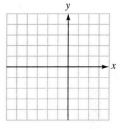

73. Find the equation of the line that is horizontal and passes through $(3, -2)$.

74. Find the equation of the line that is vertical and passes through $(3, -2)$.

75. Find the equation of the x-axis.

76. Find the equation of the y-axis.

Writing Exercises ■ *Write a paragraph using your own words.*

1. Explain why the pair (x, y) is called an ordered pair.

2. Explain why the equation $y = 2$ has no x-intercept.

3. Tell how to use the intercept method of graphing.

4. Describe the general form of the equation of a line.

Something to Think About ■ If points $P(a, b)$ and $Q(c, d)$ are two points on a rectangular coordinate system and point M is midway between them, then point M is called the **midpoint** of the line segment joining P and Q. (See Illustration 2 on the next page.) To find the coordinates of the midpoint $M(x_M, y_M)$ of the segment PQ, we find the average of the x-coordinates and the average of the y-coordinates of P and Q.

$$x_M = \frac{a + c}{2} \quad \text{and} \quad y_M = \frac{b + d}{2}$$

Find the coordinates of the midpoint of the line segment with the given endpoints.

1. $P(5, 3)$ and $Q(7, 9)$

2. $P(5, 6)$ and $Q(7, 10)$

3. $P(2, -7)$ and $Q(-3, 12)$

4. $P(-8, 12)$ and $Q(3, -9)$

5. $A(4, 6)$ and $B(10, 6)$

6. $A(8, -6)$ and the origin

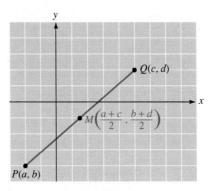

ILLUSTRATION 2

Review Exercises ■ *Solve each equation.*

1. $\dfrac{5(y - 2)}{3} = -(6 + y)$ **2.** $7(y + 14) = 35(y - 2)$

Do each division.

3. $\dfrac{8x^2y^3 + 12x^3y^2}{6x^2y^2}$ **4.** $2x + 4\overline{)6x^2 + 8x - 8}$

6.2 The Slope of a Line

■ Slope–Intercept Form of the Equation of a Line ■ Horizontal and Vertical Lines ■ Applications

GETTING READY Simplify each expression.

1. $\dfrac{6 - 2}{12 - 8}$ **2.** $\dfrac{12 - 3}{11 - 8}$ **3.** $\dfrac{4 - 16}{6 - 2}$ **4.** $\dfrac{2 - 9}{21 - 7}$

The **slope** of a nonvertical line is a number that measures the steepness of the line. In Figure 6-10, a line passes through the points $P(1, 2)$ and $Q(3, 6)$. Moving along the line from P to Q causes the value of y to change from $y = 2$ to $y = 6$, an increase of $6 - 2$, or 4 units. The value of x increases $3 - 1$, or 2 units. The slope of the line, m, is defined to be the change in y divided by the change in x.

$$m = \frac{\text{change in the } y \text{ values}}{\text{change in the } x \text{ values}}$$

$$= \frac{6 - 2}{3 - 1}$$

$$= \frac{4}{2}$$

$$= 2$$

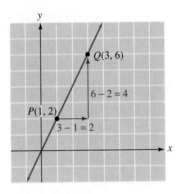

FIGURE 6-10

To distinguish between the coordinates of points P and Q in Figure 6-11, we use **subscript notation.** Point P is denoted as $P(x_1, y_1)$, read as "point P with coordinates of x sub 1 and y sub 1." Point Q is denoted as $Q(x_2, y_2)$, read as "point Q with coordinates of x sub 2 and y sub 2."

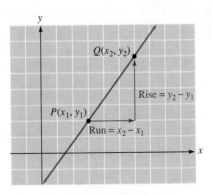

FIGURE 6-11

As a point on the line in Figure 6-11 moves from P to Q, its y-coordinate changes by the amount $y_2 - y_1$, while its x-coordinate changes by $x_2 - x_1$.

Slope of a Nonvertical Line   If $P(x_1, y_1)$ and $Q(x_2, y_2)$ are two points on a nonvertical line, the slope m of PQ is given by the formula

$$m = \frac{y_2 - y_1}{x_2 - x_1}$$

In Figure 6-11, the difference in the y-coordinates of points P and Q is the vertical distance the line *rises* between points P and Q. The difference of the x-coordinates is the *run*—the horizontal distance between points P and Q. The slope of the line is the ratio of its rise to its run.

$$m = \frac{y_2 - y_1}{x_2 - x_1} = \frac{\text{rise}}{\text{run}}$$

WARNING!   When finding the slope of a line, always subtract the y values and the x values in the same order, or your answer will have the wrong sign:

$$m = \frac{y_2 - y_1}{x_2 - x_1} \quad \text{or} \quad m = \frac{y_1 - y_2}{x_1 - x_2}$$

However,

$$m \neq \frac{y_2 - y_1}{x_1 - x_2} \quad \text{and} \quad m \neq \frac{y_1 - y_2}{x_2 - x_1}$$

EXAMPLE 1   Find the slope of the line that passes through the points $(-3, 2)$ and $(2, -5)$ and draw its graph.

Solution   Since we know the coordinates of two points on the line, we can find its slope. If (x_1, y_1) is $(-3, 2)$ and (x_2, y_2) is $(2, -5)$, then

$$x_1 = -3 \quad \text{and} \quad x_2 = 2$$
$$y_1 = 2 \qquad\qquad y_2 = -5$$

To find the slope of the line, we substitute these values into the formula for slope and simplify.

$$\text{Slope} = \frac{y_2 - y_1}{x_2 - x_1}$$

$$= \frac{-5 - 2}{2 - (-3)} \qquad \text{Substitute } -5 \text{ for } y_2, 2 \text{ for } y_1, 2 \text{ for } x_2, \text{ and } -3 \text{ for } x_1.$$

$$= -\frac{7}{5} \qquad \text{Simplify.}$$

We would obtain the same result of $-\frac{7}{5}$ if we had let $(x_1, y_1) = (2, -5)$ and $(x_2, y_2) = (-3, 2)$. The graph of the line is shown in Figure 6-12.

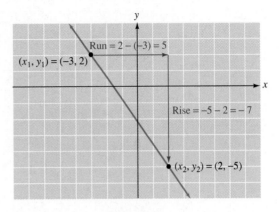

FIGURE 6-12 ■

Suppose we move from left to right along any of the lines shown in Figure 6-13. When increasing numbers x produce increasing values of y, as in Figure 6-13(a), the slope of the line is positive. Whenever increasing numbers x produce decreasing values of y, as in Figure 6-13(b), the slope is negative. If a line is horizontal, as in Figure 6-13(c), the slope is 0.

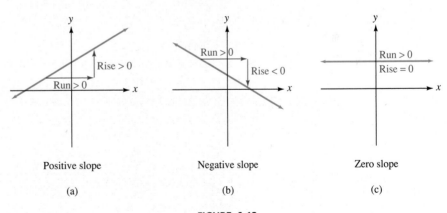

Positive slope

(a)

Negative slope

(b)

Zero slope

(c)

FIGURE 6-13

We can graph a line whenever we know the coordinates of one point on the line and its slope. For example to graph the line that passes through $P(2, 4)$ and has a slope of 3, we plot $P(2, 4)$, as in Figure 6-14. Since the slope is 3, the line rises 3 units for every 1 unit it moves to the right. Thus, we can find a second point on the line by starting at $P(2, 4)$ and moving 1 unit to the right and 3 units up. This brings us to the point Q with coordinates of $(2 + 1, 4 + 3)$, or $(3, 7)$. The required line passes through points P and Q.

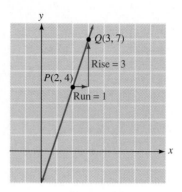

FIGURE 6-14

▪ Slope–Intercept Form of the Equation of a Line

We can find the slope of a line from its equation after first finding the coordinates of two points on the line. For example, to find the slope of the line with equation $y = 2x + 7$, we find the coordinates of two points on the graph. One point is the y-intercept, found by substituting 0 for x in the equation and solving for y.

$$y = 2x + 7$$
$$= 2(0) + 7 \qquad \text{Substitute 0 for } x.$$
$$= 7$$

The y-intercept is $(0, 7)$.

To find the coordinates of a second point, we substitute 1 for x and solve for y.

$$y = 2x + 7$$
$$= 2(1) + 7 \qquad \text{Substitute 1 for } x.$$
$$= 9$$

The line passes through the point $(1, 9)$.

To find the slope, we can let $(x_1, y_1) = (0, 7)$ and $(x_2, y_2) = (1, 9)$.

$$m = \frac{y_2 - y_1}{x_2 - x_1} \qquad \text{Use the definition of slope.}$$
$$= \frac{9 - 7}{1 - 0}$$
$$= 2$$

The slope of the line is 2, the same as the coefficient of x in the equation $y = 2x + 7$. The y-intercept is $(0, 7)$, whose y-coordinate is the same as the constant in $y = 2x + 7$. This suggests the following form of the equation of a line.

| **Slope–Intercept Form of the Equation of a Line** | If a linear equation is written in the form |

$$y = mx + b$$

where m and b are constants, then the graph of that equation is a line with slope m and with y-intercept $(0, b)$.

EXAMPLE 2 Find the slope and the y-intercept of the line determined by $3x + 5y - 15 = 0$. Then graph it.

Solution We write the equation in slope–intercept form by solving it for y.

$$3x + 5y - 15 = 0$$

$$5y = -3x + 15 \qquad \text{Add } -3x \text{ and } 15 \text{ to both sides.}$$

$$y = -\frac{3}{5}x + \frac{15}{5} \qquad \text{Divide both sides by 5.}$$

$$y = -\frac{3}{5}x + 3$$

This equation is in the form $y = mx + b$ with $m = -\frac{3}{5}$ and $b = 3$. The slope of the line is $-\frac{3}{5}$, and its y-intercept is $(0, 3)$.

To graph the equation, we use the y-intercept and plot the point $(0, 3)$, as in Figure 6-15. Since the slope is $\frac{\text{rise}}{\text{run}} = \frac{-3}{5}$, we can find a second point on the line by starting at $(0, 3)$ and moving 5 units to the right and 3 units down to reach the point

$$(0 + 5, 3 + (-3)) \qquad \text{or} \qquad (5, 0)$$

We plot the point $(5, 0)$ and draw the line shown in the figure.

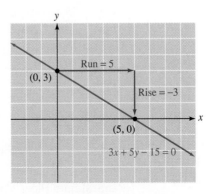

FIGURE 6-15

■ Horizontal and Vertical Lines

We can find the slope and the y-intercept of the line $y = 3$ by writing the equation in slope–intercept form.

$$y = 0x + 3$$

We can now see that the slope is 0, and the y-intercept is $(0, 3)$. Since the slope is 0, the line is horizontal. Its graph is shown in Figure 6-16(a).

The graph of the equation $x = 3$ is also a line. However, since the equation does not have a variable y, we cannot find its slope or y-intercept by writing the equation in the form $y = mx + b$. Instead, we will find two points on the line and try to find the slope by using the definition. The points $(3, 1)$ and $(3, 2)$ lie on the given line, because any point with an x-coordinate of 3 lies on the line. Thus,

$$m = \frac{y_2 - y_1}{x_2 - x_1}$$

$$m = \frac{2 - 1}{3 - 3} \qquad \text{Substitute 2 for } y_2, \text{ 1 for } y_1, \text{ 3 for } x_2, \text{ and 3 for } x_1.$$

$$= \frac{1}{0}$$

Since division by 0 is undefined, the symbol $\frac{1}{0}$ has no meaning. The slope of the line $x = 3$ is undefined. Its graph is the vertical line shown in Figure 6-16(b). Its x-intercept is 3, but it has no y-intercept.

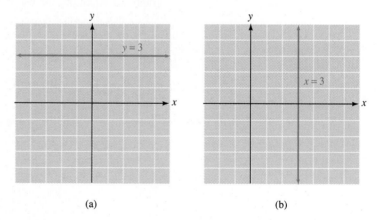

(a)   (b)

FIGURE 6-16

The previous discussion suggests the following facts.

Horizontal and Vertical Lines	Horizontal lines (lines with equations of the form $y = b$) have a slope of 0.
	Vertical lines (lines with equations of the form $x = a$) have no defined slope.

■ Applications

A linear equation often describes the relationship between the price of a product and the number of units that can be sold. For example, the owners of a bicycle shop have found that for the stock they have on hand, the number of bicycles they can sell is related to the price they charge according to a linear equation. If x is the price, in dollars, of one bicycle, and y is the number of bicycles that will be sold, the equation is

1. $\quad y = 500 - \dfrac{3}{5}x$

To find the number of bicycles they can sell at a unit price of $100, we substitute 100 for x in Equation 1 and solve for y.

$$y = 500 - \frac{3}{5}x$$

$$y = 500 - \frac{3}{5}(\mathbf{100}) \qquad \text{Substitute 100 for } x.$$

$$= 500 - 60$$

$$= 440$$

At a price of $100, the owners expect to sell 440 bicycles.

If we substitute 150 for x, we find that $y = 410$. Thus, at a price of $150, the owners would expect to sell 410 bicycles. In this case, increasing the price will cause a decrease in sales.

To graph this equation, we plot several points (x, y) that satisfy the equation and join them with a line, as in Figure 6-17. The break in the x-axis shows that we have omitted numbers less than 100. The break in the y-axis shows that we have omitted numbers less than 380.

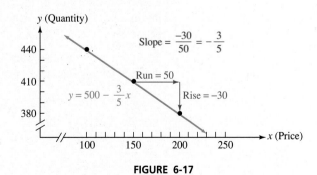

FIGURE 6-17

The slope of the line is $-\frac{3}{5}$, which is the coefficient of x in the equation $y = 500 - \frac{3}{5}x$. This slope represents the ratio of a change in the number of bicycles sold to the change in the price. Since $-\frac{3}{5} = -\frac{30}{50}$, for each increase of $50 in the cost of a bicycle, 30 fewer bicycles will be sold.

As machinery wears out, it is worth less. Accountants often estimate the decreasing value of aging equipment with **linear depreciation,** a method based on the equation of a line.

EXAMPLE 3 **Linear depreciation** A company buys a $12,500 computer with an estimated useful life of 6 years. After x years of use, the value y of the computer is linearly depreciated according to the equation

$$y = 12,500 - 2000x$$

a. Find the value of the computer after $3\frac{1}{2}$ years.

b. Find the economic meaning of the y-intercept.

c. Find the economic meaning of the slope.

d. Find the value of the computer at the end of its useful life.

Solution **a.** To find the computer's value after $3\frac{1}{2}$ years, we substitute 3.5 for x in the equation and solve for y.

$$y = 12,500 - 2000x$$
$$y = 12,500 - 2000(3.5)$$
$$= 12,500 - 7000$$
$$= 5500$$

When the computer is $3\frac{1}{2}$ years old, its value will be $5500.

b. The y-intercept is the point (0, 12,500). In this example, $12,500 is the value of a zero-year-old computer, which is the computer's original cost.

c. Each year, the value of the computer decreases by $2000, because the slope of the line is -2000. The slope of the depreciation line is the **annual depreciation rate.**

d. To find the computer's value at the end of its useful life, we substitute 6 for x in the equation and solve for y.

$$y = 12,500 - 2000x$$
$$y = 12,500 - 2000(6)$$
$$= 12,500 - 12,000$$
$$= 500$$

After 6 years, the computer will be worth $500. This is its **salvage value.** ∎

ORALS *Find the slope of the line passing through the given points.*

1. $(0, 0), (2, 3)$ **2.** $(0, 0), (6, -5)$

Find the slope and y-intercept of the graph of each equation.

3. $y = 4x + 9$ **4.** $y = -\dfrac{2}{3}x - 1$

5. $y - 1 = 3x$ **6.** $y + 4x = 2$

EXERCISE 6.2

In Exercises 1–16, find the slope of the line passing through the given points, when possible.

1. $(1, 3), (2, 4)$ **2.** $(1, 4), (2, 3)$ **3.** $(2, 5), (3, 4)$ **4.** $(3, 6), (5, 2)$

5. $(2, 6), (3, 7)$ **6.** $(5, 8), (2, 9)$ **7.** $(0, -5), (4, 3)$ **8.** $(3, -2), (3, 5)$

9. $(2, 3), (-3, 2)$ **10.** $(-7, 3), (-3, 7)$ **11.** $(-5, -7), (-4, -7)$ **12.** $(-6, -8), (-5, -8)$

13. $(-2, 0), (0, -2)$ **14.** $(10, -9), (9, 0)$ **15.** $(3, 7), (3, 8)$ **16.** $(8, 3), (7, 3)$

In Exercises 17–28, graph the line that passes through the given point and has the given slope.

17. $(0, 3), m = 1$ **18.** $(3, 2), m = 3$ **19.** $(-3, 2), m = 4$ **20.** $(-1, 0), m = 2$

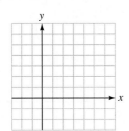 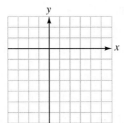 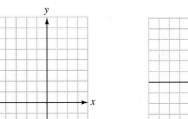 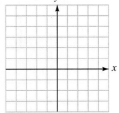

21. $(1, -3), m = -1$ **22.** $(1, -3), m = -2$ **23.** $(3, 5), m = -4$ **24.** $(0, 0), m = -5$

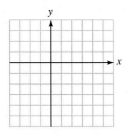 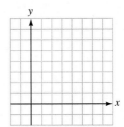 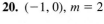 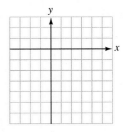

25. $(0, 0)$, $m = \dfrac{1}{2}$ **26.** $(2, 3)$, $m = -\dfrac{1}{2}$ **27.** $(-1, 3)$, $m = -\dfrac{5}{3}$ **28.** $(0, 0)$, $m = \dfrac{3}{5}$

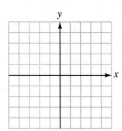 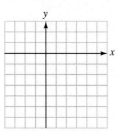 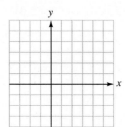 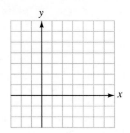

In Exercises 29–40, find the slope and the y-intercept of the line defined by each equation. Then use the slope and y-intercept to graph the line.

29. $y = 3x + 3$ **30.** $y = 4x - 5$ **31.** $y = 5x + 1$ **32.** $y = -3x + 2$

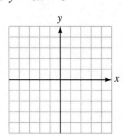 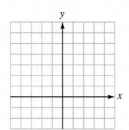 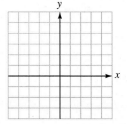

33. $y = -3x$ **34.** $y = -3x + 5$ **35.** $y = 3x - 2$ **36.** $y = -5x + 1$

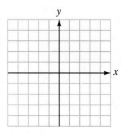 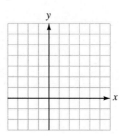 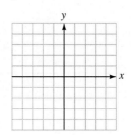 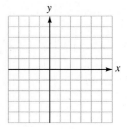

37. $y = \dfrac{x}{3}$ **38.** $y = -\dfrac{x}{2} + 2$ **39.** $y = \dfrac{5}{3}x + \dfrac{1}{2}$ **40.** $y = \dfrac{3}{5}x - \dfrac{1}{2}$

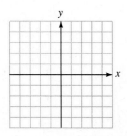 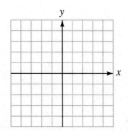 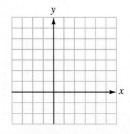 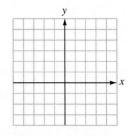

In Exercises 41–52, write each equation in slope–intercept form and then find the slope and the y-intercept of the line.

41. $3(x - 2) + y = 1$

42. $2(x + 2) - 3y = 2$

43. $5(x - 5) = 2y + 1$

44. $2(y + 7) - x = 3$

45. $2(y - 1) = y$

46. $y - 7(y + 5) = 6$

47. $\dfrac{2y + 7}{2} = x$

48. $\dfrac{2 - x}{4} = y$

49. $\dfrac{3(y - 5)}{2} = x + 3$

50. $\dfrac{5(3 + x)}{2} = y - 5$

51. $x = 3y + 5$

52. $x = -\dfrac{1}{5}y - \dfrac{1}{2}$

53. Grade of a road If the vertical rise of the road shown in Illustration 1 is 24 feet for a horizontal run of 1 mile, find the slope of the road. (*Hint:* 1 mile = 5280 feet.)

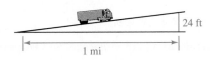

ILLUSTRATION 1

54. Demand for television sets The demand, y, for television sets depends on the price, x, according to the equation $y = 5500 - 6x$. Find the decrease in sales for each $1 increase in price.

55. Pitch of a roof If the rise of the roof shown in Illustration 2 is 5 feet for a run of 12 feet, find the pitch of the roof.

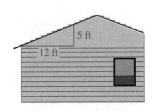

ILLUSTRATION 2

56. Slope of a ramp If a ramp rises 4 feet over a run of 12 feet, find the slope of the ramp.

57. Fixed costs An electronic company's daily cost C for manufacturing x television sets is given by the equation $C = 1200 + 130x$. The y-intercept of the graph of this equation is called the company's **fixed cost**. Find the fixed cost.

58. Marginal cost The slope of the line in Exercise 57 is called the **marginal cost**. Find the marginal cost.

59. Truck depreciation A truck is depreciated by the formula $y = 57,000 - 5500x$, where x represents its age in years. Find the value of the truck after $5\frac{1}{2}$ years.

60. Annual rate of depreciation Find the annual rate of depreciation of the truck in Exercise 59.

61. Finding the original cost A copy machine is valued annually by the formula $y = 6000 - 1200x$. Find the copier's original cost.

62. Salvage value Find the salvage value of the copy machine discussed in Exercise 61 if its useful life is $4\frac{1}{2}$ years.

63. Finding rate of depreciation A $450-drill press will have no salvage value after 3 years. Find the annual rate of depreciation.

64. Finding rate of depreciation A $750 video camera has a useful life of 5 years and a salvage value of $50. Find its annual rate of depreciation.

Writing Exercises ■ *Write a paragraph using your own words.*

1. Explain how to find the equation of a line passing through points $(1, 2)$ and $(5, 8)$.

2. Explain why a vertical line has no defined slope.

Something to Think About ■
1. Could $y = 8000 - 2000x$ be the linear depreciation equation for a computer with a useful life of 5 years?

2. Could $y = 8000 + 1000x$ be the linear depreciation equation for a color copier with a useful life of 6 years?

Review Exercises ■ *Factor each expression.*

1. $3x^2 - 6x$　　**2.** $y^2 - 25$　　**3.** $2z^2 - 5z - 3$　　**4.** $9t^2 - 15t + 6$

Solve each equation.

5. $4y^2 + 8y = 0$　　**6.** $r^2 - 36 = 0$　　**7.** $x^2 - 7x + 6 = 0$　　**8.** $12s^2 + 13s = 4$

6.3 Writing Equations of Lines

■ **Point–Slope Form of the Equation of a Line** ■ **Parallel and Perpendicular Lines** ■ **Applications**

GETTING READY　*Write each equation in general form.*

1. $3x = 2y + 4$　　　　　　　　**2.** $y = -2x + 7$
3. $y - 4 = 2x + 10$　　　　　　**4.** $y + 3 = -3(x - 5)$

Write each equation in slope–intercept form.

5. $2x - y = 12$　　**6.** $3x + 4y = 1$　　**7.** $y - 4 = 4(x - 1)$　**8.** $x + \dfrac{y}{2} = 4$

We have learned how to graph a linear equation. We now consider the reverse problem: finding a linear equation from its graph.

Since the linear equation $y = mx + b$ is written in slope–intercept form, the slope of its straight-line graph is m, and its y-intercept is $(0, b)$. If we know the slope and the y-intercept of a line, we can write its equation. For example, to write the equation of the line with slope 7 and y-intercept $(0, -2)$, we substitute 7 for m and -2 for b in the slope–intercept form and simplify.

$$y = mx + b$$
$$y = 7x + (-2) \qquad \text{Substitute 7 for } m \text{ and } -2 \text{ for } b.$$
$$y = 7x - 2$$

The equation of the line with slope 7 and y-intercept $(0, -2)$ is $y = 7x - 2$.

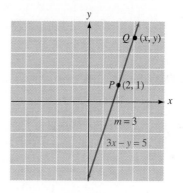

EXAMPLE 1 Find the equation of the line with slope $-\frac{3}{2}$ and y-intercept $(0, 3)$. Write the equation in general form.

Solution

$$y = mx + b \qquad \text{Use the slope–intercept form.}$$

$$y = -\frac{3}{2}x + 3 \qquad \text{Substitute } -\frac{3}{2} \text{ for } m \text{ and 3 for } b.$$

$$2y = -3x + 6 \qquad \text{Multiply both sides by 2.}$$

$$3x + 2y = 6 \qquad \text{Add } 3x \text{ to both sides.}$$

The equation of $3x + 2y = 6$ is written in the form $Ax + By = C$, which is the general form of the equation of a line. ∎

■ Point–Slope Form of the Equation of a Line

The line shown in Figure 6-18 passes through the point $P(2, 1)$ and has a slope of 3. If $Q(x, y)$ is any other point on the line, we can find the equation of the line by using the definition of slope.

FIGURE 6-18

$$m = \frac{y_2 - y_1}{x_2 - x_1} \qquad \text{The definition of slope.}$$

1. $m = \dfrac{y - 1}{x - 2}$ Substitute y for y_2, 1 for y_1, x for x_2, and 2 for x_1.

Since we are given that the slope m of the line is 3, we have

$$3 = \frac{y - 1}{x - 2} \qquad \text{Substitute 3 for } m \text{ in Equation 1.}$$

$$3(x - 2) = y - 1 \qquad \text{Multiply both sides by } x - 2.$$

$$3x - 6 = y - 1 \qquad \text{Use the distributive property to remove parentheses.}$$

$$3x - y = 5 \qquad \text{Add } -y \text{ and 6 to both sides.}$$

The equation of the required line, written in general form, is $3x - y = 5$.

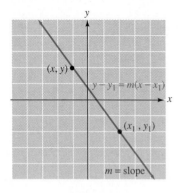

FIGURE 6-19

To find the equation of any nonvertical line with known slope passing through a known point, we refer to Figure 6-19, where the slope of the line is m and the known point has coordinates of (x_1, y_1). If (x, y) are the coordinates of any point on the line, the slope m of the line is

$$m = \frac{y - y_1}{x - x_1}$$

If we multiply both sides of this equation by $x - x_1$, we have

$$m(x - x_1) = y - y_1 \qquad \text{Multiply both sides by } x - x_1.$$

or

$$y - y_1 = m(x - x_1)$$

This equation is called the **point–slope form** of the equation of a line.

Point–Slope Form of the Equation of a Line	If a line of slope m passes through the point (x_1, y_1), then the equation of the line is $$y - y_1 = m(x - x_1)$$

 EXAMPLE 2   A line has a slope of $\dfrac{3}{4}$ and passes through the point $\left(-1, \dfrac{1}{2}\right)$. Write its equation in general form.

Solution   We substitute $\frac{3}{4}$ for m, -1 for x_1, and $\frac{1}{2}$ for y_1 in the point–slope form of a linear equation and simplify.

$$y - y_1 = m(x - x_1)$$

$$y - \frac{1}{2} = \frac{3}{4}[x - (-1)]$$

$$4y - 2 = 3(x + 1) \qquad \text{Multiply both sides by 4 and simplify.}$$

$$4y - 2 = 3x + 3 \qquad \text{Remove parentheses.}$$

$$-5 = 3x - 4y \qquad \text{Add } -3 \text{ and } -4y \text{ to both sides.}$$

In general form, the equation of the line is $3x - 4y = -5$. ∎

▪ Parallel and Perpendicular Lines

Because horizontal lines are **parallel** and horizontal lines have a slope of 0, all distinct lines with a slope of 0 are parallel.

Likewise, because vertical lines are parallel and have no defined slope, distinct lines with undefined slopes are parallel.

Because the slope of a line is a measure of the steepness of a line, parallel lines have equal slopes.

| **Slopes of Parallel Lines** | Two lines with the same slope are **parallel.** |

The equations $y = 5x + 7$ and $y = 5x - 8$ represent parallel lines, because each line has a slope of 5.

EXAMPLE 3 Graph the equation $y = -3x + 4$ and find the equation of the line parallel to it that passes through the point $(5, -2)$. Write the equation in general form and graph it.

Solution Because the equation $y = -3x + 4$ is written in slope–intercept form, we can see that the slope of its graph is -3 and its y-intercept is $(0, 4)$.

To graph the equation, we plot the y-intercept $(0, 4)$. Since the slope is -3, the line drops 3 units for every 1 unit it moves to the right. We begin at $(0, 4)$ and move 1 unit to the right and 3 units down to locate $(1, 1)$, which is another point on the line. The line through $(0, 4)$ and $(1, 1)$ is the graph of $y = -3x + 4$, shown in Figure 6-20.

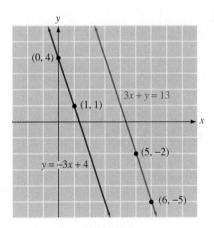

FIGURE 6-20

Since the required line is to be parallel to the graph of $y = -3x + 4$, it must also have a slope of -3. Furthermore, it passes through the point $(5, -2)$. We substitute these values into the point–slope form of the equation of a line and simplify.

$$y - y_1 = m(x - x_1)$$
$$y - (-2) = -3(x - 5) \qquad \text{Substitute } -3 \text{ for } m, 5 \text{ for } x_1, \text{ and } -2 \text{ for } y_1.$$
$$y + 2 = -3x + 15 \qquad \text{Remove parentheses.}$$
$$3x + y = 13 \qquad \text{Add } 3x \text{ and } -2 \text{ to both sides.}$$

In general form, the equation of the required line is $3x + y = 13$. To graph the line, we begin by plotting the point $(5, -2)$. Since the slope is -3, we move 1 unit to the right and 3 units down to the point $(6, -5)$. The graph, shown in Figure 6-20, is the line through these two points. ■

Lines that meet at right angles are called **perpendicular lines**. For example, a vertical line is perpendicular to a horizontal line. The slopes of other perpendicular lines are related by the following fact.

Slopes of Perpendicular Lines	The product of the slopes of perpendicular lines is -1.

Two numbers that have a product of -1 are called **negative reciprocals**. For example, 3 and $-\frac{1}{3}$ are negative reciprocals, because their product is -1:

$$3\left(-\frac{1}{3}\right) = -1$$

In general, the negative reciprocal of a $(a \neq 0)$ is $-\frac{1}{a}$.

EXAMPLE 4 One line has a slope of 5, and another has a slope of $-\frac{1}{5}$. Determine whether the lines are parallel, perpendicular, or neither.

Solution Because the slopes are not equal $\left(5 \neq -\frac{1}{5}\right)$, the lines are not parallel. Since the product of their slopes is $5\left(-\frac{1}{5}\right)$ or -1, the lines are perpendicular. ∎

EXAMPLE 5 Determine whether the graphs of $y = 2x - 5$ and $y = 2(3 - x)$ are parallel, perpendicular, or neither.

Solution The slope of the line $y = 2x - 5$ is 2, the coefficient of x. To find the slope of the second line, we solve its equation for y.

$$y = 2(3 - x)$$
$$y = 6 - 2x \qquad \text{Use the distributive law to remove parentheses.}$$
$$y = -2x + 6$$

Here the slope is -2, the coefficient of x in the equation $y = -2x + 5$. Because the slopes (2 and -2) are not equal, the lines are not parallel. Because the product of the slopes is not -1, the lines are not perpendicular. Thus, the lines are neither parallel nor perpendicular. ∎

EXAMPLE 6 Find the equation of the line that is perpendicular to the line $y = \frac{1}{3}x + 5$ and passes through $(2, -1)$. Write the equation in general form and graph both equations.

Solution The slope of the line with equation $y = \frac{1}{3}x + 5$ is $\frac{1}{3}$. Since the required line is to be perpendicular to the given line, its slope will be the negative reciprocal of $\frac{1}{3}$.

$$\text{Slope of the required line} = -\frac{1}{\frac{1}{3}} = -3$$

We can use the point–slope form to find the equation of the line with a slope of -3 and passing through the point $(2, -1)$, as follows:

$$y - y_1 = m(x - x_1)$$
$$y - (-1) = -3(x - 2) \qquad \text{Substitute } -3 \text{ for } m, 2 \text{ for } x_1, \text{ and } -1 \text{ for } y_1.$$
$$y + 1 = -3x + 6 \qquad \text{Remove parentheses.}$$
$$3x + y = 5 \qquad \text{Add } 3x \text{ and } -1 \text{ to both sides.}$$

Written in general form, the equation of the required line is $3x + y = 5$. The graphs are shown in Figure 6-21.

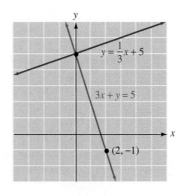

FIGURE 6-21 ∎

The various forms of a linear equation are summarized as follows.

Forms of a Line	$Ax + By = C$	**General form** of a linear equation. A and B cannot both be 0, $A \geq 0$, and A, B, and C are integers, when possible.
	$y = mx + b$	**Slope–intercept form** of a linear equation. The slope is m, and the y-intercept is $(0, b)$.
	$y - y_1 = m(x - x_1)$	**Point–slope form** of a linear equation. The slope is m, and the line passes through (x_1, y_1).
	$y = b$	**Horizontal line.** The slope is 0, and the y-intercept is $(0, b)$.
	$x = a$	**Vertical line.** The slope is undefined, and the x-intercept is $(a, 0)$.

■ Applications

EXAMPLE 7   **Water billing**   A city water department's monthly charge for water usage is related to the number of gallons used by a linear equation. If a customer is charged $12 for using 1000 gallons and $16 for using 1800 gallons, find the charge for 2000 gallons.

Solution   When the customer uses 1000 gallons, the charge is $12. When he uses 1800 gallons, the charge is $16. Since y is related to x by a linear equation, the points $P(1000, 12)$ and $Q(1800, 16)$ will lie on the line shown in Figure 6-22.

To write the equation of the line passing through P and Q, we first find the slope of the line passing through those points:

$$m = \frac{y_2 - y_1}{x_2 - x_1}$$

$$= \frac{16 - 12}{1800 - 1000}$$

$$= \frac{4}{800}$$

$$= \frac{1}{200}$$

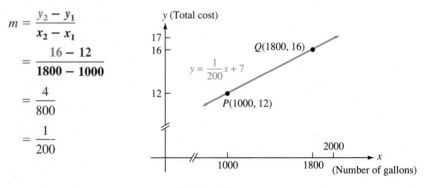

FIGURE 6-22

We then substitute $\frac{1}{200}$ for m and the coordinates of one of the known points (say, $P(1000, 12)$) into the point–slope form of the equation of a line and proceed as follows:

$$y - y_1 = m(x - x_1)$$

$$y - 12 = \frac{1}{200}(x - 1000)$$

$$y - 12 = \frac{1}{200}x - \frac{1}{200}(1000)$$

$$y - 12 = \frac{1}{200}x - 5$$

2. $y = \dfrac{1}{200}x + 7$ Add 12 to both sides.

To find the charge for 2000 gallons, we substitute 2000 for x in Equation 2 and find y.

$$y = \frac{1}{200}(2000) + 7$$

$$y = 10 + 7$$

$$y = 17$$

The charge for 2000 gallons of water is $17.   ■

ORALS *Find the equation of the line with the given slope and y-intercept.*

1. $m = 2, (0, 4)$ **2.** $m = -\dfrac{1}{2}, (0, 2)$

Tell whether the graphs of the following pairs of equations are parallel or perpendicular.

3. $y = 2x - 3, y = 2x + 7$

4. $y = \dfrac{1}{2}x - \dfrac{3}{2}, y = -2x + 45$

E X E R C I S E   6 . 3

In Exercises 1–12, find the equation of each line with the given slope and y-intercept. **Write the answer in general form.**

1. $4, (0, 5)$ **2.** $2, (0, -3)$ **3.** $-7, (0, -2)$ **4.** $-3, (0, 10)$

5. $-\dfrac{1}{2}, \left(0, -\dfrac{1}{2}\right)$ **6.** $\dfrac{2}{3}, (0, -5)$ **7.** $-\dfrac{3}{5}, \left(0, -\dfrac{2}{5}\right)$ **8.** $-\dfrac{5}{7}, \left(0, -\dfrac{3}{7}\right)$

9. $0, (0, 3)$ **10.** $3, (0, 0)$ **11.** $0,$ origin **12.** $-\dfrac{4}{5},$ origin

In Exercises 13–24, use the point–slope form to find the equation of each line with the given slope and passing through the given point. **Write the answer in general form.**

13. slope $2, (1, 1)$ **14.** slope $3, (2, 3)$

15. slope $5, (1, -1)$ **16.** slope $-3, (0, -3)$

17. slope $-2, (0, 0)$ **18.** slope $\dfrac{1}{2}, (2, 3)$

19. slope $-\dfrac{1}{2}, (-2, 4)$ **20.** slope $-\dfrac{2}{3}, (0, 3)$

21. slope $\dfrac{7}{5}, \left(3, \dfrac{2}{5}\right)$ **22.** slope $-\dfrac{3}{7}, \left(1, \dfrac{5}{7}\right)$

23. slope $0, (739, 3)$ **24.** slope $0, (0, 0)$

In Exercises 25–32, tell whether the lines with the given slopes are parallel, perpendicular, or neither.

25. slopes of 3 and -3 **26.** slopes of 2 and $\dfrac{1}{2}$

27. slopes of 4 and $\dfrac{8}{2}$

28. slopes of $-\dfrac{3}{2}$ and $\dfrac{2}{3}$

29. slopes of 3 and $-\dfrac{1}{3}$

30. slopes of 0.5 and $\dfrac{1}{2}$

31. slopes of 1.5 and $\dfrac{3}{2}$

32. slopes of 1.5 and -1.5

In Exercises 33–42, tell whether the given pairs of lines are parallel, perpendicular, or neither.

33. $y = 3x + 2$ and $y = 3x - \dfrac{1}{2}$

34. $y = 2x + 5$ and $y = 2x - 7$

35. $y = \dfrac{1}{3}x + 1$ and $y = 3x - 1$

36. $y = 5x + 5$ and $y = -\dfrac{1}{5}x + 5$

37. $y = x + \dfrac{7}{2}$ and $y = -x + \dfrac{2}{7}$

38. $y = 5$ and $x = 5$

39. $y = -\dfrac{3}{4}x$ and $y = \dfrac{4}{3}x - \dfrac{1}{2}$

40. $y = 7x + 1$ and $y = -\dfrac{1}{7}x - 7$

41. $2x + 3y = 5$ and $y = -\dfrac{2}{3}x - 5$

42. $3x - y = 8$ and $x + 3y = 8$

In Exercises 43–62, write the equation of each line with the given properties in general form.

43. $m = \dfrac{2}{15}$, passes through $(15, 10)$

44. $m = \dfrac{3}{11}$, y-intercept $\left(0, \dfrac{1}{11}\right)$

45. $m = -\dfrac{4}{9}$, y-intercept $\left(0, -\dfrac{2}{9}\right)$

46. $m = -\dfrac{13}{5}$, passes through $\left(3, -\dfrac{2}{5}\right)$

47. Parallel to the line $y = 5x - 8$, passes through $(0, 0)$

48. Parallel to the line $y = -2(7 - x)$, passes through $(0, 5)$

49. Parallel to the line $y = 5(x + 3)$, y-intercept $(0, -4)$

50. Parallel to the line $y = -6(x - 2)$, y-intercept $(0, 5)$

51. Parallel to the line $5x - 2y = 3$, passes through $(2, 1)$

52. Parallel to the line $3x + 4y = -9$, y-intercept $(0, -3)$

53. Passes through $(2, 5)$ and $(3, 7)$ (*Hint:* First find the slope.)

54. Passes through $(-3, 0)$ and $(3, -1)$ (*Hint:* First find the slope.)

55. Passes through $(-5, 2)$ and $(7, 2)$ (*Hint:* First find the slope.)

56. Passes through $(-5, 2)$ and $(-5, 7)$ (*Hint:* First try to find the slope.)

57. Passes through $(2, 5)$, and perpendicular to a line with slope $\dfrac{1}{5}$

58. Passes through $(2, 1)$, and perpendicular to a line with slope -3

59. Passes through the origin and perpendicular to the line $y = 5x + 11$

60. Passes through $(3, -5)$, and perpendicular to the line $y = -\dfrac{1}{2}x - 7$

61. Has a y-intercept of $(0, 3)$ and perpendicular to the line $y = \dfrac{1}{5}x + 12$

62. Has a y-intercept of $\left(0, -\dfrac{1}{2}\right)$ and perpendicular to the line $y = -\dfrac{1}{3}x$

63. Projecting sales In the first year of operation, the sales of a software company totaled $50 million. In the second year, the sales totaled $58 million. If a linear trend continues, find the sales in the fourth year.

64. Projecting enrollments The enrollment at Rock Valley College was 7500 students in 1993 and 7900 students in 1994. If a linear trend continues, estimate the enrollment in the year 2000.

65. Installing rain gutters Seamless aluminum rain gutters can be installed for a fixed charge, plus an additional per-foot cost. If 200 feet of gutter costs $350 and 250 feet costs $425, find the cost to install 500 feet of gutter.

66. Converting temperatures Fahrenheit temperature, F, is related to Celsius temperature, C, by the formula $F = mC + b$. If water freezes at $0°$ Celsius and $32°$ Fahrenheit, and water boils at $100°$ Celsius and $212°$ Fahrenheit, find m and b.

67. Find the purchase price The accounting department uses the linear method to depreciate a word-processing system. After 3 years, the system is worth $2000. After 4 years, it is worth nothing. Find the purchase price of the system.

68. Finding salvage value The useful life of a lathe is 10 years. If it is worth $330 after 3 years and $90 after 9 years, find its salvage value.

Writing Exercises ■ *Write a paragraph using your own words.*

1. Explain why the slopes of parallel lines are equal.

2. Explain why the equation $y - y_1 = m(x - x_1)$ is called the point–slope form of the equation of a line.

Something to Think About ■

1. Can two lines with the same slope be nonparallel?

2. Can two lines be perpendicular when the product of their slopes is not -1?

Review Exercises ■ *Do each operation.*

1. $\dfrac{x^2 - 25}{x^2 + 10x + 25} \cdot \dfrac{x^2 + 6x + 5}{x^2 - 5x}$

2. $\dfrac{x^2 - x - 2}{x^2 + 4x + 3} \div \dfrac{x^2 - 2x}{x + 3}$

3. $\dfrac{3 + x}{x + 1} + \dfrac{3 - x}{x + 1}$

4. $\dfrac{x + 1}{x - 1} - \dfrac{x - 1}{x + 1}$

6.4 Graphs of Other Equations

■ Graphing Calculators ■ Possible Pitfalls When Using Graphing Calculators ■ Using the TRACE and ZOOM Keys

GETTING READY   *If* $y = x^2 - 1$, *find* y *when*

1. $x = 0$ **2.** $x = 1$ **3.** $x = -1$ **4.** $x = 2$

If $y = |x|$, *find* y *when*

5. $x = 0$ **6.** $x = -2$ **7.** $x = 2$ **8.** $x = -3$

Many equations have graphs that are not lines. For example, the graph of the equation $y = x^2$ is a curve, called a **parabola.**

EXAMPLE 1   Graph the equation $y = x^2$.

Solution   We make a table of values by substituting numbers for x and finding the corresponding values of y. For example, if we substitute -3 for x, we get

$$y = x^2$$
$$y = (-3)^2$$
$$y = 9$$

After plotting the points listed in the table shown in Figure 6-23 and joining them with a smooth curve, we obtain the graph of the equation.

$y = x^2$

x	y
-3	9
-2	4
-1	1
0	0
1	1
2	4
3	9

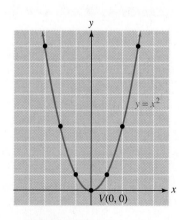

FIGURE 6-23

WARNING!   When an equation is nonlinear, its graph is not a line. To graph nonlinear equations, we must usually plot many points to recognize the shape of the graph.

EXAMPLE 2   Graph the equation $y = -|x| + 2$.

Solution   We make a table of values by substituting numbers for x and finding the corresponding values of y. For example, if we substitute -2 for x, we get

$$y = -|x| + 2$$
$$y = -|-2| + 2$$
$$y = -(2) + 2$$
$$y = 0$$

After plotting the points listed in the table shown in Figure 6-24, we obtain the graph of the equation.

$$y = -|x| + 2$$

x	y
-3	-1
-2	0
-1	1
0	2
1	1
2	0
3	-1

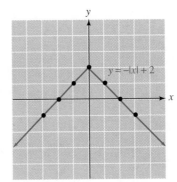

FIGURE 6-24

■ Graphing Calculators

So far, we have graphed equations by making a table of values and plotting points. This method is usually tedious and time-consuming. Fortunately, the task of graphing is made much easier when we use a graphing calculator.

Several brands of graphing calculators are available. Although we will use calculators to graph equations, we will not show the keystrokes of any specific brand. For these details, please consult your owner's manual.

All graphing calculators have a **viewing window,** used to display graphs (see Figure 6-25). To see the proper picture of a graph, we must often set the minimum and maximum values for the x- and y-coordinates. The standard RANGE settings of

$$\text{Xmin} = -10 \qquad \text{Xmax} = 10 \qquad \text{Ymin} = -10 \qquad \text{Ymax} = 10$$

indicate that -10 is the minimum x-coordinate and the minimum y-coordinate that will be used in the graph, and that 10 is the maximum x- and y-coordinate that will

FIGURE 6-25

be used. If these settings do not appear when you turn your calculator on and se-
lect the RANGE option, move the cursor to the desired position by pressing the
cursor keys ◄ , ► , ▲ , and ▼ and enter the standard RANGE settings.
To delete any unwanted digits, press the DEL key.

To graph the equation $y = -x^2 + 3$, press the Y = key and enter the right-
hand side of the equation after the symbol $Y_1 =$. The display will show the equation

$$Y_1 = -X^2 + 3$$

Then press the GRAPH key to produce the graph shown in Figure 6-26(a). To
show more detail, redraw the graph after setting other RANGE values. The settings

$$Xmin = -4 \qquad Xmax = 4 \qquad Ymin = -4 \qquad Ymax = 4$$

produce the graph shown in Figure 6-26(b).

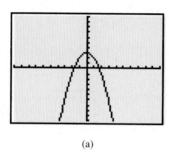
(a)

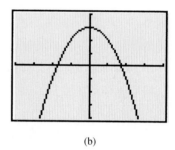
(b)

FIGURE 6-26

EXAMPLE 3   Graph the equation $y = \dfrac{x^2}{2} + 2$ in the following viewing windows:

a. $x = 10$ to 20 and $y = 10$ to 20.

b. $x = 0$ to 5 and $y = -5$ to 5.

c. $x = -6$ to 6 and $y = -2$ to 6.

Solution   **a.** We enter the appropriate x and y RANGE values and enter the right-hand side
of the equation. The viewing window will display

$$Y = X^2/2 + 2$$

Finally, we press the GRAPH key and obtain the blank window shown in
Figure 6-27(a). The blank window shows that no portion of the graph appears
on the coordinate grid seen in the viewing window.

b. We set the x and y RANGE values. There is no need to re-enter the equation.
We press the GRAPH key to obtain the graph shown in Figure 6-27(b). In
this case, only part of the graph appears, indicating that we need a larger viewing window.

c. We set the RANGE values and press the GRAPH key to obtain the more complete graph shown in Figure 6-27(c).

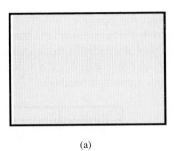
(a)

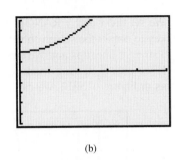
(b)

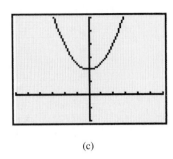
(c)

FIGURE 6-27 ■

As Example 3 shows, the choice of viewing window makes a big difference in how a graph looks on a calculator. One of the challenges of using graphing calculators is finding an appropriate viewing window.

EXAMPLE 4 Graph the equation $y = |x - 4|$.

Solution Since absolute values are always nonnegative, the minimum value of y is 0. To obtain a nice viewing window, we set the Ymin value slightly lower, at Ymin = -3. We set Ymax to be 10 greater than Ymin, at Ymax = 7.

The minimum value of y occurs when $x = 4$. To center the graph in the viewing window, we will set the Xmin and Xmax values at 5 on either side of 4:

Xmin = -1 and   Xmax = 9

After entering the right-hand side of the equation, we obtain the graph shown in Figure 6-28. Consult your owner's manual to learn how to enter an absolute value.

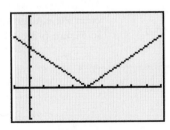

FIGURE 6-28 ■

■ Possible Pitfalls When Using Graphing Calculators

We have already seen that the choice of viewing window is extremely important when graphing equations. The next example reinforces this fact.

EXAMPLE 5   Graph the equation $y = 2x^2 - 20$ in the viewing window with x values from -1 to 5 and y values from -5 to 5.

Solution   We set the x and y RANGE values, enter the equation, and press the GRAPH key to obtain the screen shown in Figure 6-29(a). Although the graph appears to be a straight line, it is not. We are seeing a part of a parabola that appears to be straight. If we pick a viewing window with x values of -4 to 6 and y values of -20 to 2, as in Figure 6-29(b), we can see that the graph is a parabola.

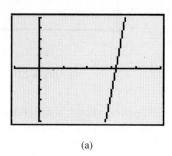   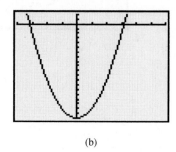

(a)   (b)

FIGURE 6-29   ∎

EXAMPLE 6   Use a graphing calculator to graph the equation $2x - 3y = 12$.

Solution   In order to use a graphing calculator, we must first solve the equation for y.

$$2x - 3y = 12$$
$$-3y = 12 - 2x \qquad \text{Add } -2x \text{ to both sides.}$$
$$y = -4 + \frac{2}{3}x \qquad \text{Divide both sides by } -3 \text{ and simplify.}$$
$$y = \frac{2}{3}x - 4$$

We can now set the standard range values of $x = -10$ to 10 and $y = -10$ to 10, enter the equation as $(2/3)x - 4$, and press the GRAPH key to obtain the graph in Figure 6-30.

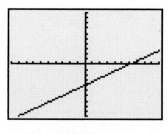

FIGURE 6-30   ∎

 WARNING!   To graph an equation with a graphing calculator, the equation must be solved for y.

■ Using the TRACE and ZOOM Keys

By using the TRACE key, it is possible to find the coordinates of any point on a graph. For example, the graph of the equation $y = x^2 - 3$ is shown in Figure 6-31(a), which uses the standard RANGE values.

To find where the parabola intersects the positive x-axis, we press the TRACE key and use the ◄ and ► keys to move the flashing cursor along the parabola until we approach the x-intercept. The coordinates $x = 1.7894737$ and $y = .20221607$ appear at the bottom of the viewing window. This means the cursor is at the point with coordinates (1.7894737, .20221607). See Figure 6-31(b).

To get better results, we can press the ZOOM key, zoom in, and press ENTER. We now see a magnified picture of the graph. We can press TRACE again and move the cursor to the point (1.7105263, −.0740997). Since y is nearly 0, the number 1.7105263 is nearly the x-intercept. We can achieve even better results with repeated zooms. See Figure 6-31(c).

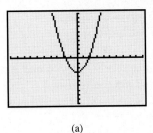
(a)

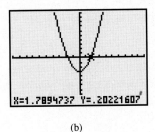
(b)

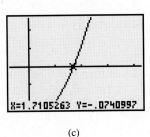
(c)

FIGURE 6-31

ORALS   *If $y = x^2 + 2$, find y when*

1. $x = 0$ **2.** $x = 1$ **3.** $x = -2$ **4.** $x = 5$

If $y = |x|$, find y when

5. $x = 0$ **6.** $x = 2$ **7.** $x = -4$ **8.** $x = -6$

E X E R C I S E   6.4

In Exercises 1–8, graph each equation by plotting points.

1. $y = x^2 - 1$

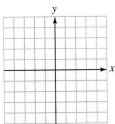

2. $y = -x^2 + 1$

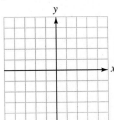

3. $y = (x - 2)^2$

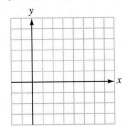

4. $y = -(x + 1)^2$

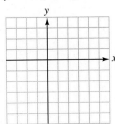

5. $y = -|x| + 3$ **6.** $y = |x - 2|$ **7.** $y = |x + 1| + 2$ **8.** $y = -|x - 2| + 5$

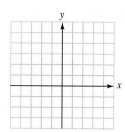 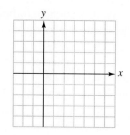 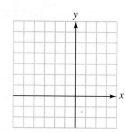 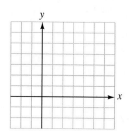

In Exercises 9–16, use a graphing calculator to graph each equation. Use a viewing window of $x = -5$ to 5 and $y = -5$ to 5. Compare the calculator graphs with your hand-made graphs in Exercises 1–8.

9. $y = x^2 - 1$ **10.** $y = -x^2 + 1$ **11.** $y = (x - 2)^2$ **12.** $y = -(x + 1)^2$

13. $y = -|x| + 3$ **14.** $y = |x - 2|$ **15.** $y = |x + 1| + 2$ **16.** $y = -|x - 2| + 5$

In Exercises 17–24, graph each equation in a viewing window of $x = -4$ to 4 and $y = -4$ to 4. The graph is not what it appears to be. Pick a better viewing window and find the true graph.

17. $y = x^2 - 6x$ **18.** $y = x^2 + 6x$ **19.** $y = |x - 4|$ **20.** $y = |x + 4|$

21. $y = x^2 - 12x + 13$ **22.** $y = x^2 + 12x + 13$ **23.** $y = x^3 - 8$ **24.** $y = x^3 + 10$

In Exercises 25–32, solve each equation for y and use a graphing calculator to graph it.

25. $2x + 5y = 10$ **26.** $3x - 4y = 12$ **27.** $4x - 8y = 12$ **28.** $x - 3y = 6$

29. $x^2 + y = 4$ **30.** $x^2 - y = -2$ **31.** $|x| + 2y = 4$ **32.** $|x| - \frac{1}{2}y = 2$

In Exercises 33–40, graph each parabola and find the coordinates of its highest or lowest point. This point is called the **vertex** of the parabola.

33. $y = x^2 + 4x$ **34.** $y = x^2 - 2x$ **35.** $y = -x^2 + 6x$ **36.** $y = -x^2 + 4x$

37. $y = x^2 + 4x - 5$ **38.** $y = x^2 - 5x - 2$ **39.** $y = -x^2 - 4x + 2$ **40.** $y = -x^2 - 4x + 1$

Writing Exercises ■ *Write a paragraph using your own words.*

1. Explain how to graph an equation by plotting points.

2. Explain why the correct choice of RANGE values is important when using a graphing calculator.

Something to Think About ■

1. Use a graphing calculator with the standard RANGE settings to graph
  a. $y = x^2$
  b. $y = x^2 + 1$
  c. $y = x^2 + 2$
  What do you notice?

2. Use a graphing calculator with the standard RANGE settings to graph
  a. $y = -|x|$
  b. $y = -|x| - 1$
  c. $y = -|x| - 2$
  What do you notice?

Review Exercises ■

1. List the prime numbers between 40 and 50.

2. State the associative property of addition.

3. State the commutative property of multiplication.

4. What is the additive identity element?

5. What is the multiplicative identity element?

6. What is the multiplicative inverse of $\dfrac{5}{3}$?

6.5 Graphing Inequalities

■ An Application of Linear Inequalities

GETTING READY   *Graph the equation $y = \frac{1}{3}x + 3$. Tell whether the given point lies on the line, above the line, or below the line.*

1. $(0, 0)$ **2.** $(0, 4)$ **3.** $(2, 2)$ **4.** $(6, 5)$
5. $(-3, 2)$ **6.** $(6, 8)$ **7.** $(-6, 0)$ **8.** $(-9, 5)$

The graph of $y = x - 5$ is a line consisting of the points whose coordinates satisfy the equation. The graph of the *inequality* $y \geq x - 5$ is not a line, but an area bounded by a line, called a **half-plane**. The half-plane consists of the points whose coordinates satisfy the inequality.

EXAMPLE 1   Graph the inequality $y \geq x - 5$.

Solution   Since $y \geq x - 5$ means that $y > x - 5$ or $y = x - 5$, we begin by graphing the equation $y = x - 5$. The graph appears in Figure 6-32(a).

Since $y \geq x - 5$ also indicates that y can be greater than $x - 5$, the coordinates of points other than those shown in Figure 6-32(a) satisfy the inequality. For example, the coordinates of the origin satisfy the inequality. We can verify this by letting x and y be 0 in the given inequality:

$$y \geq x - 5$$
$$0 \geq 0 - 5 \qquad \text{Substitute 0 for } x \text{ and 0 for } y.$$
$$0 \geq -5$$

Because $0 \geq -5$, the coordinates of the origin satisfy the original inequality. In fact, the coordinates of every point on the same side of the line as the origin satisfy the inequality. The graph of $y \geq x - 5$ is the half-plane that is shaded in Figure 6-32(b). Since the boundary line $y = x - 5$ is included, we draw it with a solid line.

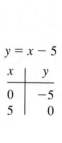

$y = x - 5$

x	y
0	-5
5	0

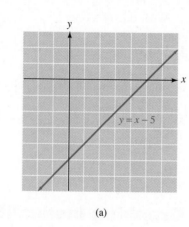

(a)

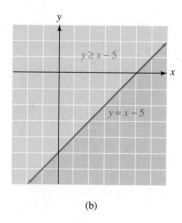

(b)

FIGURE 6-32   ■

EXAMPLE 2   Graph the inequality $2(x - 3) - (x - y) \leq -1$.

Solution   We begin by simplifying the inequality as follows:

$$2(x - 3) - (x - y) \leq -1$$
$$2x - 6 - x + y \leq -1 \qquad \text{Remove parentheses.}$$
$$x - 6 + y \leq -1 \qquad \text{Combine like terms.}$$
$$x + y \leq 5 \qquad \text{Add 6 to both sides.}$$

To graph the inequality $x + y \leq 5$, we graph the boundary line whose equation is $x + y = 5$, as in Figure 6-33(a).

Because the inequality $x + y \leq 5$ allows $x + y$ to be less than 5, the coordinates of the origin satisfy the inequality. We can verify this by substituting 0 for x and y in the given inequality.

$x + y \leq 5$

$0 + 0 \leq 5$ Substitute 0 for x and y.

$0 \leq 5$

Since $0 \leq 5$ is a true statement, the origin is in the graph. In fact, the coordinates of every point in the origin's side of the boundary line satisfy the inequality.

The graph of $x + y \leq 5$ is the half-plane that appears in color in Figure 6-33(b). Since the boundary line $x + y = 5$ is included, we draw it with a solid line.

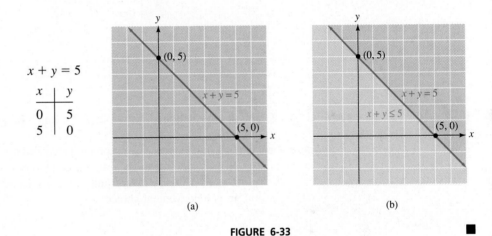

$x + y = 5$

x	y
0	5
5	0

(a)   (b)

FIGURE 6-33   ■

EXAMPLE 3   Graph the inequality $y > 2x$.

Solution   To find the boundary line, we graph the equation $y = 2x$. We draw the boundary line as a broken line to show that the points on it are not part of the solution, as in Figure 6-34(a).

We substitute into $y > 2x$ the coordinates of some point that lies on one side of the boundary line. Point $T(2, 0)$, for example, is below the boundary line in Figure 6-34(a). To see if point $T(2, 0)$ satisfies $y > 2x$, we substitute 2 for x and 0 for y in the inequality.

$y > 2x$

$0 > 2(2)$ Substitute 2 for x and 0 for y.

$0 > 4$

Since $0 > 4$ is a false statement, the coordinates of point T do not satisfy the inequality, and point T is not on the side of the line we wish to shade. Instead, we shade the other side of the boundary line. The graph of the solution set of $y > 2x$ is shown in Figure 6-34(b).

$y = 2x$

x	y
-1	-2
0	0
3	6

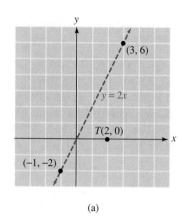

(a)

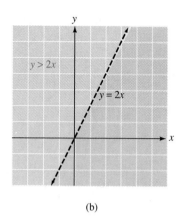

(b)

FIGURE 6-34   ∎

EXAMPLE 4   Graph the inequalities   **a.** $x + 2y < 6$ and   **b.** $y \geq 0$.

Solution   **a.** We find the boundary by graphing the equation $x + 2y = 6$. We draw the boundary as a broken line to show that it is not part of the solution. We then choose a point not on the boundary and see if its coordinates satisfy $x + 2y < 6$. The origin is a convenient choice.

$$x + 2y < 6$$
$$0 + 2(0) < 6 \qquad \text{Substitute 0 for } x \text{ and for } y.$$
$$0 < 6$$

Since $0 < 6$ is a true statement, we shade the side of the line that includes the origin. The graph is shown in Figure 6-35(a).

b. We find the boundary by graphing the equation $y = 0$. We draw the boundary as a solid line to show that it is part of the solution. We then choose a point not on the boundary and see if its coordinates satisfy $y \geq 0$. The point $T(0, 1)$ is a convenient choice.

$$y \geq 0$$
$$1 \geq 0 \qquad \text{Substitute 1 for } y.$$

Since $1 \geq 0$ is a true statement, we shade the side of the line that includes point T. The graph is shown in Figure 6-35(b).

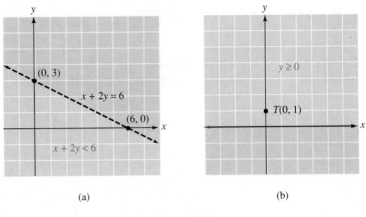

FIGURE 6-35 ■

The inequalities of Examples 1–4 are called **linear inequalities.** The following is a summary of the procedure for graphing linear inequalities.

Procedure for Graphing a Linear Inequality in Two Variables

1. Graph the boundary line of the region. If the inequality allows the possibility of equality (the symbol is either $\leq$ or $\geq$), draw the boundary line as a solid line. If equality is not allowed ($<$ or $>$), draw the boundary line as a broken line.

2. Pick a point that is on one side of the boundary line. (Use the origin if possible.) Replace x and y with the coordinates of that point. If the inequality is satisfied, shade the side that contains that point. If the inequality is not satisfied, shade the other side.

■ **An Application of Linear Inequalities**

EXAMPLE 5   **Earning money**   Carlos has two part-time jobs, one paying $5 per hour, and the other paying $6 per hour. He must earn at least $120 per week to pay his expenses while attending college. Write an inequality that shows the various ways he can schedule his time to achieve his goal.

Solution   If we let

x represent the number of hours he works on the first job and

y represent the number of hours he works on the second job we have

The hourly rate on the first job	·	the hours worked on the first job	+	the hourly rate on the second job	·	the hours worked on the second job	≥	$120.
$5	·	x	+	$6	·	y	≥	$120

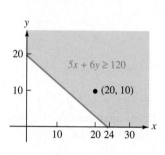

FIGURE 6-36

The graph of the inequality $5x + 6y \geq 120$ is shown in Figure 6-36. Any point in the shaded region indicates a possible way he can schedule his time and earn $120 or more per week. For example, if he works 20 hours on the first job and 10 hours on the second job, he will earn

$$\$5(20) + \$6(10) = \$100 + \$60$$
$$= \$160$$

Since Carlos cannot work a negative number of hours, the graph in the figure has no meaning when either x or y is negative. ∎

ORALS   *Tell whether the following coordinates satisfy the inequality $y > 3x + 2$.*

1. $(0, 0)$ **2.** $(5, 5)$ **3.** $(-2, 4)$ **4.** $(-3, -6)$

Tell whether the following coordinates satisfy the inequality $y \leq \frac{1}{2}x - 1$.

5. $(0, 0)$ **6.** $(2, 0)$ **7.** $(4, 3)$ **8.** $(-4, -3)$

E X E R C I S E   6.5

In Exercises 1–8, complete the graph by shading the correct side of the boundary.

1. $y \leq x + 2$

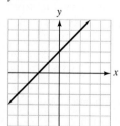

2. $y > x - 3$

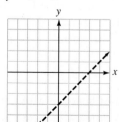

3. $y > 2x - 4$

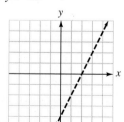

4. $y \leq -x + 1$

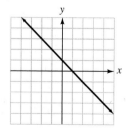

5. $x - 2y \leq 4$

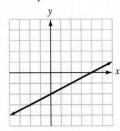

6. $3x + 2y > 12$

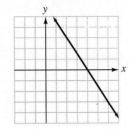

7. $y \leq 4x$

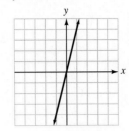

8. $y + 2x < 0$

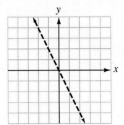

In Exercises 9–24, graph each inequality.

9. $y \geq 3 - x$

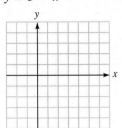

10. $y < 2 - x$

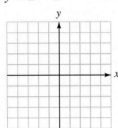

11. $y < 2 - 3x$

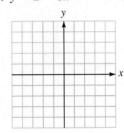

12. $y \geq 5 - 2x$

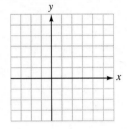

13. $y \geq 2x$

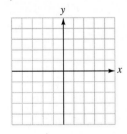

14. $y < 3x$

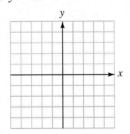

15. $2y - x < 8$

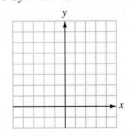

16. $y + 9x \geq 3$

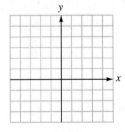

17. $y - x \geq 0$

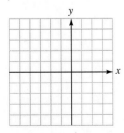

18. $y + x < 0$

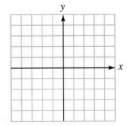

19. $2x + y > 2$

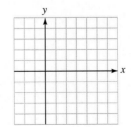

20. $3x - 2y > 6$

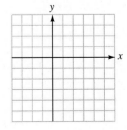

21. $3x - 4y > 12$

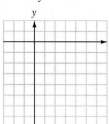

22. $4x + 3y \leq 12$

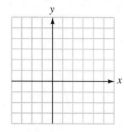

23. $5x + 4y \geq 20$

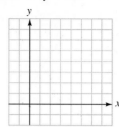

24. $7x - 2y < 21$

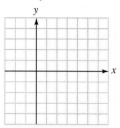

In Exercises 25–40, simplify each inequality and graph it.

25. $3(x + y) + x < 6$

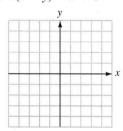

26. $2(x - y) - y \geq 4$

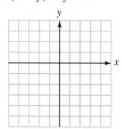

27. $4x - 3(x + 2y) \geq -6y$

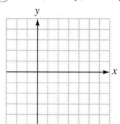

28. $3y + 2(x + y) < 5y$

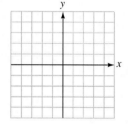

29. $7(x + 2y) - 2(x - 3y) < 50$

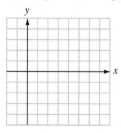

30. $3(6x + 5y) - 5(3x + 4y) \geq 15$

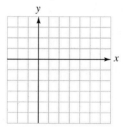

31. $5(x - 2y) > 5x$

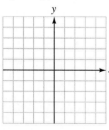

32. $3(y - x) \leq x - 3$

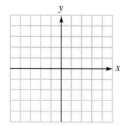

33. $x(x + 2) \leq x^2 + 3x + 1$

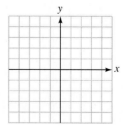

34. $y(y - 5) > y^2 - 7$

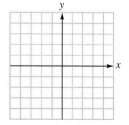

35. $x^2 + 3y \le x(x + 2) - 1$

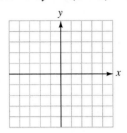

36. $x + y(y - 3) < (y + 1)(y - 2)$

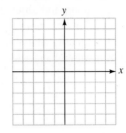

37. $3x + 7 \le 5y + 7$

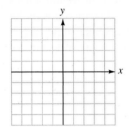

38. $5x \le x + 5(y + x)$

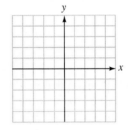

39. $(x + 1)(x - 2) + y^2 < x^2 + y(y - 2)$

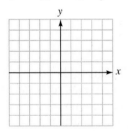

40. $(x + 2)(x - 2) + y^2 \ge (y - 2)(y + 1) + x^2$

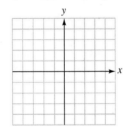

In Exercises 41–46, graph each inequality for nonnegative values of x and y. Then give some ordered pairs that satisfy the inequality.

41. Baking cakes and pies   It costs a bakery $3 to make a cake and $4 to make a pie. If production costs cannot exceed $120 per day, find an inequality that shows the possible combinations of cakes (x) and pies (y) that can be made.

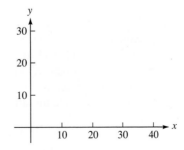

42. Hiring babysitters Mary has a choice of two babysitters. Sitter 1 charges $6 per hour, and sitter 2 charges $7 per hour. If Mary can afford no more than $42 per week for sitters, find an inequality that shows the possible ways that she can hire sitter 1 (x) and sitter 2 (y).

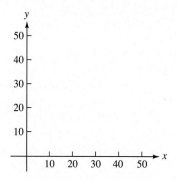

43. Inventory A clothing store advertises that it maintains an inventory of at least $4400 worth of men's jackets. If a leather jacket costs $100 and a nylon jacket costs $88, find an inequality that shows the possible ways that leather jackets (x) and nylon jackets (y) can be stocked.

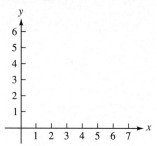

44. Making sporting goods To keep up with demand, a sporting goods manufacturer allocates at least 2400 units of time per day to make baseballs and footballs. If it takes 20 units of time to make a baseball and 30 units of time to make a football, find an inequality that shows the possible ways to schedule the time to make baseballs (x) and footballs (y).

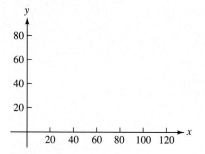

45. Investing in stocks Robert has up to $8000 to invest in two companies. If stock in Robotronics sells for $40 per share and stock in Macrocorp sells for $50 per share, find an inequality that shows the possible ways that he can buy shares of Robotronics (x) and Macrocorp (y).

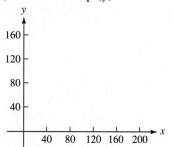

46. Buying baseball tickets Tickets to the Rockford Rox baseball games cost $6 for reserved seats and $4 for general admission. If nightly receipts must average at least $10,200 to meet expenses, find an inequality that shows the possible ways that the Rox can sell reserved seats (x) and general admission tickets (y).

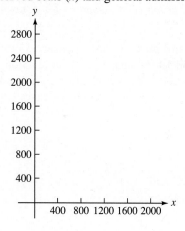

Writing Exercises ■ *Write a paragraph using your own words.*

1. Explain how to find the boundary for the graph of an inequality.

2. Explain how to decide which side of the boundary line to shade.

Something to Think About ■ **1.** Graph the inequality $y < |x|$. Is the graph a half-plane?

2. Graph the inequality $y < x^2$. Is the graph a half-plane?

Review Exercises ■ *Let $P(x) = 3x^2 - 4x + 3$. Find each quantity.*

1. $P(2)$ **2.** $P(-3)$ **3.** $P(-x)$ **4.** $P(x + 1)$

6.6 Relations and Functions

■ Relations ■ Functions ■ Finding Domains and Ranges from Graphs
■ Function Notation

GETTING READY *Let $y = x^2 - 2$. Find y when*

1. $x = 0$ **2.** $x = 2$ **3.** $x = -1$ **4.** $x = -2$

The equation $y = -x + 2$ determines ordered pairs (x, y) where a single value of y corresponds to each number x. If $x = 3$, the corresponding value of y is $-3 + 2$, or -1, and the ordered pair $(3, -1)$ indicates this correspondence. The graphs of all such ordered pairs form the line shown in Figure 6-37(a).

The inequality $y \le -x + 2$ also determines a set of ordered pairs. For this inequality, however, to each number x there correspond many values of y. Since the boundary of the graph $y \le -x + 2$ is the line $y = -x + 2$, the inequality determines the ordered pair $(3, -1)$. The inequality also determines many other pairs with an x-coordinate of 3: $(3, -2)$, $\left(3, -3\frac{1}{2}\right)$, and $(3, -5)$ are examples. The graph of the inequality appears in Figure 6-37(b).

Sonya Kovalevskaya (1850–1891)
This talented young Russian woman hoped to study mathematics at the University of Berlin, but strict rules prohibited women from attending lectures. Undaunted, she studied privately with the great mathematician Karl Weierstrauss and published several important papers.

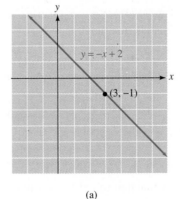

(a)

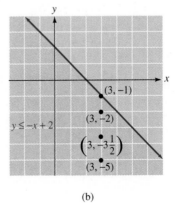

(b)

FIGURE 6-37

■ Relations

Because any set of ordered pairs of numbers indicates that certain numbers x and values y are related, any set of ordered pairs is called a **relation.** The set of ordered pairs shown in either part of Figure 6-37 is a relation.

Relation	A **relation** is a correspondence that assigns to each number x or one or more values of y.

The set of all possible numbers x is called the **domain of the relation,** and the set of all possible values of y is called the **range of the relation.** In the relations shown in Figure 6-37, both x and y could be any real number. The domain and the range of both relations are the set of real numbers.

■ Functions

The relations shown in Figure 6-37 differ in an important way. Several ordered pairs in part (b) have the same x but different values of y. The relation contains, for example, the pairs $(3, -1)$ and $(3, -2)$. In part (a), however, a single value of y corresponds to each number x. The relation in part (a) is called a **function.**

Function	A **function** is a correspondence that assigns to each number x exactly one value y. The variable x is called the **independent variable,** and the variable y is called the **dependent variable.**

WARNING! A function is always a relation, but a relation is not necessarily a function.

EXAMPLE 1 Do the equations **a.** $y = 3x + 1$ and **b.** $y^2 = x$ define y to be a function of x?

Solution **a.** For a function to exist, *each* number x must determine a *single* value of y. To find y in the equation $y = 3x + 1$, we multiply x by 3 and then add 1. Since this arithmetic always gives a single result, each choice of x determines a single value y. Thus, the equation $y = 3x + 1$ does define a function.

b. For a function to exist, *each* number x must determine a *single* value y. If we let $x = 9$, for example, y could be 3 or -3, because $3^2 = 9$ and $(-3)^2 = 9$. Since more than one value of y is determined when $x = 9$, the equation does not represent a function. ■

■ Finding Domains and Ranges from Graphs

The graph of a function enables us to see the domain and range of the function. For example, the domain of the graph in Figure 6-38 is shown on the x-axis, and the range is shown on the y-axis.

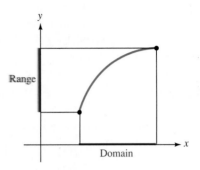

FIGURE 6-38

Since it is easy to graph a function with a graphing calculator, the calculator is a valuable tool for finding domains and ranges.

EXAMPLE 2 Find the domain and range of the function defined by $y = x^2 - 1$.

Solution The graph of $y = x^2 - 1$ is shown in Figure 6-39. Since every number x determines a corresponding y, the domain is the set of real numbers. Since the values of y are never less than -1, the range is the set of numbers greater than or equal to -1.

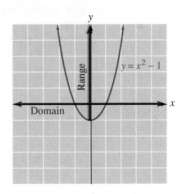

FIGURE 6-39   ■

■ Function Notation

Function Notation The notation $y = f(x)$ denotes that the variable y is a function of x.

The notation $f(x)$ is read as "f of x."

WARNING!   The notation $y = f(x)$ does not mean "f times x."

The notation $y = f(x)$ is similar to polynomial notation discussed in Section 3.4. The notation $f(x)$ provides a way of denoting the values of y that correspond to individual numbers x. If $y = f(x)$, the value of y that is determined by $x = 3$ is denoted by $f(3)$. Similarly, $f(-1)$ represents the value of y that corresponds to $x = -1$.

EXAMPLE 3   Let $f(x) = 3x + 1$. Find   **a.** $f(3)$,   **b.** $f(-1)$,   **c.** $f(0)$,   and   **d.** $f(r)$.

Solution   **a.** We replace x with 3:

$$f(x) = 3x + 1$$
$$f(3) = 3(3) + 1$$
$$= 9 + 1$$
$$= 10$$

b. We replace x with -1:

$$f(x) = 3x + 1$$
$$f(-1) = 3(-1) + 1$$
$$= -3 + 1$$
$$= -2$$

c. We replace x with 0:

$$f(x) = 3x + 1$$
$$f(0) = 3(0) + 1$$
$$= 0 + 1$$
$$= 1$$

d. We replace x with r:

$$f(x) = 3x + 1$$
$$f(r) = 3r + 1$$

The letter f used in the notation $y = f(x)$ represents the word *function*. However, other letters can be used to represent functions. The notations $y = g(x)$ and $y = h(x)$ also denote functions involving the variable x.

EXAMPLE 4   Let $g(x) = x^2 + 2x$. Find   **a.** $g\left(\frac{2}{5}\right)$,   **b.** $g(s)$,   **c.** $g(s^2)$,   and   **d.** $g(-t)$.

Solution   **a.** We replace x with $\frac{2}{5}$:

$$g(x) = x^2 + 2x$$
$$g\left(\frac{2}{5}\right) = \left(\frac{2}{5}\right)^2 + 2\left(\frac{2}{5}\right)$$
$$= \frac{4}{25} + \frac{4}{5}$$
$$= \frac{24}{25}$$

b. We replace x with s:

$$g(x) = x^2 + 2x$$
$$g(s) = s^2 + 2s$$

c. We replace x with s^2:

$$g(x) = x^2 + 2x$$
$$g(s^2) = (s^2)^2 + 2(s^2)$$
$$= s^4 + 2s^2$$

d. We replace x with $-t$:

$$g(x) = x^2 + 2x$$
$$g(-t) = (-t)^2 + 2(-t)$$
$$= t^2 - 2t$$

(b)

EXAMPLE 5   Let $f(x) = 3x + 2$. Find   **a.** $f(3) + f(2)$ and   **b.** $f(a) - f(b)$.

Solution   **a.** We find $f(3)$ and $f(2)$ separately.

$$f(x) = 3x + 2 \qquad\qquad\qquad f(x) = 3x + 2$$
$$f(3) = 3(3) + 2 \qquad\qquad\qquad f(2) = 3(2) + 2$$
$$= 9 + 2 \qquad\qquad\qquad\qquad = 6 + 2$$
$$= 11 \qquad\qquad\qquad\qquad\quad = 8$$

We then add the results to obtain

$$f(3) + f(2) = 11 + 8$$
$$= 19$$

b. We find $f(a)$ and $f(b)$ separately.

$$f(x) = 3x + 2 \qquad\qquad\qquad f(x) = 3x + 2$$
$$f(a) = 3a + 2 \qquad\qquad\qquad f(b) = 3b + 2$$

We then subtract the results to obtain

$$f(a) - f(b) = (3a + 2) - (3b + 2)$$
$$= 3a + 2 - 3b - 2 \qquad \text{Remove parentheses.}$$
$$= 3a - 3b$$

■

ORALS   *Tell whether each equation or inequality determines y to be a function of x.*

1. $y = x$ **2.** $y < 2x$ **3.** $y = x + 3$ **4.** $y^2 = -x$

If $f(x) = 2x + 1$, find

5. $f(0)$ **6.** $f(1)$ **7.** $f(-1)$ **8.** $f(-2)$

E X E R C I S E   6.6

In Exercises 1–12, tell whether each equation or inequality determines y to be a function of x. If it does not, indicate some numbers x for which there is more than one corresponding value of y.

1. $y = 2x + 3$ **2.** $y = 4x - 1$ **3.** $y = 3x^2$ **4.** $y^2 = x + 1$

5. $y = 3 + 7x^2$ **6.** $y^2 = 3 - 2x$ **7.** $y \le x$ **8.** $y > x$

9. $y = |x|$ **10.** $x = |y|$ **11.** $x = -|y|$ **12.** $y = -|x|$

In Exercises 13–16, find the domain and range of each function.

13. $y = -x^2 + 2$ **14.** $y = (x - 2)^2$ **15.** $y = |x - 2| - 2$ **16.** $y = -|x + 1|$

*In Exercises 17–24, find   **a.** $f(3)$,   **b.** $f(0)$,   and   **c.** $f(-1)$.*

17. $f(x) = 3x$ **18.** $f(x) = -4x$ **19.** $f(x) = 2x - 3$ **20.** $f(x) = 3x - 5$

21. $f(x) = 7 + 5x$ **22.** $f(x) = 3 + 3x$ **23.** $f(x) = 9 - 2x$ **24.** $f(x) = 12 + 3x$

*In Exercises 25–32, find   **a.** $f(1)$,   **b.** $f(-2)$,   and   **c.** $f(3)$.*

25. $f(x) = x^2$ **26.** $f(x) = x^2 - 2$ **27.** $f(x) = x^3 - 1$ **28.** $f(x) = x^3$

29. $f(x) = (x + 1)^2$ **30.** $f(x) = (x - 3)^2$ **31.** $f(x) = 2x^2 - x$ **32.** $f(x) = 5x^2 + 2x - 1$

*In Exercises 33–40, find   **a.** $f(2)$,   **b.** $f(1)$,   and   **c.** $f(-2)$.*

33. $f(x) = |x| + 2$ **34.** $f(x) = |x| - 5$ **35.** $f(x) = x^2 - 2$ **36.** $f(x) = x^2 + 3$

37. $f(x) = \dfrac{1}{x + 3}$ **38.** $f(x) = \dfrac{3}{x - 4}$ **39.** $f(x) = \dfrac{x}{x - 3}$ **40.** $f(x) = \dfrac{x}{x^2 + 2}$

*In Exercises 41–48, find   **a.** $g(w)$ and   **b.** $g(w + 1)$.*

41. $g(x) = 2x$ **42.** $g(x) = -3x$ **43.** $g(x) = 3x - 5$ **44.** $g(x) = 2x - 7$

45. $g(x) = -2x + 3$ **46.** $g(x) = -3x - 1$ **47.** $g(x) = |x|$ **48.** $g(x) = |x - 1|$

In Exercises 49–56, let $f(x) = 2x + 1$. Then find the requested value.

49. $f(3) + f(2)$ **50.** $f(1) - f(-1)$ **51.** $f(b) - f(a)$ **52.** $f(b) + f(a)$

53. $f(b) - 1$ **54.** $f(b) - f(1)$ **55.** $f(0) + f\left(-\dfrac{1}{2}\right)$ **56.** $f(a) + f(2a)$

57. Ballistics   A stone tossed upward is s feet above the earth after t seconds, where $s = -16t^2 + 128t$. Find the height of the stone 2 seconds after it is thrown.

58. Artillery fire   A mortar shell is s feet above the ground after t seconds, where $s = -16t^2 + 512t + 64$. Find the height of the shell 20 seconds after it is fired.

Writing Exercises   ■   *Write a paragraph using your own words.*

1. Explain why a relation is not always a function.

2. Explain why a function is always a relation.

Something to Think About ■ *Let $f(x) = 2x + 1$ and $g(x) = x^2$. Assume that $f(x) \neq 0$ and $g(x) \neq 0$.*

1. Is $f(x) + g(x) = g(x) + f(x)$?   **2.** Is $f(x) - g(x) = g(x) - f(x)$?

3. Is $f(x) \cdot g(x) = g(x) \cdot f(x)$?   **4.** Is $\dfrac{f(x)}{g(x)} = \dfrac{g(x)}{f(x)}$?

Review Exercises ■ *Solve each equation.*

1. $\dfrac{y + 2}{2} = 4(y + 2)$ **2.** $\dfrac{3z - 1}{6} - \dfrac{3z + 4}{3} = \dfrac{z + 3}{2}$

3. $\dfrac{2}{x - 3} - 1 = -\dfrac{1}{3}$ **4.** $\dfrac{5}{x} + \dfrac{6}{x^2 + 2x} = \dfrac{-3}{x + 2}$

6.7 Variation

■ Direct Variation ■ Inverse Variation ■ Joint Variation ■ Combined Variation

GETTING READY   *Solve for k.*

1. $8 = 2k$ **2.** $8 = \dfrac{k}{2}$ **3.** $A = kbh$ **4.** $P = \dfrac{kT}{V}$

We now introduce some special terminology that scientists use to describe functions.

■ Direct Variation

Direct Variation   The words *y varies directly with x* mean that

$$y = kx$$

for some constant k. The constant k is called the **constant of variation.**

     The more force that is applied to a spring, the more it will stretch. Scientists call this fact Hooke's law: *The distance a spring will stretch varies directly with the force applied.* If d represents distance and f represents force, this relationship can be expressed by the equation

1. $d = kf$

where k is the constant of variation. If a spring stretches 5 inches when a weight of 2 pounds is attached, we can find the constant of variation by substituting 5 for d and 2 for f in Equation 1 and solving for k:

$$d = kf$$
$$5 = k(2)$$
$$\frac{5}{2} = k$$

To find the distance that the spring will stretch when a weight of 6 pounds is attached, we substitute $\frac{5}{2}$ for k and 6 for f in Equation 1 and solve for d.

$$d = kf$$
$$d = \frac{5}{2}(6)$$
$$d = 15$$

The spring will stretch 15 inches when a weight of 6 pounds is attached.

EXAMPLE 1   At a constant speed, the distance traveled varies directly with time. If a bus driver can drive 105 miles in three hours, how far could he drive in 5 hours?

Solution   We let d represent distance traveled and let t represent time. We then translate the words *distance varies directly with time* into the equation

2. $d = kt$

To find the constant of variation, k, we substitute 105 for d and 3 for t in Equation 2 and solve for k.

$$d = kt$$
$$105 = k(3)$$
$$35 = k \qquad \text{Divide both sides by 3.}$$

We can now substitute 35 for k in Equation 2 to obtain Equation 3.

3. $d = 35t$

To find the distance traveled in 5 hours, we substitute 5 for t in Equation 3.

$$d = 35t$$
$$d = 35(5)$$
$$d = 175$$

In 5 hours, the bus driver could travel 175 miles.   ■

■ Inverse Variation

Inverse Variation   The words *y varies inversely with x* mean that

$$y = \frac{k}{x}$$

for some constant k. The constant k is the constant of variation.

Under constant temperature, the volume occupied by a gas varies inversely with its pressure. If V represents volume and p represents pressure, this relationship is expressed by the equation

4. $V = \dfrac{k}{p}$

EXAMPLE 2   A gas occupies a volume of 15 cubic inches when placed under 4 pounds per square inch of pressure. How much pressure is needed to compress the gas into a volume of 10 cubic inches?

Solution   To find the constant of variation, we substitute 15 for V and 4 for p in Equation 4 and solve for k.

$$V = \frac{k}{p}$$

$$15 = \frac{k}{4}$$

$$60 = k \qquad \text{Multiply both sides by 4.}$$

To find the pressure needed to compress the gas into a volume of 10 cubic inches, we substitute 60 for k and 10 for V in Equation 4 and solve for p.

$$V = \frac{k}{p}$$

$$10 = \frac{60}{p}$$

$$10p = 60 \qquad \text{Multiply both sides by } p.$$

$$p = 6 \qquad \text{Divide both sides by 10.}$$

It will take 6 pounds per square inch of pressure to compress the gas into a volume of 10 cubic inches.   ■

■ **Joint Variation**

Joint Variation   The words *y varies jointly with x and z* mean that

$$y = kxz$$

for some constant *k*. The constant *k* is the constant of variation.

The area of a rectangle depends on its length *l* and its width *w* by the formula

$$A = lw$$

We could say that the area of a rectangle varies jointly with its length and its width. In this example, the constant of variation is *k* = 1.

EXAMPLE 3   The area of a triangle varies jointly with the length of its base and its height. If a triangle with an area of 63 square inches has a base of 18 inches and a height of 7 inches, find the area of a triangle with a base of 12 inches and a height of 10 inches.

Solution   We let *A* represent the area of the triangle, let *b* represent the length of the base, and let *h* represent the height. We translate the words *area varies jointly with the length of the base and the height* into the formula

5. $A = kbh$

We are given that *A* = 63 when *b* = 18 and *h* = 7. To find *k*, we substitute these values into Equation 5 and solve for *k*.

$$A = kbh$$
$$63 = k(18)(7)$$
$$63 = k(126)$$
$$\frac{63}{126} = k \qquad \text{Divide both sides by 126.}$$
$$\frac{1}{2} = k \qquad \text{Simplify.}$$

Thus, $k = \frac{1}{2}$, and the formula for finding the area is

6. $A = \dfrac{1}{2}bh$

To find the area of a triangle with a base of 12 inches and a height of 10 inches, we substitute 12 for *b* and 10 for *h* in Equation 6.

$$A = \frac{1}{2}bh$$

$$A = \frac{1}{2}(12)(10)$$

$$A = 60$$

The area is 60 square inches. ∎

■ Combined Variation

Combined variation involves a combination of direct and inverse variation.

EXAMPLE 4 The pressure of a fixed amount of gas varies directly with its temperature and inversely with its volume. A sample of gas at a pressure of 1 atmosphere occupies a volume of 3 cubic meters when its temperature is 273 degrees Kelvin (about 0° Celsius). Find the pressure after the gas is heated to 364 K and compressed to 1 cubic meter.

Solution We let P represent the pressure of the gas, T represent its temperature, and V represent its volume. The words *the pressure varies directly with temperature and inversely with volume* translate into the equation

7. $\quad P = \dfrac{kT}{V}$

To find k, we substitute 1 for P, 273 for T, and 3 for V into Equation 7.

$$P = \frac{kT}{V}$$

$$1 = \frac{k(273)}{3}$$

$$1 = 91k$$

$$k = \frac{1}{91}$$

Since $k = \frac{1}{91}$, Equation 7 can be written as

$$P = \frac{1}{91} \cdot \frac{T}{V} \quad \text{or} \quad P = \frac{T}{91V}$$

To find the pressure under the new conditions, we substitute 364 for T and 1 for V into the previous equation and solve for P.

$$P = \frac{T}{91V}$$

$$P = \frac{364}{91(1)}$$

$$= 4$$

The pressure of the heated and compressed gas is 4 atmospheres. ∎

ORALS *Tell whether each equation defines direct variation, inverse variation, joint variation, or combined variation.*

1. $y = \dfrac{5}{x}$ **2.** $y = 3x$ **3.** $y = \dfrac{3x}{z}$ **4.** $y = 3xz$

5. $y = \dfrac{1}{2}x$ **6.** $y = \dfrac{2x}{z}$ **7.** $y = \dfrac{1}{2}xz$ **8.** $y = 4x$

EXERCISE 6.7

In Exercises 1–10, express each sentence as a formula.

1. The distance d a car can travel while moving at a constant speed varies directly with n, the number of gallons of gasoline it consumes.

2. A farmer's harvest h varies directly with a, the number of acres he plants.

3. For a fixed area, the length l of a rectangle varies inversely with its width w.

4. The value v of a car varies inversely with its age a.

5. The area A of a circle varies directly with the square of its radius r.

6. The distance s that a body falls varies directly with the square of the time t.

7. The distance d traveled varies jointly with the speed s and time t.

8. The interest i on a savings account varies jointly with the rate r and the time t.

9. The current I varies directly with the voltage V and inversely with the resistance R.

10. The force of gravity F varies directly with the product of the masses m_1 and m_2, and inversely with the square of the distance between them.

In Exercises 11–28, assume that all variables represent positive numbers.

11. Assume that y varies directly with x. If $y = 10$ when $x = 2$, find y when $x = 7$.

12. Assume that A varies directly with z. If $A = 30$ when $z = 5$, find A when $z = 9$.

13. Assume that r varies directly with s. If $r = 21$ when $s = 6$, find r when $s = 12$.

14. Assume that d varies directly with t. If $d = 15$ when $t = 3$, find t when $d = 3$.

15. Assume that s varies directly with t^2. If $s = 12$ when $t = 4$, find s when $t = 30$.

16. Assume that y varies directly with x^3. If $y = 16$ when $x = 2$, find y when $x = 3$.

17. Assume that y varies inversely with x. If $y = 8$ when $x = 1$, find y when $x = 8$.

18. Assume that V varies inversely with p. If $V = 30$ when $p = 5$, find V when $p = 6$.

19. Assume that r varies inversely with s. If $r = 40$ when $s = 10$, find r when $s = 15$.

20. Assume that J varies inversely with v. If $J = 90$ when $v = 5$, find J when $v = 45$.

21. Assume that y varies inversely with x^2. If $y = 6$ when $x = 4$, find y when $x = 2$.

22. Assume that i varies inversely with d^2. If $i = 6$ when $d = 3$, find i when $d = 2$.

23. Assume that y varies jointly with r and s. If $y = 4$ when $r = 2$ and $s = 6$, find y when $r = 3$ and $s = 4$.

24. Assume that A varies jointly with x and y. If $A = 18$ when $x = 3$ and $y = 3$, find A when $x = 7$ and $y = 9$.

25. Assume that D varies jointly with p and q. If $D = 20$ when p and q are both 5, find D when p and q are both 10.

26. Assume that z varies jointly with r and the square of s. If $z = 24$ when r and s are 2, find z when $r = 3$ and $s = 4$.

27. Assume that y varies directly with a and inversely with b. If $y = 1$ when $a = 2$ and $b = 10$, find y when $a = 7$ and $b = 14$.

28. Assume that y varies directly with the square of x and inversely with z. If $y = 1$ when $x = 2$ and $z = 10$, find y when $x = 4$ and $z = 5$.

29. Objects in free fall The distance traveled by an object in free fall varies directly with the square of the time that it falls. If the object falls 256 feet in 4 seconds, how far will it fall in 6 seconds?

30. Commuting distance The distance that a car can travel without refueling varies directly with the number of gallons of gasoline in the tank. If a car can go 360 miles on 12 gallons of gas, how far can it go on 7 gallons?

31. Computing interest For a fixed rate and principal, the interest earned in a bank account paying simple interest varies directly with the length of time the principal is left on deposit. If an investment of $5000 earns $700 in 2 years, how much will it earn in 7 years?

32. Computing forces The force of gravity acting on an object varies directly with the mass of the object. The force on a mass of 5 kilograms is 49 newtons. What is the force acting on a mass of 12 kilograms?

33. Commuting time The time it takes a car to travel a certain distance varies inversely with its rate of speed. If a certain trip takes 3 hours when the driver travels at 50 miles per hour, how long will the trip take when the driver travels at 60 miles per hour?

34. Geometry For a fixed area, the length of a rectangle is inversely proportional to its width. A rectangle has a width of 12 feet and a length of 20 feet. If the length is increased to 24 feet, find the width of the rectangle.

35. Computing pressures If the temperature of a gas is constant, the volume occupied varies inversely with the pressure. If a gas occupies a volume of 40 cubic meters under a pressure of 8 atmospheres, find the volume when the pressure is changed to 6 atmospheres.

36. Computing depreciation   Assume that the value of a machine varies inversely with its age. If a drill press is worth $300 when it is 2 years old, find its value when it is 6 years old. How much has the machine depreciated in 4 years?

37. Computing interest   The interest earned on a fixed amount of money varies jointly with the annual interest rate and the time that the money is left on deposit. If an account earns $120 at 8% annual interest when left on deposit for 2 years, how much interest would be earned in 3 years at an annual rate of 12%?

38. Cost of a well   The cost of drilling a water well is jointly proportional to the length and diameter of the steel casing. If a 30-foot well using 4-inch casing costs $1200, find the cost of a 35-foot well using 6-inch casing.

39. Electronics   The current in a circuit varies directly with the voltage and inversely with the resistance. If a current of 4 amperes flows when 36 volts is applied to a 9-ohm resistance, find the current when the voltage is 42 volts and the resistance is 11 ohms.

40. Building construction   The deflection of a beam is inversely proportional to its width and the cube of its depth. If the deflection is 2 inches when the width is 4 inches and the depth is 3 inches, find the deflection when the width is 3 inches and the depth is 4 inches.

Writing Exercises ■   *Write a paragraph using your own words.*

1. Explain why the words *y varies jointly with x and z* mean the same as the words *y varies directly with the product of x and z.*

2. Explain the meaning of combined variation.

Something to Think About ■

1. Can direct variation be defined as $\frac{y}{x} = k$, rather than $y = kx$?

2. Can inverse variation be defined as $xy = k$, rather than $y = \frac{k}{x}$?

Review Exercises ■   *Remove parentheses and simplify.*

1. $2(x + 4) + 3(2x - 1)$

2. $-3(3x + 5) - 2(2x + 4)$

3. $3x(x^2 - 2) - 6x^2(x - 1)$

4. $-5a^2(a + 1) - 3a(a^2 + 4a - 3)$

MATHEMATICS IN REAL ESTATE

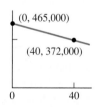

When the office building discussed at the beginning of the chapter was purchased, its value was $465,000. Forty years later, it is estimated to be worth 80% of $465,000, or $372,000. Since the points (0, 465,000) and (40, 372,000) lie on the straight-line depreciation graph of the building, we can write the depreciation equation as follows:

$$m = \frac{372{,}000 - 465{,}000}{40 - 0} = -2325$$

Since the y-intercept is (0, 465,000) and the slope is $m = -2325$, we have

$$y = mx + b$$
$$y = -2325x + 465{,}000$$

To find the value after 35 years, we substitute 35 for x and solve for y to get $383,625. Since the investor sold it for $400,000, the taxable capital gain is $400,000 − $383,625, or $16,375.

PROJECT ■ Deceptive Graphs

Graphs are often used in newspapers and magazines to convey complex information at a glance. Unfortunately, it is easy to use graphs to convey misleading information. For example, the percent of profits of a company for several years are given in the table and two graphs in Figure 6-40.

Year	Profit
1989	6.2%
1990	6.0%
1991	6.2%
1992	6.1%
1993	6.3%
1994	6.6%

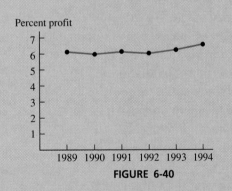

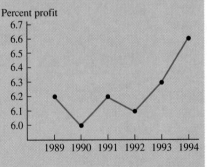

FIGURE 6-40

The first graph in the figure accurately indicates the company's steady performance over five years. But because the vertical axis of the second graph does not start at zero, the performance appears deceptively erratic.

As your college's head librarian, you spend much of your time writing reports, either trying to make the school library look good (for college promotional literature) or bad (to encourage greater funding). In 1989, the library held a collection of 17,000 volumes. Over the years, the library has acquired many new books and has retired several old books. The details appear in Table 6-2.

(continued)

P R O J E C T ■ Deceptive Graphs *(continued)*

Year	Volumes acquired	Volumes removed
1989	215	137
1990	217	145
1991	235	185
1992	257	210
1993	270	200
1994	275	180

TABLE 6-2

Using the data in the table,

• Draw a misleading graph that makes the library look good.
• Draw a misleading graph that makes the library look bad.
• Draw a graph that accurately reflects the library's condition.

Chapter Summary

KEY WORDS

abscissa (6.1)
Cartesian coordinate system (6.1)
combined variation (6.7)
constant of variation (6.7)
coordinates of a point (6.1)
dependent variable (6.6)
direct variation (6.7)
domain of a relation (6.6)
function (6.6)
general form of a linear equation (6.1)
graph (6.1)
graph of a line (6.1)
half-plane (6.5)
independent variable (6.6)
intercept method of graphing (6.1)
inverse variation (6.7)
joint variation (6.7)
linear depreciation (6.2)

linear equation (6.1)
linear inequality (6.5)
ordered pair (6.1)
ordinate (6.1)
origin (6.1)
parabola (6.4)
parallel lines (6.3)
perpendicular lines (6.3)
point–slope form of a linear equation (6.3)
quadrant (6.1)
range of a relation (6.6)
rectangular coordinate system (6.1)
relation (6.6)
slope (6.2)
slope–intercept form of a linear equation (6.2)
subscript notation (6.2)
viewing window (6.4)

***x*-axis** (6.1)   ***y*-axis** (6.1)
***x*-coordinate** (6.1)   ***y*-coordinate** (6.1)
***x*-intercept** (6.1)   ***y*-intercept** (6.1)

KEY IDEAS

(6.1) Ordered pairs of numbers are associated with points in a rectangular coordinate system.

To graph an equation in the variables x and y, choose several numbers x (at least 3), find the corresponding values of y, plot the points (x, y), and draw the line that passes through them.

General form of a linear equation: $Ax + By = C$, where A and B are not both zero.

To graph a linear equation by the intercept method, plot the points corresponding to the x- and y-intercepts and draw the line through them. Plot a third point as a check.

The equation $x = a$ represents the y-axis or a line parallel to the y-axis.

The equation $y = b$ represents the x-axis or a line parallel to the x-axis.

(6.2) The slope of a line passing through (x_1, y_1) and (x_2, y_2) is given by the formula

$$m = \frac{y_2 - y_1}{x_2 - x_1} \qquad (x_2 \neq x_1)$$

The graph $y = mx + b$ is a line with a slope of m and a y-intercept of $(0, b)$.

Slope–intercept form of a linear equation: $y = mx + b$

Horizontal lines have a slope of 0.

The slope of a vertical line is undefined.

(6.3) Point–slope form of a linear equation:

$$y - y_1 = m(x - x_1)$$

Nonvertical parallel lines have the same slope.

The product of the slopes of perpendicular lines is -1, provided neither line is vertical.

(6.4) To graph nonlinear equations, plot many points until you recognize the shape of the graph.

(6.5) To graph an inequality in the variables x and y, first graph the boundary and then use a convenient point not on that boundary to find which side of the line to shade.

(6.6) A relation is any set of ordered pairs of numbers.

A function is a correspondence that assigns to each number x exactly one value y. To indicate such a correspondence, the notation $y = f(x)$ is used.

(6.7) The equation $y = kx$ represents direct variation.

The equation $y = \dfrac{k}{x}$ represents inverse variation.

The equation $y = kxz$ represents joint variation.

Direct and inverse variation are used together in combined variation.

■ Chapter 6 Review Exercises

(6.1) *In Review Exercises 1–6, plot each point on a rectangular coordinate system.*

1. $A(1, 3)$ **2.** $B(1, -3)$

3. $C(-3, 1)$ **4.** $D(-3, -1)$

5. $E(0, 5)$ **6.** $F(-5, 0)$

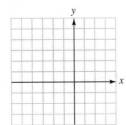

In Review Exercises 7–14, find the coordinates of each point in Illustration 1.

7. *A*

8. *B*

9. *C*

10. *D*

11. *E*

12. *F*

13. *G*

14. *H*

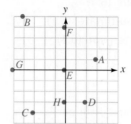

ILLUSTRATION 1

In Review Exercises 15–22, graph each equation on a rectangular coordinate system.

15. $y = x - 5$

16. $y = 2x + 1$

17. $y = \dfrac{x}{2} + 2$

18. $y = 3$

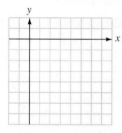
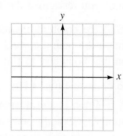
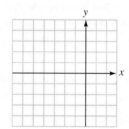
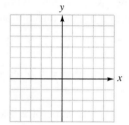

19. $x + y = 4$

20. $x - y = -3$

21. $3x + 5y = 15$

22. $7x - 4y = 28$

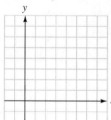
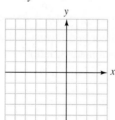
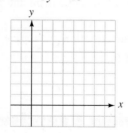
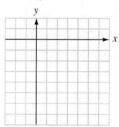

(6.2) *In Review Exercises 23–26, find the slope of the line passing through the two given points. If the slope is undefined, write "undefined slope."*

23. $(1, 4), (2, 3)$

24. $(-1, 3), (3, -2)$

25. $(-1, -1), (-3, 0)$

26. $(-8, 2), (3, 2)$

In Review Exercises 27–30, graph the line that passes through the given point and has the given slope.

27. $(-1, 4)$, $m = 2$

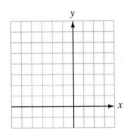

28. $(1, -2)$, $m = -2$

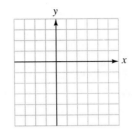

29. $\left(0, \dfrac{1}{2}\right)$, $m = \dfrac{3}{2}$

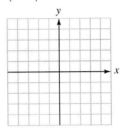

30. $(-3, 0)$, $m = -\dfrac{5}{2}$

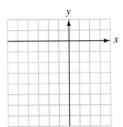

In Review Exercises 31–34, find the slope and the y-intercept of the line defined by each equation. Then graph the equation. If the line has no defined slope, write "undefined slope."

31. $y = 5x + 2$

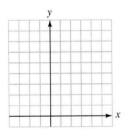

32. $y = -\dfrac{x}{2} + 4$

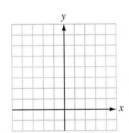

33. $y + 3 = 0$

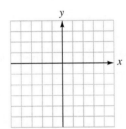

34. $x + 3y = 1$

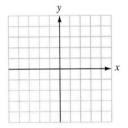

(6.3) *In Review Exercises 35–38, use the slope–intercept form of a linear equation to find the equation of each line with the given properties. Write the equation in general form.*

35. $m = -3$, y-intercept $(0, 2)$

36. $m = 0$, y-intercept $(0, -7)$

37. $m = 7$, y-intercept $(0, 0)$

38. $m = \dfrac{1}{2}$, y-intercept $\left(0, -\dfrac{3}{2}\right)$

In Review Exercises 39–42, use the point–slope form of a linear equation to find the equation of each line with the given properties. Write the equation in general form.

39. $m = 3$, passes through $(0, 0)$

40. $m = -\dfrac{1}{3}$, passes through $\left(1, \dfrac{2}{3}\right)$

41. $m = \dfrac{1}{9}$, passes through $(-27, -2)$

42. $m = -\dfrac{3}{5}$, passes through $\left(1, -\dfrac{1}{5}\right)$

In Review Exercises 43–46, tell whether the lines with the given slopes are parallel, perpendicular, or neither.

43. 5 and $\dfrac{1}{5}$

44. $\dfrac{2}{4}$ and 0.5

45. -5 and $\dfrac{1}{5}$

46. 0.25 and $\dfrac{1}{4}$

In Review Exercises 47–50, tell whether the graphs of the given equations are parallel, perpendicular, or neither.

47. $y = 3x$, $x = 3y$

48. $3x = y$, $x = -3y$

49. $x + 2y = y - x$, $2x + y = 3$

50. $3x + 2y = 7$, $2x - 3y = 8$

In Review Exercises 51–54, find the equation of each line with the given properties. Write the equation in general form.

51. parallel to $y = 7x - 18$, passes through $(2, 5)$

52. parallel to $3x + 2y = 7$, passes through $(-3, 5)$

53. perpendicular to $2x - 5y = 12$, passes through $(0, 0)$

54. perpendicular to $y = \dfrac{x}{3} + 17$, y-intercept $(0, -4)$

(6.4) *In Review Exercises 55–58, graph each equation.*

55. $y = x^2 - 4$

56. $y = -x^2 + 3$

57. $y = |x - 2|$

58. $y = |x + 2| - 1$

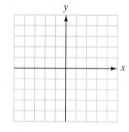 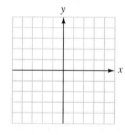 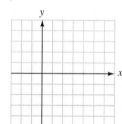 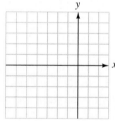

(6.5) *In Review Exercises 59–62, graph each inequality.*

59. $y \le 3x + 1$

60. $y > x - 5$

61. $2x - 3y \ge 6$

62. $2y + 3(x - y) < 5y$

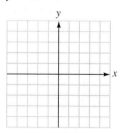 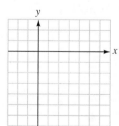 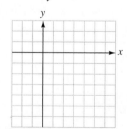 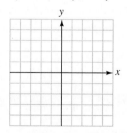

(6.6) *In Review Exercises 63–66, tell whether the equation determines y as a function of x. If it does not, give some number x for which there are two corresponding values of y.*

63. $y = 2x$ **64.** $y = 5x^2$ **65.** $|y| = x$ **66.** $y^2 = 4x^2$

In Review Exercises 67–74, $f(x) = x^2 - x + 1$. Find each value.

67. $f(0)$ **68.** $f(-2)$ **69.** $f(3)$ **70.** $f(w)$

71. $f(1) - f(-1)$ **72.** $f(2) + f(-2)$ **73.** $f(a) + f(0)$ **74.** $f(1) - f(a)$

(6.7) *In Review Exercises 75–78, express each variation as an equation. Then find the requested value.*

75. s varies directly with the square of t. Find s when $t = 10$ if $s = 64$ when $t = 4$.

76. l varies inversely with w. Find the constant of variation if $l = 30$ when $w = 20$.

77. R varies jointly with b and c. If $R = 72$ when $b = 4$ and $c = 24$, find R when $b = 6$ and $c = 18$.

78. s varies directly with w and inversely with the square of m. If $s = \frac{7}{4}$ when w and m are 4, find s when $w = 5$ and $m = 7$.

Chapter 6 Test

In Problems 1–4, graph each equation.

1. $y = \dfrac{x}{2} + 1$ **2.** $2(x + 1) - y = 4$ **3.** $x = 1$ **4.** $2y = 8$

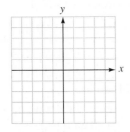 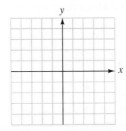 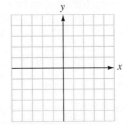 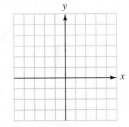

5. Find the slope of the line passing through $(0, 0)$ and $(6, 8)$.

6. Find the slope of the line passing through $(-1, 3)$ and $(3, -1)$.

7. Find the slope of the line determined by $2x + y = 3$.

8. Find the y-intercept of the line determined by $2y - 7(x + 5) = 7$.

9. Find the slope of a line parallel to a line with a slope of 2.

10. Find the slope of a line perpendicular to a line with a slope of 2.

11. In general form, write the equation of a line with a slope of $\frac{1}{2}$ and a y-intercept of $(0, 3)$.

12. In general form, write the equation of a line with a slope of 7 that passes through the point $(-2, 5)$.

13. Write the equation of a line that is parallel to the y-axis and passes through $(-3, 17)$.

14. Write the equation of a line passing through $(3, -5)$ and perpendicular to the line with the equation $y = \frac{1}{3}x + 11$.

In Problems 15–16, graph each equation.

15. $y = -x^2 + 2$

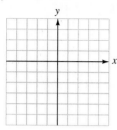

16. $y = |x| - 2$

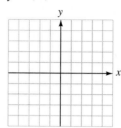

In Problems 17–18, graph each inequality.

17. $y \geq x + 2$

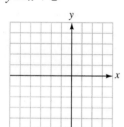

18. $x < 3$

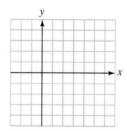

19. Does the equation $x = 7y - 8$ determine y to be a function of x?

20. Does the equation $y^2 = x$ determine y to be a function of x?

In Problems 21–24, assume that $f(x) = 3x + 2$. Find each value.

21. $f(3)$

22. $f(-2) + f(0)$

23. $f(a) - f(b)$

24. $f(t^2)$

25. If y varies directly with x and $y = 32$ when $x = 8$, find x when $y = 4$.

26. If i varies inversely with the square of d, find the constant of variation if $i = 100$ when $d = 2$.

■ Cumulative Review Exercises

In Exercises 1–4, simplify each expression.

1. $(3x^2 - 2x) + (6x^3 - 3x^2 - 1)$

2. $(4x^3 - 2x) - (2x^3 - 2x^2 - 3x + 1)$

3. $3(5x^2 - 4x + 3) + 2(-x^2 + 2x - 4)$

4. $4(3x^2 - 4x - 1) - 2(-2x^2 + 4x - 3)$

In Exercises 5–8, do each multiplication.

5. $(3x^3y^2)(-4x^2y^3)$

6. $-5x^2(7x^3 - 2x^2 - 2)$

7. $(3x + 1)(2x + 4)$

8. $(5x - 4y)(3x + 2y)$

In Exercises 9–10, do each division.

9. $x + 3 \overline{)x^2 + 7x + 12}$

10. $2x - 3 \overline{)2x^3 - x^2 - x - 3}$

In Exercises 11–20, factor each expression.

11. $3x^2y - 6xy^2$

12. $3(a + b) + x(a + b)$

13. $2a + 2b + ab + b^2$

14. $25p^4 - 16q^2$

15. $x^2 - 11x - 12$

16. $x^2 - xy - 6y^2$

17. $6a^2 - 7a - 20$

18. $8m^2 - 10mn - 3n^2$

19. $p^3 - 27q^3$

20. $8r^3 + 64s^3$

In Exercises 21–22, solve each equation.

21. $x^2 + 3x + 2 = 0$

22. $2y^2 + 5y - 12 = 0$

In Exercises 23–24, simplify each fraction.

23. $\dfrac{x^2 + 2x + 1}{x^2 - 1}$

24. $\dfrac{x^2 + 2x - 15}{x^2 + 3x - 10}$

In Exercises 25–30, do the operation(s) and simplify when possible.

25. $\dfrac{x^2 + x - 6}{5x - 5} \cdot \dfrac{5x - 10}{x + 3}$

26. $\dfrac{p^2 - p - 6}{3p - 9} \div \dfrac{p^2 + 6p + 9}{p^2 - 9}$

27. $\dfrac{3x}{x + 2} + \dfrac{5x}{x + 2} - \dfrac{7x - 2}{x + 2}$

28. $\dfrac{x - 1}{x + 1} + \dfrac{x + 1}{x - 1}$

29. $\dfrac{a + 1}{2a + 4} - \dfrac{a^2}{2a^2 - 8}$

30. $\dfrac{\dfrac{1}{x} + \dfrac{1}{y}}{\dfrac{1}{x} - \dfrac{1}{y}}$

In Exercises 31–32, solve each equation.

31. $\dfrac{4}{a} = \dfrac{6}{a} - 1$

32. $\dfrac{a + 2}{a + 3} - 1 = \dfrac{-1}{a^2 + 2a - 3}$

In Exercises 33–34, solve each proportion.

33. $\dfrac{4 - a}{13} = \dfrac{11}{26}$

34. $\dfrac{3a - 2}{7} = \dfrac{a}{28}$

In Exercises 35–36, graph each equation.

35. $3x - 4y = 12$

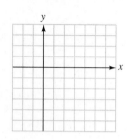

36. $y - 2 = \dfrac{1}{2}(x - 4)$

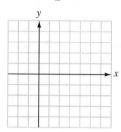

In Exercises 37–38, find the slope of the line with the given properties.

37. Passing through $(-2, 4)$ and $(6, 8)$

38. The equation of the line is $3x + 6y = 13$.

In Exercises 39–40, write the equation of the line with the following properties.

39. slope of $\dfrac{2}{3}$, y-intercept of $(0, 5)$

40. Passing through $(-2, 4)$ and $(6, 10)$

In Exercises 41–42, are the graphs of the lines parallel or perpendicular?

41. $\begin{cases} 3x + 4y = 15 \\ 4x - 3y = 25 \end{cases}$

42. $\begin{cases} 3x + 4y = 15 \\ 6x = 15 - 8y \end{cases}$

43. Graph the equation $y = x^2 - 4$.

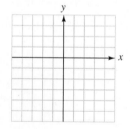

44. Graph the inequality $3x + 4y \le 12$.

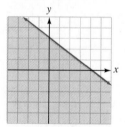

In Exercises 44–45, tell whether each equation defines a function.

45. $y = x^3 - 4$

46. $x = |y|$

In Exercises 46–50, $y = f(x) = 2x^2 - 3$. Find each value.

47. $f(0)$ **48.** $f(3)$ **49.** $f(-2)$ **50.** $f(2x)$

51. Assume that y varies directly with x. If $y = 4$ when $x = 10$, find y when $x = 30$.

52. Assume that y varies inversely with x. If $y = 8$ when $x = 2$, find y when $x = 8$.

7

Solving Systems of Equations and Inequalities

MATHEMATICS IN ECONOMICS

The number of canoes that boaters will buy is related to price. The higher the price, the fewer canoes people will buy. The equation that relates the retail price of a canoe to the number of canoes bought at that price is called a **demand equation.** Suppose that the demand equation for canoes is

1. $p = -\dfrac{1}{2}q + 1300$

where p is the price and q is the number purchased each month at that price. From the graph of Equation 1, shown in the illustration, we see that boaters will buy 1200 canoes at a price of $700.

   The number of canoes a manufacturer is willing to produce is also related to price. The higher the price, the more manufacturers are willing to produce. The equation that relates the number of canoes produced to the retail price is called a **supply equation.** Suppose that the supply equation for canoes is

2. $p = \dfrac{1}{3}q + \dfrac{1400}{3}$

where p is the retail price and q is the number produced for sale at that price. From the graph of this equation, shown in the illustration, we see that manufacturers will produce only 700 canoes at a price of $700.

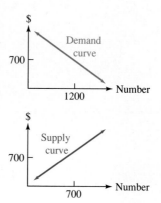

Since suppliers will produce 700 canoes and people want 1200, there will be a shortage and the price will go up. The **equilibrium price** is the price where supply equals demand. Find the equilibrium price.

After you have read this chapter, you will be able to solve this problem.

We have considered equations such as $x + y = 3$, which have two variables. Because there are infinitely many pairs of numbers whose sum is 3, there are infinitely many pairs (x, y) that will satisfy this equation. Some of these pairs are

$$x + y = 3$$

x	y
0	3
1	2
2	1
3	0

Likewise, there are infinitely many pairs (x, y) that will satisfy the equation $3x - y = 1$. Some of these pairs are

$$3x - y = 1$$

x	y
0	-1
1	2
2	5
3	8

Although there are infinitely many pairs that satisfy each of these equations, only the pair $(1, 2)$ satisfies both equations at the same time. The pair of equations

$$\begin{cases} x + y = 3 \\ 3x - y = 1 \end{cases}$$

is called a **system of equations.** Because the ordered pair $(1, 2)$ satisfies both equations simultaneously, it is called a **simultaneous solution,** or just a **solution of the system of equations.** We will discuss three methods for finding the simultaneous solution of a system of two equations, each with two variables.

7.1 Solving Systems of Equations by Graphing

■ Inconsistent Systems ■ Dependent Equations ■ Solving Systems with a Graphing Calculator

GETTING READY   *If $y = x^2 - 3$, find y when x is*

1. 0   **2.** 1   **3.** -2 **4.** 3

Find the slope of the graph of each line.

5. $y = \dfrac{3}{2}x + 2$ **6.** $y = -\dfrac{4}{3}x - 32$

To use the method of graphing to solve the system

$$\begin{cases} x + y = 3 \\ 3x - y = 1 \end{cases}$$

we graph both equations on a single set of coordinate axes, as in Figure 7-1.

Although there are infinitely many pairs (x, y) that satisfy $x + y = 3$, and infinitely many pairs (x, y) that satisfy $3x - y = 1$, only the coordinates of the point where their graphs intersect satisfy both equations simultaneously. Thus, the solution of the system is $x = 1$ and $y = 2$, or just $(1, 2)$.

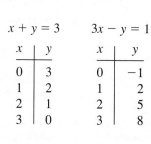

$x + y = 3$

x	y
0	3
1	2
2	1
3	0

$3x - y = 1$

x	y
0	-1
1	2
2	5
3	8

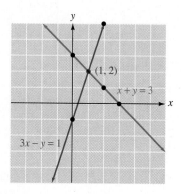

FIGURE 7-1

To check this solution, we substitute 1 for x and 2 for y in each equation and verify that the pair $(1, 2)$ satisifes each equation.

$x + y = 3$	$3x - y = 1$
$1 + 2 \overset{?}{=} 3$	$3(1) - 2 \overset{?}{=} 1$
$3 = 3$	$3 - 2 \overset{?}{=} 1$
	$1 = 1$

When the graphs of two equations in a system are different lines, the equations are called **independent equations.** When a system of equations has a solution, the system is called a **consistent system.**

To solve a system of equations in two variables by graphing, we follow these steps.

The Graphing Method
1. Carefully graph each equation.
2. When possible, find the coordinates of the point where the graphs intersect.
3. Check the solution in the equations of the original system.

■ PERSPECTIVE

A Big Computer and a Bad Prediction

For efficient scheduling of a company's workers, managers must juggle several factors to match the abilities of the workers to the demands of various tasks and to match company resources to the requirements of the job. To design bridges or office buildings, engineers must analyze the effects of thousands of forces to ensure that structures won't collapse. The telephone switching network decides which of thousands of possible routes is the most efficient, and then it rings the correct telephone—next door or around the world—in seconds. Each of these tasks requires solving systems of equations. Not just two equations in two variables, but hundreds of equations in hundreds of variables. These tasks are common to every business, industry, educational institution, and government in the world. All would be impossible without the computer.

One of the earliest of computers, and an ancestor of millions now in use, was the Mark I, which resulted from a collaboration between IBM and a Harvard mathematician, Howard Aiken. Started in 1939 and finished in 1944, using Aiken's plans and IBM's money and engineering, the Mark I was 8 feet tall, 2 feet thick, and over 50 feet long.

Mark I Relay Computer (1944)
Courtesy of IBM Corporation.

It contained 750,000 parts, performed an amazing 3 calculations per second, and sounded "like a roomful of people knitting."

Ironically, Aiken, a pioneer of the new age of computers, could not envision the potential of his invention. He advised the National Bureau of Standards that there was no point in building a better machine, because "there will never be enough work for more than one or two of these computers." How wrong he was!

EXAMPLE 1 Use the graphing method to solve the system $\begin{cases} 2x + 3y = 2 \\ 3x = 2y + 16 \end{cases}$.

Solution We graph both equations on a single set of coordinate axes, as in Figure 7-2.

Although there are infinitely many pairs (x, y) that satisfy $2x + 3y = 2$ and infinitely many pairs (x, y) that satisfy $3x = 2y + 16$, only the coordinates of the point where the graphs intersect satisfy both equations simultaneously. The solution is $x = 4$ and $y = -2$, or just $(4, -2)$.

To check this solution, we substitute 4 for x and -2 for y in each equation and verify that the pair $(4, -2)$ satisfies each equation.

$$
\begin{array}{c|c}
2x + 3y = 2 & 3x = 2y + 16 \\
2(4) + 3(-2) \overset{?}{=} 2 & 3(4) \overset{?}{=} 2(-2) + 16 \\
8 - 6 \overset{?}{=} 2 & 12 \overset{?}{=} -4 + 16 \\
2 = 2 & 12 = 12
\end{array}
$$

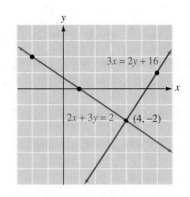

$$2x + 3y = 2 \qquad 3x = 2y + 16$$

x	y
1	0
−2	2
4	−2

x	y
6	1
0	−8
4	−2

FIGURE 7-2

The equations in this system are independent equations, and the system is a consistent system of equations. ■

■ Inconsistent Systems

Sometimes a system of equations will have no solution. Such systems are called **inconsistent systems.**

EXAMPLE 2   Solve the system $\begin{cases} 2x + y = -6 \\ 4x + 2y = 8 \end{cases}$.

Solution   We graph both equations on one set of coordinate axes, as in Figure 7-3.

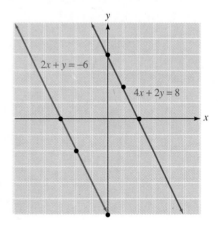

$$2x + y = -6 \qquad 4x + 2y = 8$$

x	y
−3	0
0	−6
−2	−2

x	y
2	0
0	4
1	2

FIGURE 7-3

The lines in Figure 7-3 are parallel. We can verify this by writing each equation in slope–intercept form and observing that the coefficients of x are equal.

$$2x + y = -6 \qquad\qquad 4x + 2y = 8$$
$$y = -2x - 6 \qquad\qquad 2y = -4x + 8$$
$$\qquad\qquad\qquad\qquad y = -2x + 4$$

Because parallel lines do not intersect, the system has no solution, and the system is inconsistent. Since the graphs are different lines, the equations of the system are independent. ■

■ Dependent Equations

Sometimes a system will have an infinite number of solutions. In this case, we say that the equations of the system are **dependent equations.**

EXAMPLE 3   Solve the system $\begin{cases} y - 2x = 4 \\ 4x + 8 = 2y \end{cases}$.

Solution   We graph both equations on one set of coordinate axes, as in Figure 7-4.

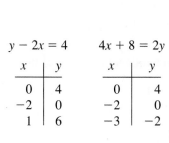

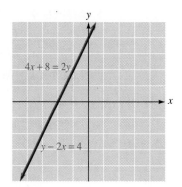

FIGURE 7-4

The lines in Figure 7-4 coincide (they are the same line). Because the lines intersect at infinitely many points, there is an infinite number of solutions. Any pair (x, y) that satisfies one of the equations satisfies the other also.

From the graph, we can see that some possible solutions are $(0, 4)$, $(1, 6)$, and $(-1, 2)$ since each of these points lies on the one line that is the graph of both equations. ■

The possibilities that can occur when two linear equations, each with two variables, are graphed are summarized as shown in Table 7-1.

Possible graph	If the	Then
	lines are different and intersect,	the equations are independent, and the system is consistent. One solution exists.
	lines are different and parallel,	the equations are independent, and the system is inconsistent. No solutions exist.
	lines coincide,	the equations are dependent, and the system is consistent. An infinite number of solutions exist.

TABLE 7-1

EXAMPLE 4 Solve the system $\begin{cases} \dfrac{2}{3}x - \dfrac{1}{2}y = 1 \\ \dfrac{1}{10}x + \dfrac{1}{15}y = 1 \end{cases}$.

Solution We can multiply both sides of the first equation by 6 to clear it of fractions.

$$\frac{2}{3}x - \frac{1}{2}y = 1$$

$$6\left(\frac{2}{3}x - \frac{1}{2}y\right) = 6(1)$$

1. $4x - 3y = 6$

We then multiply both sides of the second equation by 30 to clear it of fractions.

$$\frac{1}{10}x + \frac{1}{15}y = 1$$

$$30\left(\frac{1}{10}x + \frac{1}{15}y\right) = 30(1)$$

2. $3x + 2y = 30$

Equations 1 and 2 form the following equivalent system of equations, which has the same solution as the original system.

$$\begin{cases} 4x - 3y = 6 \\ 3x + 2y = 30 \end{cases}$$

If we graph each equation of the previous system, as in Figure 7-5, we find that their point of intersection has coordinates of $(6, 6)$. The solution of the given system is $x = 6$ and $y = 6$, or just $(6, 6)$.

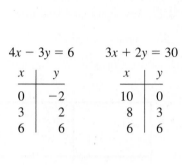

$$4x - 3y = 6 \qquad 3x + 2y = 30$$

x	y
0	−2
3	2
6	6

x	y
10	0
8	3
6	6

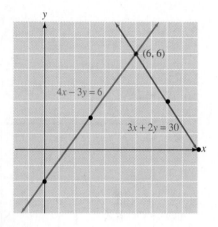

FIGURE 7-5

To verify that $(6, 6)$ satisfies each equation of the original system, we substitute 6 for x and 6 for y in each of the original equations and simplify.

$$\frac{2}{3}x - \frac{1}{2}y = 1 \qquad\qquad \frac{1}{10}x + \frac{1}{15}y = 1$$

$$\frac{2}{3}(6) - \frac{1}{2}(6) \overset{?}{=} 1 \qquad\qquad \frac{1}{10}(6) + \frac{1}{15}(6) \overset{?}{=} 1$$

$$4 - 3 \overset{?}{=} 1 \qquad\qquad \frac{3}{5} + \frac{2}{5} \overset{?}{=} 1$$

$$1 = 1 \qquad\qquad 1 = 1$$

The equations in this system are independent, and the system is consistent. ■

■ Solving Systems with a Graphing Calculator

EXAMPLE 5 Use a graphing calculator to solve the system $\begin{cases} 2x + y = 12 \\ 2x - y = -2 \end{cases}$.

Solution Before we enter the equations into a graphing calculator, we must solve them for y.

$$2x + y = 12 \qquad\qquad 2x - y = -2$$

$$y = -2x + 12 \qquad\qquad -y = -2x - 2$$

$$y = 2x + 2$$

We can enter the resulting equations into a graphing calculator and graph them on the same coordinate axes. If we use the standard RANGE settings, their graphs will look like Figure 7-6(a). We can use the TRACE key to find that the coordinates of the intersection point are

$$x = 2.4210526 \quad \text{and} \quad y = 7.1578947$$

See Figure 7-6(b). For better results, we can zoom in on the intersection point, use the TRACE key again, and find that

$$x = 2.5 \quad \text{and} \quad y = 7$$

See Figure 7-6(c). Check each solution.

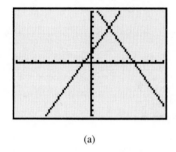

(a)

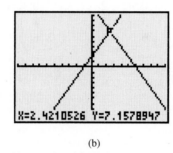

(b)

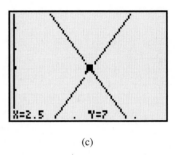

(c)

FIGURE 7-6

ORALS *Tell whether the pair is a solution of the system.*

1. $(3, 2)$, $\begin{cases} x + y = 5 \\ x - y = 1 \end{cases}$

2. $(1, 2)$, $\begin{cases} x - y = -1 \\ x + y = 3 \end{cases}$

3. $(4, 1)$ $\begin{cases} x + y = 5 \\ x - y = 2 \end{cases}$

4. $(5, 2)$, $\begin{cases} x - y = 3 \\ x + y = 6 \end{cases}$

EXERCISE 7.1

In Exercises 1–12, tell whether the ordered pair is a simultaneous solution for the given system.

1. $(1, 1)$, $\begin{cases} x + y = 2 \\ 2x - y = 1 \end{cases}$

2. $(1, 3)$, $\begin{cases} 2x + y = 5 \\ 3x - y = 0 \end{cases}$

3. $(3, -2)$, $\begin{cases} 2x + y = 4 \\ x + y = 1 \end{cases}$

4. $(-2, 4)$, $\begin{cases} 2x + 2y = 4 \\ x + 3y = 10 \end{cases}$

5. $(4, 5)$, $\begin{cases} 2x - 3y = -7 \\ 4x - 5y = 25 \end{cases}$

6. $(2, 3)$, $\begin{cases} 3x - 2y = 0 \\ 5x - 3y = -1 \end{cases}$

7. $(-2, -3)$, $\begin{cases} 4x + 5y = -23 \\ -3x + 2y = 0 \end{cases}$

8. $(-5, 1)$, $\begin{cases} -2x + 7y = 17 \\ 3x - 4y = -19 \end{cases}$

9. $\left(\dfrac{1}{2}, 3\right), \begin{cases} 2x + y = 4 \\ 4x - 3y = 11 \end{cases}$

10. $\left(2, \dfrac{1}{3}\right), \begin{cases} x - 3y = 1 \\ -2x + 6y = -6 \end{cases}$

11. $\left(-\dfrac{2}{5}, \dfrac{1}{4}\right), \begin{cases} 5x - 4y = -6 \\ 8y = 10x + 12 \end{cases}$

12. $\left(-\dfrac{1}{3}, \dfrac{3}{4}\right), \begin{cases} 3x + 4y = 2 \\ 12y = 3(2 - 3x) \end{cases}$

In Exercises 13–24, solve each system.

13. $\begin{cases} x + y = 2 \\ x - y = 0 \end{cases}$

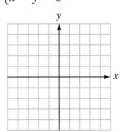

14. $\begin{cases} x + y = 4 \\ x - y = 0 \end{cases}$

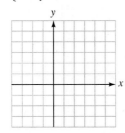

15. $\begin{cases} x + y = 2 \\ x - y = 4 \end{cases}$

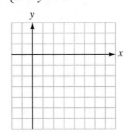

16. $\begin{cases} x + y = 1 \\ x - y = -5 \end{cases}$

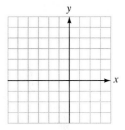

17. $\begin{cases} 3x + 2y = -8 \\ 2x - 3y = -1 \end{cases}$

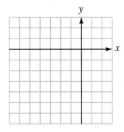

18. $\begin{cases} x + 4y = -2 \\ x + y = -5 \end{cases}$

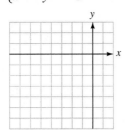

19. $\begin{cases} 4x - 2y = 8 \\ y = 2x - 4 \end{cases}$

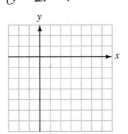

20. $\begin{cases} 3x - 6y = 18 \\ x = 2y + 3 \end{cases}$

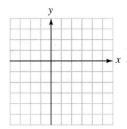

21. $\begin{cases} 2x - 3y = -18 \\ 3x + 2y = -1 \end{cases}$

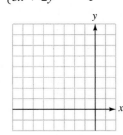

22. $\begin{cases} -x + 3y = -11 \\ 3x - y = 17 \end{cases}$

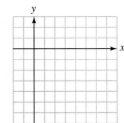

23. $\begin{cases} 4x = 3(4 - y) \\ 2y = 4(3 - x) \end{cases}$

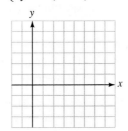

24. $\begin{cases} 2x = 3(2 - y) \\ 3y = 2(3 - x) \end{cases}$

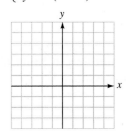

In Exercises 25–32, solve each system.

25. $\begin{cases} x + 2y = -4 \\ x - \dfrac{1}{2}y = 6 \end{cases}$

26. $\begin{cases} \dfrac{2}{3}x - y = -3 \\ 3x + y = 3 \end{cases}$

27. $\begin{cases} -\dfrac{3}{4}x + y = 3 \\ \dfrac{1}{4}x + y = -1 \end{cases}$

28. $\begin{cases} \dfrac{1}{3}x + y = 7 \\ \dfrac{2}{3}x - y = -4 \end{cases}$

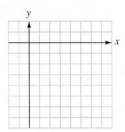
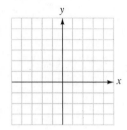
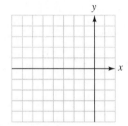
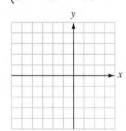

29. $\begin{cases} \dfrac{1}{2}x + \dfrac{1}{4}y = 0 \\ \dfrac{1}{4}x - \dfrac{3}{8}y = -2 \end{cases}$

30. $\begin{cases} \dfrac{1}{2}x + \dfrac{2}{3}y = -5 \\ \dfrac{3}{2}x - y = 3 \end{cases}$

31. $\begin{cases} \dfrac{1}{3}x - \dfrac{1}{2}y = \dfrac{1}{6} \\ \dfrac{2}{5}x + \dfrac{1}{2}y = \dfrac{13}{10} \end{cases}$

32. $\begin{cases} \dfrac{3}{4}x + \dfrac{2}{3}y = -\dfrac{19}{6} \\ y - x = -\dfrac{4x}{3} \end{cases}$

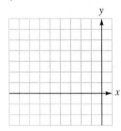
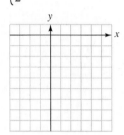
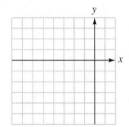

In Exercises 33–36, use a graphing calculator to solve each system, if possible.

33. $\begin{cases} y = 4 - x \\ y = 2 + x \end{cases}$

34. $\begin{cases} y = x - 2 \\ y = x + 2 \end{cases}$

35. $\begin{cases} 3x - 6y = 4 \\ 2x + y = 1 \end{cases}$

36. $\begin{cases} 4x + 9y = 4 \\ 6x + 3y = -1 \end{cases}$

Writing Exercises ■ *Write a paragraph using your own words.*

1. Explain what we mean when we say "inconsistent system."

2. Explain what we mean when we say "the equations of a system are dependent."

Something to Think About ■ **1.** Use a graphing calculator to solve the system

$$\begin{cases} 11x - 20y = 21 \\ -4x + 7y = 21 \end{cases}$$

What problems did you encounter?

2. Can the equations of an inconsistent system with two equations with two variables be dependent?

Review Exercises ■ *Write each expression as an expression with one exponent.*

1. $x^3x^4x^5$ **2.** $\dfrac{y^7}{y^2y^3}$ **3.** $\dfrac{(a^3)^2}{(a^2)^3}$ **4.** $\dfrac{(t^2t^3)^3}{(t^2t)^4}$

7.2 Solving Systems of Equations by Substitution

■ Inconsistent Systems ■ Dependent Equations

GETTING READY   *Remove parentheses.*

1. $2(3x + 2)$ **2.** $5(-5 - 2x)$

Substitute $x - 2$ for y and remove parentheses.

3. $2y$ **4.** $3(y - 2)$

We now consider the **substitution method** for solving systems of equations.
    To solve the system

$$\begin{cases} y = 3x - 2 \\ 2x + y = 8 \end{cases}$$

by the substitution method, we note that $y = 3x - 2$. Because $y = 3x - 2$, we can substitute $3x - 2$ for y in the equation $2x + y = 8$ to get

$$2x + y = 8$$
$$2x + (3x - 2) = 8$$

The resulting equation has only one variable and can be solved for x.

$$2x + (3x - 2) = 8$$
$$2x + 3x - 2 = 8 \qquad \text{Remove parentheses.}$$
$$5x - 2 = 8 \qquad \text{Combine like terms.}$$
$$5x = 10 \qquad \text{Add 2 to both sides.}$$
$$x = 2 \qquad \text{Divide both sides by 5.}$$

We can find y by substituting 2 for x in either equation of the given system. Because $y = 3x - 2$ is already solved for y, it is easiest to substitute in this equation.

$$y = 3x - 2$$
$$= 3(2) - 2$$
$$= 6 - 2$$
$$= 4$$

The solution to the given system is $x = 2$ and $y = 4$, or just $(2, 4)$.

$$Check:\ \begin{array}{l|l} y = 3x - 2 & 2x + y = 8 \\ 4 \stackrel{?}{=} 3(2) - 2 & 2(2) + 4 \stackrel{?}{=} 8 \\ 4 \stackrel{?}{=} 6 - 2 & 4 + 4 \stackrel{?}{=} 8 \\ 4 = 4 & 8 = 8 \end{array}$$

Since the pair $x = 2$ and $y = 4$ is a solution, the lines represented by the equations of the given system intersect at the point $(2, 4)$. The equations of this system are independent, and the system is consistent.

To solve a system of equations in x and y by the substitution method, we follow these steps.

The Substitution Method

1. Solve one of the equations for x or y. This step may not be necessary.
2. Substitute the resulting expression for the variable obtained in Step 1 into the other equation and solve that equation.
3. Find the value of the other variable by substituting the solution found in Step 2 in any equation containing both variables.
4. Check the solution in the equations of the original system.

EXAMPLE 1 Solve the system $\begin{cases} 2x + y = -5 \\ 3x + 5y = -4 \end{cases}$.

Solution We solve one of the equations for one of the variables. Since the term y in the first equation has a coefficient of 1, we solve the first equation for y.

$$2x + y = -5$$
$$y = -5 - 2x \qquad \text{Subtract } 2x \text{ from both sides.}$$

We then substitute $-5 - 2x$ for y in the second equation and solve for x.

$$3x + 5y = -4$$
$$3x + 5(-5 - 2x) = -4$$
$$3x - 25 - 10x = -4 \qquad \text{Remove parentheses.}$$
$$-7x - 25 = -4 \qquad \text{Combine like terms.}$$
$$-7x = 21 \qquad \text{Add 25 to both sides.}$$
$$x = -3 \qquad \text{Divide both sides by } -7.$$

We can find y by substituting -3 for x in the equation $y = -5 - 2x$.

$$y = -5 - 2x$$
$$= -5 - 2(-3)$$
$$= -5 + 6$$
$$= 1$$

The solution to the given system is $x = -3$ and $y = 1$, or just $(-3, 1)$.

Check:

$$2x + y = -5$$
$$2(-3) + 1 \overset{?}{=} -5$$
$$-6 + 1 \overset{?}{=} -5$$
$$-5 = -5$$

$$3x + 5y = -4$$
$$3(-3) + 5(1) \overset{?}{=} -4$$
$$-9 + 5 \overset{?}{=} -4$$
$$-4 = -4$$ ■

EXAMPLE 2   Solve the system $\begin{cases} 2x + 3y = 5 \\ 3x + 2y = 0 \end{cases}$.

Solution   We can solve the second equation for x:

$$3x + 2y = 0$$

$$3x = -2y \qquad \text{Subtract } 2y \text{ from both sides.}$$

$$x = \frac{-2y}{3} \qquad \text{Divide both sides by 3.}$$

We then substitute $\dfrac{-2y}{3}$ for x in the other equation and solve for y.

$$2x + 3y = 5$$

$$2\left(\frac{-2y}{3}\right) + 3y = 5$$

$$\frac{-4y}{3} + 3y = 5 \qquad \text{Remove parentheses.}$$

$$3\left(\frac{-4y}{3}\right) + 3(3y) = 3(5) \qquad \text{Multiply both sides by 3.}$$

$$-4y + 9y = 15 \qquad \text{Remove parentheses.}$$

$$5y = 15 \qquad \text{Combine like terms.}$$

$$y = 3 \qquad \text{Divide both sides by 5.}$$

We can find x by substituting 3 for y in the equation $x = \dfrac{-2y}{3}$.

$$x = \frac{-2y}{3}$$

$$= \frac{-2(3)}{3}$$

$$= -2$$

Check the solution $(-2, 3)$ in each equation of the system.   ■

■ Inconsistent Systems

EXAMPLE 3 Solve the system $\begin{cases} x = 4(3 - y) \\ 2x = 4(3 - 2y) \end{cases}$.

Solution Since $x = 4(3 - y)$, we can substitute $4(3 - y)$ for x in the second equation and solve for y.

$$2x = 4(3 - 2y)$$
$$2[4(3 - y)] = 4(3 - 2y)$$
$$8(3 - y) = 4(3 - 2y) \qquad 2 \cdot 4 = 8.$$
$$24 - 8y = 12 - 8y \qquad \text{Remove parentheses.}$$
$$24 = 12 \qquad \text{Add } 8y \text{ to both sides.}$$

This impossible result indicates that the equations in this system are independent, but that the system is inconsistent. If each equation in this system were graphed, these graphs would be parallel lines. There are no solutions to this system. ■

■ Dependent Equations

EXAMPLE 4 Solve the system $\begin{cases} 3x = 4(6 - y) \\ 4y + 3x = 24 \end{cases}$.

Solution We can substitute $4(6 - y)$ for $3x$ in the second equation and proceed as follows:

$$4y + 3x = 24$$
$$4y + 4(6 - y) = 24$$
$$4y + 24 - 4y = 24 \qquad \text{Remove parentheses.}$$
$$24 = 24 \qquad \text{Combine like terms.}$$

Although the result $24 = 24$ is true, we did not find y. This result indicates that the equations of this system are dependent. If each equation in this system were graphed, the same line would result.

Because any ordered pair that satisfies one equation of the system satisfies the other also, the system has an infinite number of solutions. To find some, we substitute 8, 0, and 4 for x in either equation and solve for y. The pairs $(8, 0)$, $(0, 6)$, and $(4, 3)$ are some of the solutions. ■

EXAMPLE 5 Solve the system $\begin{cases} 3(x - y) = 5 \\ x + 3 = -\dfrac{5}{2}y \end{cases}$.

Solution   We begin by writing each equation in general form:

$$3(x - y) = 5 \qquad\qquad x + 3 = -\frac{5}{2}y$$

1. $3x - 3y = 5$ $\qquad\qquad$ $2x + 6 = -5y$ $\qquad$ Multiply both sides by 2.

2. $\qquad\qquad\qquad\qquad$ $2x + 5y = -6$ $\qquad$ Add 5y and subtract 6 from both sides.

To solve the system formed by Equations 1 and 2, we first solve Equation 1 for x.

1. $3x - 3y = 5$

$\qquad\qquad$ $3x = 5 + 3y$ $\qquad\qquad\qquad\qquad$ Add 3y to both sides.

3. $\qquad\qquad$ $x = \dfrac{5 + 3y}{3}$ $\qquad\qquad\qquad$ Divide both sides by 3.

We then substitute $\frac{5+3y}{3}$ for x in Equation 2 and proceed as follows:

2. $\qquad\qquad$ $2x + 5y = -6$

$$2\left(\frac{5 + 3y}{3}\right) + 5y = -6$$

$\qquad$ $2(5 + 3y) + 15y = -18$ $\qquad$ Multiply both sides by 3.

$\qquad\qquad$ $10 + 6y + 15y = -18$ $\qquad$ Remove parentheses.

$\qquad\qquad\qquad$ $10 + 21y = -18$ $\qquad$ Combine like terms.

$\qquad\qquad\qquad\qquad$ $21y = -28$ $\qquad$ Subtract 10 from both sides.

$\qquad\qquad\qquad\qquad$ $y = \dfrac{-28}{21}$ $\qquad$ Divide both sides by 21.

$\qquad\qquad\qquad\qquad$ $y = -\dfrac{4}{3}$ $\qquad$ Simplify $\frac{-28}{21}$.

To find x, we substitute $-\frac{4}{3}$ for y in Equation 3 and simplify.

$$x = \frac{5 + 3y}{3}$$

$$= \frac{5 + 3\left(-\frac{4}{3}\right)}{3}$$

$$= \frac{5 - 4}{3}$$

$$= \frac{1}{3}$$

Check the solution $\left(\frac{1}{3}, -\frac{4}{3}\right)$ in each equation. $\qquad\qquad\qquad\qquad$ ■

ORALS *Let $y = x + 1$. Find y after each of the following quantities is substituted for x.*

1. $2z$　　　　　　　　　　　　　**2.** $z + 1$

3. $3t + 2$　　　　　　　　　　　**4.** $\dfrac{t}{3} + 3$

E X E R C I S E 7.2

In Exercises 1–42, use the substitution method to solve each system.

1. $\begin{cases} y = 2x \\ x + y = 6 \end{cases}$
　2. $\begin{cases} y = 3x \\ x + y = 4 \end{cases}$
　3. $\begin{cases} y = 2x - 6 \\ 2x + y = 6 \end{cases}$
　4. $\begin{cases} y = 2x - 9 \\ x + 3y = 8 \end{cases}$

5. $\begin{cases} y = 2x + 5 \\ x + 2y = -5 \end{cases}$
　6. $\begin{cases} y = -2x \\ 3x + 2y = -1 \end{cases}$
　7. $\begin{cases} 2a + 4b = -24 \\ a = 20 - 2b \end{cases}$
　8. $\begin{cases} 3a + 6b = -15 \\ a = -2b - 5 \end{cases}$

9. $\begin{cases} 2a = 3b - 13 \\ b = 2a + 7 \end{cases}$
　10. $\begin{cases} a = 3b - 1 \\ b = 2a + 2 \end{cases}$
　11. $\begin{cases} r + 3s = 9 \\ 3r + 2s = 13 \end{cases}$
　12. $\begin{cases} x - 2y = 2 \\ 2x + 3y = 11 \end{cases}$

13. $\begin{cases} 4x + 5y = 2 \\ 3x - y = 11 \end{cases}$
　14. $\begin{cases} 5u + 3v = 5 \\ 4u - v = 4 \end{cases}$
　15. $\begin{cases} 2x + y = 0 \\ 3x + 2y = 1 \end{cases}$
　16. $\begin{cases} 3x - y = 7 \\ 2x + 3y = 1 \end{cases}$

17. $\begin{cases} 3x + 4y = -7 \\ 2y - x = -1 \end{cases}$
　18. $\begin{cases} 4x + 5y = -2 \\ x + 2y = -2 \end{cases}$
　19. $\begin{cases} 9x = 3y + 12 \\ 4 = 3x - y \end{cases}$
　20. $\begin{cases} 8y = 15 - 4x \\ x + 2y = 4 \end{cases}$

21. $\begin{cases} 2x + 3y = 5 \\ 3x + 2y = 5 \end{cases}$
　22. $\begin{cases} 3x - 2y = -1 \\ 2x + 3y = -5 \end{cases}$
　23. $\begin{cases} 2x + 5y = -2 \\ 4x + 3y = 10 \end{cases}$
　24. $\begin{cases} 3x + 4y = -6 \\ 2x - 3y = -4 \end{cases}$

25. $\begin{cases} 2x - 3y = -3 \\ 3x + 5y = -14 \end{cases}$
　26. $\begin{cases} 4x - 5y = -12 \\ 5x - 2y = 2 \end{cases}$
　27. $\begin{cases} 7x - 2y = -1 \\ -5x + 2y = -1 \end{cases}$
　28. $\begin{cases} -8x + 3y = 22 \\ 4x + 3y = -2 \end{cases}$

29. $\begin{cases} 2a + 3b = 2 \\ 8a - 3b = 3 \end{cases}$
　30. $\begin{cases} 3a - 2b = 0 \\ 9a + 4b = 5 \end{cases}$
　31. $\begin{cases} y - x = 3x \\ 2(x + y) = 14 - y \end{cases}$
　32. $\begin{cases} y + x = 2x + 2 \\ 2(3x - 2y) = 21 - y \end{cases}$

33. $\begin{cases} 3(x - 1) + 3 = 8 + 2y \\ 2(x + 1) = 4 + 3y \end{cases}$
　　　　　34. $\begin{cases} 4(x - 2) = 19 - 5y \\ 3(x + 1) - 2y = 2y \end{cases}$

35. $\begin{cases} 6a = 5(3 + b + a) - a \\ 3(a - b) + 4b = 5(1 + b) \end{cases}$
　　　36. $\begin{cases} 5(x + 1) + 7 = 7(y + 1) \\ 5(y + 1) = 6(1 + x) + 5 \end{cases}$

37. $\begin{cases} \dfrac{1}{2}x + \dfrac{1}{2}y = -1 \\ \dfrac{1}{3}x - \dfrac{1}{2}y = -4 \end{cases}$
　　　　　38. $\begin{cases} \dfrac{2}{3}y + \dfrac{1}{5}z = 1 \\ \dfrac{1}{3}y - \dfrac{2}{5}z = 3 \end{cases}$

39. $\begin{cases} 5x = \dfrac{1}{2}y - 1 \\ \dfrac{1}{4}y = 10x - 1 \end{cases}$

40. $\begin{cases} \dfrac{2}{3}x = 1 - 2y \\ 2(5y - x) + 11 = 0 \end{cases}$

41. $\begin{cases} \dfrac{6x - 1}{3} - \dfrac{5}{3} = \dfrac{3y + 1}{2} \\ \dfrac{1 + 5y}{4} + \dfrac{x + 3}{4} = \dfrac{17}{2} \end{cases}$

42. $\begin{cases} \dfrac{5x - 2}{4} + \dfrac{1}{2} = \dfrac{3y + 2}{2} \\ \dfrac{7y + 3}{3} = \dfrac{x}{2} + \dfrac{7}{3} \end{cases}$

Writing Exercises ■ *Write a paragraph using your own words.*

1. Explain how to use substitution to solve a system of equations.

2. If the equations of a system are written in general form, why is it to your advantage to solve for a variable whose coefficient is 1?

Something to Think About ■

1. Could you use substitution to solve

$$\begin{cases} y = 2y + 4 \\ x = 3x - 5 \end{cases}$$

How would you solve it?

2. What are the advantages and disadvantages of the

   1. graphing method?

   2. substitution method?

Review Exercises ■ *Factor each expression completely.*

1. $8x^2y^2 - 32xy^2z + 16xyz^2$

2. $(x - y)a - (x - y)b$

3. $a^6 - 25$

4. $b^4 - 625$

5. $r^2 + 2rs - 15s^2$

6. $4m^2 - 15mn + 9n^2$

7.3 Solving Systems of Equations by Addition

■ Inconsistent Systems ■ Dependent Equations

GETTING READY   *Add the left-hand sides and the right-hand sides of the equations in each system.*

1. $\begin{cases} 2x + 3y = 4 \\ 3x - 3y = 6 \end{cases}$

2. $\begin{cases} 4x - 2y = 1 \\ -4x + 3y = 5 \end{cases}$

3. $\begin{cases} 6x - 5y = 23 \\ -4x + 5y = 10 \end{cases}$

4. $\begin{cases} -5x + 6y = 18 \\ 5x + 12y = 10 \end{cases}$

Another method used to solve systems of equations is the **addition method.** To solve the system

$$\begin{cases} x + y = 8 \\ x - y = -2 \end{cases}$$

by the addition method, we see that the coefficients of y are opposites and then add the left-hand sides and the right-hand sides of the equations to eliminate the variable y.

$$\begin{array}{rcr} x + y = & 8 \\ \underline{x - y = } & \underline{-2} \\ 2x = & 6 \end{array}$$

We can then solve the resulting equation for x.

$$2x = 6$$
$$x = 3 \qquad \text{Divide both sides by 2.}$$

To find y, we substitute 3 for x in either equation of the system and solve it for y.

$$x + y = 8 \qquad \text{The first equation of the system.}$$
$$3 + y = 8 \qquad \text{Substitute 3 for } x.$$
$$y = 5 \qquad \text{Subtract 3 from both sides.}$$

We check the solution by verifying that the pair $(3, 5)$ satisfies each equation of the original system.

To solve an equation in x and y by the addition method, we follow these steps.

The Addition Method   1. If necessary, write both equations in general form: $Ax + By = C$.

2. If necessary, multiply one or both of the equations by nonzero quantities to make the coefficients of x (or the coefficients of y) opposites.

3. Add the equations to eliminate the term involving x (or y).

4. Solve the equation resulting from Step 3.

5. Find the value of the other variable by substituting the solution found in Step 4 into any equation containing both variables.

6. Check the solution in the equations of the original system.

EXAMPLE 1   Solve the system $\begin{cases} 3y = 14 + x \\ x + 22 = 5y \end{cases}$.

Solution   We can write the equations in the form $\begin{cases} -x + 3y = 14 \\ x - 5y = -22 \end{cases}$.

When these equations are added, the terms involving x are eliminated. We solve the resulting equation for y.

$$-x + 3y = 14$$
$$\underline{x - 5y = -22}$$
$$-2y = - 8$$
$$y = 4 \qquad \text{Divide both sides by } -2.$$

To find x, we substitute 4 for y in either equation of the system. If we substitute 4 for y in the equation $-x + 3y = 14$, we have

$$-x + 3y = 14$$
$$-x + 3(4) = 14$$
$$-x + 12 = 14 \qquad \text{Simplify.}$$
$$-x = 2 \qquad \text{Subtract 12 from both sides.}$$
$$x = -2 \qquad \text{Divide both sides by } -1.$$

Verify that $(-2, 4)$ satisfies each equation. ∎

EXAMPLE 2   Solve the system $\begin{cases} 2x - 5y = 10 \\ 3x - 2y = -7 \end{cases}.$

Solution   The equations in the system must be rewritten so that one of the variables will be eliminated when the equations are added.

To eliminate the x variable, we can multiply the first equation by 3 and the second equation by -2 to get

$$\begin{array}{l} \text{3 times} \\ -2 \text{ times} \end{array} \begin{cases} 2x - 5y = 10 \\ 3x - 2y = -7 \end{cases} \longrightarrow \begin{cases} 6x - 15y = 30 \\ -6x + 4y = 14 \end{cases}$$

When these equations are added, the terms involving the variable x are eliminated.

$$6x - 15y = 30$$
$$\underline{-6x + 4y = 14}$$
$$-11y = 44$$
$$y = -4 \qquad \text{Divide both sides by } -11.$$

To find x, we substitute -4 for y in the equation $2x - 5y = 10$.

$$2x - 5y = 10$$
$$2x - 5(-4) = 10 \qquad \text{Substitute } -4 \text{ for } y.$$
$$2x + 20 = 10 \qquad \text{Simplify.}$$
$$2x = -10 \qquad \text{Subtract 20 from both sides.}$$
$$x = -5 \qquad \text{Divide both sides by 2.}$$

Check the solution $(-5, -4)$. ∎

■ Inconsistent Systems

EXAMPLE 3   Solve the system $\begin{cases} x - \dfrac{2}{3}y = \dfrac{8}{3} \\ -\dfrac{3}{2}x + y = -6 \end{cases}$.

Solution   We can multiply both sides of the first equation by 3 and both sides of the second equation by 2 to clear the equations of fractions.

$$\begin{array}{c} 3 \text{ times} \\ \\ 2 \text{ times} \end{array} \begin{cases} x - \dfrac{2}{3}y = \dfrac{8}{3} \\ -\dfrac{3}{2}x + y = -6 \end{cases} \longrightarrow \begin{cases} 3x - 2y = 8 \\ -3x + 2y = -12 \end{cases}$$

We can add the resulting equations to eliminate the term involving x.

$$\begin{array}{r} 3x - 2y = 8 \\ -3x + 2y = -12 \\ \hline 0 = -4 \end{array}$$

In this case, the terms involving both x and y are eliminated, and a false result is obtained. This shows that the equations of the system are independent, but the system itself is inconsistent. This system has no solutions.   ■

■ Dependent Equations

EXAMPLE 4   Solve the system $\begin{cases} x - \dfrac{5}{2}y = \dfrac{19}{2} \\ -\dfrac{2}{5}x + y = -\dfrac{19}{5} \end{cases}$.

Solution   We can multiply both sides of the first equation by 2 and both sides of the second equation by 5 to clear the equations of fractions.

$$\begin{array}{c} 2 \text{ times} \\ \\ 5 \text{ times} \end{array} \begin{cases} x - \dfrac{5}{2}y = \dfrac{19}{2} \\ -\dfrac{2}{5}x + y = -\dfrac{19}{5} \end{cases} \longrightarrow \begin{cases} 2x - 5y = 19 \\ -2x + 5y = -19 \end{cases}$$

We add the resulting equations to get

$$2x - 5y = 19$$
$$\underline{-2x + 5y = -19}$$
$$0 = 0$$

As in Example 3, both the x and y variables are eliminated. However, this time a true result is obtained. This shows that the equations of this system are dependent, and the system has an infinite number of solutions.

Any ordered pair that satisfies one of the equations satisfies the other also. Some solutions are $(2, -3)$, $(12, 1)$, and $\left(0, -\frac{19}{5}\right)$. ■

EXAMPLE 5   Solve the system $\begin{cases} \dfrac{5}{6}x + \dfrac{2}{3}y = \dfrac{7}{6} \\ \dfrac{10}{7}x - \dfrac{4}{9}y = \dfrac{17}{21} \end{cases}$.

Solution   To clear the equations of fractions, we multiply both sides of the first equation by 6 and both sides of the second equation by 63. This gives the system

1. $\begin{cases} 5x + 4y = 7 \\ 90x - 28y = 51 \end{cases}$
2.

We solve for x by eliminating the terms involving y. To do so, we multiply Equation 1 by 7 and add the result to Equation 2.

$$35x + 28y = 49$$
$$\underline{90x - 28y = 51}$$
$$125x = 100$$

$$x = \frac{100}{125} \qquad \text{Divide both sides by 125.}$$

$$x = \frac{4}{5} \qquad \text{Simplify.}$$

To solve for y, we substitute $\frac{4}{5}$ for x in Equation 1 and simplify.

$$5x + 4y = 7$$
$$5\left(\frac{4}{5}\right) + 4y = 7$$
$$4 + 4y = 7 \qquad \text{Simplify.}$$
$$4y = 3 \qquad \text{Subtract 4 from both sides.}$$
$$y = \frac{3}{4} \qquad \text{Divide both sides by 4.}$$

Check the solution of $\left(\frac{4}{5}, \frac{3}{4}\right)$. ■

ORALS *Use addition to solve each system for x.*

1. $\begin{cases} x + y = 1 \\ x - y = 1 \end{cases}$

2. $\begin{cases} 2x + y = 4 \\ x - y = 2 \end{cases}$

Use addition to solve each equation for y.

3. $\begin{cases} -x + y = 3 \\ x + y = 3 \end{cases}$

4. $\begin{cases} x + 2y = 4 \\ -x - y = 1 \end{cases}$

E X E R C I S E 7.3

In Exercises 1–12, use the addition method to solve each system.

1. $\begin{cases} x + y = 5 \\ x - y = -3 \end{cases}$

2. $\begin{cases} x - y = 1 \\ x + y = 7 \end{cases}$

3. $\begin{cases} x - y = -5 \\ x + y = 1 \end{cases}$

4. $\begin{cases} x + y = 1 \\ x - y = 5 \end{cases}$

5. $\begin{cases} 2x + y = -1 \\ -2x + y = 3 \end{cases}$

6. $\begin{cases} 3x + y = -6 \\ x - y = -2 \end{cases}$

7. $\begin{cases} 2x - 3y = -11 \\ 3x + 3y = 21 \end{cases}$

8. $\begin{cases} 3x - 2y = 16 \\ -3x + 8y = -10 \end{cases}$

9. $\begin{cases} 2x + y = -2 \\ -2x - 3y = -6 \end{cases}$

10. $\begin{cases} 3x + 4y = 8 \\ 5x - 4y = 24 \end{cases}$

11. $\begin{cases} 4x + 3y = 24 \\ 4x - 3y = -24 \end{cases}$

12. $\begin{cases} 5x - 4y = 8 \\ -5x - 4y = 8 \end{cases}$

In Exercises 13–42, use the addition method to solve each system of equations. If the equations of a system are dependent or if a system is inconsistent, so indicate.

13. $\begin{cases} x + y = 5 \\ x + 2y = 8 \end{cases}$

14. $\begin{cases} x + 2y = 0 \\ x - y = -3 \end{cases}$

15. $\begin{cases} 2x + y = 4 \\ 2x + 3y = 0 \end{cases}$

16. $\begin{cases} 2x + 5y = -13 \\ 2x - 3y = -5 \end{cases}$

17. $\begin{cases} 3x + 29 = 5y \\ 4y - 34 = -3x \end{cases}$

18. $\begin{cases} 3x - 16 = 5y \\ 33 - 5y = 4x \end{cases}$

19. $\begin{cases} 2x = 3(y - 2) \\ 2(x + 4) = 3y \end{cases}$

20. $\begin{cases} 3(x - 2) = 4y \\ 2(2y + 3) = 3x \end{cases}$

21. $\begin{cases} -2(x + 1) = 3(y - 2) \\ 3(y + 2) = 6 - 2(x - 2) \end{cases}$

22. $\begin{cases} 5(x - 1) = 8 - 3(y + 2) \\ 4(x + 2) - 7 = 3(2 - y) \end{cases}$

23. $\begin{cases} 4(x + 1) = 17 - 3(y - 1) \\ 2(x + 2) + 3(y - 1) = 9 \end{cases}$

24. $\begin{cases} 3(x + 3) + 2(y - 4) = 5 \\ 3(x - 1) = -2(y + 2) \end{cases}$

25. $\begin{cases} 2x + y = 10 \\ x + 2y = 10 \end{cases}$

26. $\begin{cases} 3x + 2y = 0 \\ 2x - 3y = -13 \end{cases}$

27. $\begin{cases} 2x - y = 16 \\ 3x + 2y = 3 \end{cases}$

28. $\begin{cases} 3x + 4y = -17 \\ 4x - 3y = -6 \end{cases}$

29. $\begin{cases} 4x + 5y = -20 \\ 5x - 4y = -25 \end{cases}$ **30.** $\begin{cases} 3x - 5y = 4 \\ 7x + 3y = 68 \end{cases}$ **31.** $\begin{cases} 6x = -3y \\ 5y = 2x + 12 \end{cases}$ **32.** $\begin{cases} 3y = 4x \\ 5x = 4y - 2 \end{cases}$

33. $\begin{cases} 4(2x - y) = 18 \\ 3(x - 3) = 2y - 1 \end{cases}$ **34.** $\begin{cases} 2(2x + 3y) = 5 \\ 8x = 3(1 + 3y) \end{cases}$ **35.** $\begin{cases} \dfrac{3}{5}x + \dfrac{4}{5}y = 1 \\ -\dfrac{1}{4}x + \dfrac{3}{8}y = 1 \end{cases}$ **36.** $\begin{cases} \dfrac{1}{2}x - \dfrac{1}{4}y = 1 \\ \dfrac{1}{3}x + y = 3 \end{cases}$

37. $\begin{cases} \dfrac{3}{5}x + y = 1 \\ \dfrac{4}{5}x - y = -1 \end{cases}$ **38.** $\begin{cases} \dfrac{1}{2}x + \dfrac{4}{7}y = -1 \\ 5x - \dfrac{4}{5}y = -10 \end{cases}$

39. $\begin{cases} \dfrac{x}{2} - \dfrac{y}{3} = -2 \\ \dfrac{2x - 3}{2} + \dfrac{6y + 1}{3} = \dfrac{17}{6} \end{cases}$ **40.** $\begin{cases} \dfrac{x + 2}{4} + \dfrac{y - 1}{3} = \dfrac{1}{12} \\ \dfrac{x + 4}{5} - \dfrac{y - 2}{2} = \dfrac{5}{2} \end{cases}$

41. $\begin{cases} \dfrac{x - 3}{2} + \dfrac{y + 5}{3} = \dfrac{11}{6} \\ \dfrac{x + 3}{3} - \dfrac{5}{12} = \dfrac{y + 3}{4} \end{cases}$ **42.** $\begin{cases} \dfrac{x + 2}{3} = \dfrac{3 - y}{2} \\ \dfrac{x + 3}{2} = \dfrac{2 - y}{3} \end{cases}$

Writing Exercises ■ *Write a paragraph using your own words.*

1. Why is it usually to your advantage to write the equations of a system in general form before using the addition method to solve it?

2. How would you decide whether to use substitution or addition to solve a system of equations?

Something to Think About ■

1. If possible, find a solution to the system

$$\begin{cases} x + y = 5 \\ x - y = -3 \\ 2x - y = -2 \end{cases}$$

2. If possible, find a solution to the system

$$\begin{cases} x + y = 5 \\ x - y = -3 \\ x - 2y = 0 \end{cases}$$

Review Exercises ■ *Solve each equation.*

1. $8(3x - 5) - 12 = 4(2x + 3)$

2. $3y + \dfrac{y + 2}{2} = \dfrac{2(y + 3)}{3} + 16$

3. $3z^2 - 5z + 2 = 0$

4. $3t^2 + 4 = 8t$

5. $10y^2 + 21y = 10$

6. $x(9x - 24) + 12 = 0$

7.4 Applications of Systems of Equations

GETTING READY *Use an algebraic expression to denote each phrase.*

1. The sum of x and y

2. The difference when y is subtracted from x

3. Give the formula for the area of a rectangle.

4. Give the formula for the perimeter of a rectangle.

5. If $\$a$ is invested at 6% annual interest and $\$b$ is invested at 7%, find the interest earned in 1 year.

6. If x liters of 5% alcohol is mixed with y liters of 12% alcohol, find the total amount of alcohol in the mixture.

We have previously set up equations involving one variable to solve problems. In this section, we consider ways to solve problems by using two variables.

The following steps are helpful when solving problems involving two unknown quantities.

Solving Application Problems

1. Read the problem several times and analyze the facts. Occasionally, a sketch, chart, or diagram will help you visualize the facts of the problem.

2. Pick different variables to represent two unknown quantities, and write a sentence stating what each variable represents.

3. Find two equations involving each of the two variables. This will give a system of two equations in two variables.

4. Solve the system using the most convenient method.

5. Check the solution in the words of the problem.

EXAMPLE 1 **Farming** A farmer raises wheat and soybeans on 215 acres. If he wants to plant 31 more acres in wheat than in soybeans, how many acres of each should he plant?

Solution If w represents the number of acres of wheat and s represents the number of acres of soybeans to be planted, we have

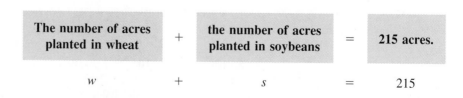

The number of acres planted in wheat	$+$	the number of acres planted in soybeans	$=$	215 acres.
w	$+$	s	$=$	215

Since the farmer wants to plant 31 more acres in wheat than in soybeans, we have

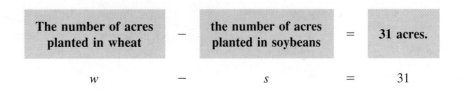

The number of acres planted in wheat	−	the number of acres planted in soybeans	=	31 acres.
w	−	s	=	31

We can now solve the system

1. $\begin{cases} w + s = 215 \\ w - s = 31 \end{cases}$
2.

by using the addition method.

$$w + s = 215$$
$$\underline{w - s = 31}$$
$$2w = 246$$
$$w = 123 \qquad \text{Divide both sides by 2.}$$

To find s, we substitute 123 for w in Equation 1.

$$w + s = 215$$
$$\mathbf{123} + s = 215 \qquad \text{Substitute 123 for } w.$$
$$s = 92 \qquad \text{Add } -123 \text{ to both sides.}$$

The farmer should plant 123 acres of wheat and 92 acres of soybeans.

Check: The total acreage planted is $123 + 92$, or 215 acres. The area planted in wheat is 31 acres greater than that planted in soybeans, because $123 - 92 = 31$.

■

EXAMPLE 2

Lawn care   An installer of underground irrigation systems wants to cut a 20-foot length of plastic tubing into two pieces. The longer piece is to be 2 feet longer than twice the shorter piece. Find the length of each piece.

Solution   We can let s represent the length of the shorter piece and l represent the length of the longer piece (see Figure 7-7).

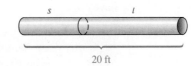

FIGURE 7-7

Since the length of tubing is 20 feet long, we have

The length of the shorter piece	+	the length of the longer piece	=	20 feet.
s	+	l	=	20

Since the longer piece is 2 feet longer than twice the shorter piece, we have

The length of the longer piece	=	2	·	the length of the shorter piece	+	2 feet.
l	=	2	·	s	+	2

We can use the substitution method to solve the system.

1. $\begin{cases} s + l = 20 \\ l = 2s + 2 \end{cases}$
2.

$$s + 2s + 2 = 20 \qquad \text{Substitute } 2s + 2 \text{ for } l.$$
$$3s + 2 = 20 \qquad \text{Combine like terms.}$$
$$3s = 18 \qquad \text{Subtract 2 from both sides.}$$
$$s = 6 \qquad \text{Divide both sides by 3.}$$

The shorter piece should be 6 feet long.

To find the length of the longer piece, we substitute 6 for s in Equation 1 and solve for l.

$$s + l = 20$$
$$6 + l = 20$$
$$l = 14 \qquad \text{Subtract 6 from both sides.}$$

The longer piece should be 14 feet long.

Check: The sum of 6 and 14 is 20.
14 is 2 more than twice 6. ∎

EXAMPLE 3

Gardening Tom has 150 feet of fencing to enclose a rectangular garden. If the length is to be 5 feet less than 3 times the width, find the area of the garden.

Solution We can let l represent the length of the garden and w represent the width (see Figure 7-8). Since the perimeter of a rectangle is two lengths plus two widths, we have

2	·	the length of the garden	+	2	·	the width of the garden	=	150 feet.
2	·	l	+	2	·	w	=	150

Since the length is 5 feet less than 3 times the width,

The length of the garden	=	3	·	the width of the garden	−	5 feet.

$$l \qquad = \quad 3 \quad \cdot \qquad w \qquad - \qquad 5$$

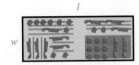

FIGURE 7-8

We can use the substitution method to solve this system.

1. $\begin{cases} 2l + 2w = 150 \\ l = 3w - 5 \end{cases}$
2.

$\quad 2(3w - 5) + 2w = 150$ Substitute $3w - 5$ for l in Equation 1.

$\quad\quad 6w - 10 + 2w = 150$ Remove parentheses.

$\quad\quad\quad\quad 8w - 10 = 150$ Combine like terms.

$\quad\quad\quad\quad\quad\quad 8w = 160$ Add 10 to both sides.

$\quad\quad\quad\quad\quad\quad\quad w = 20$ Divide both sides by 8.

The width is 20 feet.

To find the length, we substitute 20 for w in Equation 2 and simplify.

$$l = 3w - 5$$
$$= 3(20) - 5$$
$$= 60 - 5$$
$$= 55$$

Since the dimensions of the rectangle are 55 feet by 20 feet, and the area of a rectangle is given by the formula

$$A = l \cdot w \qquad \text{Area = length times width.}$$

we have

$$A = 55 \cdot 20$$
$$= 1100$$

The garden covers an area of 1100 square feet.

Check: Because the dimensions of the garden are 55 feet by 20 feet, the perimeter is

$$P = 2l + 2w$$
$$= 2(55) + 2(20)$$
$$= 110 + 40$$
$$= 150$$

It is also true that 55 feet is 5 feet less than 3 times 20 feet. ∎

EXAMPLE 4 **Manufacturing** The setup cost of a machine that mills brass plates is $750. After setup, it costs $0.25 to mill each plate. Management is considering the purchase of a larger machine that can produce the same plate at a cost of $0.20 per plate. If the setup cost of the larger machine is $1200, how many plates would the company have to produce to make the purchase worthwhile?

Solution We begin by finding the number of plates (called the **break point**) that will cost equal amounts to produce on either machine. We can let c represent this cost of milling p plates. Then we have

The cost of making p plates on machine 1	=	the setup cost of machine 1	+	the cost per plate on machine 1	$\cdot$	the number of plates to be made.
c	=	750	+	0.25	$\cdot$	p

The cost of making p plates on machine 2	=	the setup cost of machine 2	+	the cost per plate on machine 2	$\cdot$	the number of plates to be made.
c	=	1200	+	0.20	$\cdot$	p

Since the costs are equal, we can use the substitution method to solve the system

$$\begin{cases} c = 750 + 0.25p \\ c = 1200 + 0.20p \end{cases}$$

$750 + 0.25p = 1200 + 0.20p$	Substitute $750 + 0.25p$ for c in the second equation.
$0.25p = 450 + 0.20p$	Subtract 750 from both sides.
$0.05p = 450$	Subtract $0.20p$ from both sides.
$p = 9000$	Divide both sides by 0.05.

If 9000 plates are milled, the cost will be the same on either machine. If more than 9000 plates are milled, the cost will be cheaper on the newer machine, and it should be purchased. ∎

EXAMPLE 5 **Investments** Terri and Juan earned $1150 from a one-year investment of $15,000 to purchase a large screen television set. If Terri invested some of the money at 8% interest and Juan invested the rest at 7%, how much did each invest?

Solution We can let x represent the amount of money invested by Terri and y represent the amount of money invested by Juan. Because the total investment is $15,000, we have

| The amount invested by Terri | + | the amount invested by Juan | = | **$15,000.** |

$$x \qquad + \qquad y \qquad = \qquad 15,000$$

Since the income on x dollars invested at 8% is $0.08x$, the income on y dollars invested at 7% is $0.07y$, and the combined income is $1150, we have

| The income on the 8% investment | + | the income on the 7% investment | = | **$1150.** |

$$0.08x \qquad + \qquad 0.07y \qquad = \qquad 1150$$

Thus, we have the system

1. $\begin{cases} x + y = 15,000 \\ 0.08x + 0.07y = 1150 \end{cases}$
2.

To solve the system, we use the addition method.

$$-8x - 8y = -120,000 \qquad \text{Multiply both sides of Equation 1 by } -8.$$
$$\underline{8x + 7y = 115,000} \qquad \text{Multiply both sides of Equation 2 by } 100.$$
$$-y = -5000$$
$$y = 5000 \qquad \text{Multiply both sides by } -1.$$

To find x, we substitute 5000 for y in Equation 1 and simplify.

$$x + y = 15,000$$
$$x + \mathbf{5000} = 15,000$$
$$x = 10,000 \qquad \text{Subtract 5000 from both sides.}$$

Terri invested $10,000 and Juan invested $5000.

Check: $10,000 + $5000 = $15,000
$$0.08(\$10,000) = \$800$$
$$0.07(\$5000) = \$350$$

The combined interest is $1150. ■

EXAMPLE 6 **Boating** A boat traveled 30 kilometers downstream in 3 hours and made the return trip in 5 hours. Find the speed of the boat in still water.

Solution We can let s represent the speed of the boat in still water and let c represent the speed of the current. Then the rate of speed of the boat while going downstream is $s + c$. The rate of the boat while going upstream is $s - c$. We can organize the information of the problem as in Figure 7-9.

Distance =	Rate	·	Time
Downstream	30	$s + c$	3
Upstream	30	$s - c$	5

FIGURE 7-9

Because $d = r \cdot t$, the information in the table gives two equations in two variables.

$$\begin{cases} 30 = 3(s + c) \\ 30 = 5(s - c) \end{cases}$$

After removing parentheses and rearranging terms, we have

1. $\begin{cases} 3s + 3c = 30 \\ 5s - 5c = 30 \end{cases}$
2.

To solve this system by addition, we multiply Equation 1 by 5 and Equation 2 by 3, add the equations, and solve for s.

$$\begin{aligned} 15s + 15c &= 150 \\ \underline{15s - 15c = 90} \\ 30s &= 240 \\ s &= 8 \end{aligned}$$ Divide both sides by 30.

The speed of the boat in still water is 8 kilometers per hour. Check the result.

∎

EXAMPLE 7   **Medical technology**   A laboratory technician has one batch of antiseptic that is 40% alcohol and a second batch that is 60% alcohol. She would like to make 8 liters of solution that is 55% alcohol. How many liters of each batch should she use?

Solution   We can let x represent the number of liters to be used from batch 1, let y represent the number of liters to be used from batch 2, and organize the information of the problem as in Figure 7-10.

	Fractional part that is alcohol	·	Number of liters of solution	=	Number of liters of alcohol
Batch 1	0.40		x		$0.40x$
Batch 2	0.60		y		$0.60y$
Mixture	0.55		8		$0.55(8)$

FIGURE 7-10

The information in Figure 7-10 provides two equations.

1. $x + y = 8$ — The number of liters of batch 1 plus the number of liters of batch 2 equals the total number of liters in the mixture.

2. $0.40x + 0.60y = 0.55(8)$ — The amount of alcohol in batch 1 plus the amount of alcohol in batch 2 equals the amount of alcohol in the mixture.

We can use addition to solve this system.

$$
\begin{array}{ll}
-40x - 40y = -320 & \text{Multiply both sides of Equation 1 by } -40. \\
\underline{40x + 60y = 440} & \text{Multiply both sides of Equation 2 by } 100. \\
20y = 120 & \\
y = 6 & \text{Divide both sides by 20.}
\end{array}
$$

To find x, we substitute 6 for y in Equation 1 and simplify.

$$
\begin{array}{ll}
x + y = 8 & \\
x + 6 = 8 & \\
x = 2 & \text{Subtract 6 from both sides.}
\end{array}
$$

The technician should use 2 liters of the 40% solution and 6 liters of the 60% solution. Check the result. ■

ORALS *If x and y are integers, express each quantity.*

1. Twice x

2. One more than y

3. The sum of twice x and three times y

If a book costs $x and a calculator costs $y, find

4. The cost of 3 books and 2 calculators

5. The cost of 4 books and 5 calculators

EXERCISE 7.4

Use two equations in two variables to solve each problem.

1. Integer problem One integer is twice another, and their sum is 96. Find the integers.

2. Integer problem The sum of two integers is 38, and their difference is 12. Find the integers.

3. Integer problem Three times one integer plus another integer is 29. If the first integer plus twice the second is 18, find the integers.

4. Integer problem Twice one integer plus another integer is 21. If the first integer plus 3 times the second is 33, find the integers.

5. Raising livestock A rancher raises five times as many cows as horses. If he has 168 animals, how many head of cattle does he have?

6. **Grass seed mixture** A landscaper used 100 pounds of grass seed containing twice as much bluegrass as rye. She adds 15 more pounds of bluegrass to the mixture before seeding a lawn. How many pounds of bluegrass did she use?

7. **Buying painting supplies** Two partial receipts for paint supplies appear in Illustration 1. How much does each gallon of paint and each brush cost?

ILLUSTRATION 1

8. **Buying baseball equipment** One catcher's mitt and ten outfielder's gloves cost $239.50. How much does each cost if one catcher's mitt and five outfielder's gloves cost $134.50?

9. **Buying contact lens cleaner** Two bottles of contact lens cleaner and three bottles of soaking solution cost $29.40, and three bottles of cleaner and two bottles of soaking solution cost $28.60. Find the cost of each.

10. **Buying clothing** Two pairs of shoes and four pairs of socks cost $109, and three pairs of shoes and five pairs of socks cost $160. Find the cost of a pair of socks.

11. **Cutting pipe** A plumber wants to cut the pipe shown in Illustration 2 into two pieces so that one piece is 5 feet longer than the other. How long should each piece be?

ILLUSTRATION 2

12. **Cutting lumber** A carpenter wants to cut a 20-foot board into two pieces so that one piece is 4 times as long as the other. How long should each piece be?

13. **Splitting the lottery** Maria and Susan pool their resources to buy several lottery tickets. They agree that Susan should get $50,000 more than Maria, because she gave most of the money. If they win $250,000, how much will Susan get?

14. **Figuring inheritances** According to a will, a man left his older son $10,000 more than twice as much as he left his younger son. If the estate is worth $497,500, how much will the younger son get?

15. **Geometry** The perimeter of the rectangle shown in Illustration 3 is 110 feet. Find its dimensions.

ILLUSTRATION 3

16. **Geometry** A rectangle is 3 times as long as it is wide, and its perimeter is 80 centimeters. Find its dimensions.

17. **Geometry** A rectangle has a length that is 2 feet more than twice its width. If its perimeter is 34 feet, find its area.

18. **Geometry** A 50-meter path surrounds the rectangular garden shown in Illustration 4. The width of the garden is two-thirds its length. Find its area.

ILLUSTRATION 4

19. **Choosing a furnace** A high-efficiency 90+ furnace costs $2250 and costs an average of $412 per year to operate in Rockford, Ill. An 80+ furnace costs only $1715 but costs $466 per year to operate. Find the break point.

20. **Making tires**   A company has two molds to form tires. One mold has a setup cost of $600 and the other a setup cost of $1100. The cost to make each tire on the first machine is $15, and the cost to make each tire on the second machine is $13. Find the break point.

21. **Choosing a furnace**   If you intended to live in a house for seven years, which furnace would you choose in Exercise 19?

22. **Making tires**   If you planned a production run of 500 tires, which mold would you use in Exercise 20?

23. **Investing money**   Bill invested some money at 5% annual interest, and Janette invested some at 7%. If their combined interest was $310 on a total investment of $5000, how much did Bill invest?

24. **Investing money**   Peter invested some money at 6% annual interest, and Martha invested some at 12%. If their combined investment was $6000 and their combined interest was $540, how much money did Martha invest?

25. **Buying tickets**   Students can buy tickets to a basketball game for $1. However, the admission for nonstudents is $2. If 350 tickets are sold and the total receipts are $450, how many student tickets were sold?

26. **Buying tickets**   If receipts for the movie advertised in Illustration 5 were $720 for an audience of 190 people, how many senior citizens attended?

ILLUSTRATION 5

27. **Boating**   A boat can travel 24 miles downstream in 2 hours and can make the return trip in 3 hours. Find the speed of the boat in still water.

28. **Aviation**   With the wind, a plane can fly 3000 miles in 5 hours. Against the same wind, the trip takes 6 hours. Find the airspeed of the plane (the speed in still air).

29. **Aviation**   An airplane can fly downwind a distance of 600 miles in 2 hours. However, the return trip against the same wind takes 3 hours. Find the speed of the wind.

30. **Finding the speed of a current**   It takes a motorboat 4 hours to travel 56 miles down a river, and takes 3 hours longer to make the return trip. Find the speed of the current.

31. **Mixing chemicals**   A chemist has one solution that is 40% alcohol and another that is 55% alcohol. How much of each must she use to make 15 liters of a solution that is 50% alcohol?

32. **Mixing pharmaceuticals**   A nurse has a solution that is 25% alcohol and another that is 50% alcohol. How much of each must he use to make 20 liters of a solution that is 40% alcohol?

33. **Mixing nuts**   A merchant wants to mix the peanuts with the cashews shown in Illustration 6 to get 48 pounds of mixed nuts to sell at $4 per pound. How many pounds of each should the merchant use?

ILLUSTRATION 6

34. **Mixing peanuts and candy**   A merchant wants to mix peanuts worth $3 per pound with jelly beans worth $1.50 per pound to make 30 pounds of a mixture worth $2.10 per pound. How many pounds of each should he use?

35. Markdown A set of golf clubs has been marked down 40% to a sale price of $384. Let r represent the retail price and d represent the discount. Then use the following equations to find the original retail price.

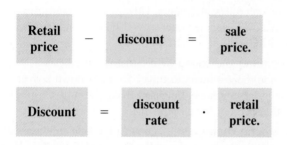

36. Markup A stereo system retailing at $565.50 has been marked up 45% from wholesale. Let w represent the wholesale cost and m represent the markup. Then use the following equations to find the wholesale cost.

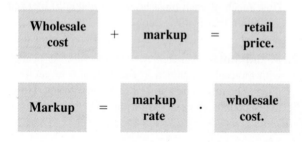

37. Selling radios An electronics store put two types of car radio on sale. One model sold for $87, and the other sold for $119. During the sale, the receipts for the 25 radios sold were $2495. How many cheap radios were sold?

38. Selling ice cream At a store, ice cream cones cost $.90 and sundaes cost $1.65. One day the receipts for a total of 148 cones and sundaes were $180.45. How many cones were sold?

39. Investing money An investment of $950 at one rate of interest and $1200 at another rate together generate an annual income of $205.50. If the investment rates differ by 1%, find the lower rate. (*Hint:* Treat 1% as .01).

40. Motion problem A man drives for a while at 45 miles per hour. Realizing that he is running late, he increases his speed to 60 miles per hour and completes his 405-mile trip in 8 hours. How long did he drive at 45 miles per hour?

Writing Exercises ■ *Write a paragraph using your own words.*

1. Which problem in the preceding set did you find the hardest? Why?

2. Which problem in the preceding set did you find the easiest? Why?

Something to Think About ■

1. Work Exercise 11 using one variable. Do you prefer using one or two variables?

2. See Illustration 7 (pages 446 and 447). How many nails will balance one nut?

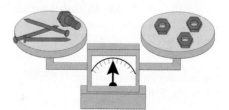

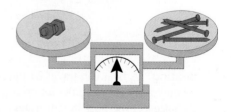

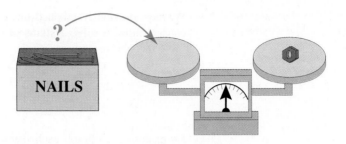

ILLUSTRATION 7

Review Exercises ■ *Graph each inequality.*

1. $2x - y \leq 4$

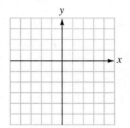

2. $x + 3y > -2$

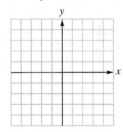

3. $2x + 3y > 0$

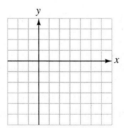

4. $\frac{1}{3}x + 2y \geq 1$

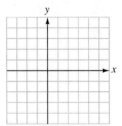

7.5 Solving Systems of Linear Inequalities

■ An Application of Systems of Linear Inequalities

GETTING READY *Solve each inequality.*

1. $2x < 12$

2. $-2x > 6$

3. $3x + 4 \geq 13$

4. $-3x - 2 \leq 13$

We have previously graphed linear inequalities containing two variables. We now consider how to solve **systems of linear inequalities** containing two variables. To solve the system

$$\begin{cases} x + y \geq 1 \\ x - y \geq 1 \end{cases}$$

for example, we graph each inequality on a set of coordinate axes, as in Figure 7-11.

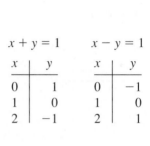

$x + y = 1$			$x - y = 1$	
x	y		x	y
0	1		0	−1
1	0		1	0
2	−1		2	1

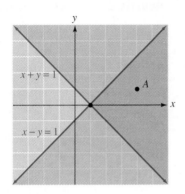

FIGURE 7-11

The graph of $x + y \geq 1$ includes the graph of the equation $x + y = 1$ and all points above it. Because the boundary line is included, we draw it with a solid line. The graph of the inequality $x - y \geq 1$ includes the graph of the equation $x - y = 1$ and all points below it. Because the boundary line is included, we draw it with a solid line.

The area that is shaded twice represents the set of simultaneous solutions of the given system of inequalities. Any point in the doubly shaded region has coordinates that satisfy both of the inequalities of the system.

To see that this is true, we can pick a point, such as point A, that lies in the doubly shaded region and show that its coordinates satisfy both inequalities. Because point A has coordinates $(4, 1)$, we have

$$
\begin{array}{ccc}
x + y \geq 1 & \text{and} & x - y \geq 1 \\
4 + 1 \geq 1 & & 4 - 1 \geq 1 \\
5 \geq 1 & & 3 \geq 1
\end{array}
$$

Since the coordinates of point A satisfy each equation, point A is a solution of the system. If we pick a point that is not in the doubly shaded region, its coordinates will not satisfy both of the inequalities.

In general, to solve systems of linear inequalities, we will do the following.

Solving Systems of Inequalities	1. Graph each inequality in the system on the same coordinate axes.
	2. Find the region that is common to every graph.
	3. Pick a test point from the region to verify the solution.

EXAMPLE 1 Graph the solution set of the system $\begin{cases} 2x + y < 4 \\ -2x + y > 2 \end{cases}$.

Solution We graph each inequality on the same set of coordinate axes, as in Figure 7-12.

$$2x + y = 4 \qquad -2x + y = 2$$

x	y		x	y
0	4		-1	0
1	2		0	2
2	0		2	6

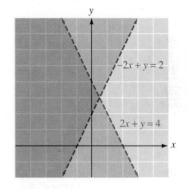

FIGURE 7-12

We then find the graph of each inequality.

- The graph of $2x + y < 4$ includes all points below the line $2x + y = 4$. Since the boundary is not included, we draw it as a broken line.
- The graph of $-2x + y > 2$ includes all points above the line $-2x + y = 2$. Since the boundary is not included, we draw it as a broken line.

The area that is shaded twice represents the set of simultaneous solutions of the given system of inequalities. Any point in the doubly shaded region has coordinates that will satisfy both inequalities of the system.

Pick a point in the doubly shaded region and show that it satisfies both inequalities. ■

EXAMPLE 2 Graph the solution set of the system $\begin{cases} x \le 2 \\ y > 3 \end{cases}$.

Solution We graph each inequality on the same set of coordinate axes, as in Figure 7-13.

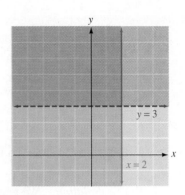

$x = 2$		$y = 3$	
x	y	x	y
2	0	0	3
2	2	1	3
2	4	4	3

FIGURE 7-13

We then find the graph of each inequality.

- The graph of $x \leq 2$ includes all points on the line $x = 2$ and all points to the left of the line. Since the boundary line is included, we draw it as a solid line.
- The graph $y > 3$ includes all points above the line $y = 3$. Since the boundary is not included, we draw it as a broken line.

The area that is shaded twice represents the set of simultaneous solutions of the given system of inequalities. Any point in the doubly shaded region has coordinates that will satisfy both inequalities of the system. Pick a point in the doubly shaded region and show that this is true. ∎

EXAMPLE 3   Graph the solution set of the system $\begin{cases} y < 3x - 1 \\ y \geq 3x + 1 \end{cases}$.

Solution   We graph each inequality, as in Figure 7-14.

- The graph of $y < 3x - 1$ includes all of the points below the broken line $y = 3x - 1$.
- The graph of $y \geq 3x + 1$ includes all of the points on and above the solid line $y = 3x + 1$.

Because the graphs of these inequalities do not intersect, the solution set is empty. There are no solutions.

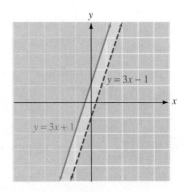

FIGURE 7-14   ∎

EXAMPLE 4 Graph the solution set of the system $\begin{cases} x \geq 0 \\ y \geq 0 \\ x + 2y \leq 6 \end{cases}$.

Solution We graph each inequality, as in Figure 7-15.

- The graph of $x \geq 0$ includes all of the points on the y-axis and to the right.
- The graph of $y \geq 0$ includes all of the points on the x-axis and above.
- The graph of $x + 2y \leq 6$ includes all of the points on the line $x + 2y = 6$ and below.

The solution is the region that is shaded three times. This includes triangle OPQ and the triangular region it encloses.

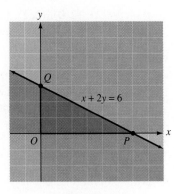

FIGURE 7-15   ■

■ An Application of Systems of Linear Inequalities

EXAMPLE 5 **Landscaping** A homeowner budgets from $300 to $600 for trees and bushes to landscape his yard. After shopping around, he finds that good trees cost $150 and mature bushes cost $75. What combinations of trees and bushes can he afford to buy?

Solution We can let

> x represent the number of trees purchased and
> y represent the number of bushes purchased.

We can then form the following system of inequalities:

The cost of a tree	·	the number of trees purchased	+	the cost of a bush	·	the number of bushes purchased	≥	$300.
$150	·	x	+	$75	·	y	≥	$300

The cost of a tree	·	the number of trees purchased	+	the cost of a bush	·	the number of bushes purchased	≤	**$600.**
$150	·	x	+	$75	·	y	≤	$600

We graph the system $\begin{cases} 150x + 75y \geq 300 \\ 150x + 75y \leq 600 \end{cases}$ as in Figure 7-16. The coordinates of each point shown in the graph gives a possible combination of trees (x) and bushes (y) that can be purchased while remaining within budget. These possibilities are

$(0, 4), (0, 5), (0, 6), (0, 7), (0, 8)$

$(1, 2), (1, 3), (1, 4), (1, 5), (1, 6)$

$(2, 0), (2, 1), (2, 2), (2, 3), (2, 4)$

$(3, 0), (3, 1), (3, 2)$

$(4, 0)$

Only these points (with integer coordinates) are possible solutions, since the homeowner cannot buy a portion of a tree or bush.

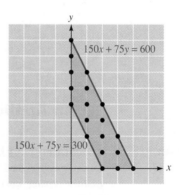

FIGURE 7-16   ■

ORALS   *Tell whether each point is a solution of the inequality* $3x + 2y < 12$.

1. $(0, 0)$ **2.** $(2, 3)$

3. $(-2, 4)$ **4.** $(3, -2)$

Tell whether each point is a solution of the inequality $3x - 2y \geq 6$.

5. $(0, 0)$ **6.** $(2, 0)$

E X E R C I S E   7 . 5

In Exercises 1–26, find the solution set of each system of inequalities, when possible.

1. $\begin{cases} x + 2y \le 3 \\ 2x - y \ge 1 \end{cases}$

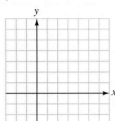

2. $\begin{cases} 2x + y \ge 3 \\ x - 2y \le -1 \end{cases}$

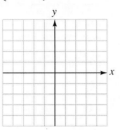

3. $\begin{cases} x + y < -1 \\ x - y > -1 \end{cases}$

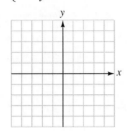

4. $\begin{cases} x + y > 2 \\ x - y < -2 \end{cases}$

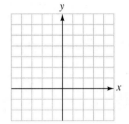

5. $\begin{cases} 2x - y < 4 \\ x + y \ge -1 \end{cases}$

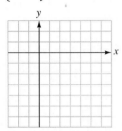

6. $\begin{cases} x - y \ge 5 \\ x + 2y < -4 \end{cases}$

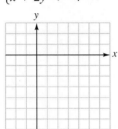

7. $\begin{cases} x > 2 \\ y \le 3 \end{cases}$

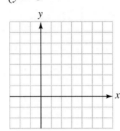

8. $\begin{cases} x \ge -1 \\ y > -2 \end{cases}$

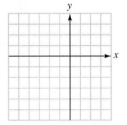

9. $\begin{cases} 2x - 3y < 0 \\ y > x - 1 \end{cases}$

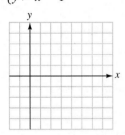

10. $\begin{cases} 3x - y \ge -1 \\ y \ge 3x + 1 \end{cases}$

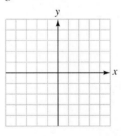

11. $\begin{cases} x + y < 1 \\ x + y > 3 \end{cases}$

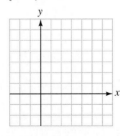

12. $\begin{cases} x + y > 2 \\ x + y < 4 \end{cases}$

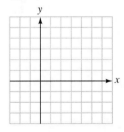

13. $\begin{cases} x > 0 \\ y > 0 \end{cases}$

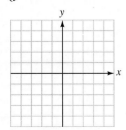

14. $\begin{cases} x \le 0 \\ y < 0 \end{cases}$

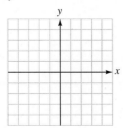

15. $\begin{cases} 3x + 4y > -7 \\ 2x - 3y \ge 1 \end{cases}$

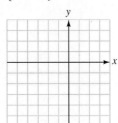

16. $\begin{cases} 3x + y \le 1 \\ 4x - y > -8 \end{cases}$

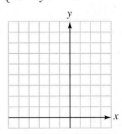

17. $\begin{cases} x < 3y - 1 \\ y \geq 2x - 3 \end{cases}$

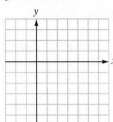

18. $\begin{cases} y \geq x + 2 \\ x \leq y - 2 \end{cases}$

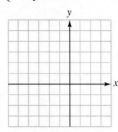

19. $\begin{cases} 2x + y < 7 \\ y > 2(1 - x) \end{cases}$

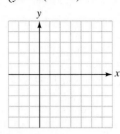

20. $\begin{cases} 2x + y \geq 6 \\ y \leq 2(2x - 3) \end{cases}$

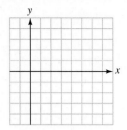

21. $\begin{cases} 2x - 4y > -6 \\ 3x + y \geq 5 \end{cases}$

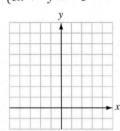

22. $\begin{cases} 2x - 3y < 0 \\ 2x + 3y \geq 12 \end{cases}$

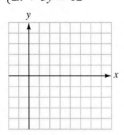

23. $\begin{cases} 3x - y \leq -4 \\ 3y > -2(x + 5) \end{cases}$

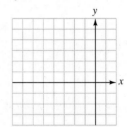

24. $\begin{cases} 3x + y < -2 \\ y > 3(1 - x) \end{cases}$

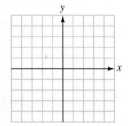

25. $\begin{cases} \dfrac{x}{2} + \dfrac{y}{3} \geq 2 \\ \dfrac{x}{2} - \dfrac{y}{2} < -1 \end{cases}$

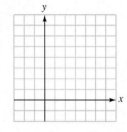

26. $\begin{cases} \dfrac{x}{3} - \dfrac{y}{2} < -3 \\ \dfrac{x}{3} + \dfrac{y}{2} > -1 \end{cases}$

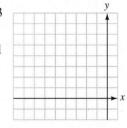

In Exercises 27–30, use the graphing method to find the region that satisfies all of the inequalities of the system.

27. $\begin{cases} x \geq 0 \\ y \geq 0 \\ x + y \leq 3 \end{cases}$

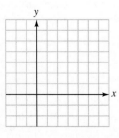

28. $\begin{cases} x - y \leq 6 \\ x + 2y \leq 6 \\ x \geq 0 \end{cases}$

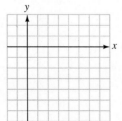

29. $\begin{cases} x \geq 0 \\ y \geq 0 \\ x \leq 5 \\ y \leq x \end{cases}$

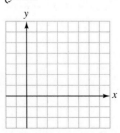

30. $\begin{cases} x \geq 0 \\ y \geq 0 \\ y \leq 2 + x \\ y \geq 4x - 2 \end{cases}$

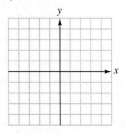

In Exercises 31–34, graph each system of inequalities and give two possible solutions to each problem.

31. Buying compact disks   Melodic Music has compact disks on sale for either $10 or $15. If a customer wants to spend at least $30 but no more than $60 on CDs, find a system of inequalities whose graph will show the possible combinations of $10 CDs (*x*) and $15 CDs (*y*) that the customer can buy.

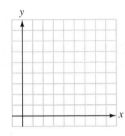

33. Buying furniture   A distributor wholesales desk chairs for $150 and side chairs for $100. Best Furniture wants to order no more than $900 worth of chairs and wants to order more side chairs than desk chairs. Find a system of inequalities whose graph will show the possible combinations of desk chairs (*x*) and side chairs (*y*) that can be ordered.

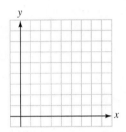

32. Buying boats   Dry Boatworks wholesales aluminum boats for $800 and fiberglass boats for $600. Northland Marina wants to order at least $2400 but no more than $4800 worth of boats. Find a system of inequalities whose graph will show the possible combination of aluminum boats (*x*) and fiberglass boats (*y*) that can be ordered.

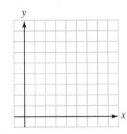

34. Ordering furnace equipment   J. Bolden Heating Company wants to order no more than $2000 worth of electronic air cleaners and humidifiers from a wholesaler that charges $500 for air cleaners and $200 for humidifiers. If Bolden wants more humidifiers than air cleaners, find a system of inequalities whose graph will show the possible combinations of air cleaners (*x*) and humidifiers (*y*) that can be ordered.

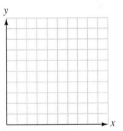

Writing Exercises ■ *Write a paragraph using your own words.*

1. Explain how to use graphing to solve a system of inequalities.

2. Explain when a system of inequalities will have no solutions.

Something to Think About ■

1. What are the limitations of the graphing method for solving inequalities?

2. Solve the system $\begin{cases} y < x^2 \\ y > x - 2 \end{cases}$.

Review Exercises ■ *Simplify each expression.*

1. $(z^3 z^{-2})^{-3}$

2. $\dfrac{t^2 t^{-1}}{t^3}$

3. $\left(\dfrac{3m^2}{n^3}\right)^4$

4. $\left(\dfrac{y^4}{y^{-2}}\right)^3$

5. $\dfrac{\dfrac{x^2}{y^3 z}}{\dfrac{x}{yz^2}}$

6. $\dfrac{\dfrac{x}{y} + \dfrac{y}{x}}{\dfrac{y}{x} - \dfrac{x}{y}}$

7.6 Solving Systems of Three Equations in Three Variables

■ Problem Solving

GETTING READY *Tell whether each point (x, y, z) is a solution of $x + 2y - z = 2$.*

1. $(0, 0, 0)$ **2.** $(1, 1, 1)$ **3.** $(1, 0, 1)$ **4.** $(0, 1, 1)$

Add the left-hand and right-hand sides of each equation to get a new equation.

5. $3x - 3y + 2z = 5$
    $2x + 3y - 3z = 7$

6. $2x - 4y + 3z = 10$
    $3x + 8y - 3z = 12$

We have seen that a solution to the system of equations

$$\begin{cases} 2x + 3y = 13 \\ 3x + 2y = 12 \end{cases}$$

is an ordered pair of real numbers (x, y) that satisfies both of the given equations simultaneously. Likewise, a solution to the system

$$\begin{cases} 2x + 3y + 4z = 9 \\ 3x + 2y + 7z = 12 \\ 5x + 12y + z = 9 \end{cases}$$

is an ordered triple of numbers (x, y, z) that satisfies each of the three given equations simultaneously.

A linear equation in two variables has a graph that is a straight line. A system of two linear equations in two variables is consistent or inconsistent, depending on whether a pair of lines intersect or are parallel.

The graph of an equation in three variables of the form $ax + by + cz = j$ is a flat surface called a **plane.** A system of three equations in three variables is consistent or inconsistent, depending on how the three planes corresponding to the three equations intersect. The drawings in Figure 7-17 illustrate some of the possibilities.

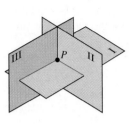

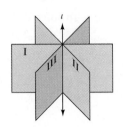

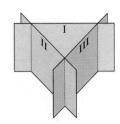

The three planes intersect at a single point P: One solution

The three planes have a line l in common: An infinite number of solutions

The three planes have no point in common: No solutions

(a)

(b)

(c)

FIGURE 7-17

Example 1 discusses a consistent system of three equations in three variables. Example 2 discusses a system that is inconsistent.

EXAMPLE 1 Solve the system $\begin{cases} 2x + y + 4z = 12 \\ x + 2y + 2z = 9 \\ 3x - 3y - 2z = 1 \end{cases}$.

Solution   We are given the following system of equations in three variables:

1. $\begin{cases} 2x + y + 4z = 12 \\ \ \ x + 2y + 2z = 9 \\ 3x - 3y - 2z = 1 \end{cases}$
2.
3.

We can use the addition method to eliminate the variable z and thereby obtain a system of two equations in two variables. If Equations 2 and 3 are added, the variable z is eliminated:

2. $x + 2y + 2z = 9$
3. $\dfrac{3x - 3y - 2z = 1}{}$
4. $4x - y \qquad = 10$

We now pick a different pair of equations and eliminate the variable z again. If each side of Equation 3 is multiplied by 2 and the resulting equation is added to Equation 1, the variable z is eliminated again:

1. $2x + y + 4z = 12$
    $\dfrac{6x - 6y - 4z = 2}{}$
5. $8x - 5y \qquad = 14$

Equations 4 and 5 form a system of two equations in two variables:

4. $\begin{cases} 4x - y = 10 \\ 8x - 5y = 14 \end{cases}$
5.

To solve this system, we multiply Equation 4 by -5, add the resulting equation to Equation 5 to eliminate the variable y, and solve for x:

$$-20x + 5y = -50$$
5. $\underline{\quad 8x - 5y = \quad 14}$
6. $\quad -12x \qquad = -36$
$$x = 3 \qquad \text{Divide both sides by } -12.$$

To find y, we substitute 3 for x in an equation containing the variables x and y, such as Equation 5, and solve for y:

5. $\quad 8x - 5y = 14$
$$8(3) - 5y = 14$$
$$24 - 5y = 14 \qquad \text{Simplify.}$$
$$-5y = -10 \qquad \text{Subtract 24 from both sides.}$$
$$y = 2 \qquad \text{Divide both sides by } -5.$$

To find z, we substitute 3 for x and 2 for y in an equation that contains the variables x, y, and z, such as Equation 1, and solve for z:

1. $\quad 2x + y + 4z = 12$
$$2(3) + 2 + 4z = 12$$
$$8 + 4z = 12 \qquad \text{Simplify.}$$
$$4z = 4 \qquad \text{Subtract 8 from both sides.}$$
$$z = 1 \qquad \text{Divide both sides by 4.}$$

Verify that the solution $(3, 2, 1)$ satisfies each equation in the system. ■

We can follow these steps to solve a system of three equations, each containing three variables.

Solving Three Equations with Three Variables	1. Pick any pair of equations and eliminate a variable to produce an equation with two variables.
	2. Pick a different pair of equations and eliminate the same variable as in Step 1. This produces a second equation with the same variables as the equation found in Step 1.
	3. Solve the system containing the two equations with two variables found in Steps 1 and 2.
	4. To find the value of the third variable, substitute the solutions found in Step 3 into any of the equations of the original system. The triple of numbers found is the solution of the system.
	5. Check the solution in all three equations of the original system.

EXAMPLE 2 Solve the system $\begin{cases} 2x + y - 3z = -3 \\ 3x - 2y + 4z = 2. \\ 4x + 2y - 6z = -7 \end{cases}$

Solution We are given the following system of equations in three variables:

1. $\begin{cases} 2x + y - 3z = -3 \\ 3x - 2y + 4z = 2 \\ 4x + 2y - 6z = -7 \end{cases}$
2.
3.

We begin by multiplying Equation 1 by 2 and adding the resulting equation to Equation 2 to eliminate the variable y:

$$4x + 2y - 6z = -6$$
2. $3x - 2y + 4z = 2$
4. $\overline{7x - 2z = -4}$

We now add Equations 2 and 3 to eliminate the variable y again:

2. $3x - 2y + 4z = 2$
3. $4x + 2y - 6z = -7$
5. $\overline{7x - 2z = -5}$

Equations 4 and 5 form the system

4. $\begin{cases} 7x - 2z = -4 \\ 7x - 2z = -5 \end{cases}$
5.

Because no values of x and z can cause $7x - 2z$ to equal both -4 and -5 at the same time, this system must be inconsistent. The original system has no solutions, either; it is inconsistent. ■

■ Problem Solving

EXAMPLE 3 **Passengers on an airline** To attract more customers, an airline added an economy class to its usual first and business class services. On one recent flight with 200 passengers, 13 more passengers flew economy class than business class, and 59 fewer flew first class than business class. How many passengers flew in each class?

Solution We can let f, b, and e represent the number of passengers in first class, business class, and economy class, respectively. Because their sum is 200, we have

1. $f + b + e = 200$

Because those who flew economy class exceeded those who flew business class by 13, we know that $e = b + 13$, or

2. $-b + e = 13$

Because those who flew business class exceeded those who flew first class by 59, we know that $f + 59 = b$, or

3. $f - b = -59$

Equations 1, 2, and 3 form a system of three equations in three variables, which can be solved as follows:

1.
2.
3.
$$\begin{cases} f + b + e = 200 \\ -b + e = 13 \\ f - b = -59 \end{cases}$$

We add Equations 1 and 3 to get Equation 4:

4. $2f + e = 141$

We add Equations 1 and 2 to get Equation 5:

5. $f + 2e = 213$

Equations 4 and 5 form a system of two equations in two variables:

4.
5.
$$\begin{cases} 2f + e = 141 \\ f + 2e = 213 \end{cases}$$

We multiply Equation 5 by -2 and add it to Equation 4 to get

$$-3e = -285$$
$$e = 95$$

To find f, we substitute 95 for e in Equation 5.

5.
$$\begin{aligned} f + 2e &= 213 \\ f + 2(\mathbf{95}) &= 213 \\ f + 190 &= 213 &&\text{Simplify.} \\ f &= 23 &&\text{Subtract 190 from both sides.} \end{aligned}$$

Finally, we substitute 23 for f in Equation 3 to find b:

3.
$$\begin{aligned} f - b &= -59 \\ \mathbf{23} - b &= -59 \\ -b &= -82 &&\text{Subtract 23 from both sides.} \\ b &= 82 &&\text{Divide both sides by } -1. \end{aligned}$$

Check the solution: 23 passengers flew first class, 82 flew business class, and 95 flew economy class. ∎

ORALS *If $x = 1$ and $y = 2$, find z.*

1. $x + y + z = 5$ **2.** $x - y + z = 7$

3. $2x + y + z = 4$ **4.** $3x + 2y + z = -2$

EXERCISE 7.6

In Exercises 1–12, solve each system of equations.

1. $\begin{cases} x + y + z = 4 \\ 2x + y - z = 1 \\ 2x - 3y + z = 1 \end{cases}$

2. $\begin{cases} x + y + z = 4 \\ x - y + z = 2 \\ x - y - z = 0 \end{cases}$

3. $\begin{cases} 2x + 2y + 3z = 10 \\ 3x + y - z = 0 \\ x + y + 2z = 6 \end{cases}$

4. $\begin{cases} x - y + z = 4 \\ x + 2y - z = -1 \\ x + y - 3z = -2 \end{cases}$

5. $\begin{cases} x + y + 2z = 7 \\ x + 2y + z = 8 \\ 2x + y + z = 9 \end{cases}$

6. $\begin{cases} x + 2y + 2z = 10 \\ 2x + y + 2z = 9 \\ 2x + 2y + z = 11 \end{cases}$

7. $\begin{cases} 2x + y - z = 1 \\ x + 2y + 2z = 2 \\ 4x + 5y + 3z = 3 \end{cases}$

8. $\begin{cases} 4x + 3z = 4 \\ 2y - 6z = -1 \\ 8x + 4y + 3z = 9 \end{cases}$

9. $\begin{cases} 2x + 3y + 4z = 6 \\ 2x - 3y - 4z = -4 \\ 4x + 6y + 8z = 12 \end{cases}$

10. $\begin{cases} x - 3y + 4z = 2 \\ 2x + y + 2z = 3 \\ 4x - 5y + 10z = 7 \end{cases}$

11. $\begin{cases} x + \dfrac{1}{3}y + z = 13 \\ \dfrac{1}{2}x - y + \dfrac{1}{3}z = -2 \\ x + \dfrac{1}{2}y - \dfrac{1}{3}z = 2 \end{cases}$

12. $\begin{cases} x - \dfrac{1}{5}y - z = 9 \\ \dfrac{1}{4}x + \dfrac{1}{5}y - \dfrac{1}{2}z = 5 \\ 2x + y + \dfrac{1}{6}z = 12 \end{cases}$

In Exercises 13–22, solve each problem.

13. **Integer problem** The sum of three numbers is 18. If the third number is four times the second, and the second number is 6 more than the first, find the numbers.

14. **Integer problem** The sum of three numbers is 48. If the first number is doubled, the sum is 60. If the second number is doubled, the sum is 63. Find the numbers.

15. **Integer problem** Three numbers have a sum of 30. The third number is 8 less than the sum of the first and second, and the second number is one-half the sum of the first and third. Find the numbers.

16. **Angles in a triangle** The sum of the three angles in any triangle is 180°. In the triangular brace shown in Illustration 1, angle A is 100° less than the sum of angles B and C, and angle C is 40° less than twice angle B. Find each angle.

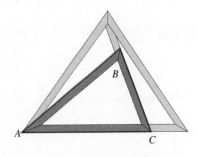

ILLUSTRATION 1

17. Making computer chips It cost an electronics company $20,000 to produce three types of computer chip. The chips cost $2, $3, and $4 to produce and were sold for $4, $5, and $6, respectively. If the company produced a total of 6000 chips and received $32,000 for their sale, how many chips of each type were produced?

18. Nutrition Refer to Table 1. How many units of each food source will be needed to provide exactly 11 grams of fat, 6 grams of carbohydrate, and 10 grams of protein?

Food source	Grams supplied by each unit		
	Fat	Carbohydrates	Protein
1	1	1	2
2	2	1	1
3	2	1	2

TABLE 1

19. Manufacturing footballs A factory makes three types of footballs at a monthly cost of $2425 for 1125 footballs. The unit costs are shown in Table 2. How many of each type are made if the monthly profit is $9275? (*Hint:* Profit = income − cost.)

Football type	Unit manufacturing cost	Unit selling price
Model A	$4	$16
Model B	$3	$12
Model C	$2	$10

TABLE 2

20. Retailing A retailer purchased 105 radios from sources A, B, and C. Five fewer units were purchased from C than from A and B combined. If twice as many had been purchased from A, the total would have been 130. Find the number purchased from each source.

21. Selling tickets Tickets for a concert cost $5, $3, and $2. Twice as many $5 tickets were sold as $2 tickets. The receipts for 750 tickets were $2625. How many of each price ticket were sold?

22. Mixing nuts The owner of a candy store wants to mix some peanuts worth $3 per pound, some cashews worth $9 per pound, and some Brazil nuts worth $9 per pound to get 50 pounds of a mixture that will sell for $6 per pound. She used 15 fewer pounds of cashews than peanuts. How many pounds of each did she use?

Writing Exercises ■ *Write a paragraph using your own words.*

1. Explain how to solve a system of three linear equations in three variables.

2. When can you tell whether a system of three equations in three variables is inconsistent?

Something to Think About ■

1. When will the equations of a system of three equations in three variables be dependent?

2. How would you construct an inconsistent system of three equations in three variables?

Review Exercises ■

1. A line with slope 3 and y-intercept $(0, -5)$ passes through the point $(a, 1)$. Find a.

2. A lines passes through the points $(2, 3)$, $(5, y)$, and $(-7, 3)$. Find y.

3. Find the slope of the line that passes through the points $(-3, 5)$ and $(3, -5)$.

4. Find the y-intercept of the line that passes through $(-3, 5)$ and $(3, -5)$.

MATHEMATICS IN ECONOMICS

In the problem discussed at the beginning of the chapter, demand will equal supply at the point where the graphs of the demand equation and the supply equation intersect. We can find this point by solving the following system.

$$\begin{cases} p = -\dfrac{1}{2}q + 1300 \\ p = \dfrac{1}{3}q + \dfrac{1400}{3} \end{cases}$$

The demand equation.

The supply equation.

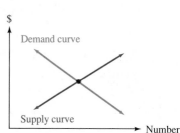

$

If we substitute the expression for p in the first equation for p in the second equation, we have

$$-\frac{1}{2}q + 1300 = \frac{1}{3}q + \frac{1400}{3}$$

$$-3q + 7800 = 2q + 2800 \qquad \text{Multiply both sides by 6.}$$

$$5000 = 5q \qquad \text{Subtract 2800 from both sides and add } 3q \text{ to both sides.}$$

$$1000 = q \qquad \text{Divide both sides by 5.}$$

The quantity q that will determine the equilibrium price is 1000 canoes. If we substitute 1000 for q in either of the original equations, we get

$$p = \frac{1}{3}q + \frac{1400}{3}$$

$$p = \frac{1}{3}(1000) + \frac{1400}{3}$$

$$p = \frac{2400}{3}$$

$$p = 800$$

The equilibrium price is \$800. When 1000 canoes are manufactured each month and priced at \$800, supply will exactly equal demand.

PROJECT   ■ Missing the Point

The graphing method of solving a system of equations is not as accurate as algebraic methods, and some systems are more difficult than others to solve accurately. For example, the two lines in Figure 7-18(a) could be drawn, in error, with slopes slightly different from their true values, and the point of intersec-

(continued)

P R O J E C T   ▪   Missing the Point *(continued)*

tion would not be far from the correct location. If the lines in Figure 7-18(b) were drawn with slopes slightly different from their correct values, the point of intersection could move substantially from its correct location.

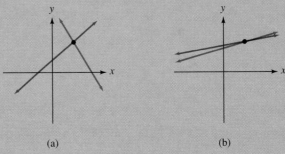

(a)   (b)

FIGURE 7-18

- Carefully solve each of these systems of equations graphically (by hand, not with a graphing calculator). Indicate your best estimate of the solution of each system.

$$\begin{cases} 2x - 4y = -7 \\ 4x + 2y = 11 \end{cases} \qquad \begin{cases} 5x - 4y = -1 \\ 12x - 10y = -3 \end{cases}$$

- Solve each system algebraically. How close were your graphical solutions to the actual solutions? Write a paragraph explaining any differences.
- Create a system of equations with the solutions $x = 3$, $y = 2$ for which an accurate solution could be obtained graphically.
- Create a system of equations with the solutions $x = 3$, $y = 2$ that is more difficult to solve accurately than the previous system, and write a paragraph explaining why.

Chapter Summary

KEY WORDS

addition method (7.3)

break point (7.4)

consistent system of equations (7.1)

dependent equations (7.1)

equilibrium price (7.1)

inconsistent system of equations (7.1)

independent equations (7.1)

plane (7.6)

simultaneous solution (7.1)

solution of a system of equations (7.1)

substitution method (7.2)

system of equations (7.1)

system of linear inequalities (7.5)

(7.1) To solve a system of equations graphically, carefully graph each equation of the system. If the lines intersect, the coordinates of the point of intersection give the solution of the system.

(7.2) To solve a system of equations by substitution, solve one of the equations of the system for one of its variables, substitute the resulting expression into the other equation, and solve. Then substitute the value obtained back into one of the original equations to solve for the other variable.

(7.3) To solve a system of equations by addition, first multiply one or both of the equations by suitable constants, if necessary, to eliminate one of the variables when the equations are added. The equation that results can be solved for its single variable. Then substitute the value obtained back into one of the original equations and solve for the other variable.

(7.4) Systems of equations are useful in solving many types of problems.

(7.5) To graph a system of inequalities, first graph the individual inequalities of the system on the same coordinate axes. The final solution, if one exists, is that region where all the individual graphs intersect.

(7.6) To solve a system of three equations in three variables, eliminate one variable from two of the equations. From another pair, eliminate the same variable. Solve the resulting system of two equations in two variables using the methods of Sections 7.2 or 7.3. Find the third value by substituting back into an original equation.

■ Chapter 7 Review Exercises

(7.1) *In Review Exercises 1–4, tell whether the ordered pair is a solution of the given system.*

1. $(1, 5)$, $\begin{cases} 3x - y = -2 \\ 2x + 3y = 17 \end{cases}$

2. $(-2, 4)$, $\begin{cases} 5x + 3y = 2 \\ -3x + 2y = 16 \end{cases}$

3. $\left(14, \dfrac{1}{2}\right)$, $\begin{cases} 2x + 4y = 30 \\ \dfrac{x}{4} - y = 3 \end{cases}$

4. $\left(\dfrac{7}{2}, -\dfrac{2}{3}\right)$, $\begin{cases} 4x - 6y = 18 \\ \dfrac{x}{3} + \dfrac{y}{2} = \dfrac{5}{6} \end{cases}$

In Review Exercises 5–8, use the graphing method to solve each system.

5. $\begin{cases} x + y = 7 \\ 2x - y = 5 \end{cases}$

6. $\begin{cases} \dfrac{x}{3} + \dfrac{y}{5} = -1 \\ x - 3y = -3 \end{cases}$

7. $\begin{cases} 3x + 6y = 6 \\ x + 2y = 2 \end{cases}$

8. $\begin{cases} 6x + 3y = 12 \\ 2x + y = 2 \end{cases}$

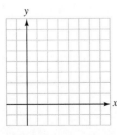

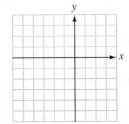

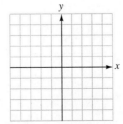

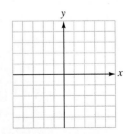

(7.2) In Review Exercises 9–12, use the substitution method to solve each system.

9. $\begin{cases} x = 3y + 5 \\ 5x - 4y = 3 \end{cases}$

10. $\begin{cases} 3x - \dfrac{2y}{5} = 2(x - 2) \\ 2x - 3 = 3 - 2y \end{cases}$

11. $\begin{cases} 8x + 5y = \ 3 \\ 5x - 8y = 13 \end{cases}$

12. $\begin{cases} 6(x + 2) = y - 1 \\ 5(y - 1) = x + 2 \end{cases}$

(7.3) In Review Exercises 13–20, use the addition method to solve each system.

13. $\begin{cases} 2x + y = \ 1 \\ 5x - y = 20 \end{cases}$

14. $\begin{cases} x + 8y = 7 \\ x - 4y = 1 \end{cases}$

15. $\begin{cases} 5x + \ y = \ 2 \\ 3x + 2y = 11 \end{cases}$

16. $\begin{cases} x + y = 3 \\ 3x = 2 - y \end{cases}$

17. $\begin{cases} 11x + 3y = 27 \\ \ 8x + 4y = 36 \end{cases}$

18. $\begin{cases} 9x + 3y = 5 \\ 3x = 4 - y \end{cases}$

19. $\begin{cases} 9x + 3y = 5 \\ 3x + \ y = \dfrac{5}{3} \end{cases}$

20. $\begin{cases} \dfrac{x}{3} + \dfrac{y + 2}{2} = 1 \\ \dfrac{x + 8}{8} + \dfrac{y - 3}{3} = 0 \end{cases}$

(7.4) In Review Exercises 21–26, use a system of equations to solve each word problem.

21. Integer problem   One number is 5 times another, and their sum is 18. Find the numbers.

22. Geometry   The length of a rectangle is 3 times its width, and its perimeter is 24 feet. Find its dimensions.

23. Buying grapefruit   A grapefruit costs 15 cents more than an orange. Together, they cost 85 cents. Find the cost of a grapefruit.

24. Utility bills   A man's electric bill for January was $23 less than his gas bill. The two utilities charged him a total of $109. Find the amount of his gas bill.

25. Buying groceries   Two gallons of milk and 3 dozen eggs cost $6.80. Three gallons of milk and 2 dozen eggs cost $7.35. How much does each gallon of milk cost?

26. Investing money   Carlos invested part of $3000 in a 10% certificate account and the rest in a 6% passbook account. The total annual interest from both accounts is $270. How much did he invest at 6%?

(7.5) In Review Exericses 27–30, solve each system of inequalities.

27. $\begin{cases} 5x + 3y < 15 \\ 3x - \ y > \ 3 \end{cases}$

28. $\begin{cases} 5x - 3y \geq 5 \\ 3x + 2y \geq 3 \end{cases}$

29. $\begin{cases} x \geq 3y \\ y < 3x \end{cases}$

30. $\begin{cases} x > 0 \\ x \leq 3 \end{cases}$

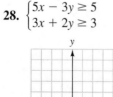

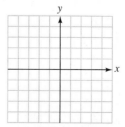

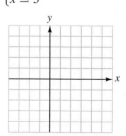

(7.6) *In Review Exercises 31–34, solve each system.*

31. $\begin{cases} x - 2y - 2z = 7 \\ 2x - y + z = -1 \\ x + 5y + 3z = -8 \end{cases}$

32. $\begin{cases} x + 2z = 4 \\ y - 3z = -4 \\ 2x + 3y + z = 2 \end{cases}$

33. $\begin{cases} x + y - z = 1 \\ 3x - y + z = 11 \\ x - 2y + 2z = 7 \end{cases}$

34. $\begin{cases} 2x - y + z = 1 \\ x + 2y + 2z = 2 \\ x - 3y + 2z = 2 \end{cases}$

■ Chapter 7 Test

In Problems 1–2, tell whether the given ordered pair is a solution of the given system.

1. $(2, -3)$, $\begin{cases} 3x - 2y = 12 \\ 2x + 3y = -5 \end{cases}$

2. $(-2, -1)$, $\begin{cases} 4x + y = -9 \\ 2x - 3y = -7 \end{cases}$

In Problems 3–4, solve each system by graphing.

3. $\begin{cases} 3x + y = 7 \\ x - 2y = 0 \end{cases}$

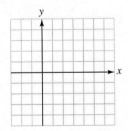

4. $\begin{cases} x + \dfrac{y}{2} = 1 \\ y = 1 - 3x \end{cases}$

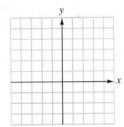

In Problems 5–6, solve each system by substitution.

5. $\begin{cases} y = x - 1 \\ 2x + y = -7 \end{cases}$

6. $\begin{cases} \dfrac{x}{6} + \dfrac{y}{10} = 3 \\ \dfrac{5x}{16} - \dfrac{3y}{16} = \dfrac{15}{8} \end{cases}$

In Problems 7–8, solve each system by addition.

7. $\begin{cases} 3x - y = 2 \\ 2x + y = 8 \end{cases}$

8. $\begin{cases} 4x + 3 = -3y \\ \dfrac{-x}{7} + \dfrac{4y}{21} = 1 \end{cases}$

In Problems 9–10, identify each system as consistent or inconsistent.

9. $\begin{cases} 2x + 3(y - 2) = 0 \\ -3y = 2(x - 4) \end{cases}$

10. $\begin{cases} \dfrac{x}{3} + y - 4 = 0 \\ -3y = x - 12 \end{cases}$

In Problems 11–12, use a system of equations in two variables to solve each problem.

11. Integer problem The sum of two numbers is -18. One number is 2 greater than 3 times the other. Find the product of the numbers.

12. Investing money A woman invested some money at 8% and some at 9%. The interest on the combined investment of $10,000 was $840. How much was invested at 9%?

In Problems 13–14, solve each system of inequalities by graphing.

13. $\begin{cases} x + y < 3 \\ x - y > 1 \end{cases}$

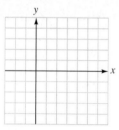

14. $\begin{cases} 2x + 3y \le 6 \\ x \ge 2 \end{cases}$

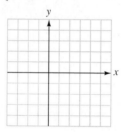

In Problems 15–16, solve each system of equations.

15. $\begin{cases} 2x + y - 3z = 5 \\ x + 2y - z = 7 \\ x + y + 5z = 4 \end{cases}$

16. $\begin{cases} 3x + 2y - 2z = 1 \\ x + 2y - 3z = 5 \\ -x + y - 3z = 7 \end{cases}$

8

Roots and Radical Expressions

MATHEMATICS IN CARPENTRY

To span the 16-foot-by-28-foot room shown in the illustration, a carpenter will use a scissors truss, with the ridge of the vaulted ceiling at the center of the room. The house plans call for the outside walls to be 8 feet high and the ridge of the room to be 12 feet high.

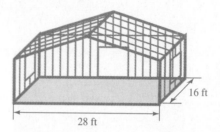

How many 4-foot-by-8-foot sheets of plaster board will be needed to drywall the entire ceiling?

After reading this chapter, you will be able to answer this question.

The product $b \cdot b$ is called the **square of b** and is usually denoted by the exponential expression b^2. For example,

- The square of 3 is 9 because $3^2 = 9$.
- The square of -3 is 9 because $(-3)^2 = 9$.
- The square of 12 is 144 because $12^2 = 144$.
- The square of -12 is 144 because $(-12)^2 = 144$.
- The square of 0 is 0 because $0^2 = 0$.

In this chapter, we will reverse the squaring process and find **square roots** of numbers. We will also discuss how to find other roots of numbers.

8.1 Radicals

■ Using a Calculator to Find Square Roots ■ Cube Roots

GETTING READY *Find each value.*

1. 3^2 **2.** 4^2 **3.** 2^3 **4.** 5^3

5. 3^4 **6.** 4^4 **7.** $(-3)^3$ **8.** $(-2)^5$

A number b is called a **square root of** a if the square of b is equal to a.

- 3 is a square root of 9 because $3^2 = 9$.
- -3 is a square root of 9 because $(-3)^2 = 9$.
- 12 is a square root of 144 because $12^2 = 144$.
- -12 is a square root of 144 because $(-12)^2 = 144$.
- 0 is a square root of 0 because $0^2 = 0$.

In general, the following is true.

Square Root of a Number   The number b is a **square root of** a if $b^2 = a$.

All positive numbers have two square roots, one that is positive and one that is negative. For example, the two square roots of 9 are 3 and -3, and the two square roots of 144 are 12 and -12. The number 0 is the only number that has a single square root, which is 0.

The symbol $\sqrt{}$, called a **radical sign,** is used to represent the *positive* square root of a number.

Principal Square Root   If $a > 0$, the **principal square root of** a, denoted by $\sqrt{a}$, is the positive square root of a.

The principal square root of 0 is 0: $\sqrt{0} = 0$.

The expression under a radical sign is called a **radicand.**

The principal square root of a positive number a is a positive number also. Although 3 and -3 are both square roots of 9, only 3 is the principal square root. The symbol $\sqrt{9}$ represents 3. To designate -3, we place a $-$ sign in front of the radical:

$$\sqrt{9} = 3 \qquad \text{and} \qquad -\sqrt{9} = -3$$

Likewise,

$$\sqrt{144} = 12 \qquad \text{and} \qquad -\sqrt{144} = -12$$

EXAMPLE 1 (c, d, f, j)

a. $\sqrt{0} = 0$ **b.** $\sqrt{1} = 1$ **c.** $\sqrt{49} = 7$ **d.** $\sqrt{121} = 11$
e. $-\sqrt{4} = -2$ **f.** $-\sqrt{81} = -9$ **g.** $\sqrt{225} = 15$ **h.** $\sqrt{169} = 13$
i. $-\sqrt{625} = -25$ **j.** $-\sqrt{900} = -30$ **k.** $\sqrt{576} = 24$ **l.** $\sqrt{1600} = 40$ ∎

Square roots of certain numbers like 7 are hard to compute by hand. However, we can easily find $\sqrt{7}$ with a calculator.

■ Using a Calculator to Find Square Roots

To find the principal square root of 7, we enter 7 into a calculator and press the $\sqrt{}$ key. The approximate value of $\sqrt{7}$ will appear on the calculator's display.

$$\sqrt{7} \approx 2.6457513 \qquad \text{Read} \approx \text{as "is approximately equal to."}$$

Numbers such as 4, 9, 16, and 49 are called **integer squares,** because each number is the square of an integer. The square root of any integer square is an integer, and therefore a **rational number:**

$$\sqrt{4} = 2, \qquad \sqrt{9} = 3, \qquad \sqrt{16} = 4, \qquad \text{and} \qquad \sqrt{49} = 7$$

The square roots of many positive integers are not rational numbers. For example, $\sqrt{7}$ is an **irrational number.** Recall that the union of the set of rational numbers and the set of irrational numbers is the set of **real numbers.**

WARNING! Square roots of negative numbers are not real numbers. For example, $\sqrt{-4}$ is nonreal, because the square of no real number is -4. The number $\sqrt{-4}$ is an example from a set of numbers called the **imaginary numbers,** discussed in Chapter 9. Remember: *The square root of a negative number is not a real number.*

In this chapter, we will assume that *all radicands under the square root symbols are either positive numbers or 0.* Thus, all square roots will be real numbers.

■ Cube Roots

The **cube root of a** is any number whose cube is a.

Cube Root   The **cube root of a** is denoted by $\sqrt[3]{a}$, and
$$\sqrt[3]{a} = b \qquad \text{if} \qquad b^3 = a$$

EXAMPLE 2   **a.** $\sqrt[3]{8} = 2$ because $2^3 = 8$

b. $\sqrt[3]{27} = 3$ because $3^3 = 27$

c. $\sqrt[3]{-8} = -2$ because $(-2)^3 = 8$

The cube root of a negative number is a negative number. It is the square root of a negative number that is imaginary.   ■

(a, b, c, d)   **EXAMPLE 3**   **a.** $\sqrt[3]{0} = 0$ **b.** $\sqrt[3]{-1} = -1$ **c.** $\sqrt[3]{64} = 4$

d. $\sqrt[3]{-64} = -4$ **e.** $\sqrt[3]{125} = 5$ **f.** $\sqrt[3]{-125} = -5$ ■

Just as there are square roots and cube roots, there are also fourth roots, fifth roots, sixth roots, and so on. In general, the following is true.

nth Root of a The **nth root of a** is denoted by $\sqrt[n]{a}$, and
$$\sqrt[n]{a} = b \quad \text{if} \quad b^n = a$$
The number n is called the **index** of the radical.
If n is an even natural number, a must be positive or 0.

In the square root symbol $\sqrt{}$, the unwritten index is understood to be 2. Thus,
$$\sqrt{a} = \sqrt[2]{a}$$

EXAMPLE 4 **a.** $\sqrt[4]{81} = 3$ because $3^4 = 81$

b. $\sqrt[5]{32} = 2$ because $2^5 = 32$

c. $\sqrt[5]{-32} = -2$ because $(-2)^5 = -32$

d. $\sqrt[4]{-81}$ is a nonreal number because no number raised to the fourth power is -81 ■

We can find square roots of many quantities that contain variables, provided that these variables represent positive numbers or 0.

EXAMPLE 5 Assume that each variable represents a positive number and find each root.

(a, b, d)

a. $\sqrt{x^2} = x$ because $x^2 = x^2$

b. $\sqrt{x^4} = x^2$ because $(x^2)^2 = x^4$

c. $\sqrt{x^4 y^2} = x^2 y$ because $(x^2 y)^2 = x^4 y^2$

d. $\sqrt[3]{x^6 y^3} = x^2 y$ because $(x^2 y)^3 = x^6 y^3$

e. $\sqrt[4]{x^{12} y^8} = x^3 y^2$ since $(x^3 y^2)^4 = x^{12} y^8$

f. $\sqrt[5]{x^{10} y^5} = x^2 y$ since $(x^2 y)^5 = x^{10} y^5$ ■

EXAMPLE 6 Assume that each variable represents a positive number and find each root.

(a, b)

a. $\sqrt{(x + 1)^2} = x + 1$ since $(x + 1)^2 = (x + 1)^2$

b. $\sqrt{x^2 + 4x + 4} = \sqrt{(x + 2)^2}$ Factor $x^2 + 4x + 4$.
$$= x + 2$$

c. $\sqrt{x^2 + 2xy + y^2} = \sqrt{(x + y)^2}$ Factor $x^2 + 2xy + y^2$.
$$= x + y$$ ■

ORALS *Find each root. Assume x > 0.*

1. $\sqrt{25}$	**2.** $\sqrt{4}$	**3.** $\sqrt{4x^2}$	**4.** $\sqrt{36x^4}$
5. $\sqrt[3]{64}$	**6.** $\sqrt[3]{27}$	**7.** $\sqrt[3]{8}$	**8.** $\sqrt[3]{125}$
9. $\sqrt[4]{1}$	**10.** $\sqrt[4]{256}$	**11.** $\sqrt[5]{32}$	**12.** $\sqrt[5]{1}$

EXERCISE 8.1

In Exercises 1–24, find each value.

1. $\sqrt{9}$	**2.** $\sqrt{16}$	**3.** $\sqrt{49}$	**4.** $\sqrt{100}$
5. $\sqrt{36}$	**6.** $\sqrt{4}$	**7.** $\sqrt{81}$	**8.** $\sqrt{121}$
9. $-\sqrt{25}$	**10.** $-\sqrt{49}$	**11.** $-\sqrt{81}$	**12.** $-\sqrt{36}$
13. $\sqrt{196}$	**14.** $\sqrt{169}$	**15.** $\sqrt{256}$	**16.** $\sqrt{225}$
17. $-\sqrt{289}$	**18.** $\sqrt{400}$	**19.** $\sqrt{10,000}$	**20.** $-\sqrt{2500}$
21. $\sqrt{324}$	**22.** $-\sqrt{625}$	**23.** $-\sqrt{3600}$	**24.** $\sqrt{1600}$

In Exercises 25–48, use a calculator to find each square root to three decimal places.

25. $\sqrt{2}$	**26.** $\sqrt{3}$	**27.** $\sqrt{5}$	**28.** $\sqrt{10}$
29. $\sqrt{6}$	**30.** $\sqrt{8}$	**31.** $\sqrt{11}$	**32.** $\sqrt{17}$
33. $\sqrt{23}$	**34.** $\sqrt{53}$	**35.** $\sqrt{95}$	**36.** $\sqrt{99}$
37. $\sqrt{6428}$	**38.** $\sqrt{4444}$	**39.** $-\sqrt{9876}$	**40.** $-\sqrt{3619}$
41. $\sqrt{21.35}$	**42.** $\sqrt{13.78}$	**43.** $\sqrt{0.3588}$	**44.** $\sqrt{0.9999}$
45. $\sqrt{0.9925}$	**46.** $\sqrt{0.12345}$	**47.** $-\sqrt{0.8372}$	**48.** $-\sqrt{0.4279}$

In Exercises 49–56, tell whether each number is rational, irrational, or imaginary.

49. $\sqrt{9}$	**50.** $\sqrt{17}$	**51.** $\sqrt{49}$	**52.** $\sqrt{-49}$
53. $-\sqrt{5}$	**54.** $\sqrt{0}$	**55.** $\sqrt{-100}$	**56.** $-\sqrt{225}$

In Exercises 57–72, find each value.

57. $\sqrt[3]{1}$	**58.** $\sqrt[3]{8}$	**59.** $\sqrt[3]{27}$	**60.** $\sqrt[3]{0}$
61. $\sqrt[3]{-8}$	**62.** $\sqrt[3]{-1}$	**63.** $\sqrt[3]{-64}$	**64.** $\sqrt[3]{-27}$
65. $\sqrt[3]{125}$	**66.** $\sqrt[3]{1000}$	**67.** $-\sqrt[3]{-1}$	**68.** $-\sqrt[3]{-27}$
69. $-\sqrt[3]{64}$	**70.** $-\sqrt[3]{343}$	**71.** $\sqrt[3]{729}$	**72.** $\sqrt[3]{512}$

In Exercises 73–80, find each value.

73. $\sqrt[4]{16}$	**74.** $\sqrt[4]{81}$	**75.** $-\sqrt[5]{32}$	**76.** $-\sqrt[5]{243}$
77. $\sqrt[6]{1}$	**78.** $\sqrt[6]{0}$	**79.** $\sqrt[5]{-32}$	**80.** $\sqrt[7]{-1}$

In Exercises 81–104, write each expression without a radical sign. All variables represent positive numbers.

81. $\sqrt{x^2y^2}$

82. $\sqrt{x^2y^4}$

83. $\sqrt{x^4z^4}$

84. $\sqrt{y^6z^8}$

85. $-\sqrt{x^4y^2}$

86. $-\sqrt{x^6y^4}$

87. $\sqrt{4z^2}$

88. $\sqrt{9t^6}$

89. $-\sqrt{9x^4y^2}$

90. $-\sqrt{16x^2y^4}$

91. $\sqrt{x^2y^2z^2}$

92. $\sqrt{x^4y^6z^8}$

93. $-\sqrt{x^2y^2z^4}$

94. $-\sqrt{a^8b^6c^2}$

95. $-\sqrt{25x^4z^{12}}$

96. $-\sqrt{100a^6b^4}$

97. $\sqrt{36z^{36}}$

98. $\sqrt{64y^{64}}$

99. $-\sqrt{16z^2}$

100. $-\sqrt{729x^8y^2}$

101. $\sqrt[3]{27y^3z^6}$

102. $\sqrt[3]{64x^3y^6z^9}$

103. $\sqrt[3]{-8p^6q^3}$

104. $\sqrt[3]{-r^{12}s^3t^6}$

In Exercises 105–114, write each expression without a radical sign. All variables represent positive numbers.

105. $\sqrt{(x+9)^2}$

106. $\sqrt{(x+6)^2}$

107. $\sqrt{(x+11)^2}$

108. $\sqrt{(x+13)^2}$

109. $\sqrt{x^2+6x+9}$

110. $\sqrt{x^2+10x+25}$

111. $\sqrt{x^2+20x+100}$

112. $\sqrt{x^2+16x+64}$

113. $\sqrt{x^2+14xy+49y^2}$

114. $\sqrt{x^2+50xy+625y^2}$

Writing Exercises ■ *Write a paragraph using your own words.*

1. Explain why a negative number cannot have a real number for its square root.

2. Explain why negative numbers always have real-number cube roots.

Something to Think About ■

1. Is $\sqrt{x^2-4x+4}=x-2$? What are the exceptions?

2. When is $\sqrt{x^2}\neq x$?

Review Exercises ■ *If $y=f(x)=2x^2-x-1$, find each value.*

1. $f(0)$

2. $f(2)$

3. $f(-2)$

4. $f(-t)$

8.2 Simplifying Radical Expressions

■ **The Multiplication Property of Radicals** ■ **Simplifying Radicals** ■ **The Division Property of Radicals** ■ **Simplifying Cube Roots**

GETTING READY *Simplify each radical.*

1. $\sqrt{100}$

2. $\sqrt{4}$

3. $\sqrt{25}$

4. $\sqrt{144}$

5. $\sqrt{9}$

6. $\sqrt{16}$

7. $\sqrt[3]{27x^3y^6}$

8. $\sqrt[3]{-8x^6y^9}$

■ The Multiplication Property of Radicals

We introduce the first of two properties of radicals with the following examples:

$$\sqrt{4 \cdot 25} = \sqrt{100} \qquad \text{and} \qquad \sqrt{4}\sqrt{25} = 2 \cdot 5$$
$$= 10 \qquad\qquad\qquad\qquad\qquad = 10$$

In each case, the answer is 10. Thus, $\sqrt{4 \cdot 25} = \sqrt{4}\sqrt{25}$. Likewise,

$$\sqrt{9 \cdot 16} = \sqrt{144} \qquad \text{and} \qquad \sqrt{9}\sqrt{16} = 3 \cdot 4$$
$$= 12 \qquad\qquad\qquad\qquad\qquad = 12$$

In each case, the answer is 12. Thus, $\sqrt{9 \cdot 16} = \sqrt{9}\sqrt{16}$. These results suggest the **multiplication property of radicals.**

Multiplication Property of Radicals	If a and b are positive or 0, then $$\sqrt{ab} = \sqrt{a}\sqrt{b}$$

In words, *the square root of the product of two nonnegative numbers is equal to the product of their square roots.*

■ Simplifying Radicals

A radical is in **simplified form** when each of the following statements is true.

Simplified Form of a Radical	**1.** Except for 1, the radicand has no perfect square factors.
	2. No fraction appears in a radicand.
	3. No radical appears in the denominator of a fraction.

We can use the multiplication property of radicals to simplify radicals that have perfect square factors. For example, we can simplify $\sqrt{12}$ as follows:

$$\sqrt{12} = \sqrt{4 \cdot 3} \qquad \text{Factor 12 as } 4 \cdot 3.$$
$$= \sqrt{4}\sqrt{3} \qquad \text{Use the multiplication property of radicals.}$$
$$= 2\sqrt{3} \qquad \text{Write } \sqrt{4} \text{ as 2.}$$

To simplify more difficult radicals, we need to know the integers that are perfect squares. The number 81, for example, is a perfect square because $9^2 = 81$. The first twenty integer squares are

1, 4, 9, 16, 25, 36, 49, 64, 81, 100, 121, 144, 169, 196, 225, 256, 289, 324, 361, 400

Expressions with variables can also be perfect squares. For example, $9x^4y^2$ is a perfect square because

$$9x^4y^2 = (3x^2y)^2$$

EXAMPLE 1 Simplify $\sqrt{72x^3}$. Assume that $x \geq 0$.

Solution We factor $72x^3$ into two factors, one of which is the greatest perfect square that divides $72x^3$. Because the greatest perfect square that divides $72x^3$ is $36x^2$, such a factorization is $72x^3 = 36x^2 \cdot 2x$. We can now use the multiplication property of radicals and simplify to get

$$
\begin{aligned}
\sqrt{72x^3} &= \sqrt{36x^2 \cdot 2x} \\
&= \sqrt{36x^2}\sqrt{2x} \qquad \text{Use the multiplication property of radicals.} \\
&= 6x\sqrt{2x} \qquad\qquad \sqrt{36x^2} = 6x
\end{aligned}
$$ ∎

EXAMPLE 2 Simplify $\sqrt{45x^2y^3}$. Assume that $x \geq 0$ and $y \geq 0$.

Solution We look for the greatest perfect square that divides $45x^2y^3$. Because

- 9 is the greatest perfect square that divides 45,
- x^2 is the greatest perfect square that divides x^2, and
- y^2 is the greatest perfect square that divides y^3,

the factor $9x^2y^2$ is the greatest perfect square that divides $45x^2y^3$.
We can now use the multiplication property of radicals and simplify to get

$$
\begin{aligned}
\sqrt{45x^2y^3} &= \sqrt{9x^2y^2 \cdot 5y} \\
&= \sqrt{9x^2y^2}\sqrt{5y} \qquad \text{Use the multiplication property of radicals.} \\
&= 3xy\sqrt{5y} \qquad\qquad \sqrt{9x^2y^2} = 3xy
\end{aligned}
$$ ∎

EXAMPLE 3 Simplify $3a\sqrt{288a^5b^7}$. Assume that $a \geq 0$ and $b \geq 0$.

Solution We look for the greatest perfect square that divides $288a^5b^7$. Because

- 144 is the greatest perfect square that divides 288,
- a^4 is the greatest perfect square that divides a^5, and
- b^6 is the greatest perfect square that divides b^7,

the factor $144a^4b^6$ is the greatest perfect square that divides $288a^5b^7$.
We can now use the multiplication property of radicals and simplify to get

$$
\begin{aligned}
3a\sqrt{288a^5b^7} &= 3a\sqrt{144a^4b^6 \cdot 2ab} \\
&= 3a\sqrt{144a^4b^6}\sqrt{2ab} \qquad \text{Use the multiplication property} \\
&\qquad\qquad\qquad\qquad\qquad\quad \text{of radicals.} \\
&= 3a\left(12a^2b^3\sqrt{2ab}\right) \qquad \sqrt{144a^4b^6} = 12a^2b^3. \\
&= 36a^3b^3\sqrt{2ab}
\end{aligned}
$$ ∎

■ The Division Property of Radicals

To find the second property of radicals, we consider these examples.

$$\sqrt{\frac{100}{25}} = \sqrt{4} \qquad \text{and} \qquad \frac{\sqrt{100}}{\sqrt{25}} = \frac{10}{5}$$
$$= 2 \qquad\qquad\qquad\qquad = 2$$

In each case, the answer is 2. Thus, $\sqrt{\frac{100}{25}} = \frac{\sqrt{100}}{\sqrt{25}}$. Likewise,

$$\sqrt{\frac{36}{4}} = \sqrt{9} \qquad \text{and} \qquad \frac{\sqrt{36}}{\sqrt{4}} = \frac{6}{2}$$
$$= 3 \qquad\qquad\qquad\qquad = 3$$

In each case, the answer is 3. Thus, $\sqrt{\frac{36}{4}} = \frac{\sqrt{36}}{\sqrt{4}}$. These results suggest the **division property of radicals.**

Division Property of Radicals	If $a \geq 0$ and $b > 0$, then $$\sqrt{\frac{a}{b}} = \frac{\sqrt{a}}{\sqrt{b}}$$

In words, *the square root of the quotient of two numbers is the quotient of their square roots.*

We can use the division property of radicals to simplify many radicals that have fractions in their radicands. For example,

$$\sqrt{\frac{59}{49}} = \frac{\sqrt{59}}{\sqrt{49}}$$
$$= \frac{\sqrt{59}}{7} \qquad \sqrt{49} = 7.$$

EXAMPLE 4 Simplify $\sqrt{\frac{108}{25}}$.

Solution

$$\sqrt{\frac{108}{25}} = \frac{\sqrt{108}}{\sqrt{25}} \qquad \text{Use the division property of radicals.}$$

$$= \frac{\sqrt{36 \cdot 3}}{5} \qquad \text{Factor 108 using the factorization involving 36, the largest perfect square factor of 108, and write } \sqrt{25} \text{ as 5.}$$

$$= \frac{\sqrt{36}\sqrt{3}}{5} \qquad \text{Use the multiplication property of radicals.}$$

$$= \frac{6\sqrt{3}}{5} \qquad \sqrt{36} = 6.$$

■

EXAMPLE 5   Simplify $\sqrt{\dfrac{44x^3}{9xy^2}}$. Assume that $x > 0$ and $y > 0$.

Solution

$\sqrt{\dfrac{44x^3}{9xy^2}} = \sqrt{\dfrac{44x^2}{9y^2}}$ Simplify the fraction by dividing out the common factor of x.

$\phantom{\sqrt{\dfrac{44x^3}{9xy^2}}} = \dfrac{\sqrt{44x^2}}{\sqrt{9y^2}}$ Use the division property of radicals.

$\phantom{\sqrt{\dfrac{44x^3}{9xy^2}}} = \dfrac{\sqrt{4x^2}\sqrt{11}}{\sqrt{9y^2}}$ Use the multiplication property of radicals.

$\phantom{\sqrt{\dfrac{44x^3}{9xy^2}}} = \dfrac{2x\sqrt{11}}{3y}$

■

■ Simplifying Cube Roots

The multiplication and division properties of radicals are also true for cube roots and higher. To simplify a cube root, we must recall the following perfect cube integers:

1, 8, 27, 64, 125, 216, 343, 512, 729, 1000

Expressions with variables can also be perfect cubes. For example, $27x^6y^3$ is a perfect cube because

$27x^6y^3 = (3x^2y)^3$

EXAMPLE 6   Simplify   **a.** $\sqrt[3]{16x^3y^4}$ and   **b.** $\sqrt[3]{\dfrac{64n^4}{27m^3}}$.

Solution   **a.** We factor $16x^3y^4$ into two factors, one of which is the greatest perfect cube that divides $16x^3y^4$. Since $8x^3y^3$ is the greatest perfect cube that divides $16x^3y^4$, such a factorization is

$16x^3y^4 = 8x^3y^3 \cdot 2y$

We can now use the multiplication property of radicals and simplify to get

$\sqrt[3]{16x^3y^4} = \sqrt[3]{8x^3y^3 \cdot 2y}$

$\phantom{\sqrt[3]{16x^3y^4}} = \sqrt[3]{8x^3y^3}\sqrt[3]{2y}$ Use the multiplication property of radicals.

$\phantom{\sqrt[3]{16x^3y^4}} = 2xy\sqrt[3]{2y}$ $\sqrt[3]{8x^3y^3} = 2xy$.

b. $\sqrt[3]{\dfrac{64n^4}{27m^3}} = \dfrac{\sqrt[3]{64n^4}}{\sqrt[3]{27m^3}}$ Use the division property of radicals.

$\phantom{\sqrt[3]{\dfrac{64n^4}{27m^3}}} = \dfrac{\sqrt[3]{64n^3}\sqrt[3]{n}}{3m}$ Use the multiplication property of radicals and write $\sqrt[3]{27m^3}$ as $3m$.

$\phantom{\sqrt[3]{\dfrac{64n^4}{27m^3}}} = \dfrac{4n\sqrt[3]{n}}{3m}$

■

WARNING! $\sqrt{a+b} \neq \sqrt{a} + \sqrt{b}$ and $\sqrt{a-b} \neq \sqrt{a} - \sqrt{b}$. To see that this is true, we consider these correct simplifications:

$$\sqrt{9+16} = \sqrt{25} = 5 \quad \text{and} \quad \sqrt{25-16} = \sqrt{9} = 3$$

It is incorrect to write

$$\sqrt{9+16} = \sqrt{9} + \sqrt{16} \qquad\qquad \sqrt{25-16} = \sqrt{25} - \sqrt{16}$$
$$= 3 + 4 \qquad\qquad\qquad\qquad = 5 - 4$$
$$= 7 \qquad\qquad\qquad\qquad\qquad = 1$$

ORALS *Simplify each radical ($x > 0$).*

1. $\sqrt{8}$ **2.** $\sqrt{12}$ **3.** $\sqrt{\dfrac{5}{9}}$ **4.** $\sqrt{\dfrac{7}{25}}$

5. $\sqrt{4x^2}$ **6.** $\sqrt{9a^4}$

EXERCISE 8.2

In Exercises 1–48, simplify each radical. Assume that all variables represent positive numbers.

1. $\sqrt{20}$ 　　　　 **2.** $\sqrt{18}$ 　　　　 **3.** $\sqrt{50}$ 　　　　 **4.** $\sqrt{75}$

5. $\sqrt{45}$ 　　　　 **6.** $\sqrt{54}$ 　　　　 **7.** $\sqrt{98}$ 　　　　 **8.** $\sqrt{27}$

9. $\sqrt{48}$ 　　　　 **10.** $\sqrt{128}$ 　　　 **11.** $\sqrt{200}$ 　　 **12.** $\sqrt{300}$

13. $\sqrt{192}$ 　　　 **14.** $\sqrt{250}$ 　　　 **15.** $\sqrt{88}$ 　　　 **16.** $\sqrt{275}$

17. $\sqrt{324}$ 　　　 **18.** $\sqrt{405}$ 　　　 **19.** $\sqrt{147}$ 　　 **20.** $\sqrt{722}$

21. $\sqrt{180}$ 　　　 **22.** $\sqrt{320}$ 　　　 **23.** $\sqrt{432}$ 　　 **24.** $\sqrt{720}$

25. $4\sqrt{288}$ 　　 **26.** $2\sqrt{800}$ 　　 **27.** $-7\sqrt{1000}$ 　 **28.** $-3\sqrt{252}$

29. $2\sqrt{245}$ 　　 **30.** $3\sqrt{196}$ 　　 **31.** $-5\sqrt{162}$ 　 **32.** $-4\sqrt{243}$

33. $\sqrt{25x}$ 　　　 **34.** $\sqrt{36y}$ 　　　 **35.** $\sqrt{a^2b}$ 　　 **36.** $\sqrt{rs^2}$

37. $\sqrt{9x^2y}$ 　　 **38.** $\sqrt{16xy^2}$ 　 **39.** $\dfrac{1}{5}x^2y\sqrt{50x^2y^2}$ 　 **40.** $\dfrac{1}{5}x^5y\sqrt{75x^3y^2}$

41. $12x\sqrt{16x^2y^3}$ 　 **42.** $-4x^5y^3\sqrt{36x^3y^3}$ 　 **43.** $-3xyz\sqrt{18x^3y^5}$ 　 **44.** $15xy^2\sqrt{72x^2y^3}$

45. $\dfrac{3}{4}\sqrt{192a^3b^5}$ 　 **46.** $-\dfrac{2}{9}\sqrt{162r^3s^3t}$ 　 **47.** $-\dfrac{2}{5}\sqrt{80mn^2}$ 　 **48.** $\dfrac{5}{6}\sqrt{180ab^2c}$

In Exercises 49–64, write each quotient as the quotient of two radicals and simplify.

49. $\sqrt{\dfrac{25}{9}}$

50. $\sqrt{\dfrac{36}{49}}$

51. $\sqrt{\dfrac{81}{64}}$

52. $\sqrt{\dfrac{121}{144}}$

53. $\sqrt{\dfrac{26}{25}}$

54. $\sqrt{\dfrac{17}{169}}$

55. $\sqrt{\dfrac{20}{49}}$

56. $\sqrt{\dfrac{50}{9}}$

57. $\sqrt{\dfrac{48}{81}}$

58. $\sqrt{\dfrac{27}{64}}$

59. $\sqrt{\dfrac{32}{25}}$

60. $\sqrt{\dfrac{75}{16}}$

61. $\sqrt{\dfrac{125}{121}}$

62. $\sqrt{\dfrac{250}{49}}$

63. $\sqrt{\dfrac{245}{36}}$

64. $\sqrt{\dfrac{500}{81}}$

In Exercises 65–72, simplify each expression. All variables represent positive numbers.

65. $\sqrt{\dfrac{72x^3}{y^2}}$

66. $\sqrt{\dfrac{108a^3b^2}{c^2d^4}}$

67. $\sqrt{\dfrac{125m^2n^5}{64n}}$

68. $\sqrt{\dfrac{72p^5q^7}{16pq^3}}$

69. $\sqrt{\dfrac{128m^3n^5}{36mn^7}}$

70. $\sqrt{\dfrac{75p^3q^2}{9p^5q^4}}$

71. $\sqrt{\dfrac{12r^7s^6t}{81r^5s^2t}}$

72. $\sqrt{\dfrac{36m^2n^9}{100mn^3}}$

In Exercises 73–86, simplify each cube root.

73. $\sqrt[3]{8x^3}$

74. $\sqrt[3]{27x^3y^3}$

75. $\sqrt[3]{-64x^5}$

76. $\sqrt[3]{-16x^4y^3}$

77. $\sqrt[3]{54x^3y^4z^6}$

78. $\sqrt[3]{-24x^5y^5z^4}$

79. $\sqrt[3]{-81x^2y^3z^4}$

80. $\sqrt[3]{1600xy^2z^3}$

81. $\sqrt[3]{\dfrac{27m^3}{8n^6}}$

82. $\sqrt[3]{\dfrac{125t^9}{27s^6}}$

83. $\sqrt[3]{\dfrac{16r^4s^5}{1000t^3}}$

84. $\sqrt[3]{\dfrac{54m^4n^3}{r^3s^6}}$

85. $\sqrt[3]{\dfrac{250a^3b^4}{16b}}$

86. $\sqrt[3]{\dfrac{81p^5q^3}{1000p^2q^6}}$

Writing Exercises ■

1. State the multiplication and division properties of radicals.

2. Explain the terms *perfect square* and *perfect cube*.

Something to Think About ■

1. Find the mistake in this work.

$$\begin{aligned} \sqrt{13} &= \sqrt{9+4} \\ &= \sqrt{9} + \sqrt{4} \\ &= 3 + 2 \\ &= 5 \end{aligned}$$

2. Use scientific notation to find $\sqrt{0.00000004}$.

Review Exercises ■

Simplify each fraction.

1. $\dfrac{5xy^2z^3}{10x^2y^2z^4}$

2. $\dfrac{35a^3b^2c}{63a^2b^3c^2}$

3. $\dfrac{a^2 - a - 2}{a^2 + a - 6}$

4. $\dfrac{y^2 + 3y - 18}{y^2 - 9}$

Do the operations and simplify the result when possible.

5. $\dfrac{t^2 - t - 6}{t^2 - 3t} \cdot \dfrac{t^2 - t}{t^2 + t - 2}$

6. $\dfrac{x + 3}{x^2 + x - 6} \div \dfrac{x - 2}{x^2 - 5x + 6}$

7. $\dfrac{2r}{r + 3} + \dfrac{6}{r + 3}$

8. $\dfrac{2(u - 4)}{u + 1} - \dfrac{2(u + 4)}{u + 1}$

8.3 Adding and Subtracting Radical Expressions

■ Combining Expressions with Higher-Order Radicals

GETTING READY   *Combine like terms.*

1. $3x + 4x$ **2.** $5y - 2y$ **3.** $7xy - xy$ **4.** $7t^2 + 2t^2$

Simplify each radical.

5. $\sqrt{20}$ **6.** $\sqrt{45}$ **7.** $\sqrt{8x^2y}$ **8.** $\sqrt{18x^2y}$

When adding monomials, we can combine **like terms.** For example,

$$3x + 5x = (3 + 5)x \qquad \text{Use the distributive property.}$$
$$= 8x$$

 WARNING!   The expression $3x + 5y$ cannot be simplified, because $3x$ and $5y$ are not like terms.

It is often possible to combine terms that contain **like radicals.**

Like Radicals   Radicals are called **like radicals** when they have the same index and the same radicand.

Since the terms $3\sqrt{2}$ and $5\sqrt{2}$ have like radicals, they are like terms and can be combined.

$$3\sqrt{2} + 5\sqrt{2} = (3 + 5)\sqrt{2} \qquad \text{Use the distributive property.}$$
$$= 8\sqrt{2}$$

Likewise,

$$5x\sqrt{3y} - 2x\sqrt{3y} = (5 - 2)x\sqrt{3y} \qquad \text{Use the distributive property.}$$
$$= 3x\sqrt{3y}$$

WARNING!   The terms in $3\sqrt{2} + 5\sqrt{7}$ have unlike radicals and cannot be combined.

We cannot combine the terms in the expression $2x\sqrt{5z} + 3y\sqrt{5z}$, because the terms have different variables.

Radicals such as $3\sqrt{18}$ and $5\sqrt{8}$ can be simplified so that they contain like radicals. They can then be combined.

EXAMPLE 1   Simplify $3\sqrt{18} + 5\sqrt{8}$.

Solution   The radical $\sqrt{18}$ is not in simplified form, because 18 has a perfect square factor of 9. The radical $\sqrt{8}$ is not in simplified form either, because 8 has perfect square factor of 4. To simplify the radicals and combine like terms, we proceed as follows.

$$3\sqrt{18} + 5\sqrt{8} = 3\sqrt{9 \cdot 2} + 5\sqrt{4 \cdot 2}$$ Factor 18 and 8. Look for perfect square factors.

$$= 3\sqrt{9}\sqrt{2} + 5\sqrt{4}\sqrt{2}$$ Use the multiplication property of radicals.

$$= 3(3)\sqrt{2} + 5(2)\sqrt{2}$$ Simplify.

$$= 9\sqrt{2} + 10\sqrt{2}$$ Simplify.

$$= 19\sqrt{2}$$ Combine like terms.   ∎

EXAMPLE 2   Simplify $\sqrt{20} + \sqrt{45} + 3\sqrt{5}$.

Solution   We simplify the first two radicals and combine like terms.

$$\sqrt{20} + \sqrt{45} + 3\sqrt{5} = \sqrt{4 \cdot 5} + \sqrt{9 \cdot 5} + 3\sqrt{5}$$ Factor.

$$= \sqrt{4}\sqrt{5} + \sqrt{9}\sqrt{5} + 3\sqrt{5}$$ Use the multiplication property of radicals.

$$= 2\sqrt{5} + 3\sqrt{5} + 3\sqrt{5}$$ Simplify.

$$= 8\sqrt{5}$$ Combine like terms.   ∎

EXAMPLE 3   Simplify $\sqrt{8x^2y} + \sqrt{18x^2y}$. Assume $x > 0$ and $y > 0$.

Solution   We simplify each radical and combine like terms.

$$\sqrt{8x^2y} + \sqrt{18x^2y} = \sqrt{4 \cdot 2x^2y} + \sqrt{9 \cdot 2x^2y}$$ Factor.

$$= \sqrt{4x^2}\sqrt{2y} + \sqrt{9x^2}\sqrt{2y}$$ Use the multiplication property of radicals.

$$= 2x\sqrt{2y} + 3x\sqrt{2y}$$ Simplify.

$$= 5x\sqrt{2y}$$ Combine like terms.   ∎

EXAMPLE 4   Simplify $\sqrt{28x^2y} - 2\sqrt{63y^3}$. Assume $x > 0$ and $y > 0$.

Solution   We simplify each radical and combine like terms.

$$\sqrt{28x^2y} - 2\sqrt{63y^3} = \sqrt{4 \cdot 7x^2y} - 2\sqrt{9 \cdot 7y^2y} \qquad \text{Factor.}$$

$$= \sqrt{4x^2}\sqrt{7y} - 2\sqrt{9y^2}\sqrt{7y} \qquad \begin{array}{l}\text{Use the multiplication}\\ \text{property of radicals.}\end{array}$$

$$= 2x\sqrt{7y} - 2 \cdot 3y\sqrt{7y} \qquad \text{Simplify.}$$

$$= 2x\sqrt{7y} - 6y\sqrt{7y} \qquad \text{Simplify.}$$

Since the variables in the terms are different, the expression does not simplify further. ■

EXAMPLE 5   Simplify $\sqrt{27xy} + \sqrt{20xy}$. Assume $x > 0$ and $y > 0$.

Solution
$$\sqrt{27xy} + \sqrt{20xy} = \sqrt{9 \cdot 3xy} + \sqrt{4 \cdot 5xy} \qquad \text{Factor.}$$

$$= \sqrt{9}\sqrt{3xy} + \sqrt{4}\sqrt{5xy} \qquad \begin{array}{l}\text{Use the multiplication property}\\ \text{of radicals.}\end{array}$$

$$= 3\sqrt{3xy} + 2\sqrt{5xy} \qquad \text{Simplify.}$$

Since the terms have unlike radicals, the expression does not simplify further. ■

EXAMPLE 6   Simplify $\sqrt{8x} + \sqrt{3y} - \sqrt{50x} + \sqrt{27y}$. Assume $x > 0$ and $y > 0$.

Solution   We simplify the radicals and combine like terms, when possible.

$$\sqrt{8x} + \sqrt{3y} - \sqrt{50x} + \sqrt{27y} = \sqrt{4 \cdot 2x} + \sqrt{3y} - \sqrt{25 \cdot 2x} + \sqrt{9 \cdot 3y}$$

$$= \sqrt{4}\sqrt{2x} + \sqrt{3y} - \sqrt{25}\sqrt{2x} + \sqrt{9}\sqrt{3y}$$

$$= 2\sqrt{2x} + \sqrt{3y} - 5\sqrt{2x} + 3\sqrt{3y}$$

$$= -3\sqrt{2x} + 4\sqrt{3y} \qquad ■$$

■ Combining Expressions with Higher-Order Radicals

It is often possible to combine terms containing like radicals other than square roots.

EXAMPLE 7 Simplify $\sqrt[3]{81x^4} - x\sqrt[3]{24x}$.

Solution We simplify each radical and combine like terms, when possible.

$$\sqrt[3]{81x^4} - x\sqrt[3]{24x} = \sqrt[3]{27x^3 \cdot 3x} - x\sqrt[3]{8 \cdot 3x}$$
$$= \sqrt[3]{27x^3}\sqrt[3]{3x} - x\sqrt[3]{8}\sqrt[3]{3x}$$
$$= 3x\sqrt[3]{3x} - 2x\sqrt[3]{3x}$$
$$= x\sqrt[3]{3x} \qquad \blacksquare$$

ORALS *Combine like radicals.*

1. $3\sqrt{5} + 2\sqrt{5}$ **2.** $6\sqrt{7} - \sqrt{7}$

3. $4\sqrt{xy} - 2\sqrt{xy}$ **4.** $\sqrt{5yz} + 3\sqrt{5yz}$

5. $\sqrt{8} + \sqrt{2}$ **6.** $\sqrt{3} - \sqrt{12}$

E X E R C I S E   8 . 3

In Exercises 1–24, find each sum.

1. $\sqrt{12} + \sqrt{27}$ **2.** $\sqrt{20} + \sqrt{45}$ **3.** $\sqrt{48} + \sqrt{75}$ **4.** $\sqrt{48} + \sqrt{108}$

5. $\sqrt{45} + \sqrt{80}$ **6.** $\sqrt{80} + \sqrt{125}$ **7.** $\sqrt{125} + \sqrt{245}$ **8.** $\sqrt{36} + \sqrt{196}$

9. $\sqrt{20} + \sqrt{180}$ **10.** $\sqrt{80} + \sqrt{245}$ **11.** $\sqrt{160} + \sqrt{360}$ **12.** $\sqrt{12} + \sqrt{147}$

13. $3\sqrt{45} + 4\sqrt{245}$ **14.** $2\sqrt{28} + 7\sqrt{63}$ **15.** $2\sqrt{28} + 2\sqrt{112}$ **16.** $4\sqrt{63} + 6\sqrt{112}$

17. $5\sqrt{32} + 3\sqrt{72}$ **18.** $3\sqrt{72} + 2\sqrt{128}$ **19.** $3\sqrt{98} + 8\sqrt{128}$ **20.** $5\sqrt{90} + 7\sqrt{250}$

21. $\sqrt{20} + \sqrt{45} + \sqrt{80}$ **22.** $\sqrt{48} + \sqrt{27} + \sqrt{75}$ **23.** $\sqrt{24} + \sqrt{150} + \sqrt{240}$ **24.** $\sqrt{28} + \sqrt{63} + \sqrt{112}$

In Exercises 25–44, find each difference, when possible.

25. $\sqrt{18} - \sqrt{8}$ **26.** $\sqrt{32} - \sqrt{18}$ **27.** $\sqrt{9} - \sqrt{50}$ **28.** $\sqrt{50} - \sqrt{32}$

29. $\sqrt{72} - \sqrt{32}$ **30.** $\sqrt{98} - \sqrt{72}$ **31.** $\sqrt{12} - \sqrt{48}$ **32.** $\sqrt{48} - \sqrt{75}$

33. $\sqrt{108} - \sqrt{75}$ **34.** $\sqrt{147} - \sqrt{48}$ **35.** $\sqrt{1000} - \sqrt{360}$ **36.** $\sqrt{180} - \sqrt{125}$

37. $2\sqrt{80} - 3\sqrt{125}$ **38.** $3\sqrt{245} - 2\sqrt{180}$ **39.** $8\sqrt{96} - 5\sqrt{24}$ **40.** $3\sqrt{216} - 3\sqrt{150}$

41. $\sqrt{288} - 3\sqrt{200}$ **42.** $\sqrt{392} - 2\sqrt{128}$ **43.** $5\sqrt{250} - 3\sqrt{160}$ **44.** $4\sqrt{490} - 3\sqrt{360}$

In Exercises 45–56, simplify each expression.

45. $\sqrt{12} + \sqrt{18} - \sqrt{27}$ **46.** $\sqrt{8} - \sqrt{50} + \sqrt{72}$ **47.** $\sqrt{200} - \sqrt{75} + \sqrt{48}$ **48.** $\sqrt{20} + \sqrt{80} - \sqrt{125}$

49. $\sqrt{24} - \sqrt{150} - \sqrt{54}$ **50.** $\sqrt{98} - \sqrt{300} + \sqrt{800}$ **51.** $\sqrt{200} + \sqrt{300} - \sqrt{75}$ **52.** $\sqrt{175} + \sqrt{125} - \sqrt{28}$

53. $\sqrt{48} - \sqrt{8} + \sqrt{27} - \sqrt{32}$ **54.** $\sqrt{162} + \sqrt{50} - \sqrt{75} - \sqrt{108}$

55. $\sqrt{147} + \sqrt{216} - \sqrt{108} - \sqrt{27}$ **56.** $\sqrt{180} - \sqrt{112} + \sqrt{45} - \sqrt{700}$

In Exercises 57–70, simplify each expression. All variables represent positive numbers.

57. $\sqrt{2x^2} + \sqrt{8x^2}$ **58.** $\sqrt{3y^2} - \sqrt{12y^2}$ **59.** $\sqrt{2x^3} + \sqrt{8x^3}$ **60.** $\sqrt{3y^3} - \sqrt{12y^3}$

61. $\sqrt{18x^2y} - \sqrt{27x^2y}$ **62.** $\sqrt{49xy} + \sqrt{xy}$ **63.** $\sqrt{32x^5} - \sqrt{18x^5}$ **64.** $\sqrt{27xy^3} - \sqrt{48xy^3}$

65. $3\sqrt{54x^2} + 5\sqrt{24x^2}$ **66.** $3\sqrt{24x^4y^3} + 2\sqrt{54x^4y^3}$ **67.** $y\sqrt{490y} - 2\sqrt{360y^3}$ **68.** $3\sqrt{20x} + 2\sqrt{63y}$

69. $\sqrt{20x^3y} + \sqrt{45x^5y^3} - \sqrt{80x^7y^5}$ **70.** $x\sqrt{48xy^2} - y\sqrt{27x^3} + \sqrt{75x^3y^2}$

In Exercises 71–82, simplify each expression.

71. $\sqrt[3]{16} + \sqrt[3]{54}$ **72.** $\sqrt[3]{24} - \sqrt[3]{81}$ **73.** $\sqrt[3]{81} - \sqrt[3]{24}$ **74.** $\sqrt[3]{32} + \sqrt[3]{108}$

75. $\sqrt[3]{40} + \sqrt[3]{125}$ **76.** $\sqrt[3]{3000} - \sqrt[3]{192}$ **77.** $\sqrt[3]{x^4} - \sqrt[3]{x^7}$ **78.** $\sqrt[3]{8x^5} + \sqrt[3]{27x^8}$

79. $\sqrt[3]{192x^4y^5} - \sqrt[3]{24x^4y^5}$ **80.** $\sqrt[3]{24a^5b^4} + \sqrt[3]{81a^5b^4}$

81. $\sqrt[3]{135x^7y^4} - \sqrt[3]{40x^7y^4}$ **82.** $\sqrt[3]{56a^4b^5} + \sqrt[3]{7a^4b^5}$

Writing Exercises ■ *Write a paragraph using your own words.*

1. Explain why $\sqrt{3x}$ and $\sqrt{2y}$ cannot be combined.

2. Explain why $\sqrt{4x}$ and $\sqrt[3]{4x}$ cannot be combined.

Something to Think About ■

1. What is wrong with the following work?

$$7\sqrt{5} - 3\sqrt{2} = 4\sqrt{3}$$

2. What is wrong with the following work?

$$\frac{6 + \sqrt{8}}{3 + \sqrt{2}} = 2 + \sqrt{4} = 2 + 2 = 4$$

Review Exercises ■ *Solve each proportion.*

1. $\dfrac{a - 2}{8} = \dfrac{a + 10}{24}$ 2. $\dfrac{6}{t + 12} = \dfrac{18}{4t}$ 3. $\dfrac{-2}{x + 14} = \dfrac{6}{x - 6}$ 4. $\dfrac{y - 4}{4} = \dfrac{y + 2}{12}$

8.4 Multiplying and Dividing Radical Expressions

■ Multiplying Radical Expressions ■ Dividing Radical Expressions
■ Rationalizing the Denominator

GETTING READY   *Do the operation and simplify, when possible.*

1. $x^2 x^3$ 2. $y^3 y^4$ 3. $\dfrac{x^5}{x^2}$ 4. $\dfrac{y^8}{y^5}$

5. $x(x + 2)$ 6. $2y^3(3y^2 - 4y)$

7. $(x + 2)(x - 3)$ 8. $(2x + 3y)(3x + 2y)$

The definition of square root implies that $\sqrt{5}$ is the positive number whose square is 5:

$$\left(\sqrt{5}\right)^2 = 5 \qquad \text{and} \qquad \sqrt{5}\sqrt{5} = 5$$

In general, we have

$$\left(\sqrt{x}\right)^2 = x \qquad \text{and} \qquad \sqrt{x}\sqrt{x} = x$$

Because of the multiplication property of radicals, the *product of the square roots of two nonnegative numbers is equal to the square root of the product of those numbers.* For example,

$$\sqrt{2}\sqrt{8} = \sqrt{2 \cdot 8} = \sqrt{16} = 4$$
$$\sqrt{3}\sqrt{27} = \sqrt{3 \cdot 27} = \sqrt{81} = 9$$
$$\sqrt{x}\sqrt{x^3} = \sqrt{x \cdot x^3} = \sqrt{x^4} = x^2$$

Likewise, the *product of the cube roots of two numbers is equal to the cube root of the product of those numbers.* For example,

$$\sqrt[3]{2} \cdot \sqrt[3]{4} = \sqrt[3]{2 \cdot 4} = \sqrt[3]{8} = 2$$
$$\sqrt[3]{4} \cdot \sqrt[3]{16} = \sqrt[3]{4 \cdot 16} = \sqrt[3]{64} = 4$$
$$\sqrt[3]{3x^2} \cdot \sqrt[3]{9x} = \sqrt[3]{3x^2 \cdot 9x} = \sqrt[3]{27x^3} = 3x$$

■ Multiplying Radical Expressions

To multiply monomials containing radicals, we multiply the coefficients and multiply the radicals separately and simplify the result, when possible.

 (a)

EXAMPLE 1 Multiply **a.** $3\sqrt{6}$ by $4\sqrt{3}$ and **b.** $-2\sqrt[3]{7}$ by $6\sqrt[3]{49}$.

Solution The commutative and associative properties enable us to multiply the integers and the radicals separately.

a. $3\sqrt{6} \cdot 4\sqrt{3} = 3(4)\sqrt{6}\sqrt{3}$ **b.** $-2\sqrt[3]{7} \cdot 6\sqrt[3]{49} = -2(6)\sqrt[3]{7}\sqrt[3]{49}$
$\qquad\qquad = 12\sqrt{18}$ $\qquad\qquad\qquad = -12\sqrt[3]{7 \cdot 49}$
$\qquad\qquad = 12\sqrt{9}\sqrt{2}$ $\qquad\qquad\qquad = -12\sqrt[3]{343}$
$\qquad\qquad = 12(3)\sqrt{2}$ $\qquad\qquad\qquad = -12(7)$
$\qquad\qquad = 36\sqrt{2}$ $\qquad\qquad\qquad = -84$ ■

To multiply a polynomial by a monomial, we use the distributive property to remove parentheses and combine like terms.

(b)

EXAMPLE 2 Multiply **a.** $\sqrt{2}(\sqrt{6} + \sqrt{8})$ and **b.** $\sqrt[3]{3}(\sqrt[3]{9} - 2)$.

Solution **a.** $\sqrt{2}(\sqrt{6} + \sqrt{8}) = \sqrt{2}\sqrt{6} + \sqrt{2}\sqrt{8}$ Use the distributive property to remove parentheses.

$\qquad\qquad\qquad = \sqrt{12} + \sqrt{16}$ Use the multiplication property of radicals.

$\qquad\qquad\qquad = \sqrt{4 \cdot 3} + \sqrt{16}$ Factor 12.

$\qquad\qquad\qquad = \sqrt{4}\sqrt{3} + \sqrt{16}$ Use the multiplication property of radicals.

$\qquad\qquad\qquad = 2\sqrt{3} + 4$ Simplify.

b. $\sqrt[3]{3}\left(\sqrt[3]{9} - 2\right) = \sqrt[3]{3}\sqrt[3]{9} - 2\sqrt[3]{3}$ Use the distributive property to remove parentheses.

$= \sqrt[3]{27} - 2\sqrt[3]{3}$ Use the multiplication property of radicals.

$= 3 - 2\sqrt[3]{3}$ Simplify. ∎

We can use the FOIL method to multiply one binomial by another.

EXAMPLE 3   Multiply and simplify $\left(\sqrt{3} + \sqrt{2}\right)\left(\sqrt{3} - \sqrt{2}\right)$.

Solution   We can find the product with the FOIL method.

$\left(\sqrt{3} + \sqrt{2}\right)\left(\sqrt{3} - \sqrt{2}\right) = \sqrt{3}\sqrt{3} - \sqrt{3}\sqrt{2} + \sqrt{2}\sqrt{3} - \sqrt{2}\sqrt{2}$ Use the FOIL method.

$= 3 - 2$ Combine like terms and simplify.

$= 1$ Simplify. ∎

EXAMPLE 4   Multiply and simplify $\left(\sqrt{3x} + 1\right)\left(\sqrt{3x} + 2\right)$.

Solution $\left(\sqrt{3x} + 1\right)\left(\sqrt{3x} + 2\right) = \sqrt{3x}\sqrt{3x} + 2\sqrt{3x} + 1\sqrt{3x} + 2$ Use the FOIL method.

$= 3x + 3\sqrt{3x} + 2$ Combine like terms and simplify. ∎

EXAMPLE 5   Multiply and simplify $\left(\sqrt[3]{4x} - 3\right)\left(\sqrt[3]{2x^2} + 1\right)$.

Solution $\left(\sqrt[3]{4x} - 3\right)\left(\sqrt[3]{2x^2} + 1\right) = \sqrt[3]{4x}\sqrt[3]{2x^2} + 1\sqrt[3]{4x} - 3\sqrt[3]{2x^2} - 3$ Use the FOIL method.

$= \sqrt[3]{8x^3} + \sqrt[3]{4x} - 3\sqrt[3]{2x^2} - 3$ Use the multiplication property of radicals.

$= 2x + \sqrt[3]{4x} - 3\sqrt[3]{2x^2} - 3$ Simplify. ∎

■ Dividing Radical Expressions

To divide radical expressions, we use the division property of radicals. For example, to divide $\sqrt{108}$ by $\sqrt{36}$, we proceed as follows:

$$\frac{\sqrt{108}}{\sqrt{36}} = \sqrt{\frac{108}{36}}$$ Use the division property of radicals.

$$= \sqrt{3}$$ $\frac{108}{36} = \frac{36 \cdot 3}{36} = 3.$

EXAMPLE 6   Simplify $\dfrac{\sqrt{22a^2b^7}}{\sqrt{99a^4b^3}}$. Assume that $a > 0$ and $b > 0$.

Solution

$$\dfrac{\sqrt{22a^2b^7}}{\sqrt{99a^4b^3}} = \sqrt{\dfrac{22a^2b^7}{99a^4b^3}}$$

$$= \sqrt{\dfrac{2b^4}{9a^2}} \qquad \text{Simplify the radicand.}$$

$$= \dfrac{\sqrt{2b^4}}{\sqrt{9a^2}} \qquad \text{Use the division property of radicals.}$$

$$= \dfrac{\sqrt{b^4}\sqrt{2}}{\sqrt{9a^2}} \qquad \text{Use the multiplication property of radicals.}$$

$$= \dfrac{b^2\sqrt{2}}{3a} \qquad \text{Simplify the radicals.}$$ ■

■ Rationalizing the Denominator

We can **rationalize the denominator** to simplify fractions with radicals in their denominators. In the following example, we can eliminate the radical in the denominator by multiplying both the numerator and the denominator by $\sqrt{2}$, since $\sqrt{2}\sqrt{2}$ is the rational number 2.

$$\dfrac{1}{\sqrt{2}} = \dfrac{1\sqrt{2}}{\sqrt{2}\sqrt{2}} \qquad \text{Multiply both numerator and denominator by } \sqrt{2}.$$

$$= \dfrac{\sqrt{2}}{2} \qquad \text{Simplify the denominator.}$$

EXAMPLE 7   Rationalize the denominator of the fractions   **a.** $\dfrac{3}{\sqrt{3}}$ and   **b.** $\dfrac{2}{\sqrt[3]{3}}$.

Solution   **a.** We multiply both the numerator and denominator by $\sqrt{3}$ and simplify.

$$\dfrac{3}{\sqrt{3}} = \dfrac{3\sqrt{3}}{\sqrt{3}\sqrt{3}} \qquad \text{Multiply both numerator and denominator by } \sqrt{3}.$$

$$= \dfrac{3\sqrt{3}}{3} \qquad \sqrt{3}\sqrt{3} = 3.$$

$$= \sqrt{3} \qquad \text{Divide out the common factor of 3.}$$

WARNING! $\dfrac{3}{\sqrt{3}}$ is not $\dfrac{\overset{1}{\cancel{3}}}{\sqrt{\cancel{3}}} = \dfrac{1}{\sqrt{1}} = 1$

b. Since $\sqrt[3]{3}\sqrt[3]{9} = \sqrt[3]{27}$ and 27 is a perfect integer cube, we multiply both the numerator and denominator of the fraction by $\sqrt[3]{9}$ and simplify.

$$\frac{2}{\sqrt[3]{3}} = \frac{2\sqrt[3]{9}}{\sqrt[3]{3}\sqrt[3]{9}}$$

$$= \frac{2\sqrt[3]{9}}{\sqrt[3]{27}}$$

$$= \frac{2\sqrt[3]{9}}{3}$$

■

If a radical expression contains any radicals in a denominator, it is not in simplified form. The following examples show how to simplify fractions with radicals in their denominators.

EXAMPLE 8 Divide 5 by $\sqrt{20}$ by simplifying the fraction $\dfrac{5}{\sqrt{20}}$.

Solution To rationalize the denominator, it is not necessary to multiply numerator and denominator by $\sqrt{20}$. To keep the numbers small, we can multiply by $\sqrt{5}$, because $5 \cdot 20 = 100$, which is a perfect integer square.

$$\frac{5}{\sqrt{20}} = \frac{5\sqrt{5}}{\sqrt{20}\sqrt{5}} \qquad \text{Multiply both numerator and denominator by } \sqrt{5}.$$

$$= \frac{5\sqrt{5}}{\sqrt{100}} \qquad \text{Use the multiplication property of radicals.}$$

$$= \frac{5\sqrt{5}}{10} \qquad \sqrt{100} = 10.$$

$$= \frac{\sqrt{5}}{2} \qquad \text{Simplify the fraction.}$$

■

EXAMPLE 9 Simplify the fraction $\dfrac{\sqrt{72x^5}}{\sqrt{45}}$. Assume $x > 0$.

Solution

$$\frac{\sqrt{72x^5}}{\sqrt{45}} = \frac{\sqrt{36 \cdot x^4 \cdot 2x}}{\sqrt{9 \cdot 5}}$$

Factor the radicands, looking for perfect square factors.

$$= \frac{\sqrt{36} \cdot \sqrt{x^4} \cdot \sqrt{2x}}{\sqrt{9} \cdot \sqrt{5}}$$

Use the multiplication property of radicals.

$$= \frac{6x^2\sqrt{2x}}{3\sqrt{5}}$$

Simplify.

$$= \frac{6x^2\sqrt{2x}\sqrt{5}}{3\sqrt{5}\sqrt{5}}$$

Multiply both numerator and denominator by $\sqrt{5}$ to rationalize the denominator.

$$= \frac{2x^2\sqrt{10x}}{5}$$

Simplify.   ∎

EXAMPLE 10 Simplify the fraction $\sqrt{\dfrac{3x^3y^2}{27xy^3}}$ $(x > 0$ and $y > 0)$.

Solution

$$\sqrt{\frac{3x^3y^2}{27xy^3}} = \sqrt{\frac{x^2}{9y}}$$

Simplify the fraction within the radical.

$$= \sqrt{\frac{x^2 \cdot y}{9y \cdot y}}$$

Multiply both numerator and denominator by y.

$$= \frac{\sqrt{x^2y}}{\sqrt{9y^2}}$$

Use the division property of radicals.

$$= \frac{x\sqrt{y}}{3y}$$

Simplify.   ∎

EXAMPLE 11 Simplify the fraction $\dfrac{2}{\sqrt{3} - 1}$.

Solution Since the denominator contains a radical, the fraction is not in simplified form. The denominator is a *binomial,* so multiplying the numerator and denominator by $\sqrt{3}$ will not make the denominator a rational number. The key is to multiply the numerator and denominator by the binomial $\sqrt{3} + 1$. This works because the product $(\sqrt{3} + 1)(\sqrt{3} - 1)$ has no radicals. Radical expressions such as $\sqrt{3} + 1$ and $\sqrt{3} - 1$ are called **conjugates** of each other.

$$\frac{2}{\sqrt{3} - 1} = \frac{2(\sqrt{3} + 1)}{(\sqrt{3} - 1)(\sqrt{3} + 1)}$$

Multiply both numerator and denominator by the conjugate of the denominator.

$$= \frac{2(\sqrt{3} + 1)}{3 - 1}$$

Multiply the binomials in the denominator.

$$= \frac{2(\sqrt{3} + 1)}{2} \qquad \text{Simplify.}$$

$$= \sqrt{3} + 1 \qquad \text{Divide out the common factor of 2.} \qquad \blacksquare$$

EXAMPLE 12 Simplify the fraction $\dfrac{10\sqrt{7}}{\sqrt{7x} + \sqrt{2x}}$.

Solution We multiply the numerator and denominator by the conjugate of the denominator, which is $\sqrt{7x} - \sqrt{2x}$. The product $\left(\sqrt{7x} + \sqrt{2x}\right)\left(\sqrt{7x} - \sqrt{2x}\right)$ has no radicals.

$$\frac{10\sqrt{7}}{\sqrt{7x} + \sqrt{2x}} = \frac{10\sqrt{7}\left(\sqrt{7x} - \sqrt{2x}\right)}{\left(\sqrt{7x} + \sqrt{2x}\right)\left(\sqrt{7x} - \sqrt{2x}\right)} \qquad \begin{array}{l}\text{Multiply both numerator}\\ \text{and denominator by}\\ \sqrt{7x} - \sqrt{2x}.\end{array}$$

$$= \frac{10\sqrt{7}\left(\sqrt{7x} - \sqrt{2x}\right)}{7x - 2x} \qquad \text{Multiply the binomials.}$$

$$= \frac{10\sqrt{7}\left(\sqrt{7x} - \sqrt{2x}\right)}{5x} \qquad \text{Simplify.}$$

$$= \frac{2\sqrt{7}\left(\sqrt{7x} - \sqrt{2x}\right)}{x} \qquad \text{Simplify the fraction.}$$

$$= \frac{14\sqrt{x} - 2\sqrt{14x}}{x} \qquad \text{Remove parentheses.} \qquad \blacksquare$$

ORALS *Do each multiplication. All variables represent positive numbers.*

1. $\sqrt{5}\sqrt{5}$

2. $\sqrt{2}\sqrt{50}$

3. $\sqrt{x}\left(\sqrt{x} + 2\right)$

4. $\sqrt{y}\left(4 - \sqrt{y}\right)$

5. $\left(\sqrt{x} + 1\right)\left(\sqrt{x} - 1\right)$

6. $\left(1 + \sqrt{z}\right)\left(1 - \sqrt{z}\right)$

Rationalize each denominator.

7. $\dfrac{1}{\sqrt{5}}$

8. $\dfrac{x}{\sqrt{x}}$

EXERCISE 8.4

In Exercises 1–62, do each multiplication. All variables represent positive numbers.

1. $\sqrt{3}\sqrt{3}$

2. $\sqrt{7}\sqrt{7}$

3. $\sqrt{2}\sqrt{8}$

4. $\sqrt{27}\sqrt{3}$

5. $\sqrt{16}\sqrt{4}$

6. $\sqrt{32}\sqrt{2}$

7. $\sqrt[3]{8}\sqrt[3]{8}$

8. $\sqrt[3]{4}\sqrt[3]{250}$

9. $\sqrt{x^3}\sqrt{x^3}$

10. $\sqrt{a^7}\sqrt{a^3}$

11. $\sqrt{b^8}\sqrt{b^6}$

12. $\sqrt{y^4}\sqrt{y^8}$

13. $(2\sqrt{5})(2\sqrt{3})$

14. $(4\sqrt{3})(2\sqrt{2})$

15. $(-5\sqrt{6})(4\sqrt{3})$

16. $(6\sqrt{3})(-7\sqrt{3})$

17. $(2\sqrt[3]{4})(3\sqrt[3]{16})$

18. $(-3\sqrt[3]{100})(\sqrt[3]{10})$

19. $(4\sqrt{x})(-2\sqrt{x})$

20. $(3\sqrt{y})(15\sqrt{y})$

21. $(-14\sqrt{50x})(-5\sqrt{20x})$

22. $(12\sqrt{24y})(-16\sqrt{2y})$

23. $\sqrt{8x}\sqrt{2x^3y}$

24. $\sqrt{27y}\sqrt{3y^3}$

25. $\sqrt{2}(\sqrt{2}+1)$

26. $\sqrt{3}(\sqrt{3}-2)$

27. $\sqrt{3}(\sqrt{27}-1)$

28. $\sqrt{2}(\sqrt{8}-1)$

29. $\sqrt{7}(\sqrt{7}-3)$

30. $\sqrt{5}(\sqrt{5}+2)$

31. $\sqrt{5}(3-\sqrt{5})$

32. $\sqrt{7}(2+\sqrt{7})$

33. $\sqrt{3}(\sqrt{6}+1)$

34. $\sqrt{2}(\sqrt{6}-2)$

35. $\sqrt[3]{7}(\sqrt[3]{49}-2)$

36. $\sqrt[3]{5}(\sqrt[3]{25}+3)$

37. $\sqrt{x}(\sqrt{3x}-2)$

38. $\sqrt{y}(\sqrt{y}+5)$

39. $2\sqrt{x}(\sqrt{9x}+3)$

40. $3\sqrt{z}(\sqrt{4z}-\sqrt{z})$

41. $3\sqrt{x}(2+\sqrt{x})$

42. $5\sqrt{y}(5-\sqrt{5y})$

43. $\sqrt{21x}(\sqrt{3x}+\sqrt{2x})$

44. $\sqrt{35y}(\sqrt{7y}-\sqrt{5y})$

45. $(\sqrt{2}+1)(\sqrt{2}-1)$

46. $(\sqrt{3}-1)(\sqrt{3}+1)$

47. $(\sqrt{5}+2)(\sqrt{5}-2)$

48. $(\sqrt{7}+5)(\sqrt{7}-5)$

49. $(\sqrt[3]{2}+1)(\sqrt[3]{2}+1)$

50. $(\sqrt[3]{5}-2)(\sqrt[3]{5}-2)$

51. $(\sqrt{7}-x)(\sqrt{7}+x)$

52. $(\sqrt{2}-\sqrt{x})(\sqrt{x}+\sqrt{2})$

53. $(\sqrt{2}-\sqrt{x})^2$

54. $(\sqrt{a}+\sqrt{3})^2$

55. $(\sqrt{6x}+\sqrt{7})(\sqrt{6x}-\sqrt{7})$

56. $(\sqrt{8y}+\sqrt{2z})(\sqrt{8y}-\sqrt{2z})$

57. $(\sqrt{2x}+3)(\sqrt{8x}-6)$

58. $(\sqrt{5y}-3)(\sqrt{20y}+6)$

59. $(\sqrt{8xy}+1)(\sqrt{8xy}+1)$

60. $(\sqrt{5x}+3\sqrt{y})(\sqrt{5x}-3\sqrt{y})$

61. $(\sqrt{16x}-\sqrt{8})(\sqrt{16x}+\sqrt{8})$

62. $(\sqrt{9xz}+2\sqrt{y})(\sqrt{25xz}-\sqrt{y})$

In Exercises 63–74, simplify each expression. Assume that all variables represent positive numbers.

63. $\dfrac{\sqrt{12x^3}}{\sqrt{27x}}$

64. $\dfrac{\sqrt{32}}{\sqrt{98x^2}}$

65. $\dfrac{\sqrt{18xy^2}}{\sqrt{25x}}$

66. $\dfrac{\sqrt{27y^3}}{\sqrt{75x^2y}}$

67. $\dfrac{\sqrt{196xy^3}}{\sqrt{49x^3y}}$

68. $\dfrac{\sqrt{50xyz^4}}{\sqrt{98xyz^2}}$

69. $\dfrac{\sqrt[3]{16x^6}}{\sqrt[3]{54x^3}}$

70. $\dfrac{\sqrt[3]{128a^6b^3}}{\sqrt[3]{16a^3b^6}}$

71. $\dfrac{\sqrt{3x^2y^3}}{\sqrt{27x}}$

72. $\dfrac{\sqrt{44x^2y^5}}{\sqrt{99x^4y}}$

73. $\dfrac{\sqrt{5x}\sqrt{10y^2}}{\sqrt{x^3y}}$

74. $\dfrac{\sqrt{7y}\sqrt{14x}}{\sqrt{8xy}}$

In Exercises 75–106, do each division by rationalizing a denominator and simplifying. All variables represent positive numbers.

75. $\dfrac{1}{\sqrt{3}}$

76. $\dfrac{1}{\sqrt{5}}$

77. $\dfrac{2}{\sqrt{7}}$

78. $\dfrac{3}{\sqrt{11}}$

79. $\dfrac{5}{\sqrt[3]{5}}$

80. $\dfrac{7}{\sqrt[3]{7}}$

81. $\dfrac{9}{\sqrt{27}}$

82. $\dfrac{4}{\sqrt{20}}$

83. $\dfrac{3}{\sqrt{32}}$

84. $\dfrac{5}{\sqrt{18}}$

85. $\dfrac{4}{\sqrt[3]{4}}$

86. $\dfrac{7}{\sqrt[3]{10}}$

87. $\dfrac{\sqrt{5}}{\sqrt{3}}$

88. $\dfrac{\sqrt{3}}{\sqrt{5}}$

89. $\dfrac{10}{\sqrt{x}}$

90. $\dfrac{12}{\sqrt{y}}$

91. $\dfrac{\sqrt{9}}{\sqrt{2x}}$

92. $\dfrac{\sqrt{4}}{\sqrt{3z}}$

93. $\dfrac{\sqrt{2x}}{\sqrt{9y}}$

94. $\dfrac{\sqrt{3xy}}{\sqrt{4x}}$

95. $\dfrac{2\sqrt{3}}{\sqrt{8x^2}}$

96. $\dfrac{3\sqrt{5}}{\sqrt{27y^2}}$

97. $\dfrac{5\sqrt{6x}}{\sqrt{50}}$

98. $\dfrac{8\sqrt{10y}}{\sqrt{40}}$

99. $\dfrac{\sqrt[3]{5}}{\sqrt[3]{2}}$

100. $\dfrac{\sqrt[3]{2}}{\sqrt[3]{5}}$

101. $\dfrac{\sqrt[3]{2x^2}}{\sqrt[3]{2x}}$

102. $\dfrac{\sqrt[3]{3y^4}}{\sqrt[3]{3y}}$

103. $\dfrac{2}{\sqrt[3]{4x^2y}}$

104. $\dfrac{3}{\sqrt[3]{9xy^2}}$

105. $\dfrac{-5}{\sqrt[3]{25a^2b^2}}$

106. $\dfrac{-4}{\sqrt[3]{4ab^2c^2}}$

In Exercises 107–130, do each division by rationalizing the denominator and simplifying. All variables represent positive numbers.

107. $\dfrac{3}{\sqrt{3}-1}$

108. $\dfrac{3}{\sqrt{5}-2}$

109. $\dfrac{3}{\sqrt{7}+2}$

110. $\dfrac{5}{\sqrt{8}+3}$

111. $\dfrac{12}{3-\sqrt{3}}$

112. $\dfrac{10}{5-\sqrt{5}}$

113. $\dfrac{\sqrt{2}}{\sqrt{2}+1}$

114. $\dfrac{\sqrt{3}}{\sqrt{3}-1}$

115. $\dfrac{-\sqrt{3}}{\sqrt{3}+1}$

116. $\dfrac{-\sqrt{2}}{\sqrt{2}-1}$

117. $\dfrac{5}{\sqrt{3}+\sqrt{2}}$

118. $\dfrac{3}{\sqrt{3}-\sqrt{2}}$

119. $\dfrac{\sqrt{8}}{\sqrt{5}-\sqrt{3}}$

120. $\dfrac{\sqrt{32}}{\sqrt{7}-\sqrt{3}}$

121. $\dfrac{\sqrt{3}-\sqrt{2}}{\sqrt{3}+\sqrt{2}}$

122. $\dfrac{\sqrt{5}+\sqrt{3}}{\sqrt{5}-\sqrt{3}}$

123. $\dfrac{\sqrt{3x}-1}{\sqrt{3x}+1}$

124. $\dfrac{\sqrt{5x}+3}{\sqrt{5x}-3}$

125. $\dfrac{\sqrt{2x}+5}{\sqrt{2x}+3}$

126. $\dfrac{\sqrt{3y}+3}{\sqrt{3y}-2}$

127. $\dfrac{1 - \sqrt{2z}}{\sqrt{2z} + 1}$ **128.** $\dfrac{1 + \sqrt{3z}}{\sqrt{3z} - 1}$ **129.** $\dfrac{y - \sqrt{15}}{y + \sqrt{15}}$ **130.** $\dfrac{x + \sqrt{17}}{x - \sqrt{17}}$

Writing Exercises ■ *Write a paragraph using your own words.*

1. How do you know when a radical has been simplified?

2. Explain how to rationalize a denominator.

Something to Think About ■ **1.** How would you make the numerator of $\dfrac{\sqrt{3}}{2}$ a rational number?

2. Rationalize the numerator of $\dfrac{\sqrt{5} + 2}{5}$.

Review Exercises ■ *Factor each polynomial.*

1. $x^2 - 4x - 21$ **2.** $y^2 + 6y - 27$ **3.** $6x^2y - 15xy$ **4.** $x^3 + 8$

8.5 Solving Equations Containing Radicals

■ Solving Equations Containing Square Roots ■ Solving Equations Containing Cube Roots ■ Applications

GETTING READY *Find each power.*

1. $\left(\sqrt{x}\right)^2$ **2.** $\left(\sqrt{3y}\right)^2$ **3.** $\left(\sqrt{x - 2}\right)^2$ **4.** $\left(\sqrt[3]{y + 3}\right)^3$

5. $\left(\sqrt[4]{x}\right)^4$ **6.** $\left(\sqrt[5]{x - 2}\right)^5$ **7.** $(x + 1)^2$ **8.** $(2y - 3)^2$

In this section, we will solve equations that contain radicals. To do so, we note that if two numbers are equal, then their squares are equal.

Squaring Property of Equality If $a = b$, then $a^2 = b^2$.

The only solution of the equation $x = 2$ is 2. However, if we square both sides, we get $x^2 = 4$. This equation has two solutions, 2 and -2. Thus, squaring both sides of $x = 2$ leads to another equation with more solutions than the first. No solutions were lost, however. The number 2 is still a solution of $x^2 = 4$.

■ Solving Equations Containing Square Roots

To solve the equation $\sqrt{x + 2} = 3$, we must square both sides to eliminate the radical. Since this might produce an equation with more solutions than the original equation, we must check each solution of the squared equation in the original equation.

EXAMPLE 1 Solve $\sqrt{x + 2} = 3$.

Solution We square both sides to eliminate the radical and proceed as follows:

$$\sqrt{x + 2} = 3$$
$$\left(\sqrt{x + 2}\right)^2 = 3^2 \qquad \text{Square both sides.}$$
$$x + 2 = 9 \qquad \text{Simplify.}$$
$$x = 7 \qquad \text{Subtract 2 from both sides.}$$

We check the solution by substituting 7 for x in the original equation.

$$\sqrt{x + 2} = 3$$
$$\sqrt{7 + 2} \overset{?}{=} 3 \qquad \text{Replace } x \text{ with 7.}$$
$$\sqrt{9} \overset{?}{=} 3$$
$$3 = 3$$

The solution checks. Since no solutions are lost in this process, 7 is the only solution of the original equation. ■

EXAMPLE 2 Solve the equation $\sqrt{x + 1} + 5 = 3$.

Solution We isolate the radical on one side of the equation and proceed as follows:

$$\sqrt{x + 1} + 5 = 3$$
$$\sqrt{x + 1} = -2 \qquad \text{Subtract 5 from both sides.}$$
$$\left(\sqrt{x + 1}\right)^2 = (-2)^2 \qquad \text{Square both sides.}$$
$$x + 1 = 4 \qquad \text{Simplify.}$$
$$x = 3 \qquad \text{Subtract 1 from both sides.}$$

We check the solution by substituting 3 for x in the original equation.

$$\sqrt{x + 1} + 5 = 3$$
$$\sqrt{3 + 1} + 5 \overset{?}{=} 3 \qquad \text{Replace } x \text{ with 3.}$$
$$\sqrt{4} + 5 \overset{?}{=} 3$$
$$2 + 5 \overset{?}{=} 3$$
$$7 \neq 3$$

Since $7 \neq 3$, 3 is not a solution. Because no solutions are lost, the original equation has no solution. ■

Example 2 shows that squaring both sides of an equation can lead to **extraneous solutions.** These solutions do not satisfy the original equation and must be discarded.

We follow these steps to solve an equation containing radical expressions.

Steps for Solving Radical Equations	1. Whenever possible, isolate a single radical on one side of the equation.
	2. Square both sides of the equation and solve the resulting equation.
	3. Check the solution in the original equation. This step is required.

EXAMPLE 3   Solve the equation $\sqrt{x + 12} = 3\sqrt{x + 4}$.

Solution   We square both sides to eliminate the radical expressions and proceed as follows:

$$\sqrt{x + 12} = 3\sqrt{x + 4}$$
$$\left(\sqrt{x + 12}\right)^2 = \left(3\sqrt{x + 4}\right)^2 \qquad \text{Square both sides.}$$
$$x + 12 = 9(x + 4) \qquad \text{Simplify.}$$
$$x + 12 = 9x + 36 \qquad \text{Remove parentheses.}$$
$$-24 = 8x \qquad \text{Subtract } x \text{ and } 36 \text{ from both sides.}$$
$$-3 = x \qquad \text{Divide both sides by 8.}$$

We check the solution by substituting -3 for x in the original equation.

$$\sqrt{x + 12} = 3\sqrt{x + 4}$$
$$\sqrt{-3 + 12} \stackrel{?}{=} 3\sqrt{-3 + 4} \qquad \text{Replace } x \text{ with } -3.$$
$$\sqrt{9} \stackrel{?}{=} 3\sqrt{1}$$
$$3 = 3$$

The solution checks.   ■

EXAMPLE 4   Solve the equation $x = \sqrt{2x + 10} - 1$.

Solution   We add 1 to both sides to isolate the radical on the right-hand side and then square both sides to eliminate the radical.

WARNING!   Be sure to square the entire left-hand side.

$$x = \sqrt{2x + 10} - 1$$
$$x + 1 = \sqrt{2x + 10} \qquad \text{Add 1 to both sides to isolate the radical.}$$
$$(x + 1)^2 = \left(\sqrt{2x + 10}\right)^2 \qquad \text{Square both sides.}$$
$$x^2 + 2x + 1 = 2x + 10 \qquad \text{Remove parentheses.}$$
$$x^2 - 9 = 0 \qquad \text{Subtract } 2x \text{ and } 10 \text{ from both sides.}$$
$$(x - 3)(x + 3) = 0 \qquad \text{Factor.}$$

$$x - 3 = 0 \quad \text{or} \quad x + 3 = 0 \qquad \text{Set each factor equal to 0.}$$
$$x = 3 \quad | \quad x = -3 \qquad \text{Solve each linear equation.}$$

We check each possible solution.

For $x = 3$	For $x = -3$
$x = \sqrt{2x + 10} - 1$	$x = \sqrt{2x + 10} - 1$
$3 \stackrel{?}{=} \sqrt{2(3) + 10} - 1$	$-3 \stackrel{?}{=} \sqrt{2(-3) + 10} - 1$
$3 \stackrel{?}{=} \sqrt{16} - 1$	$-3 \stackrel{?}{=} \sqrt{4} - 1$
$3 \stackrel{?}{=} 4 - 1$	$-3 \stackrel{?}{=} 2 - 1$
$3 = 3$	$-3 \neq 1$

Since 3 is the only number that checks, it is the only solution. The false solution -3 is extraneous. ■

▓ Solving Equations Containing Cube Roots

We can use the power theorem to solve equations whose radicals have indexes greater than 2.

The Power Theorem Let n be a real number.
  If $a = b$, then $a^n = b^n$.

EXAMPLE 5 Solve the equation $\sqrt[3]{2x + 10} = 2$.

Solution We cube both sides and proceed as follows:

$$\sqrt[3]{2x + 10} = 2$$
$$\left(\sqrt[3]{2x + 10}\right)^3 = (2)^3 \qquad \text{Cube both sides.}$$
$$2x + 10 = 8 \qquad \text{Simplify.}$$
$$2x = -2 \qquad \text{Subtract 10 from both sides.}$$
$$x = -1 \qquad \text{Divide both sides by 2.}$$

Check the result. ■

▓ Applications

EXAMPLE 6 The time t (in seconds) required for the pendulum in Figure 8-1 to swing through one complete cycle is given by the formula

$$t = 1.11\sqrt{L}$$

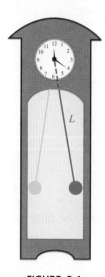

Find the length (in feet) of the pendulum if it completes one cycle in 2 seconds.

Solution We substitute 2 for t in the formula and solve for L.

$$t = 1.11\sqrt{L}$$

$$2 = 1.11\sqrt{L}$$

$$\frac{2}{1.11} = \sqrt{L} \qquad \text{Divide both sides by 1.11.}$$

$$\left(\frac{2}{1.11}\right)^2 = L \qquad \text{Square both sides.}$$

$$L = 3.2464897 \qquad \text{Simplify.}$$

The pendulum is approximately 3.2 feet long. ■

FIGURE 8-1

ORALS *Solve each equation and check the solution.*

1. $\sqrt{x} = 4$ **2.** $\sqrt{x-1} = 2$

3. $\sqrt{x+1} = 2$ **4.** $\sqrt{2x} = 1$

5. $\sqrt{\dfrac{x}{2}} = 1$ **6.** $\sqrt{2x-1} = 1$

E X E R C I S E 8.5

In Exercises 1–48, solve each equation. Check all solutions. If an equation has no solutions, so indicate.

1. $\sqrt{x} = 3$ **2.** $\sqrt{x} = 5$ **3.** $\sqrt{x} = 7$ **4.** $\sqrt{x} = 2$

5. $\sqrt{x} = -4$ **6.** $\sqrt{x} = -1$ **7.** $\sqrt{x+3} = 2$ **8.** $\sqrt{x-2} = 3$

9. $\sqrt{x-5} = 5$ **10.** $\sqrt{x+8} = 12$ **11.** $\sqrt{3-x} = -2$ **12.** $\sqrt{5-x} = 10$

13. $\sqrt{6+2x} = 4$ **14.** $\sqrt{7+x} = -4$ **15.** $\sqrt{5x-5} = 5$ **16.** $\sqrt{6x+19} = 7$

17. $\sqrt{4x-3} = 3$ **18.** $\sqrt{11x-2} = 3$ **19.** $\sqrt{13x+14} = 1$ **20.** $\sqrt{8x+9} = 1$

21. $\sqrt{x+3} + 5 = 12$ **22.** $\sqrt{x-5} - 3 = 4$ **23.** $\sqrt{2x+10} + 3 = 5$ **24.** $\sqrt{3x+4} + 7 = 12$

25. $\sqrt{5x+9} + 4 = 7$ **26.** $\sqrt{9x+25} - 2 = 3$ **27.** $\sqrt{7-5x} + 4 = 3$ **28.** $\sqrt{7+6x} - 4 = -3$

29. $\sqrt{3x+3} = 3\sqrt{x-1}$ **30.** $2\sqrt{4x+5} = 5\sqrt{x+4}$

31. $2\sqrt{3x + 4} = \sqrt{5x + 9}$

32. $\sqrt{10 - 3x} = \sqrt{2x + 20}$

33. $\sqrt{3x + 6} = 2\sqrt{2x - 11}$

34. $2\sqrt{9x + 16} = \sqrt{3x + 64}$

35. $\sqrt{x + 1} = x - 1$

36. $\sqrt{x + 4} = x - 2$

37. $\sqrt{x + 1} = x + 1$

38. $\sqrt{x + 9} = x + 7$

39. $\sqrt{7x + 2} - 2x = 0$

40. $\sqrt{3x + 3} + 5 = x$

41. $x - 1 = \sqrt{x - 1}$

42. $x - 2 = \sqrt{x + 10}$

43. $x = \sqrt{3 - x} + 3$

44. $x = \sqrt{x - 4} + 4$

45. $\sqrt[3]{x - 1} = 4$

46. $\sqrt[3]{2x + 5} = 3$

47. $\sqrt[3]{\frac{1}{2}x - 3} = 2$

48. $\sqrt[4]{x + 4} = 1$

49. Integer problem The square root of the sum of a certain integer and 8 is 13. Find the integer.

50. Integer problem The square root of the sum of a certain integer and 12 is 15. Find the integer.

51. Falling objects The distance s (in feet) that an object will fall in t seconds is given by the formula

$$t = \frac{\sqrt{s}}{4}$$

How deep is a shaft if a stone dropped down it hits bottom in 4 seconds?

52. Falling objects How deep would the shaft in Exercise 51 be if the stone hit bottom in 3 seconds?

53. Horology Find the length of a pendulum that completes one cycle in $\frac{3}{2}$ seconds.

54. Foucault pendulum A very long pendulum in Chicago's Museum of Science and Industry completes one cycle in 8.91 seconds. How long is it? (See Illustration 1.)

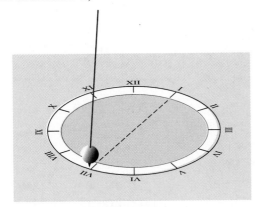

ILLUSTRATION 1

55. Electronics The current I (measured in amperes), the resistance R (measured in ohms), and the power P (measured in watts) are related by the formula

$$I = \sqrt{\frac{P}{R}}$$

Find the power used by an electrical appliance that draws 7 amps when the resistance is 20 ohms.

56. Electronics Find the resistance of a 500-watt space heater that draws 7 amperes. (See Exercise 55.)

57. Law enforcement Police sometimes use the formula $s = k\sqrt{d}$ to estimate the speed s (in miles per hour) of a car involved in an accident. In this formula, d represents the distance of a skid, and k is a constant that depends on the condition of the pavement. For wet pavement, $k = 3.24$. How fast was a car going if its skid was 300 feet on wet pavement?

58. Law enforcement How fast was a car going if its skid was 300 feet on dry pavement? For dry pavement, $k = 5.34$. (See Exercise 57.)

59. Satellite orbits The orbital speed, s, of an earth satellite is related to its distance, r, from the earth's center by the formula

$$s = \frac{2.029 \times 10^7}{\sqrt{r}}$$

(continued on page 502)

If the satellite's orbital speed is 7×10^3 meters per second, find its altitude, a, above the earth's surface. (See Illustration 2.)

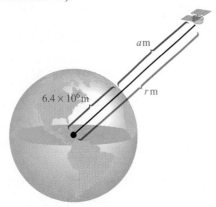

ILLUSTRATION 2

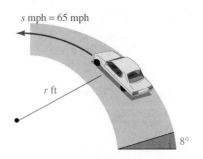

s mph = 65 mph

r ft

8°

ILLUSTRATION 3

60. Banked curves A highway curve banked at 8° will accommodate traffic traveling s miles per hour if the radius of the curve is r feet, according to the formula

$$s = 1.45\sqrt{r}$$

If highway engineers expect 65-mph traffic, what radius should they specify? (See Illustration 3.)

61. Einstein's Theory of Relativity predicts that an object moving at speed v will be shortened in the direction of its motion by a factor f given by

$$f = \sqrt{1 - \frac{v^2}{c^2}}$$

where c is the speed of light. Solve this formula for v^2.

62. Einstein's Theory of Relativity predicts that a clock moving at speed v will run slower by a factor f given by

$$f = \frac{1}{\sqrt{1 - \frac{v^2}{c^2}}}$$

where c is the speed of light. Solve this formula for v^2.

Writing Exercises ■ *Write a paragraph using your own words.*

1. Explain why a check is necessary when solving radical equations.

2. How would you know, without solving it, that the equation $\sqrt{x + 2} = -4$ has no solutions?

Something to Think About ■ *Find the error in each solution.*

1. $\sqrt{x - 2} + 2 = 3$
$x - 2 + 4 = 9$
$x + 2 = 9$
$x = 7$

2. $2 = \sqrt{x^2 - 9}$
$2 = x - 3$
$5 = x$

Review Exercises ■ *Solve each system of equations.*

1. $\begin{cases} x + y = 5 \\ x - y = -1 \end{cases}$
2. $\begin{cases} 2x + y = 0 \\ x + 3y = 5 \end{cases}$
3. $\begin{cases} 2x + 3y = 0 \\ 3x - 2y = 13 \end{cases}$
4. $\begin{cases} 3x - 4y = 11 \\ 4x + y = -17 \end{cases}$

8.6 Rational Exponents

■ Fractional Exponents with Numerators of 1 ■ Fractional Exponents with Numerators Other Than 1

GETTING READY *Simplify each expression.*

1. x^2x^4 **2.** $(x^3)^4$ **3.** $\dfrac{x^7}{x^3}$ **4.** x^0

5. x^{-3} **6.** $(xy^2)^3$ **7.** $\left(\dfrac{x^4}{y^3}\right)^2$ **8.** $(x^2x^3)^3$

We have seen that a positive integer exponent indicates the number of times that a base is to be used as a factor in a product. For example, x^4 means that x is to be used as a factor four times.

$$x^4 = x \cdot x \cdot x \cdot x$$

Furthermore, we recall the following rules of exponents.

Rules of Exponents   If m and n are natural numbers and there are no divisions by 0, then

$$x^m x^n = x^{m+n} \qquad (x^m)^n = x^{mn} \qquad (xy)^n = x^n y^n \qquad \left(\frac{x}{y}\right)^n = \frac{x^n}{y^n}$$

$$x^0 = 1 \qquad x^{-1} = \frac{1}{x} \qquad \frac{x^m}{x^n} = x^{m-n}$$

■ Fractional Exponents with Numerators of 1

Exponents do not need to be integers. It is possible to raise certain bases to fractional powers. To give meaning to rational (fractional) exponents, we consider the number $\sqrt{7}$. Because $\sqrt{7}$ is the positive number whose square is 7, we have

$$\left(\sqrt{7}\right)^2 = 7$$

We now consider the symbol $7^{1/2}$. If we demand that fractional exponents obey the same rules as integral exponents, the square of $7^{1/2}$ must be 7, because

$$(7^{1/2})^2 = 7^{(1/2)2}$$
$$= 7^1$$
$$= 7$$

Since $(7^{1/2})^2 = 7$, and $\left(\sqrt{7}\right)^2 = 7$, we define $7^{1/2}$ to be $\sqrt{7}$. Similarly, we define

$$7^{1/3} \quad \text{to be} \quad \sqrt[3]{7}$$
$$7^{1/7} \quad \text{to be} \quad \sqrt[7]{7}$$

and so on. In general, we define the **rational exponent** $\frac{1}{n}$ as follows.

Rational Exponent   If n is a positive integer greater than 1 and $\sqrt[n]{x}$ is a real number, then
$$x^{1/n} = \sqrt[n]{x}$$

EXAMPLE 1   Simplify   **a.** $64^{1/2}$,   **b.** $64^{1/3}$,   **c.** $(-64)^{1/3}$,   and   **d.** $64^{1/6}$.

Solution   **a.** $64^{1/2} = \sqrt{64} = 8$ **b.** $64^{1/3} = \sqrt[3]{64} = 4$

c. $(-64)^{1/3} = \sqrt[3]{-64} = -4$ **d.** $64^{1/6} = \sqrt[6]{64} = 2$ ∎

■ Fractional Exponents with Numerators Other Than 1

We can extend the definition of $x^{1/n}$ to cover fractional exponents for which the numerator is not 1. For example, because $4^{3/2}$ can be written as $(4^{1/2})^3$, we have

$$4^{3/2} = (4^{1/2})^3 = \left(\sqrt{4}\right)^3 = 2^3 = 8$$

Similarly, because $4^{3/2}$ can be written as $(4^3)^{1/2}$, we have

$$4^{3/2} = (4^3)^{1/2} = 64^{1/2} = \sqrt{64} = 8$$

In general, $x^{m/n}$ can be written as $(x^{1/n})^m$ or as $(x^m)^{1/n}$. Since $(x^{1/n})^m = \left(\sqrt[n]{x}\right)^m$ and $(x^m)^{1/n} = \sqrt[n]{x^m}$, we make the following definition.

Changing from Rational Exponents to Radicals   If m and n are positive integers, x is nonnegative, and the fraction m/n cannot be simplified, then
$$x^{m/n} = \sqrt[n]{x^m} = \left(\sqrt[n]{x}\right)^m$$

EXAMPLE 2   Simplify   **a.** $8^{2/3}$ and   **b.** $(-27)^{3/4}$.

Solution   **a.**
$$\begin{aligned} 8^{2/3} &= \left(\sqrt[3]{8}\right)^2 \\ &= 2^2 \\ &= 4 \end{aligned} \qquad \text{or} \qquad \begin{aligned} 8^{2/3} &= \sqrt[3]{8^2} \\ &= \sqrt[3]{64} \\ &= 4 \end{aligned}$$

b.
$$\begin{aligned} (-27)^{4/3} &= \left(\sqrt[3]{-27}\right)^4 \\ &= (-3)^4 \\ &= 81 \end{aligned} \qquad \text{or} \qquad \begin{aligned} (-27)^{4/3} &= \sqrt[3]{(-27)^4} \\ &= \sqrt[3]{531,441} \\ &= 81 \end{aligned}$$ ∎

The work in Example 2 suggests that in order to avoid large numbers, it is usually easier to take the root of the base first.

EXAMPLE 3   Simplify   **a.** $125^{4/3}$,   **b.** $9^{5/2}$,   **c.** $-25^{3/2}$,   and   **d.** $(-27)^{2/3}$.

Solution   **a.** $125^{4/3} = \left(\sqrt[3]{125}\right)^4$ **b.** $9^{5/2} = \left(\sqrt{9}\right)^5$
$$= (5)^4$$
$$= 625$$
$$= (3)^5$$
$$= 243$$

c. $-25^{3/2} = -\left(\sqrt{25}\right)^3$ **d.** $(-27)^{2/3} = \left(\sqrt[3]{-27}\right)^2$
$$= -(5)^3$$
$$= (-3)^2$$
$$= -125$$
$$= 9$$ ■

Because of the definition of $x^{1/n}$, the familiar rules of exponents are valid for rational exponents. The following example illustrates the use of each rule.

EXAMPLE 4   **a.** $4^{2/5}4^{1/5} = 4^{2/5+1/5} = 4^{3/5}$ $x^m x^n = x^{m+n}$

b. $(5^{2/3})^{1/2} = 5^{(2/3)(1/2)} = 5^{1/3}$ $(x^m)^n = x^{mn}$

c. $(3x)^{2/3} = 3^{2/3}x^{2/3}$ $(xy)^m = x^m y^m$

d. $\dfrac{4^{3/5}}{4^{2/5}} = 4^{3/5-2/5} = 4^{1/5}$ $\dfrac{x^m}{x^n} = x^{m-n}$

e. $\left(\dfrac{3}{2}\right)^{2/5} = \dfrac{3^{2/5}}{2^{2/5}}$ $\left(\dfrac{x}{y}\right)^n = \dfrac{x^n}{y^n}$

f. $4^{-2/3} = \dfrac{1}{4^{2/3}}$ $x^{-n} = \dfrac{1}{x^n}$

g. $5^0 = 1$ $x^0 = 1$ ■

We can often use the rules of exponents to simplify expressions containing rational exponents.

EXAMPLE 5   Simplify   **a.** $64^{-2/3}$,   **b.** $(x^2)^{1/2}$,   **c.** $(x^6y^4)^{1/2}$,   and   **d.** $(27x^{12})^{-1/3}$. $(x > 0$ and $y > 0.)$

Solution   **a.** $64^{-2/3} = \dfrac{1}{64^{2/3}}$ **b.** $(x^2)^{1/2} = x^{2(1/2)}$
$$= x^1$$
$$= \dfrac{1}{(64^{1/3})^2}$$
$$= x$$
$$= \dfrac{1}{4^2}$$
$$= \dfrac{1}{16}$$

c. $(x^6y^4)^{1/2} = x^{6(1/2)}y^{4(1/2)}$
$= x^3y^2$

d. $(27x^{12})^{-1/3} = \dfrac{1}{(27x^{12})^{1/3}}$

$= \dfrac{1}{27^{1/3}x^{12(1/3)}}$

$= \dfrac{1}{3x^4}$ ∎

EXAMPLE 6 Simplify **a.** $x^{1/3}x^{1/2}$, **b.** $\dfrac{3x^{2/3}}{6x^{1/5}}$, and **c.** $\dfrac{2x^{-1/2}}{x^{3/4}}$. Assume $x > 0$.

Solution **a.** $x^{1/3}x^{1/2} = x^{2/6}x^{3/6}$ Get a common denominator in the fractional exponents.
$= x^{5/6}$ Keep the base and add the exponents.

b. $\dfrac{3x^{2/3}}{6x^{1/5}} = \dfrac{3x^{10/15}}{6x^{3/15}}$ Get a common denominator in the fractional exponents.

$= \dfrac{1}{2}x^{10/15-3/15}$ Simplify $\frac{3}{6}$ and keep the base and subtract the exponents.

$= \dfrac{1}{2}x^{7/15}$

c. $\dfrac{2x^{-1/2}}{x^{3/4}} = \dfrac{2x^{-2/4}}{x^{3/4}}$ Get a common denominator in the fractional exponents.

$= 2x^{-2/4-3/4}$ Keep the base and subtract the exponents.

$= 2x^{-5/4}$ Simplify.

$= \dfrac{2}{x^{5/4}}$ $x^{-5/4} = \dfrac{1}{x^{5/4}}$. ∎

ORALS *Find each value.*

1. $16^{1/2}$ **2.** $25^{1/2}$ **3.** $27^{1/3}$ **4.** $81^{1/4}$
5. $8^{2/3}$ **6.** $32^{3/5}$ **7.** $9^{-1/2}$ **8.** $64^{-1/3}$

E X E R C I S E 8.6

In Exercises 1–24, simplify each expression.

1. $81^{1/2}$ **2.** $100^{1/2}$ **3.** $-144^{1/2}$ **4.** $-400^{1/2}$

5. $\left(\dfrac{1}{4}\right)^{1/2}$ **6.** $\left(\dfrac{1}{25}\right)^{1/2}$ **7.** $\left(\dfrac{4}{49}\right)^{1/2}$ **8.** $\left(\dfrac{9}{64}\right)^{1/2}$

9. $27^{1/3}$ **10.** $8^{1/3}$ **11.** $-125^{1/3}$ **12.** $-1000^{1/3}$

13. $(-8)^{1/3}$

14. $(-125)^{1/3}$

15. $\left(\dfrac{1}{64}\right)^{1/3}$

16. $\left(\dfrac{1}{1000}\right)^{1/3}$

17. $\left(\dfrac{27}{64}\right)^{1/3}$

18. $\left(\dfrac{64}{125}\right)^{1/3}$

19. $16^{1/4}$

20. $81^{1/4}$

21. $32^{1/5}$

22. $-32^{1/5}$

23. $-243^{1/5}$

24. $\left(-\dfrac{1}{32}\right)^{1/5}$

In Exercises 25–44, simplify each expression.

25. $81^{3/2}$

26. $16^{3/2}$

27. $25^{3/2}$

28. $4^{5/2}$

29. $125^{2/3}$

30. $8^{4/3}$

31. $1000^{2/3}$

32. $27^{2/3}$

33. $(-8)^{2/3}$

34. $(-125)^{2/3}$

35. $32^{3/5}$

36. $-243^{3/5}$

37. $81^{3/4}$

38. $256^{3/4}$

39. $(-32)^{3/5}$

40. $243^{2/5}$

41. $\left(\dfrac{8}{27}\right)^{2/3}$

42. $\left(\dfrac{27}{64}\right)^{2/3}$

43. $\left(\dfrac{16}{625}\right)^{3/4}$

44. $\left(\dfrac{49}{64}\right)^{3/2}$

In Exercises 45–68, simplify each expression. Write each answer without using negative exponents.

45. $6^{3/5}6^{2/5}$

46. $3^{4/7}3^{3/7}$

47. $5^{2/3}5^{4/3}$

48. $2^{7/8}2^{9/8}$

49. $(7^{2/5})^{5/2}$

50. $(8^{1/3})^3$

51. $(5^{2/7})^7$

52. $(3^{3/8})^8$

53. $\dfrac{8^{3/2}}{8^{1/2}}$

54. $\dfrac{11^{9/7}}{11^{2/7}}$

55. $\dfrac{5^{11/3}}{5^{2/3}}$

56. $\dfrac{27^{13/15}}{27^{8/15}}$

57. $(2^{1/2}3^{1/2})^2$

58. $(3^{2/3}5^{1/3})^3$

59. $(4^{3/4}3^{1/4})^4$

60. $(2^{1/5}3^{2/5})^5$

61. $4^{-1/2}$

62. $8^{-1/3}$

63. $27^{-2/3}$

64. $36^{-3/2}$

65. $16^{-3/2}$

66. $100^{-5/2}$

67. $(-27)^{-4/3}$

68. $(-8)^{-4/3}$

In Exercises 69–88, simplify each expression. Assume that all variables represent positive numbers.

69. $(x^{1/2})^2$

70. $(x^9)^{1/3}$

71. $(x^{12})^{1/6}$

72. $(x^{18})^{1/9}$

73. $(x^{18})^{2/9}$

74. $(x^{12})^{3/4}$

75. $x^{5/6}x^{7/6}$

76. $x^{2/3}x^{7/3}$

77. $y^{4/7}y^{10/7}$

78. $y^{5/11}y^{6/11}$

79. $\dfrac{x^{3/5}}{x^{1/5}}$

80. $\dfrac{x^{4/3}}{x^{2/3}}$

81. $\dfrac{x^{1/7}x^{3/7}}{x^{2/7}}$

82. $\dfrac{x^{5/6}x^{5/6}}{x^{7/6}}$

83. $\left(\dfrac{x^{3/5}}{x^{2/5}}\right)^5$

84. $\left(\dfrac{x^{2/9}}{x^{1/9}}\right)^9$

85. $\left(\dfrac{y^{2/7}y^{3/7}}{y^{4/7}}\right)^{49}$

86. $\left(\dfrac{z^{3/5}z^{6/5}}{z^{2/5}}\right)^5$

87. $\left(\dfrac{y^{5/6}y^{7/6}}{y^{1/3}y}\right)^3$

88. $\left(\dfrac{t^{4/9}t^{5/9}}{t^{1/9}t^{2/9}}\right)^9$

In Exercises 89–100, simplify each expression. Assume that all variables represent positive numbers.

89. $x^{2/3}x^{3/4}$

90. $a^{3/5}a^{1/2}$

91. $(b^{1/2})^{3/5}$

92. $(x^{2/5})^{4/7}$

93. $\dfrac{t^{2/3}}{t^{2/5}}$

94. $\dfrac{p^{3/4}}{p^{1/3}}$

95. $\dfrac{x^{4/5}x^{1/3}}{x^{2/15}}$

96. $\dfrac{y^{2/3}y^{3/5}}{y^{1/5}}$

97. $\dfrac{a^{2/5}a^{1/5}}{a^{-1/3}}$

98. $\dfrac{q^{3/4}q^{4/5}}{q^{-2/3}}$

99. $\dfrac{12b^{-1/3}b^{-3/4}}{4b^{3/5}}$

100. $\dfrac{4c^{-3/4}c^{1/6}}{8c^{1/4}}$

Writing Exercises ■ *Write a paragraph using your own words.*

1. Is $(-4)^{1/2}$ a real number? Explain.   **2.** Is $(-8)^{1/3}$ a real number? Explain.

Something to Think About ■ *If $x > y$, pick the larger number in each pair.*

1. 2^x, 2^y

2. $\left(\dfrac{1}{2}\right)^x$, $\left(\dfrac{1}{2}\right)^y$

Review Exercises ■ *Factor each expression.*

1. $3z^2 - 15tz + 12t^2$ **2.** $a^4 - b^4$

Solve each equation.

3. $\dfrac{x-5}{7} + \dfrac{2}{5} = \dfrac{7-x}{5}$ **4.** $\dfrac{t}{t+2} - 1 = \dfrac{1}{1-t}$

8.7 The Distance Formula

■ The Pythagorean Theorem ■ The Distance Formula

GETTING READY *In each set of numbers, verify that $a^2 + b^2 = c^2$.*

1. $a = 3, b = 4, c = 5$ **2.** $a = 6, b = 8, c = 10$
3. $a = 5, b = 12, c = 13$ **4.** $a = 7, b = 24, c = 25$

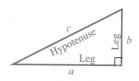

FIGURE 8-2

A triangle that contains a 90° angle is called a **right triangle.** The longest side of a right triangle is the **hypotenuse,** which is the side opposite the right angle. The remaining two sides are the **legs** of the triangle. In the right triangle shown in Figure 8-2, side c is the hypotenuse and sides a and b are the legs.

■ The Pythagorean Theorem

The **Pythagorean theorem** provides a formula relating the lengths of the three sides of any right triangle.

Pythagorean Theorem If the length of the hypotenuse of a right triangle is c and the lengths of the two legs are a and b, then
$$c^2 = a^2 + b^2$$

The Pythagorean theorem is useful because equal positive numbers have equal positive square roots.

Square Root Property of Equality	If a and b are positive numbers, then
	If $a = b$, then $\sqrt{a} = \sqrt{b}$.

Since the lengths of the sides of a triangle are positive numbers, we can use the square root property of equality and the Pythagorean theorem to find the lengths of the sides of any right triangle.

EXAMPLE 1   **Constructing a high-ropes adventure course**   A builder of a high-ropes course wants to stabilize the pole shown in Figure 8-3 by attaching a cable from a ground anchor 20 feet from its base to a point 15 feet up the pole. How long will the cable be?

■ P E R S P E C T I V E

A 350-Year-Old Problem

Because of the Pythagorean theorem, the ancient Greeks knew that the lengths of the sides of a right triangle could be natural numbers. For example, a right triangle could have sides of lengths 3, 4, and 5 because $3^2 + 4^2 = 5^2$ (check it: $9 + 16 = 25$). Similarly, a right triangle could have sides of 5, 12, and 13 because $5^2 + 12^2 = 13^2$. Natural numbers a, b, and c that satisfy the equation $a^2 + b^2 = c^2$ are called **Pythagorean triples**. The triples 3, 4, 5 and 5, 12, 13 are two of infinitely many possibilities.

In 1637, French mathematician Pierre de Fermat wrote a note in the margin of a book: There are no natural-number solutions a, b, and c to the equation $a^n + b^n = c^n$ if n is greater than 2. Fermat also mentioned that he had found a marvelous proof which wouldn't fit in the margin. Mathematicians have been trying to prove Fermat's last theorem ever since. Until recently, they had little success.

Princeton University mathematician Dr. Andrew Wiles first learned of Fermat's last theorem when he was 10 years old. He was so intrigued by the problem that he decided that he would study mathematics.

Dr. Andrew Wiles, Princeton University

Dr. Wiles worked on the problem, isolating himself in a barren attic room in his Princeton home. "The problem was on my mind all the time," said Dr. Wiles. "When you are really desperate to find an answer, you can't let go." After seven years of concentrated work, Dr. Wiles announced an apparent solution in June of 1993. If it is confirmed, Fermat's 350-year-old theorem is proved.

Solution We can use the Pythagorean theorem, with $a = 20$ and $b = 15$.

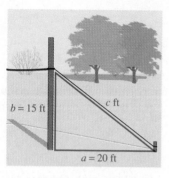

FIGURE 8-3

$$c^2 = a^2 + b^2$$
$$c^2 = 20^2 + 15^2$$
$$c^2 = 400 + 225$$
$$c^2 = 625$$
$$\sqrt{c^2} = \sqrt{625} \qquad \text{Take the square root of both sides.}$$
$$c = 25$$

The cable will be 25 feet long.

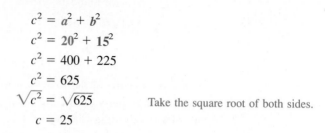

EXAMPLE 2 **Saving cable** The builder of a high-ropes course wants to use a previously cut 25-foot cable to stabilize the pole shown in Figure 8-4. To be safe, the ground anchor must be greater than 16 feet from the base of the pole. Is the cable long enough to use?

Solution We can use the Pythagorean theorem, with $b = 16$ and $c = 25$.

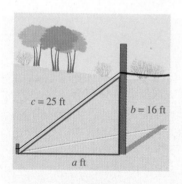

FIGURE 8-4

$$c^2 = a^2 + b^2$$
$$25^2 = a^2 + 16^2$$
$$625 = a^2 + 256$$
$$369 = a^2 \qquad \text{Subtract 256 from both sides.}$$
$$\sqrt{369} = \sqrt{a^2} \qquad \text{Take the square root of both sides.}$$
$$a \approx 19.209372 \qquad \text{Use a calculator to find the approximate value of } \sqrt{369}.$$

Since the anchor will be more than 16 feet from the base, the cable is long enough.

EXAMPLE 3 A 26-foot ladder rests against the side of a building. The base of the ladder is 10 feet from the wall. How far up the side of the building does the ladder reach?

Solution The wall, the ground, and the ladder form a right triangle, as shown in Figure 8-5. In this triangle, the hypotenuse is 26 feet and one of the legs is the base-to-wall distance of 10 feet. We can let d represent the other leg, which is the distance that the ladder reaches up the wall.

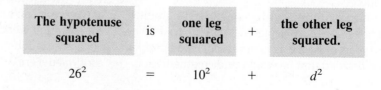

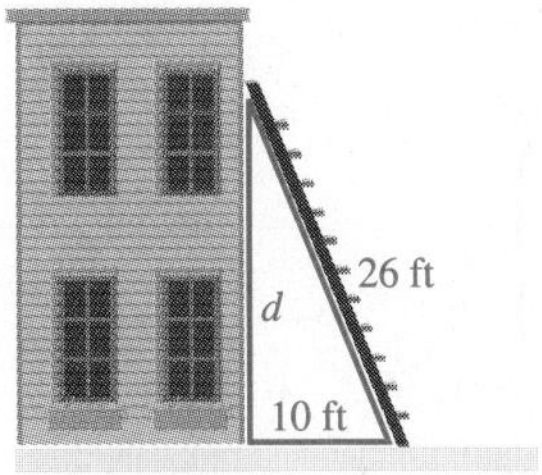

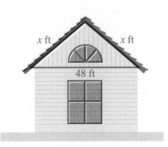

$$26^2 = 10^2 + d^2$$
$$676 = 100 + d^2$$
$$676 - 100 = d^2 \quad \text{Subtract 100 from both sides.}$$
$$576 = d^2$$
$$\sqrt{576} = \sqrt{d^2} \quad \text{Take the square root of both sides.}$$
$$24 = d$$

FIGURE 8-5

The ladder will reach 24 feet up the side of the building. ■

EXAMPLE 4 The gable end of the roof shown in Figure 8-6 is an isosceles right triangle with a span of 48 feet. Find the distance from the eaves to the peak.

Solution The two equal sides of the isosceles right triangle are the two legs of the right triangle, and the span of 48 is the length of the hypotenuse. We can let x represent the length of each of the legs, the distance from eaves to peak.

The hypotenuse squared	is	one leg squared	+	the other leg squared.
48^2	$=$	x^2	$+$	x^2

$$48^2 = x^2 + x^2$$
$$2304 = 2x^2 \quad \text{Simplify and combine like terms.}$$
$$1152 = x^2 \quad \text{Divide both sides by 2.}$$
$$\sqrt{1152} = \sqrt{x^2} \quad \text{Take the square root of both sides.}$$
$$33.941125 \approx x \quad \text{Use a calculator to find the approximate value of } \sqrt{1152}.$$

FIGURE 8-6

The eaves-to-peak distance of the roof is approximately 34 feet. ■

▪ The Distance Formula

We can use the Pythagorean theorem to derive a formula for finding the distance between two points $P(x_1, y_1)$ and $Q(x_2, y_2)$ plotted on a rectangular coordinate system. The distance d between points P and Q is the length of the hypotenuse of the triangle in Figure 8-7. The two legs have lengths $x_2 - x_1$ and $y_2 - y_1$.

By the Pythagorean theorem, we have

1. $d^2 = (x_2 - x_1)^2 + (y_2 - y_1)^2$

To solve for d, we can take the square root of both sides of Equation 1 to get

$$d = \sqrt{(x_2 - x_1)^2 + (y_2 - y_1)^2}$$

The result is called the **distance formula.**

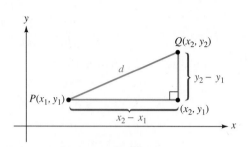

FIGURE 8-7

Distance Formula   The distance d between points $P(x_1, y_1)$ and $Q(x_2, y_2)$ is given by the formula

$$d = \sqrt{(x_2 - x_1)^2 + (y_2 - y_1)^2}$$

EXAMPLE 5   Find the distance between the points $P(1, 5)$ and $Q(4, 9)$.

Solution   We can let $(x_1, y_1) = (1, 5)$ and let $(x_2, y_2) = (4, 9)$. In other words, we let

$$x_1 = 1, \qquad y_1 = 5, \qquad x_2 = 4, \qquad \text{and} \qquad y_2 = 9$$

We substitute these values into the distance formula and simplify.

$$\begin{aligned}
d &= \sqrt{(x_2 - x_1)^2 + (y_2 - y_1)^2} \\
&= \sqrt{(4 - 1)^2 + (9 - 5)^2} \\
&= \sqrt{3^2 + 4^2} \\
&= \sqrt{9 + 16} \\
&= \sqrt{25} \\
&= 5
\end{aligned}$$

The distance between points P and Q is 5 units (see Figure 8-8).

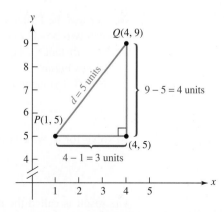

FIGURE 8-8

EXAMPLE 6 Find the distance between the points $P(-3, 6)$ and $Q(2, -6)$.

Solution We can let $(x_1, y_1) = (-3, 6)$ and let $(x_2, y_2) = (2, -6)$. In other words, we let

$$x_1 = -3, \qquad y_1 = 6, \qquad x_2 = 2, \qquad \text{and} \qquad y_2 = -6$$

We substitute these values into the distance formula and simplify.

$$\begin{aligned} d &= \sqrt{(x_2 - x_1)^2 + (y_2 - y_1)^2} \\ &= \sqrt{[2 - (-3)]^2 + (-6 - 6)^2} \\ &= \sqrt{(5)^2 + (-12)^2} \\ &= \sqrt{25 + 144} \\ &= \sqrt{169} \\ &= 13 \end{aligned}$$

The distance between points P and Q is 13 units. ∎

ORALS *If a, b, and c are lengths of the sides of a right triangle, find the length of the hypotenuse c.*

1. $a = 4$ and $b = 3$ | **2.** $a = 5$ and $b = 12$
3. $a = 6$ and $b = 8$ | **4.** $a = 15$ and $b = 8$

E X E R C I S E 8.7

In Exercises 1–10, refer to the right triangle of Illustration 1. Find the length of the unknown side.

1. $a = 4$ and $b = 3$. Find c.

2. $a = 6$ and $b = 8$. Find c.

3. $a = 5$ and $b = 12$. Find c.

4. $a = 15$ and $c = 17$. Find b.

5. $a = 21$ and $c = 29$. Find b.

6. $b = 16$ and $c = 34$. Find a.

7. $b = 45$ and $c = 53$. Find a.

9. $a = 5$ and $c = 9$. Find b.

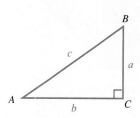

ILLUSTRATION 1

8. $a = 7$ and $b = 1$. Find c.

10. $a = 1$ and $c = \sqrt{2}$. Find b.

In Exercises 11–20, find the distance between the points.

11. (1, 2), (4, 6)   **12.** (2, 2), (5, 6)   **13.** (−2, 5), (6, −1)   **14.** (1, −8), (5, −12)

15. (−1, 4), (4, 16)   **16.** (−5, 7), (10, −1)   **17.** (−17, −3), (−23, 5)   **18.** (0, 0), (21, 20)

19. $\left(-\dfrac{1}{2}, 0\right), \left(\dfrac{5}{2}, -4\right)$ **20.** $(-2\sqrt{3}, \sqrt{3}), (\sqrt{3}, -2\sqrt{3})$

21. Adjusting a ladder A 20-foot ladder reaches a window 16 feet above the ground. How far from the wall is the base of the ladder?

22. Length of guy wires A 150-foot-tall tower is secured by three guy wires fastened at the top and to anchors 15 feet from the base of the tower. How long is each guy wire?

23. Height of a pole A 34-foot-long wire reaches from the top of a telephone pole to a point on the ground 16 feet from the base of the pole. Find the height of the pole.

24. Length of a path A rectangular garden has sides of 28 and 45 feet. Find the length of a path that extends from one corner to the opposite corner.

25. Baseball A baseball diamond is a square, with each side 90 feet long. (See Illustration 2.) How far is it from home plate to second base?

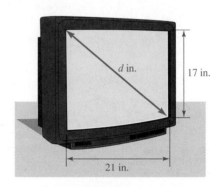

ILLUSTRATION 3

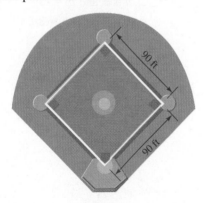

ILLUSTRATION 2

26. Television The size of the television screen shown in Illustration 3 is the diagonal measure of its rectangular screen. How large is the screen if it is 21 inches wide and 17 inches high?

27. Finding location A woman drives 4.2 miles east and then 4.0 miles north. How far is she from her starting point?

28. Taking a shortcut Instead of walking on the sidewalk, students take a diagonal shortcut across the vacant lot shown in Illustration 4. How much distance do they save?

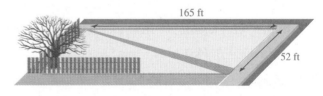

ILLUSTRATION 4

29. Carpentry A square-headed bolt is countersunk into a circular hole drilled in a wooden beam. The corners of the bolt must have $\frac{3}{8}$ inch clearance, as shown in Illustration 5. Find the diameter of the hole.

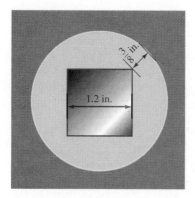

ILLUSTRATION 5

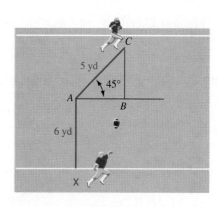

ILLUSTRATION 7

30. Designing a tunnel The entrance to a one-way tunnel is a rectangle with a semicircular roof. Its dimensions are given in Illustration 6. How tall can a 10-foot-wide truck be, without getting stuck in the tunnel?

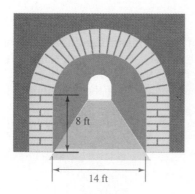

ILLUSTRATION 6

31. Football On first and ten, a coach tells his tight end to go out 6 yards, cut 45° to the right, and run 5 yards. (See Illustration 7.) The tight end follows instructions, catches a pass, and is tackled immediately. Will he gain the necessary 10 yards for a first down?

32. Geometry The legs of a right triangle are equal, and the hypotenuse is $2\sqrt{2}$ units long. Find the length of each leg.

33. Geometry The sides of a square are 3 feet long. Find the length of each diagonal of the square.

34. Perimeter of a square The diagonal of a square is 3 feet long. Find its perimeter.

35. Altitude of a triangle Find the altitude of the isosceles triangle shown in Illustration 8.

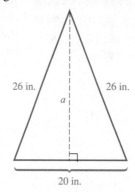

ILLUSTRATION 8

36. Geometry The square in Illustration 9 is inscribed in a circle. The sides of the square are 6 inches long. Find the area of the circle.

ILLUSTRATION 9

Writing Exercises ■ *State the theorem or formula using your own words.*

1. The Pythagorean theorem

2. The distance formula

Something to Think About ■

1. To generate Pythagorean triples, pick natural numbers for x and y ($x > y$). Let $a = 2xy$, $b = x^2 - y^2$, and $c = x^2 + y^2$. Why do you always get a Pythagorean triple?

2. When using the distance formula, why doesn't it matter which point is (x_1, y_1) and which point is (x_2, y_2)?

Review Exercises ■ *Graph each equation or inequality.*

1. $x = 3$

2. $y = -3$

3. $-2x + y = 4$

4. $4x - y > 4$

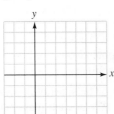
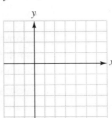
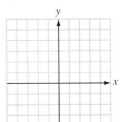
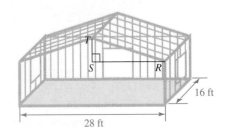

MATHEMATICS IN CARPENTRY

To find the area of the ceiling of the room discussed at the beginning of the chapter, we must find the length of the rafter *RT* in right triangle *RST*. Since the ridge is at the center of the room, the length of *RS* is 14 feet. Since the height of the ridge is 12 feet and the height of the outside wall is 8 feet, the length of *ST* is 4 feet.

We can use the Pythagorean theorem to find the length of *RT*.

$$(RT)^2 = (RS)^2 + (ST)^2$$
$$(RT)^2 = 14^2 + 4^2$$
$$= 212$$
$$RT = \sqrt{212} \quad \text{Take the square root of both sides.}$$

Since the area, *A*, of the ceiling is the sum of its two rectangular parts, we have
$$A = 2\left(16 \cdot \sqrt{212}\right) \approx 465.9270329$$

To find the number of sheets of plaster board needed to drywall the ceiling, we divide the area of the ceiling by the area of one 4 × 8-foot sheet of plaster board.

$$465.9270329 \div 32 \approx 14.56021978$$

The carpenter will need at least 15 sheets of plasterboard.

P R O J E C T   ■   Getting into the Swing of Things

The Italian mathematician and physicist Galileo Galilei (1564–1642) is best known as the inventor of the telescope and for his discovery of four of the moons of Jupiter, still known as the *Galilean satellites*. Less known is his discovery that a pendulum could be used to keep accurate time. While praying one day in the cathedral, Galileo noticed a suspended candle left swinging after it had been lit. Using his own pulse as a timer, Galileo discovered that the time for one swing remained unchanged as the swings themselves became smaller. By more experimenting, Galileo discovered the relationship between the length of a pendulum and its **period,** the time it takes to complete one swing.

You can discover the relationship, too. You will need a stopwatch, a calculator, a meter stick, and a pendulum. To make the pendulum, try tying a length of string to a rubber band and wrapping the band tightly about a small rock. By tying the free end of the string to a support, you can change the length of the pendulum.

- Start the pendulum swinging and use the stopwatch to determine its period— the time it takes to go from left to right and back to left again. (You might time 10 complete swings and then divide by 10.) Do this for pendulums of at least eight different lengths. Let t be the period (measured in seconds) and let l be the length (measured in centimeters), and record your results in the table of Figure 8-9.

- Plot the points (l, t) on the axes in the figure. Does the graph appear to be a line?

- The pendulum's period t and length l are related by the formula $t = a\sqrt{l}$ for some number, a. From your experimental data, find the approximate value of a.

- Use your formula to predict the period of a pendulum 2 meters long.

- Time the period of a 2-meter pendulum. How close was your prediction?

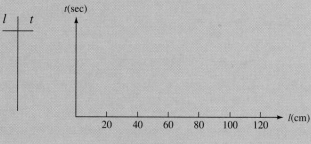

FIGURE 8-9

Chapter Summary

conjugate (8.4)

cube root (8.1)

distance formula (8.7)

division property of radicals (8.2)

extraneous solution (8.5)

hypotenuse (8.7)

imaginary number (8.1)

index (8.1)

irrational number (8.1)

leg of a right triangle (8.7)

like radicals (8.3)

multiplication property of radicals (8.2)

*n*th root of a number (8.1)

perfect integer square (8.1)

principal square root (8.1)

Pythagorean theorem (8.7)

radical sign (8.1)

radicand (8.1)

rational exponent (8.6)

rationalizing the denominator (8.4)

rational number (8.1)

real number (8.1)

right triangle (8.7)

simplified form of a radical (8.2)

square root (8.1)

square root property of equality (8.7)

squaring property of equality (8.5)

(8.1) The number b is a square root of a if $b^2 = a$.

If a is a positive number, then the principal square root of a, denoted by $\sqrt{a}$, is the positive square root of a. The principal square root of 0 is 0.

If a is a positive integer and not a perfect square, then $\sqrt{a}$ is an irrational number.

The square root of a negative number is not a real number.

The cube root of a is denoted by $\sqrt[3]{a}$, and $\sqrt[3]{a} = b$ if $b^3 = a$.

The nth root of a is denoted by $\sqrt[n]{a}$, and $\sqrt[n]{a} = b$ if $b^n = a$.

(8.2) If a and b are nonnegative numbers, then

$$\sqrt{ab} = \sqrt{a}\sqrt{b} \quad \text{and if } b \neq 0, \text{ then} \quad \sqrt{\frac{a}{b}} = \frac{\sqrt{a}}{\sqrt{b}}$$

To simplify an expression involving square roots (or cube roots) use the multiplication and division properties of radicals to remove perfect square factors (or perfect cube factors) from the radicands.

(8.3) Radical expressions can be added or subtracted if they contain like radicals. Often radicals can be converted to like radicals and then added.

(8.4) If a square root appears as a monomial in the denominator of a fraction, rationalize the denominator by multiplying both the numerator and the denominator of the fraction by some appropriate square root.

If the denominator of a fraction contains radicals within a binomial, multiply numerator and denominator by the conjugate of the denominator.

(8.5) If $a = b$, then $a^2 = b^2$.

To solve an equation that involves square roots, rearrange the terms of the equation so that no more than one radical appears on one side of the equation. Then square both sides of the equation and solve the resulting equation. Finally, *check the solution.*

(8.6) $x^{1/n} = \sqrt[n]{x}$ $x^{m/n} = \sqrt[n]{x^m} = \left(\sqrt[n]{x}\right)^m$

(8.7) **The Pythagorean theorem.** In any right triangle, the sum of the squares of the two legs is equal to the square of the hypotenuse.

Let a and b be positive numbers. If $a = b$, then $\sqrt{a} = \sqrt{b}$.

The distance formula. The distance between the points (x_1, y_1) and (x_2, y_2) is given by the formula

$$d = \sqrt{(x_2 - x_1)^2 + (y_2 - y_1)^2}$$

Chapter 8 Review Exercises

(8.1) *In Review Exercises 1–12, write each expression without a radical sign.*

1. $\sqrt{25}$ **2.** $\sqrt{64}$ **3.** $-\sqrt{144}$ **4.** $-\sqrt{289}$

5. $\sqrt{256}$ **6.** $-\sqrt{64}$ **7.** $\sqrt{169}$ **8.** $-\sqrt{225}$

9. $-\sqrt[3]{-27}$ **10.** $-\sqrt[3]{125}$ **11.** $\sqrt[4]{81}$ **12.** $\sqrt[5]{32}$

In Review Exercises 13–16, use a calculator to find each value to three decimal places.

13. $\sqrt{21}$ **14.** $-\sqrt{15}$ **15.** $-\sqrt{57.3}$ **16.** $\sqrt{751.9}$

(8.2) *In Review Exercises 17–36, simplify each expression. All variables represent positive numbers.*

17. $\sqrt{32}$ **18.** $\sqrt{50}$ **19.** $\sqrt{500}$ **20.** $\sqrt{112}$

21. $\sqrt{80x^2}$ **22.** $\sqrt{63y^2}$ **23.** $-\sqrt{250t^3}$ **24.** $-\sqrt{700z^5}$

25. $\sqrt{200x^2y}$ **26.** $\sqrt{75y^2z}$ **27.** $\sqrt[3]{8x^2y^3}$ **28.** $\sqrt[3]{250x^4y^3}$

29. $\sqrt{\dfrac{16}{25}}$ **30.** $\sqrt{\dfrac{100}{49}}$ **31.** $\sqrt[3]{\dfrac{1000}{27}}$ **32.** $\sqrt[3]{\dfrac{16}{64}}$

33. $\sqrt{\dfrac{60}{49}}$ **34.** $\sqrt{\dfrac{80}{225}}$ **35.** $\sqrt{\dfrac{242x^4}{169x^2}}$ **36.** $\sqrt{\dfrac{450a^6}{196a^2}}$

(8.3–8.4) *In Review Exercises 37–60, do the operations. All variables represent positive numbers.*

37. $\sqrt{2}+\sqrt{8}-\sqrt{18}$ **38.** $\sqrt{3}+\sqrt{27}-\sqrt{12}$ **39.** $3\sqrt{5}+5\sqrt{45}$ **40.** $5\sqrt{28}-3\sqrt{63}$

41. $3\sqrt{2x^2y}+2x\sqrt{2y}$ **42.** $3y\sqrt{5xy^3}-y^2\sqrt{20xy}$ **43.** $\sqrt[3]{16}+\sqrt[3]{54}$ **44.** $\sqrt[3]{2000x^3}-\sqrt[3]{128x^3}$

45. $\left(3\sqrt{2}\right)\left(-2\sqrt{3}\right)$ **46.** $\left(-5\sqrt{x}\right)\left(-2\sqrt{x}\right)$ **47.** $\left(3\sqrt{3x}\right)\left(4\sqrt{6x}\right)$ **48.** $\left(-2\sqrt{27y^3}\right)\left(y\sqrt{2y}\right)$

49. $\left(\sqrt[3]{4}\right)\left(2\sqrt[3]{4}\right)$ **50.** $\left(-2\sqrt[3]{32x^2}\right)\left(3\sqrt[3]{2x^2}\right)$ **51.** $\sqrt{2}\left(\sqrt{8}-\sqrt{18}\right)$ **52.** $\sqrt{6y}\left(\sqrt{2y}+\sqrt{75}\right)$

53. $\left(\sqrt{3}+\sqrt{5}\right)\left(\sqrt{3}-\sqrt{5}\right)$ **54.** $\left(\sqrt{15}+3x\right)\left(\sqrt{15}+3x\right)$

55. $\left(\sqrt[3]{3}+2\right)\left(\sqrt[3]{3}-1\right)$ **56.** $\left(\sqrt[3]{5}-1\right)\left(\sqrt[3]{5}+1\right)$

57. $\left(3\sqrt{5}+2\right)^2$ **58.** $\left(2\sqrt{3}-1\right)^2$ **59.** $\left(\sqrt{x}-\sqrt{2}\right)^2$ **60.** $\left(\sqrt{7}+\sqrt{a}\right)^2$

In Review Exercises 61–68, rationalize each denominator. All variables represent positive numbers.

61. $\dfrac{1}{\sqrt{7}}$ **62.** $\dfrac{3}{\sqrt{18}}$ **63.** $\dfrac{8}{\sqrt[3]{16}}$ **64.** $\dfrac{10}{\sqrt[3]{32}}$

65. $\dfrac{\sqrt{7}}{\sqrt{35}}$ **66.** $\dfrac{3}{\sqrt{3} - 1}$ **67.** $\dfrac{2\sqrt{5}}{\sqrt{5} + \sqrt{3}}$ **68.** $\dfrac{\sqrt{7x} + \sqrt{x}}{\sqrt{7x} - \sqrt{x}}$

(8.5) In Review Exercises 69–76, solve each equation and check all solutions.

69. $\sqrt{x + 3} = 3$

70. $\sqrt{2x + 10} = 2$

71. $\sqrt{3x + 4} = -2\sqrt{x}$

72. $\sqrt{2(x + 4)} - \sqrt{4x} = 0$

73. $\sqrt{x + 5} = x - 1$

74. $\sqrt{2x + 9} = x - 3$

75. $\sqrt{2x + 5} - 1 = x$

76. $\sqrt{4a + 13} + 2 = a$

(8.6) In Review Exercises 77–92, simplify each expression. Write answers without using negative exponents. Assume variables represent positive numbers.

77. $49^{1/2}$ **78.** $(-1000)^{1/3}$ **79.** $36^{3/2}$ **80.** $\left(\dfrac{4}{9}\right)^{5/2}$

81. $8^{2/3}8^{4/3}$ **82.** $\dfrac{5^{17/7}}{5^{3/7}}$ **83.** $\dfrac{x^{4/5}x^{3/5}}{(x^{1/5})^2}$ **84.** $\left(\dfrac{r^{1/3}r^{2/3}}{r^{4/3}}\right)^3$

85. $6^{5/3}6^{-2/3}$ **86.** $\dfrac{5^{2/3}}{5^{-1/3}}$ **87.** $\dfrac{x^{2/5}x^{1/5}}{x^{-2/5}}$ **88.** $(a^4b^8)^{-1/2}$

89. $x^{1/3}x^{2/5}$ **90.** $\dfrac{t^{3/4}}{t^{2/3}}$ **91.** $\dfrac{x^{-4/5}x^{1/3}}{x^{1/3}}$ **92.** $\dfrac{r^{1/4}r^{1/3}}{r^{5/6}}$

(8.7) In Review Exericses 93–96, refer to the right triangle shown in Illustration 1.

93. $a = 21$ and $b = 28$. Find c.

94. $a = 25$ and $c = 65$. Find b.

95. $a = 1$ and $c = \sqrt{2}$. Find b.

96. $b = 5$ and $c = 7$. Find a.

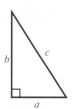

ILLUSTRATION 1

In Review Exercises 97–100, find the distance between the points.

97. $(-7, 12), (-4, 8)$ **98.** $(-15, -3), (-10, -15)$ **99.** $(1, 1), (-1, 1)$ **100.** $(-10, 11), (10, -10)$

101. Installing windows The window frame shown in Illustration 2 is 32 inches by 60 inches. It is to be shipped with a temporary brace attached diagonally. Find the length of the brace.

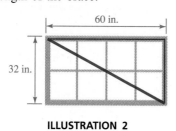

60 in.

32 in.

ILLUSTRATION 2

102. Height of a mast A 53-foot rope runs from the top of the mast shown in Illustration 3 to a point 28 feet from its base. Find the height of the mast.

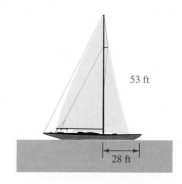

53 ft

28 ft

ILLUSTRATION 3

Chapter 8 Test

In Problems 1–4, write each expression without a radical sign. Assume $x > 0$.

1. $\sqrt{100}$ **2.** $-\sqrt{400}$ **3.** $\sqrt[3]{-27}$ **4.** $\sqrt{3x}\sqrt{27x}$

In Problems 5–10, simplify each expression. Assume $x > 0$ and $y > 0$.

5. $\sqrt{8x^2}$ **6.** $\sqrt{54x^3y}$ **7.** $\sqrt{\dfrac{320}{10}}$ **8.** $\sqrt{\dfrac{18x^2y^3}{2xy}}$

9. $\sqrt[3]{x^6y^6}$ **10.** $\sqrt[4]{\dfrac{16x^8}{y^4}}$

In Problems 11–16, do each operation and simplify.

11. $\sqrt{12} + \sqrt{27}$ **12.** $\sqrt{8x^3} - x\sqrt{18x}$

13. $\left(-2\sqrt{8x}\right)\left(3\sqrt{12x}\right)$ **14.** $\sqrt{3}\left(\sqrt{8} + \sqrt{6}\right)$

15. $\left(\sqrt{2} + \sqrt{3}\right)\left(\sqrt{2} - \sqrt{3}\right)$ **16.** $\left(2\sqrt{x} + 2\right)\left(\sqrt{x} - 3\right)$

In Problems 17–20, rationalize each denominator.

17. $\dfrac{2}{\sqrt{2}}$ **18.** $\sqrt{\dfrac{3xy^3}{48x^2}}$ **19.** $\dfrac{2}{\sqrt{5} - 2}$ **20.** $\dfrac{\sqrt{3x}}{\sqrt{x} + 2}$

In Problems 21–26, solve each equation.

21. $\sqrt{x} + 3 = 9$ **22.** $\sqrt{x - 2} - 2 = 6$

23. $\sqrt{3x + 9} = 2\sqrt{x + 1}$

24. $3\sqrt{x - 3} = \sqrt{2x + 8}$

25. $\sqrt{3x + 1} = x - 1$

26. $\sqrt[3]{x - 2} = 3$

In Problems 27–32, simplify each expression and write all answers without using negative exponents. All variables represent positive numbers.

27. $121^{1/2}$

28. $27^{-4/3}$

29. $(y^{15})^{2/5}$

30. $\left(\dfrac{a^{5/3} a^{4/3}}{(a^{1/3})^2 a^{2/3}} \right)^6$

31. $p^{2/3} p^{3/4}$

32. $\dfrac{x^{2/3} x^{-4/5}}{x^{2/15}}$

33. Find the length of the hypotenuse of a right triangle with legs of 5 inches and 12 inches.

34. Find the distance between the points $(1, 4)$ and $(7, 12)$.

35. Find the distance between the points $(-2, -3)$ and $(-5, 1)$.

36. A 26-foot ladder reaches a point on a wall 24 feet above the ground. How far from the wall is the ladder's base?

9

Quadratic Equations

American Appliance has found that it can sell more custom refrigerators each month if it decreases the price. Over the years, the sales department has found that the store will sell x refrigerators at a price of $\$\left(1800 - \frac{3}{2}x\right)$. What price should the store charge to maximize its revenue?

At that price, how many refrigerators should the store order from the distributor each month?

After reading this chapter, you will be able to answer these questions.

In this chapter, we shall develop general techniques for solving quadratic equations—equations of the form $ax^2 + bx + c = 0$, where a, b, and c are real numbers and $a \neq 0$.

9.1 Solving Equations of the Form $x^2 = c$

■ Solving Equations by Factoring ■ The Square Root Method

GETTING READY

Factor each polynomial.

1. $4x^2 - 2x$ **2.** $x^2 - 9$ **3.** $x^2 + x - 6$ **4.** $2x^2 + 3x - 9$

Find each square root.

5. $\sqrt{25}$ **6.** $-\sqrt{36}$ **7.** $\sqrt{20}$ **8.** $-\sqrt{50}$

■ Solving Equations by Factoring

Factoring Method   To solve an equation by factoring, we
1. write the equation in $ax^2 + bx + c = 0$ form (called **quadratic form**),
2. factor the trinomial on the left-hand side of the equation,
3. use the zero-factor property to set each factor equal to 0, and
4. solve each resulting linear equation.

We review the factoring method in Example 1.

 (b)

EXAMPLE 1   Solve the equations   **a.** $6x^2 - 3x = 0$ and   **b.** $6x^2 - x - 2 = 0$.

Solution   **a.**

$$6x^2 - 3x = 0$$
$$3x(2x - 1) = 0 \qquad \text{Factor out } 3x.$$
$$3x = 0 \quad \text{or} \quad 2x - 1 = 0 \qquad \text{Set each factor equal to 0.}$$
$$x = 0 \qquad\qquad 2x = 1 \qquad \text{Solve each linear equation.}$$
$$x = \frac{1}{2}$$

Check: **For $x = 0$**

$$6x^2 - 3x = 0$$
$$6(0)^2 - 3(0) \stackrel{?}{=} 0$$
$$6(0) - 0 \stackrel{?}{=} 0$$
$$0 - 0 \stackrel{?}{=} 0$$
$$0 = 0$$

For $x = \dfrac{1}{2}$

$$6x^2 - 3x = 0$$
$$6\left(\frac{1}{2}\right)^2 - 3\left(\frac{1}{2}\right) \stackrel{?}{=} 0$$
$$6\left(\frac{1}{4}\right) - \frac{3}{2} \stackrel{?}{=} 0$$
$$\frac{3}{2} - \frac{3}{2} \stackrel{?}{=} 0$$
$$0 = 0$$

WARNING!   Do not make the following error.

$$6x^2 - 3x = 0$$
$$6x^2 = 3x \qquad \text{Add } 3x \text{ to both sides.}$$
$$2x = 1 \qquad \text{Divide both sides by } 3x.$$
$$x = \frac{1}{2} \qquad \text{Divide both sides by 2.}$$

By dividing by $3x$, the solution $x = 0$ is lost.

b. $6x^2 - x - 2 = 0$

$\qquad (3x - 2)(2x + 1) = 0$ Factor the trinomial.

$\qquad 3x - 2 = 0$ or $2x + 1 = 0$ Set each factor equal to 0.

$\qquad\qquad 3x = 2$ $2x = -1$ Solve each linear equation.

$$x = \frac{2}{3} \qquad\qquad x = -\frac{1}{2}$$

Check: **For $x = \dfrac{2}{3}$**   **For $x = -\dfrac{1}{2}$**

$$6x^2 - x - 2 = 0 \qquad\qquad\qquad 6x^2 - x - 2 = 0$$

$$6\left(\frac{2}{3}\right)^2 - \frac{2}{3} - 2 \overset{?}{=} 0 \qquad\qquad 6\left(-\frac{1}{2}\right)^2 - \left(-\frac{1}{2}\right) - 2 \overset{?}{=} 0$$

$$6\left(\frac{4}{9}\right) - \frac{6}{9} - \frac{18}{9} \overset{?}{=} 0 \qquad\qquad 6\left(\frac{1}{4}\right) + \frac{2}{4} - \frac{8}{4} \overset{?}{=} 0$$

$$\frac{24}{9} - \frac{24}{9} \overset{?}{=} 0 \qquad\qquad\qquad \frac{6}{4} - \frac{6}{4} \overset{?}{=} 0$$

$$0 = 0 \qquad\qquad\qquad\qquad 0 = 0 \qquad\blacksquare$$

The factoring method doesn't work with some quadratic equations. For example, the trinomial in the equation $x^2 + 5x + 1 = 0$ cannot be factored using integer coefficients. To solve such equations, we need other methods.

■ The Square Root Method

If $x^2 = c$ $(c \geq 0)$, then x is a number whose square is c. Since $\left(\sqrt{c}\right)^2 = c$ and $\left(-\sqrt{c}\right)^2 = c$, the equation $x^2 = c$ has two solutions.

Square Root Method   If $c > 0$, the equation $x^2 = c$ has two solutions:

$$x = \sqrt{c} \qquad \text{and} \qquad x = -\sqrt{c}$$

We can write the previous result with double sign notation. The equation $x = \pm\sqrt{c}$ $\left(\text{read as } x \text{ equals plus or minus } \sqrt{c}\right)$ means that $x = \sqrt{c}$ or $x = -\sqrt{c}$.

EXAMPLE 2 Solve the equation $x^2 = 16$.

Solution The equation $x^2 = 16$ has two solutions:

$$x = \sqrt{16} \quad \text{or} \quad x = -\sqrt{16}$$
$$= 4 \quad \quad \quad = -4$$

Using double sign notation, we have $x = \pm 4$.

Check: **For $x = 4$** | **For $x = -4$**
$$x^2 = 16 \quad\quad\quad\quad x^2 = 16$$
$$4^2 \stackrel{?}{=} 16 \quad\quad\quad (-4)^2 \stackrel{?}{=} 16$$
$$16 = 16 \quad\quad\quad\quad 16 = 16$$

∎

The equation in Example 2 can also be solved by factoring.

$$x^2 = 16$$
$$x^2 - 16 = 0 \quad\quad\quad \text{Add } -16 \text{ to both sides.}$$
$$(x + 4)(x - 4) = 0 \quad\quad\quad \text{Factor the difference of two squares.}$$
$$x + 4 = 0 \quad \text{or} \quad x - 4 = 0$$
$$x = -4 \quad | \quad x = 4$$

EXAMPLE 3 Solve the equation $3x^2 - 9 = 0$.

Solution We can solve the equation by the square root method.

$$3x^2 - 9 = 0$$
$$3x^2 = 9 \quad\quad \text{Add 9 to both sides.}$$
$$x^2 = 3 \quad\quad \text{Divide both sides by 3.}$$

This equation has two solutions:

$$x = \sqrt{3} \quad \text{or} \quad x = -\sqrt{3}$$

The solutions can be written as $x = \pm \sqrt{3}$. To the nearest tenth, the solutions are ± 1.7.

Check: **For $x = \sqrt{3}$** | **For $x = -\sqrt{3}$**
$$3x^2 - 9 = 0 \quad\quad\quad\quad 3x^2 - 9 = 0$$
$$3\left(\sqrt{3}\right)^2 - 9 \stackrel{?}{=} 0 \quad\quad 3\left(-\sqrt{3}\right)^2 - 9 \stackrel{?}{=} 0$$
$$3(3) - 9 \stackrel{?}{=} 0 \quad\quad\quad 3(3) - 9 \stackrel{?}{=} 0$$
$$9 - 9 \stackrel{?}{=} 0 \quad\quad\quad\quad 9 - 9 \stackrel{?}{=} 0$$
$$0 = 0 \quad\quad\quad\quad\quad 0 = 0$$

∎

EXAMPLE 4 Solve the equation $(x + 1)^2 = 9$.

Solution The two solutions are

$$x + 1 = \sqrt{9} \quad \text{or} \quad x + 1 = -\sqrt{9}$$
$$x + 1 = 3 \quad\quad\quad\quad\; x + 1 = -3$$
$$x = 2 \quad\quad\quad\quad\quad\quad x = -4$$

Check: **For $x = 2$**   **For $x = -4$**

$$(x + 1)^2 = 9 \quad\quad\quad (x + 1)^2 = 9$$
$$(2 + 1)^2 \stackrel{?}{=} 9 \quad\quad\quad (-4 + 1)^2 \stackrel{?}{=} 9$$
$$3^2 \stackrel{?}{=} 9 \quad\quad\quad\quad (-3)^2 \stackrel{?}{=} 9$$
$$9 = 9 \quad\quad\quad\quad\quad\; 9 = 9$$

■

EXAMPLE 5 Solve the equation $(x - 2)^2 - 18 = 0$.

Solution
$$(x - 2)^2 - 18 = 0$$
$$(x - 2)^2 = 18 \quad\quad \text{Add 18 to both sides.}$$

The two solutions are

$$x - 2 = \sqrt{18} \quad\quad \text{or} \quad\quad x - 2 = -\sqrt{18}$$
$$x = 2 + \sqrt{18} \quad\quad\quad\quad x = 2 - \sqrt{18}$$
$$x = 2 + 3\sqrt{2} \quad\quad\quad\quad x = 2 - 3\sqrt{2} \quad\quad \sqrt{18} = \sqrt{9}\sqrt{2} = 3\sqrt{2}$$

Check: **For $x = 2 + 3\sqrt{2}$**   **For $x = 2 - 3\sqrt{2}$**

$$(x - 2)^2 - 18 = 0 \quad\quad\quad\quad (x - 2)^2 - 18 = 0$$
$$(2 + 3\sqrt{2} - 2)^2 - 18 \stackrel{?}{=} 0 \quad\quad (2 - 3\sqrt{2} - 2)^2 - 18 \stackrel{?}{=} 0$$
$$(3\sqrt{2})^2 - 18 \stackrel{?}{=} 0 \quad\quad\quad\quad (-3\sqrt{2})^2 - 18 \stackrel{?}{=} 0$$
$$18 - 18 \stackrel{?}{=} 0 \quad\quad\quad\quad\quad\quad 18 - 18 \stackrel{?}{=} 0$$
$$0 = 0 \quad\quad\quad\quad\quad\quad\quad\quad\; 0 = 0$$

To the nearest tenth, the solutions are 6.2 and -2.2.

■

EXAMPLE 6 Solve the equation $3x^2 - 4 = 2(x^2 + 2)$.

Solution
$$3x^2 - 4 = 2(x^2 + 2)$$
$$3x^2 - 4 = 2x^2 + 4 \quad\quad \text{Remove parentheses.}$$
$$3x^2 = 2x^2 + 8 \quad\quad\quad \text{Add 4 to both sides.}$$
$$x^2 = 8 \quad\quad\quad\quad\quad \text{Subtract } 2x^2 \text{ from both sides.}$$
$$x = \sqrt{8} \quad \text{or} \quad x = -\sqrt{8}$$
$$= 2\sqrt{2} \quad\quad\quad\; = -2\sqrt{2} \quad\quad \text{Simplify the radical.}$$

Both solutions check. To the nearest tenth, the solutions are ± 2.8.

■

Solve each equation.

1. $(x + 5)(x - 5) = 0$ **2.** $(x - 2)(x - 3) = 0$

Solve by the square root method.

3. $x^2 = 100$ **4.** $x^2 = 49$

E X E R C I S E   9.1

In Exercises 1–12, use the factoring method to solve each equation.

1. $x^2 - 9 = 0$ **2.** $x^2 + x = 0$ **3.** $3x^2 + 9x = 0$ **4.** $2x^2 - 8 = 0$

5. $x^2 - 5x + 6 = 0$ **6.** $x^2 + 7x + 12 = 0$ **7.** $3x^2 + x - 2 = 0$ **8.** $2x^2 - x - 6 = 0$

9. $6x^2 + 11x + 3 = 0$ **10.** $5x^2 + 13x - 6 = 0$ **11.** $10x^2 + x - 2 = 0$ **12.** $6x^2 + 37x + 6 = 0$

In Exercises 13–24, use the square root method to solve each equation.

13. $x^2 = 1$ **14.** $x^2 = 4$ **15.** $x^2 = 9$ **16.** $x^2 = 32$

17. $x^2 = 20$ **18.** $x^2 = 0$ **19.** $3x^2 = 27$ **20.** $4x^2 = 64$

21. $4x^2 = 16$ **22.** $5x^2 = 125$ **23.** $x^2 = a$ **24.** $x^2 = 4b$

In Exercises 25–34, use the square root method to solve each equation for x.

25. $(x + 1)^2 = 25$ **26.** $(x - 1)^2 = 49$ **27.** $(x + 2)^2 = 81$ **28.** $(x + 3)^2 = 16$

29. $(x - 2)^2 = 8$ **30.** $(x + 2)^2 = 50$ **31.** $(x - a)^2 = 4a^2$ **32.** $(x + y)^2 = 9y^2$

33. $(x + b)^2 = 16c^2$ **34.** $(x - c)^2 = 25b^2$

In Exercises 35–40, factor the trinomial square and use the square root method to solve each equation.

35. $x^2 + 4x + 4 = 4$ **36.** $x^2 - 6x + 9 = 9$

37. $9x^2 - 12x + 4 = 16$ **38.** $4x^2 - 20x + 25 = 36$

39. $4x^2 + 4x + 1 = 20$ **40.** $9x^2 + 12x + 4 = 12$

In Exercises 41–46, solve each equation.

41. $6(x^2 - 1) = 4(x^2 + 3)$ **42.** $5(x^2 - 2) = 2(x^2 + 1)$

43. $8(x^2 - 6) = 4(x^2 + 13)$ **44.** $8(x^2 - 1) = 5(x^2 + 10) + 50$

45. $5(x + 1)^2 = (x + 1)^2 + 32$ **46.** $6(x - 4)^2 = 4(x - 4)^2 + 36$

Writing Exercises ■ *Write a paragraph using your own words.*

1. Explain how to solve a quadratic equation with the factoring method.

2. Explain how to solve the equation $x^2 = 81$ with the square root method.

Something to Think About ■ **1.** Find the error.

$$2x^2 + 5x = 0$$
$$2x^2 = -5x \qquad \text{Subtract } 5x \text{ from both sides.}$$
$$2x = -5 \qquad \text{Divide both sides by } x.$$
$$x = -\frac{5}{2} \qquad \text{Divide both sides by 2.}$$

2. What would happen if you solved $x^2 = c$ ($c < 0$) by the square root method? Would the roots be real numbers?

Review Exercises ■ *Write each expression without parentheses.*

1. $(y - 1)^2$ **2.** $(z + 2)^2$ **3.** $(x + y)^2$ **4.** $(a - b)^2$

5. $(2r - s)^2$ **6.** $(m + 3n)^2$

9.2 Completing the Square

■ Completing the Square ■ Solving Equations with Lead Coefficients of 1
■ Solving Equations with Lead Coefficients Other Than 1

GETTING READY *Find one-half of each number and square it.*

1. 6   **2.** 10   **3.** 2   **4.** 5

5. -8 **6.** -12 **7.** $\dfrac{1}{2}$ **8.** $\dfrac{2}{3}$

When the polynomial in a quadratic equation factors easily, the factoring method is usually the best way to solve the equation. However, when the polynomial doesn't factor easily, we can use a method called **completing the square.**

To complete the square, we change the left-hand side of a quadratic equation such as $x^2 - 4x - 12 = 0$ into a trinomial square. Since a trinomial square can be factored, the resulting equation is easy to solve.

■ Completing the Square

The method of completing the square is based on the following special products:

$$x^2 + 2bx + b^2 = (x + b)^2 \qquad \text{and} \qquad x^2 - 2bx + b^2 = (x - b)^2$$

The trinomials $x^2 + 2bx + b^2$ and $x^2 - 2bx + b^2$ are both trinomial squares, since each factors as the square of a binomial. In each trinomial, if we take one-half of the coefficient of the x in the middle term and square it, we get the third term.

$$x^2 + 2b \cdot x + b^2 \qquad\qquad x^2 - 2b \cdot x + b^2$$

$$\left[\frac{1}{2}(2b)\right]^2 = (b)^2 = b^2 \qquad \left[\frac{1}{2}(-2b)\right]^2 = (-b)^2 = b^2$$

To change a binomial such as $x^2 + 12x$ into a trinomial square, we take one-half of 12, square it, and add it to $x^2 + 12x$.

$$x^2 + 12x + \left[\frac{1}{2}(12)\right]^2 = x^2 + 12x + (6)^2$$

$$= x^2 + 12x + 36$$

This result is a trinomial square, because $x^2 + 12x + 36 = (x + 6)^2$.

EXAMPLE 1 (a, c) Add the square of one-half of the coefficient of x to change **a.** $x^2 + 4x$, **b.** $x^2 - 6x$, and **c.** $x^2 - 5x$ into trinomial squares.

Solution **a.** $x^2 + 4x + \left[\frac{1}{2}(4)\right]^2 = x^2 + 4x + (2)^2$

$$= x^2 + 4x + 4$$
$$= (x + 2)^2$$

b. $x^2 - 6x + \left[\frac{1}{2}(-6)\right]^2 = x^2 - 6x + (-3)^2$

$$= x^2 - 6x + 9$$
$$= (x - 3)^2$$

c. $x^2 - 5x + \left[\frac{1}{2}(-5)\right]^2 = x^2 - 5x + \left(-\frac{5}{2}\right)^2$

$$= x^2 - 5x + \frac{25}{4}$$
$$= \left(x - \frac{5}{2}\right)^2$$

■ Solving Equations with Lead Coefficients of 1

If the quadratic equation $ax^2 + bx + c = 0$ has a lead coefficient of 1 ($a = 1$), it is easy to solve by completing the square.

EXAMPLE 2 Use completing the square to solve the equation $x^2 - 4x - 12 = 0$.

Solution Since the coefficient of x^2 is 1, we can complete the square as follows:

$$x^2 - 4x - 12 = 0$$
$$x^2 - 4x = 12 \qquad \text{Add 12 to both sides.}$$

We then find one-half of the coefficient of x, square it, and add the result to both sides of the equation to make the left-hand side a trinomial square.

$$x^2 - 4x + \left[\frac{1}{2}(-4)\right]^2 = 12 + \left[\frac{1}{2}(-4)\right]^2$$

$$x^2 - 4x + 4 = 12 + 4 \qquad \text{Simplify.}$$

$$(x - 2)^2 = 16 \qquad \text{Factor } x^2 - 4x + 4 \text{ and simplify.}$$

$$x - 2 = \pm\sqrt{16} \qquad \begin{array}{l}\text{Use the square root method to solve} \\ \text{the quadratic equation for } x - 2.\end{array}$$

$$x = 2 \pm 4 \qquad \text{Add 2 to both sides and simplify.}$$

Because of the $\pm$ sign, there are two solutions.

$$\begin{array}{ccc} x = 2 + 4 & \text{or} & x = 2 - 4 \\ = 6 & | & = -2 \end{array}$$

Check each solution. This equation can also be solved by factoring. ∎

■ Solving Equations with Lead Coefficients Other Than 1

If the quadratic equation $ax^2 + bx + c = 0$ has a lead coefficient other than 1 ($a \neq 0, 1$), we can make it 1 by dividing both sides of the equation by a.

EXAMPLE 3 Use completing the square to solve the equation $4x^2 + 4x - 3 = 0$.

Solution We divide both sides of the equation by 4 so that the coefficient of x^2 is 1. We then proceed as follows.

$$4x^2 + 4x - 3 = 0$$

$$x^2 + x - \frac{3}{4} = 0 \qquad \text{Divide both sides by 4.}$$

$$x^2 + x = \frac{3}{4} \qquad \text{Add } \tfrac{3}{4} \text{ to both sides.}$$

$$x^2 + x + \left(\frac{1}{2}\right)^2 = \frac{3}{4} + \left(\frac{1}{2}\right)^2 \qquad \begin{array}{l}\text{Add } \left(\tfrac{1}{2}\right)^2 \text{ to both sides to complete the} \\ \text{square.}\end{array}$$

$$\left(x + \frac{1}{2}\right)^2 = 1 \qquad \text{Factor and simplify.}$$

$$x + \frac{1}{2} = \pm 1 \qquad \text{Solve the quadratic equation for } x + \frac{1}{2}.$$

$$x = -\frac{1}{2} \pm 1 \qquad \text{Add } -\frac{1}{2} \text{ to both sides.}$$

$$x = -\frac{1}{2} + 1 \quad \text{or} \quad x = -\frac{1}{2} - 1$$

$$= \frac{1}{2} \qquad \qquad = -\frac{3}{2}$$

Check each solution. This equation can also be solved by factoring. ■

EXAMPLE 4   Use completing the square to solve the equation $2x^2 - 5x - 3 = 0$.

Solution   We divide both sides of the equation by 2 so that the coefficient of x^2 is 1. Then we proceed as follows.

$$2x^2 - 5x - 3 = 0$$

$$x^2 - \frac{5}{2}x - \frac{3}{2} = 0 \qquad \text{Divide both sides by 2.}$$

$$x^2 - \frac{5}{2}x = \frac{3}{2} \qquad \text{Add } \frac{3}{2} \text{ to both sides.}$$

$$x^2 - \frac{5}{2}x + \left[\frac{1}{2}\left(-\frac{5}{2}\right)\right]^2 = \frac{3}{2} + \left[\frac{1}{2}\left(-\frac{5}{2}\right)\right]^2 \qquad \text{Add } \left[\frac{1}{2}\left(-\frac{5}{2}\right)\right]^2 \text{ to both sides to complete the square.}$$

$$x^2 - \frac{5}{2}x + \frac{25}{16} = \frac{3}{2} + \frac{25}{16} \qquad \text{Simplify.}$$

$$\left(x - \frac{5}{4}\right)^2 = \frac{24}{16} + \frac{25}{16} \qquad \begin{array}{l}\text{Factor on the left-hand side}\\\text{and get a common denomi-}\\\text{nator on the right-hand side.}\end{array}$$

$$\left(x - \frac{5}{4}\right)^2 = \frac{49}{16} \qquad \text{Add the fractions.}$$

$$x - \frac{5}{4} = \pm\frac{7}{4} \qquad \begin{array}{l}\text{Solve the quadratic equation}\\\text{for } x - \frac{5}{4}.\end{array}$$

$$x = \frac{5}{4} \pm \frac{7}{4} \qquad \text{Add } \frac{5}{4} \text{ to both sides.}$$

$$x = \frac{5}{4} + \frac{7}{4} \quad \text{or} \quad x = \frac{5}{4} - \frac{7}{4}$$

$$= \frac{12}{4} \qquad \qquad = -\frac{2}{4}$$

$$= 3 \qquad \qquad = -\frac{1}{2}$$

Check each solution. This equation can also be solved by factoring. ■

EXAMPLE 5   Solve the equation $2x^2 - 2 = -4x$.

Solution   We add $4x$ to both sides of the equation to see if it can be solved by factoring.

$$2x^2 + 4x - 2 = 0 \qquad \text{Add } 4x \text{ to both sides.}$$

1. $x^2 + 2x - 1 = 0 \qquad \text{Divide both sides by 2.}$

Since Equation 1 cannot be solved by factoring, we complete the square.

$$x^2 + 2x = 1 \qquad \text{Add 1 to both sides.}$$

$$x^2 + 2x + (1)^2 = 1 + (1)^2 \qquad \text{Add } 1^2 \text{ to both sides to complete the square.}$$

$$(x + 1)^2 = 2 \qquad \text{Simplify and factor.}$$

$$x + 1 = \pm \sqrt{2} \qquad \text{Solve the quadratic equation for } x + 1.$$

$$x = -1 \pm \sqrt{2} \qquad \text{Subtract 1 from both sides.}$$

$$x = -1 + \sqrt{2} \qquad \text{or} \qquad x = -1 - \sqrt{2}$$

Both solutions check. To the nearest tenth, the solutions are 0.4 and -2.4.   ■

To solve an equation by completing the square, we follow these steps.

Method of Completing the Square

1. If the coefficient of x^2 is not 1, make it 1 by dividing both sides of the equation by the coefficient of x^2.

2. If necessary, add a number to both sides of the equation to get the constant term on the right-hand side.

3. Complete the square.

   a. Find half the coefficient of x and square it.

   b. Add that square to both sides of the equation.

4. Factor the trinomial square and combine like terms.

5. Solve the resulting quadratic equation.

6. Check each solution.

ORALS   *What number must be added to each binomial to make a trinomial square?*

1. $x^2 + 4x$ **2.** $x^2 + 6x$

3. $x^2 - 8x$ **4.** $x^2 - 10x$

5. $x^2 + 5x$ **6.** $x^2 - 3x$

EXERCISE 9.2

In Exercises 1–12, complete the square to make a trinomial square.

1. $x^2 + 2x$

2. $x^2 + 12x$

3. $x^2 - 4x$

4. $x^2 - 14x$

5. $x^2 + 7x$

6. $x^2 + 21x$

7. $a^2 - 3a$

8. $b^2 - 13b$

9. $b^2 + \dfrac{2}{3}b$

10. $a^2 + \dfrac{8}{5}a$

11. $c^2 - \dfrac{5}{2}c$

12. $c^2 - \dfrac{11}{3}c$

In Exercises 13–30, solve each equation by completing the square.

13. $x^2 + 6x + 8 = 0$

14. $x^2 + 8x + 12 = 0$

15. $x^2 - 8x + 12 = 0$

16. $x^2 - 4x + 3 = 0$

17. $x^2 - 2x - 15 = 0$

18. $x^2 - 2x - 8 = 0$

19. $x^2 - 7x + 12 = 0$

20. $x^2 - 7x + 10 = 0$

21. $x^2 + 5x - 6 = 0$

22. $x^2 = 14 - 5x$

23. $2x^2 = 4 - 2x$

24. $3x^2 + 9x + 6 = 0$

25. $3x^2 + 48 = -24x$

26. $3x^2 = 3x + 6$

27. $2x^2 = 3x + 2$

28. $3x^2 = 2 - 5x$

29. $4x^2 = 2 - 7x$

30. $2x^2 = 5x + 3$

In Exercises 31–38, solve each equation.

31. $x^2 + 4x + 1 = 0$

32. $x^2 + 6x + 2 = 0$

33. $x^2 - 2x - 4 = 0$

34. $x^2 - 4x - 2 = 0$

35. $x^2 = 4x + 3$

36. $x^2 = 6x - 3$

37. $2x^2 = 2 - 4x$

38. $3x^2 = 12 - 6x$

In Exercises 39–44, write each equation in the form $ax^2 + bx + c = 0$ and solve it by completing the square.

39. $2x(x + 3) = 8$

40. $3x(x - 2) = 9$

41. $6(x^2 - 1) = 5x$

42. $2(3x^2 - 2) = 5x$

43. $x(x + 3) - \dfrac{1}{2} = -2$

44. $x[(x - 2) + 3] = 3\left(x - \dfrac{2}{9}\right)$

Writing Exercises ■ *Write a paragraph using your own words.*

1. Explain how to complete the square.

2. Explain why the coefficient of x^2 should be 1 before completing the square.

Something to Think About ■ *Consider this method of completing the square on x in the binomial* $ax^2 + bx$: *Multiply the binomial by 4a and then add* b^2. *Complete the square on x in each binomial and factor the resulting trinomial.*

1. $2x^2 + 6x$

2. $3x^2 - 4x$

Review Exercises ■ *Solve each equation.*

1. $\dfrac{3t(2t + 1)}{2} + 6 = 3t^2$

2. $20r^2 - 11r - 3 = 0$

3. $\dfrac{2}{3x} - \dfrac{5}{9} = -\dfrac{1}{x}$

4. $\sqrt{x + 12} = \sqrt{3x}$

9.3 The Quadratic Formula

■ **Quadratic Equations with No Real Roots**

GETTING READY *Evaluate* $b^2 - 4ac$ *when a, b, and c have the following values.*

1. $a = 1, b = 2, c = 3$

2. $a = 4, b = 3, c = 1$

3. $a = 1, b = 0, c = -2$

4. $a = 2, b = 4, c = 2$

We can solve the **general quadratic equation** $ax^2 + bx + c = 0$ $(a \neq 0)$ by completing the square.

$$ax^2 + bx + c = 0$$

$$\frac{ax^2}{a} + \frac{bx}{a} + \frac{c}{a} = \frac{0}{a} \qquad \text{Divide both sides by } a.$$

$$x^2 + \frac{b}{a}x + \frac{c}{a} = 0 \qquad \text{Simplify.}$$

$$x^2 + \frac{b}{a}x = -\frac{c}{a} \qquad \text{Add } -\frac{c}{a} \text{ to both sides.}$$

We can now complete the square on x by adding $\left(\dfrac{1}{2} \cdot \dfrac{b}{a}\right)^2$, or $\dfrac{b^2}{4a^2}$, to both sides:

$$x^2 + \frac{b}{a}x + \frac{b^2}{4a^2} = \frac{b^2}{4a^2} - \frac{c}{a}$$

After factoring the trinomial on the left-hand side and adding the fractions on the right-hand side, we have

$$\left(x + \frac{b}{2a}\right)\left(x + \frac{b}{2a}\right) = \frac{b^2}{4a^2} - \frac{4ac}{4aa}$$

1. $$\left(x + \frac{b}{2a}\right)^2 = \frac{b^2 - 4ac}{4a^2}$$

Equation 1 can be solved by the square root method to obtain

$$x + \frac{b}{2a} = \sqrt{\frac{b^2 - 4ac}{4a^2}} \qquad \text{and} \qquad x + \frac{b}{2a} = -\sqrt{\frac{b^2 - 4ac}{4a^2}}$$

$$x + \frac{b}{2a} = \frac{\sqrt{b^2 - 4ac}}{\sqrt{4a^2}} \qquad\qquad x + \frac{b}{2a} = -\frac{\sqrt{b^2 - 4ac}}{\sqrt{4a^2}}$$

$$x = -\frac{b}{2a} + \frac{\sqrt{b^2 - 4ac}}{2a} \qquad\qquad x = -\frac{b}{2a} - \frac{\sqrt{b^2 - 4ac}}{2a}$$

$$x = \frac{-b + \sqrt{b^2 - 4ac}}{2a} \qquad\qquad x = \frac{-b - \sqrt{b^2 - 4ac}}{2a}$$

These solutions are usually written in one expression called the **quadratic formula.**

Quadratic Formula The solutions of the quadratic equation $ax^2 + bx + c = 0$, where $a \neq 0$, are

$$x = \frac{-b \pm \sqrt{b^2 - 4ac}}{2a}$$

■ PERSPECTIVE

Equations in Babylon

Clay tablets that survive from the early period of the Babylonian civilization, 1800 to 1600 B.C., show that the Babylonians were accomplished mathematicians. For use by the Babylonian merchants, many of these tablets contain multiplication tables and, for division, lists of reciprocals. For more abstract mathematical purposes, other tablets provide tables of squares, cubes, square roots, and cube roots. Still other tablets contain lists of problems and exercises. Some of these problems are practical, but many are puzzle problems, just for fun. Several problems and their solutions indicate that the Babylonians knew what we now call the Pythagorean theorem, centuries before the Greeks discovered it.

The Babylonians could also solve certain quadratic equations. For example, one puzzle problem from a Babylonian tablet asks, "What number added to its reciprocal is 5?" In modern notation, we would solve the equation $x + \frac{1}{x} = 5$. Can you show that this is equivalent to the quadratic equation $x^2 - 5x + 1 = 0$?

WARNING! When you write the quadratic formula, be careful to draw the fraction bar so that it underlines the complete numerator. Do not write

$$x = -b \pm \frac{\sqrt{b^2 - 4ac}}{2a}$$

EXAMPLE 1 Use the quadratic formula to solve $x^2 + 5x + 6 = 0$.

Solution In this equation, $a = 1$, $b = 5$, and $c = 6$. We substitute these values into the quadratic formula and simplify.

$$x = \frac{-b \pm \sqrt{b^2 - 4ac}}{2a}$$

$$= \frac{-5 \pm \sqrt{5^2 - 4(1)(6)}}{2(1)} \qquad \text{Substitute 1 for } a, 5 \text{ for } b, \text{ and } 6 \text{ for } c.$$

$$= \frac{-5 \pm \sqrt{25 - 24}}{2}$$

$$= \frac{-5 \pm \sqrt{1}}{2}$$

$$= \frac{-5 \pm 1}{2}$$

Thus,

$$x = \frac{-5 + 1}{2} \quad \text{and} \quad x = \frac{-5 - 1}{2}$$

$$= \frac{-4}{2} \qquad\qquad\quad = \frac{-6}{2}$$

$$= -2 \qquad\qquad\qquad = -3$$

Check both solutions. ■

 WARNING! Be sure to write a quadratic equation in quadratic form before identifying the values a, b, and c.

EXAMPLE 2 Use the quadratic formula to solve the equation $2x^2 = 5x + 3$.

Solution We begin by writing the equation in quadratic form.

$$2x^2 = 5x + 3$$

$$2x^2 - 5x - 3 = 0 \qquad \text{Add } -5x \text{ and } -3 \text{ to both sides.}$$

In this equation, $a = 2$, $b = -5$, and $c = -3$. We substitute these values into the quadratic formula and simplify.

$$x = \frac{-b \pm \sqrt{b^2 - 4ac}}{2a}$$

$$= \frac{-(-5) \pm \sqrt{(-5)^2 - 4(2)(-3)}}{2(2)}$$ Substitute 2 for a, -5 for b, and -3 for c.

$$= \frac{5 \pm \sqrt{25 + 24}}{4}$$

$$= \frac{5 \pm \sqrt{49}}{4}$$

$$= \frac{5 \pm 7}{4}$$

Thus,

$$x = \frac{5 + 7}{4} \quad \text{or} \quad x = \frac{5 - 7}{4}$$

$$= \frac{12}{4} \qquad\qquad\quad = \frac{-2}{4}$$

$$= 3 \qquad\qquad\qquad = -\frac{1}{2}$$

Check both solutions. ∎

EXAMPLE 3   Use the quadratic formula to solve $3x^2 = 2x + 4$.

Solution   We begin by writing the given equation in quadratic form.

$$3x^2 = 2x + 4$$

$$3x^2 - 2x - 4 = 0$$ Add $-2x$ and -4 to both sides.

In this equation, $a = 3$, $b = -2$, and $c = -4$. We substitute these values into the quadratic formula and simplify.

$$x = \frac{-b \pm \sqrt{b^2 - 4ac}}{2a}$$

$$= \frac{-(-2) \pm \sqrt{(-2)^2 - 4(3)(-4)}}{2(3)}$$ Substitute 3 for a, -2 for b, and -4 for c.

$$= \frac{2 \pm \sqrt{4 + 48}}{6}$$

$$= \frac{2 \pm \sqrt{52}}{6}$$

$$= \frac{2 \pm 2\sqrt{13}}{6}$$ $\sqrt{52} = \sqrt{4 \cdot 13} = \sqrt{4}\sqrt{13} = 2\sqrt{13}.$

$$= \frac{2\left(1 \pm \sqrt{13}\right)}{6}$$ Factor out 2.

$$= \frac{1 \pm \sqrt{13}}{3}$$ Divide out the common factor of 2.

Thus,

$$x = \frac{1}{3} + \frac{\sqrt{13}}{3} \qquad \text{or} \qquad x = \frac{1}{3} - \frac{\sqrt{13}}{3}$$

Both solutions check. To the nearest tenth, the solutions are 1.5 and -0.9. ∎

■ Quadratic Equations with No Real Roots

EXAMPLE 4   Use the quadratic formula to solve $x^2 + 2x + 5 = 0$.

Solution   In this equation $a = 1$, $b = 2$, and $c = 5$. We substitute these values into the quadratic formula.

$$x = \frac{-b \pm \sqrt{b^2 - 4ac}}{2a}$$

$$= \frac{-2 \pm \sqrt{2^2 - 4(1)(5)}}{2(1)}$$ Substitute 1 for a, 2 for b, and 5 for c.

$$= \frac{-2 \pm \sqrt{4 - 20}}{6}$$

$$= \frac{-2 \pm \sqrt{-16}}{6}$$

Since $\sqrt{-16}$ is not a real number, there can be no real-number solutions. ∎

ORALS   *Find the values of a, b, and c in each equation.*

1. $3x^2 + 4x - 12 = 0$ **2.** $-2x^2 - 4x + 10 = 0$

3. $5x^2 - x = 1$ **4.** $x^2 - 9 = -3x$

E X E R C I S E   9.3

In Exercises 1–12, change each equation into quadratic form, if necessary, and find the values of a, b, and c. **Do not solve the equation.**

1. $x^2 + 4x + 3 = 0$

2. $x^2 - x - 4 = 0$

3. $3x^2 - 2x + 7 = 0$

4. $4x^2 + 7x - 3 = 0$

5. $4y^2 = 2y - 1$

6. $2x = 3x^2 + 4$

7. $x(3x - 5) = 2$

8. $y(5y + 10) = 8$

9. $7(x^2 + 3) = -13x$

10. $5(a^2 + 5) = -4a$

11. $(2a + 3)(a - 2) = (a + 1)(a - 1)$

12. $(3a + 2)(a - 1) = (2a + 7)(a - 1)$

In Exercises 13–40, use the quadratic formula to find all real solutions of each equation.

13. $x^2 - 5x + 6 = 0$

14. $x^2 + 5x + 4 = 0$

15. $x^2 + 7x + 12 = 0$

16. $x^2 - x - 12 = 0$

17. $2x^2 - x - 1 = 0$

18. $2x^2 + 3x - 2 = 0$

19. $3x^2 + 5x + 2 = 0$

20. $3x^2 - 4x + 1 = 0$

21. $4x^2 + 4x - 3 = 0$

22. $4x^2 + 3x - 1 = 0$

23. $5x^2 - 8x - 4 = 0$

24. $6x^2 - 8x + 2 = 0$

25. $x^2 + 3x + 1 = 0$

26. $x^2 + 3x - 2 = 0$

27. $x^2 + 5x - 3 = 0$

28. $x^2 + 5x + 3 = 0$

29. $x^2 + 2x + 7 = 0$

30. $2x^2 - x + 2 = 0$

31. $2x^2 + x = 5$

32. $3x^2 - x = 1$

33. $x^2 + 1 = -4x$

34. $x^2 + 1 = -8x$

35. $x^2 + 5 = 2x$

36. $2x^2 + 3x = -3$

37. $x^2 = 1 - 2x$

38. $x^2 = 2 - 2x$

39. $3x^2 = 6x + 2$

40. $3x^2 = -8x - 2$

In Exercises 41–42, use these facts. The two solutions of the equation $ax^2 + bx + c = 0$ ($a \neq 0$) are

$$x_1 = \frac{-b + \sqrt{b^2 - 4ac}}{2a} \quad \text{and} \quad x_2 = \frac{-b - \sqrt{b^2 - 4ac}}{2a}$$

41. Show that $x_1 + x_2 = -\dfrac{b}{a}$.

42. Show that $x_1 x_2 = \dfrac{c}{a}$.

Writing Exercises ■ *Write a paragraph using your own words.*

1. Explain how to use the quadratic formula.

2. Explain the meaning of the $\pm$ symbol.

Something to Think About ■ **1.** What advantages does the following proof have over the one shown on pages 536–537?

$$ax^2 + bx + c = 0$$

$$ax^2 + bx = -c \qquad \text{Subtract } c \text{ from both sides.}$$

$$(4a)ax^2 + (4a)bx = -(4a)c \qquad \text{Multiply both sides by } 4a.$$

$$4a^2x^2 + 4abx + b^2 = b^2 - 4ac \qquad \text{Add } b^2 \text{ to both sides.}$$

$$(2ax + b)^2 = b^2 - 4ac \qquad \text{Factor the left-hand side.}$$

$$2ax + b = \pm \sqrt{b^2 - 4ac} \qquad \text{Take the square root of both sides.}$$

$$2ax = -b \pm \sqrt{b^2 - 4ac} \qquad \text{Subtract } b \text{ from both sides.}$$

$$x = \frac{-b \pm \sqrt{b^2 - 4ac}}{2a} \qquad \text{Divide both sides by } 2a.$$

2. The binomial $b^2 - 4ac$ is called the **discriminant.** From its value, you can predict whether the solutions of a given quadratic equation are real or nonreal numbers. Explain.

Review Exercises ■ *Solve each equation for the indicated variable.*

1. $A = p + prt$; for r

2. $F = \dfrac{GMm}{d^2}$, for M

In general form, write the equation of the line with the given properties.

3. Slope of $\frac{3}{5}$ and passing through $(0, 12)$

4. Passes through $(6, 8)$ and the origin

9.4 Applications of Quadratic Equations

■ Integer Problems ■ Geometric Problems ■ Ballistics Problems
■ Pythagorean Theorem Problems ■ Business Problems

GETTING READY **1.** Find the area of a rectangle with dimensions of 6 cm by 12 cm.

2. Factor the binomial $144t - 16t^2$.

3. Two sides of a right triangle are 18 feet and 24 feet. Find the length of the hypotenuse.

4. Find the interest earned in one year on a deposit of $5000 at 6% annual interest.

The solutions of many problems involve quadratic equations.

■ Integer Problems

EXAMPLE 1 The product of the first and third of three consecutive positive odd integers is 77. Find the integers.

Solution We can let x represent the first odd integer.
Then $x + 2$ will represent the second odd integer.
Then $x + 4$ will represent the third odd integer.

Since the product of the first and third is 77, we have

$$x(x + 4) = 77$$
$$x^2 + 4x - 77 = 0$$
$$(x + 11)(x - 7) = 0$$
$$x + 11 = 0 \quad \text{or} \quad x - 7 = 0$$
$$x = -11 \quad | \quad x = 7$$

Since we are looking for three positive odd integers, we ignore the solution of -11. Thus, the first of the integers is 7, the second is 9, and the third is 11. ■

■ Geometric Problems

EXAMPLE 2 The oriental rug shown in Figure 9-1 is 3 feet longer than it is wide. Find the dimensions of the rug if its area is 180 ft².

Solution We can let w represent the width of the rug.
Then $w + 3$ will represent its length.

Since the area of a rectangle is given by the formula $A = lw$ (Area = length × width), the area of the rug is $(w + 3)w$, which is equal to 180. This gives the equation

The length of the rug	·	the width of the rug	=	the area of the rug.
$(w + 3)$	·	w	=	180

We can solve this equation as follows:

$$w^2 + 3w = 180 \qquad \text{Use the distributive property to remove parentheses.}$$
$$w^2 + 3w - 180 = 0 \qquad \text{Subtract 180 from both sides.}$$
$$(w - 12)(w + 15) = 0 \qquad \text{Factor } w^2 + 3w - 180.$$
$$w - 12 = 0 \quad \text{or} \quad w + 15 = 0$$
$$w = 12 \quad | \quad w = -15$$

FIGURE 9-1

When $w = 12$, the length, $w + 3$, is 15. The dimensions of the rug are 12 feet by 15 feet. The solution $w = -15$ is irrelevant, since the rug cannot have a negative width. ∎

Ballistics Problems

EXAMPLE 3 If an object is thrown straight up with a velocity of 160 feet per second, its height h (in feet) above the ground is given by the formula $h = 160t - 16t^2$, where t represents the time (in seconds) since it was thrown. How long will it take for the object to hit the ground?

Solution When the object hits the ground, its height will be 0 feet. Thus, we can let $h = 0$ and solve the formula for t.

$$h = 160t - 16t^2$$
$$0 = 160t - 16t^2 \qquad \text{Substitute 0 for } h.$$
$$0 = 16t(10 - t) \qquad \text{Factor out the common monomial of } 16t.$$
$$16t = 0 \quad \text{or} \quad 10 - t = 0 \qquad \text{Set each factor equal to 0.}$$
$$t = 0 \quad | \quad \qquad t = 10$$

At $t = 0$, the object's height is 0, because it was just released. When $t = 10$, the height is again 0, because the object has returned to its starting point. ∎

Pythagorean Theorem Problems

EXAMPLE 4 A manufacturer of television parts received an order for 52-inch picture tubes (measured along the diagonal). The tubes are to be rectangular in shape and 4 inches wider than they are high. Find the dimensions of each tube.

Solution We can let h represent the height of a picture tube, as shown in Figure 9-2. Then $h + 4$ will represent the width.

Since a 52-inch picture tube measures 52 inches along its diagonal, we can use the Pythagorean theorem to form the equation

$$h^2 + (h + 4)^2 = 52^2$$

$$h^2 + h^2 + 8h + 16 = 2704 \qquad \text{Expand } (h + 4)^2.$$

$$2h^2 + 8h - 2688 = 0 \qquad \text{Subtract 2704 from both sides.}$$

$$h^2 + 4h - 1344 = 0 \qquad \text{Divide both sides by 2.}$$

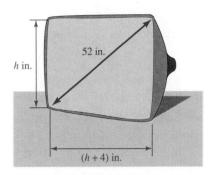

FIGURE 9-2

We can solve this equation with the quadratic formula.

$$h = \frac{-b \pm \sqrt{b^2 - 4ac}}{2a}$$

$$h = \frac{-4 \pm \sqrt{(4)^2 - 4(1)(-1344)}}{2(1)}$$

$$h = \frac{-4 \pm \sqrt{16 + 5376}}{2}$$

$$h = \frac{-4 \pm \sqrt{5392}}{2}$$

$$h = \frac{-4 \pm 73.430239}{2}$$

$$h = \frac{-4 + 73.430239}{2} \qquad \text{or} \qquad h = \frac{-4 - 73.430239}{2}$$

$$= \frac{69.430239}{2} \qquad\qquad\qquad = \frac{-77.430239}{2}$$

$$= 34.715119 \qquad\qquad\qquad = -38.715119$$

The height of each tube will be approximately 34.7 inches, and the width will be approximately 38.7 inches. The second solution, which is negative, must be discarded, because the height of a TV picture tube cannot be negative. ∎

Business Problems

EXAMPLE 5 If $P is invested at an annual rate r, it will grow to an amount of $A in n years according to the formula $A = P(1 + r)^n$.

What interest rate is needed for a $5000 investment to grow to $5618 after 2 years?

Solution We can substitute 5000 for P, 5618 for A, and 2 for n in the formula and solve for r.

$$A = P(1 + r)^n$$
$$5618 = 5000(1 + r)^2$$
$$5618 = 5000(1 + 2r + r^2) \qquad \text{Expand } (1 + r)^2.$$
$$5618 = 5000 + 10,000r + 5000r^2 \qquad \text{Remove parentheses.}$$
$$5000r^2 + 10,000r - 618 = 0 \qquad \text{Subtract 5618 from both sides.}$$

We can use a calculator and solve this equation with the quadratic formula.

$$r = \frac{-b \pm \sqrt{b^2 - 4ac}}{2a}$$
$$r = \frac{-10,000 \pm \sqrt{10,000^2 - 4(5000)(-618)}}{2(5000)}$$
$$r = \frac{-10,000 \pm \sqrt{100,000,000 + 12,360,000}}{10,000}$$
$$r = \frac{-10,000 \pm \sqrt{112,360,000}}{10,000}$$
$$r = \frac{-10,000 \pm 10,600}{10,000}$$

$$r = \frac{-10,000 + 10,600}{10,000} \qquad \text{or} \qquad r = \frac{-10,000 - 10,600}{10,000}$$
$$= \frac{600}{10,000} \qquad\qquad\qquad = \frac{-20,600}{10,000}$$
$$= 6\% \qquad\qquad\qquad = -206\%$$

The required rate is 6%. The rate of -206% has no meaning in this problem. ∎

ORALS 1. Denote three consecutive integers.
2. Denote three consecutive odd integers.
3. Give the formula for the area of a rectangle.
4. Give the formula for the perimeter of a rectangle.
5. State the Pythagorean theorem.

E X E R C I S E   9.4

In Exercises 1–14, solve each problem.

1. **Finding consecutive integers** The product of two consecutive positive even integers is 48. Find the integers.

2. **Finding consecutive integers** The product of the first and second of three consecutive positive odd integers is 35. Find the sum of the three integers.

3. **Integer problem** The sum of a positive integer and its square is 90. Find the integer.

4. **Finding the sum of two squares** The sum of the squares of two consecutive even negative integers is 52. Find the integers.

5. **Integer problems** Can the sum of the squares of two consecutive integers equal 100? Explain.

6. **Number problem** The sum of an integer and four times its reciprocal is 8.5. Find the integer.

7. **Geometric problem** A rectangular mural is 4 feet longer than it is wide. Find its dimensions if its area is 32 square feet.

8. **Geometric problem** The length of a 220-square-foot rectangular garden is 2 feet more than twice its width. Find its perimeter.

9. **Finding the height of a triangle** The triangle shown in Illustration 1 has an area of 30 square inches. Find its height.

10. **Finding the base of a triangle** Find the length of the base of the triangle shown in Illustration 1.

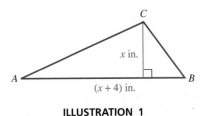

ILLUSTRATION 1

11. **Finding the area of a garden** The rectangular garden shown in Illustration 2 is surrounded by a walk of uniform width. Find the dimensions of the garden if its area is 180 square feet.

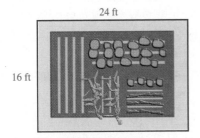

ILLUSTRATION 2

12. **Finding the area of a pool** The owner of the pool in Illustration 3 wants to surround it with a deck of uniform width. If he can afford 368 square feet of decking, how wide can he make the deck?

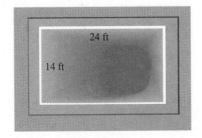

ILLUSTRATION 3

13. **Falling objects** An object will fall s feet in t seconds, where $s = 16t^2$. If a workman 1454 feet above the ground at the top of the Sears Tower drops a hammer, how long will it take for the hammer to hit the ground?

14. **Falling coins** A tourist drops a penny from the observation deck of the World Trade Center, 1377 feet above the ground. How long will it take for the penny to hit the ground? (See Exercise 13.)

In Exercises 15–17, the height, h, of a toy rocket in flight is given by the formula $h = -16t^2 + 144t$, where t is the time of the flight in seconds.

15. Height of a rocket Find the height of the rocket at $t = 3$ seconds.

16. Flight of a rocket How long will it take for the rocket to hit the ground?

17. Maximum height of a rocket If the maximum height of the rocket occurs halfway through its flight, how high will the rocket go?

In Exercises 18–20, a handgun is fired straight up with a muzzle velocity of 1088 feet per second. The height h of the bullet is given by the formula $h = -16t^2 + 1088t$, where t is the time in seconds.

18. Height of a bullet Find the height of the bullet after 10 seconds.

19. Height of a bullet When will the bullet hit the ground?

20. Maximum height of a bullet How high will the bullet go?

In Exercises 21–22, a soldier fires a rifle straight up. If the muzzle velocity of the rifle is 3488 feet per second, the height h of the bullet is given by the formula $h = -16t^2 + 3488t$, where t is the time in seconds.

21. Height of a bullet When will the bullet hit the ground?

22. Military application Could the bullet hit an airplane flying at an altitude of 40,000 feet?

23. Finding dimensions The picture frame in Illustration 4 is 2 inches wider than it is high. Find its dimensions.

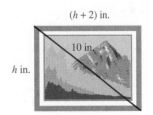

$(h + 2)$ in.

10 in.

h in.

ILLUSTRATION 4

24. Length of a sidewalk A 170-meter long sidewalk from the mathematics building M to the student center C is shown in Illustration 5. However, students prefer to walk directly from M to C. How long are the two pieces of the existing sidewalk?

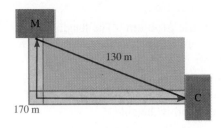

M

130 m

170 m

C

ILLUSTRATION 5

25. Navigation Two boats left port at the same time, one sailing east and one sailing south. If one boat sailed 10 nautical miles more than the other and they are 50 nautical miles apart, how far did each boat sail?

26. Navigation One plane heads west from an airport, flying at 200 mph. One hour later, a second plane heads north from same airport, flying at the same speed. When will the planes be 1000 miles apart?

In Exercises 27–28, use the formula $A = P(1 + r)^2$ that relates the amount \$A that \$P will become when invested at an annual rate r for 2 years.

27. Investing money What interest rate is needed for \$5000 to grow to \$5724.50 at the end of 2 years?

28. Investing money What interest rate is needed for \$7000 to grow to \$8470 at the end of 2 years?

29. Manufacturing television sets An electronics firm has found that its revenue for manufacturing and selling x television sets is given by the formula $R = -\frac{1}{6}x^2 + 450x$. How much revenue will be earned by manufacturing 600 television sets?

30. Selling CD players When a wholesaler sells n CD players, his revenue R is given by the formula $R = 150n - \frac{1}{2}n^2$. How many players would he have to sell to receive \$11,250?

31. Metal fabrication A piece of tin, 12 inches on a side, is to have four equal squares cut from its corners, as in Illustration 6. If the edges are then to be folded up to make a box with a floor area of 64 square inches, find the depth of the box.

32. Making gutters A piece of sheet metal, 18 inches wide, is bent to form the gutter shown in Illustration 7. If the cross-sectional area is 36 square inches, find the depth of the gutter.

ILLUSTRATION 7

33. Filling a storage tank Two pipes are used to fill a water tank. The first pipe can fill the tank in 4 hours, and the two pipes together can fill the tank in 2 hours less time than the second pipe alone. How long would it take for the second pipe to fill the tank?

34. Filling a swimming pool A small hose requires 6 more hours to fill a swimming pool than a larger hose. If the two hoses can fill the pool in 4 hours, how long would it take the larger hose alone?

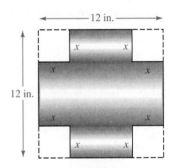

ILLUSTRATION 6

Writing Exercises ■ *Write a paragraph using your own words.*

1. Choose one of the previous application problems and list the steps you followed as you worked it.

2. What advice would you give students who were about to solve that problem?

Something to Think About ■ **1.** Construct an integer problem that requires a quadratic equation to solve it.

2. Construct a geometric problem that uses the Pythagorean theorem.

Review Exercises ■ *Simplify each radical expression.*

1. $\sqrt{80}$

2. $12\sqrt{x^3y^2}$

3. $\dfrac{x}{\sqrt{7x}}$

4. $\dfrac{\sqrt{x}+2}{\sqrt{x}-2}$

9.5 Complex Numbers

■ Complex Numbers ■ Operations with Complex Numbers ■ Complex Conjugates ■ Rationalizing Denominators ■ Absolute Value of a Complex Number

GETTING READY *Do the following operations.*

1. $(2x+4)+(3x-5)$

2. $(3x-4)-(2x+3)$

3. $(x+4)(x-5)$

4. $(3x-1)(2x-1)$

We have seen that the solutions to some quadratic equations are not real numbers.

EXAMPLE 1 Solve $x^2 + x + 1 = 0$.

Solution We can use the quadratic formula, with $a = 1$, $b = 1$, and $c = 1$.

$$x = \frac{-b \pm \sqrt{b^2 - 4ac}}{2a}$$

$$= \frac{-1 \pm \sqrt{1^2 - 4(1)(1)}}{2(1)}$$

$$= \frac{-1 \pm \sqrt{1 - 4}}{2}$$

$$= \frac{-1 \pm \sqrt{-3}}{2}$$

$$x = \frac{-1 + \sqrt{-3}}{2} \quad \text{or} \quad x = \frac{-1 - \sqrt{-3}}{2} \qquad ■$$

Each solution in Example 1 involves $\sqrt{-3}$, which is not a real number, because the square of no real number equals -3. For years, mathematicians believed that numbers like $\sqrt{-3}$, $\sqrt{-1}$, and $\sqrt{-9}$ were nonsense. Even the great English mathematician Sir Isaac Newton (1642–1727) called them impossible. These numbers were called **imaginary numbers** by René Descartes (1596–1650). Today they have important uses such as describing alternating current in electronics.

The imaginary number $\sqrt{-1}$ is usually denoted by the letter i. Since

$$i = \sqrt{-1}$$

it follows that

$$i^2 = -1$$

The powers of i produce an interesting pattern:

$$i = \sqrt{-1} = i \qquad\qquad i^5 = i^4 \cdot i = 1 \cdot i = i$$
$$i^2 = \sqrt{-1}\sqrt{-1} = -1 \qquad\qquad i^6 = i^4 \cdot i^2 = 1(-1) = -1$$
$$i^3 = i^2 \cdot i = -1 \cdot i = -i \qquad\qquad i^7 = i^4 \cdot i^3 = 1(-i) = -i$$
$$i^4 = i^2 \cdot i^2 = (-1)(-1) = 1 \qquad\qquad i^8 = i^4 \cdot i^4 = (1)(1) = 1$$

The pattern continues: $i, -1, -i, 1, \ldots$.

If we assume that multiplication of imaginary numbers is commutative and associative, then

$$(2i)^2 = 2^2 i^2$$
$$= 4(-1)$$
$$= -4$$

Since $(2i)^2 = -4$, it follows that $2i$ is a square root of -4, and we write

$$\sqrt{-4} = 2i$$

This result could have been obtained by the following process:

$$\sqrt{-4} = \sqrt{4(-1)}$$
$$= \sqrt{4}\sqrt{-1}$$
$$= 2i$$

Likewise, we have

$$\sqrt{-25} = \sqrt{25(-1)} = \sqrt{25}\sqrt{-1} = 5i$$

$$\sqrt{-\frac{1}{9}} = \sqrt{\frac{1}{9}(-1)} = \sqrt{\frac{1}{9}}\sqrt{-1} = \frac{1}{3}i$$

$$\sqrt{\frac{-100}{49}} = \sqrt{\frac{100}{49}(-1)} = \frac{\sqrt{100}}{\sqrt{49}}\sqrt{-1} = \frac{10}{7}i$$

In general, we have the following rule.

Leonhard Euler (1707–1783)
Euler first used the letter i to represent $\sqrt{-1}$, the letter e for the base of natural logarithms, and the symbol Σ for summation. Euler was one of the most prolific mathematicians of all time, contributing to almost all areas of mathematics. Much of his work was accomplished after he became blind.

Rules of Radicals   If at least one of a and b is a nonnegative real number, then

$$\sqrt{ab} = \sqrt{a}\sqrt{b} \qquad \text{and} \qquad \sqrt{\frac{a}{b}} = \frac{\sqrt{a}}{\sqrt{b}} \quad (b \neq 0)$$

■ Complex Numbers

Imaginary numbers such as $\sqrt{-3}$, $\sqrt{-1}$, and $\sqrt{-9}$ form a subset of a broader set of numbers called **complex numbers.**

Complex Number A **complex number** is any number that can be written in the form $a + bi$, where a and b are real numbers, and $i = \sqrt{-1}$.

The number a is called the **real part** and the number b is called the **imaginary part** of the complex number $a + bi$.

If $b = 0$, the complex number $a + bi$ is the real number a. If $b \neq 0$ and $a = 0$, the complex number $0 + bi$ (or just bi) is an imaginary number.

Figure 9-3 shows the relationship of the real numbers to the imaginary and complex numbers.

Complex Numbers	
Real Numbers	Imaginary Numbers
2, π, $\sqrt{5}$	i, $3i$, $-5i$

$$3 + 4i, \ 2 - 3i, \ \frac{1}{2} - \sqrt{3}\,i$$

FIGURE 9-3

Equality of Complex Numbers The complex numbers $a + bi$ and $c + di$ are equal if and only if
$$a = c \quad \text{and} \quad b = d$$

EXAMPLE 2
a. $2 + 3i = \sqrt{4} + \frac{6}{2}i$, because $2 = \sqrt{4}$ and $3 = \frac{6}{2}$.

b. $4 - 5i = \frac{12}{3} - \sqrt{25}i$, because $4 = \frac{12}{3}$ and $-5 = -\sqrt{25}$.

c. $x + yi = 4 + 7i$ if and only if $x = 4$ and $y = 7$. ■

■ Operations with Complex Numbers

Addition and Subtraction of Complex Numbers Complex numbers are added and subtracted as if they were binomials:
$$(a + bi) + (c + di) = (a + c) + (b + d)i$$

(a, c)   **EXAMPLE 3**   **a.** $(8 + 4i) + (12 + 8i) = 8 + 4i + 12 + 8i$
$$= 20 + 12i$$

b. $(7 - 4i) + (9 + 2i) = 7 - 4i + 9 + 2i$
$$= 16 - 2i$$

c. $(-6 + i) - (3 - 4i) = -6 + i - 3 + 4i$
$$= -9 + 5i$$

d. $(2 - 4i) - (-4 + 3i) = 2 - 4i + 4 - 3i$
$$= 6 - 7i$$ ∎

To multiply a complex number by an imaginary number, we use the distributive property to remove parentheses and then simplify. For example,

$$-5i(4 - 8i) = -5i(4) - (-5i)(8i)$$
$$= -20i + 40i^2$$
$$= -40 - 20i \qquad \text{Remember that } i^2 = -1.$$

Multiplication of Complex Numbers   Complex numbers are multiplied as if they were binomials, with $i^2 = -1$:
$$(a + bi)(c + di) = ac + adi + bci + bdi^2$$
$$= (ac - bd) + (ad + bc)i$$

(a, c)   **EXAMPLE 4**   **a.** $(2 + 3i)(3 - 2i) = 6 - 4i + 9i - 6i^2$
$$= 6 + 5i + 6$$
$$= 12 + 5i$$

b. $(3 + i)(1 + 2i) = 3 + 6i + i + 2i^2$
$$= 3 + 7i - 2$$
$$= 1 + 7i$$

c. $(-4 + 2i)(2 + i) = -8 - 4i + 4i + 2i^2$
$$= -8 - 2$$
$$= -10$$

d. $(-1 - i)(4 - i) = -4 + i - 4i + i^2$
$$= -4 - 3i - 1$$
$$= -5 - 3i$$ ∎

The next example shows how to write complex numbers in $a + bi$ form. When writing answers, it is acceptable to use $a - bi$ as a substitute for the form $a + (-b)i$.

EXAMPLE 5 **a.** $7 = 7 + 0i$ **b.** $3i = 0 + 3i$

c. $4 - \sqrt{-16} = 4 - \sqrt{-1(16)}$ **d.** $5 + \sqrt{-11} = 5 + \sqrt{-1(11)}$

$\qquad\qquad\quad = 4 - \sqrt{16}\sqrt{-1}$ $\qquad\qquad\qquad = 5 + \sqrt{11}\sqrt{-1}$

$\qquad\qquad\quad = 4 - 4i$ $\qquad\qquad\qquad = 5 + \sqrt{11}\,i$

e. $2i^2 + 4i^3 = 2(-1) + 4(-i)$ **f.** $\dfrac{3}{2i} = \dfrac{3}{2i} \cdot \dfrac{i}{i}$

$\qquad\qquad\quad = -2 - 4i$

$\qquad\qquad\qquad\qquad\qquad\qquad = \dfrac{3i}{2i^2}$

$\qquad\qquad\qquad\qquad\qquad\qquad = \dfrac{3i}{2(-1)}$

$\qquad\qquad\qquad\qquad\qquad\qquad = \dfrac{3i}{-2}$

$\qquad\qquad\qquad\qquad\qquad\qquad = 0 - \dfrac{3}{2}i$

g. $-\dfrac{5}{i} = -\dfrac{5}{i} \cdot \dfrac{i^3}{i^3} = -\dfrac{5(-i)}{1} = 5i = 0 + 5i$ ∎

■ Complex Conjugates

Complex Conjugates The complex numbers $a + bi$ and $a - bi$ are called **complex conjugates** of each other.

For example,

- $3 + 4i$ and $3 - 4i$ are complex conjugates.
- $5 - 7i$ and $5 + 7i$ are complex conjugates.
- $8 + 17i$ and $8 - 17i$ are complex conjugates.

EXAMPLE 6 Find the product of $3 + i$ and its complex conjugate.

Solution The complex conjugate of $3 + i$ is $3 - i$. We find their product as follows:

$(3 + i)(3 - i) = 9 - 3i + 3i - i^2$

$\qquad\qquad\quad = 9 - i^2$ Combine like terms.

$\qquad\qquad\quad = 9 - (-1)$ Because $i^2 = -1$.

$\qquad\qquad\quad = 10$ ∎

In general, the product of the complex number $a + bi$ and its complex conjugate $a - bi$ is the real number $a^2 + b^2$.

$$(a + bi)(a - bi) = a^2 - abi + abi - b^2i^2$$
$$= a^2 - b^2(\mathbf{-1})$$
$$= a^2 + b^2$$

■ Rationalizing Denominators

To write complex numbers such as $\dfrac{1}{3 + i}, \dfrac{3 - i}{2 + i},$ and $\dfrac{5 + i}{5 - i}$ in $a + bi$ form, we rationalize their denominators.

EXAMPLE 7   Write $\dfrac{1}{3 + i}$ in $a + bi$ form.

Solution   Since the product of $3 + i$ and its conjugate is a real number, we can rationalize the denominator by multiplying both the numerator and the denominator of the fraction by the complex conjugate of the denominator and simplify.

$$\frac{1}{3 + i} = \frac{1}{3 + i} \cdot \frac{\mathbf{3 - i}}{\mathbf{3 - i}}$$

$$= \frac{3 - i}{9 - 3i + 3i - i^2}$$

$$= \frac{3 - i}{9 - (-1)}$$

$$= \frac{3 - i}{10}$$

$$= \frac{3}{10} - \frac{1}{10}i$$ ■

EXAMPLE 8   Write $\dfrac{3 - i}{2 + i}$ in $a + bi$ form.

Solution   We rationalize the denominator by multiplying the numerator and denominator by the complex conjugate of the denominator and simplify.

$$\frac{3 - i}{2 + i} = \frac{3 - i}{2 + i} \cdot \frac{\mathbf{2 - i}}{\mathbf{2 - i}}$$

$$= \frac{6 - 3i - 2i + i^2}{4 - 2i + 2i - i^2}$$

$$= \frac{5 - 5i}{4 - (-1)}$$

$$= \frac{5(1 - i)}{5}$$ Factor out 5 in the numerator.

$$= 1 - i$$ Divide out the common factor of 5.   ■

EXAMPLE 9   Divide $5 + i$ by $5 - i$ and express the quotient in $a + bi$ form.

Solution   The quotient obtained when dividing $5 + i$ by $5 - i$ is expressed by the fraction $\frac{5+i}{5-i}$. To express this quotient in $a + bi$ form, we rationalize the denominator by multiplying both the numerator and the denominator by the complex conjugate of the denominator and simplify.

$$\frac{5+i}{5-i} = \frac{5+i}{5-i} \cdot \frac{5+i}{5+i}$$

$$= \frac{25 + 5i + 5i + i^2}{25 + 5i - 5i - i^2}$$

$$= \frac{25 + 10i - 1}{25 - (-1)}$$

$$= \frac{24 + 10i}{26}$$

$$= \frac{2(12 + 5i)}{26} \qquad \text{Factor out 2 in the numerator.}$$

$$= \frac{12 + 5i}{13} \qquad \text{Divide out a common factor of 2.}$$

$$= \frac{12}{13} + \frac{5}{13}i \qquad ■$$

Dr. Bernoit B. Mandelbrot has used mathematics to study unpredictable and irregular events in the physical, behavioral, and biological sciences. His study uses properties of the complex numbers to produce intricate images like the one above.

WARNING!   Complex numbers are not always written in $a + bi$ form. To avoid mistakes, always put complex numbers in $a + bi$ form before doing any arithmetic involving the numbers.

EXAMPLE 10   Write $\dfrac{4 + \sqrt{-16}}{2 + \sqrt{-4}}$ in $a + bi$ form.

Solution

$$\frac{4 + \sqrt{-16}}{2 + \sqrt{-4}} = \frac{4 + 4i}{2 + 2i}$$

$$= \frac{2(2 + 2i)}{2 + 2i} \qquad \text{Factor out 2 in the numerator.}$$

$$= 2 + 0i \qquad \text{Divide out } 2 + 2i. \qquad ■$$

■ Absolute Value of a Complex Number

Absolute Value of a Complex Number

The **absolute value** of the complex number $a + bi$ is $\sqrt{a^2 + b^2}$. In symbols,

$$|a + bi| = \sqrt{a^2 + b^2}$$

(b)

EXAMPLE 11

a. $|3 + 4i| = \sqrt{3^2 + 4^2}$
$= \sqrt{9 + 16}$
$= \sqrt{25}$
$= 5$

b. $|5 - 12i| = \sqrt{5^2 + (-12)^2}$
$= \sqrt{25 + 144}$
$= \sqrt{169}$
$= 13$ ∎

EXAMPLE 12 If a and b are both negative numbers, is the formula $\sqrt{a}\sqrt{b} = \sqrt{ab}$ still true?

Solution We can let $a = -4$ and $b = -1$ and compute $\sqrt{a}\sqrt{b}$ and $\sqrt{ab}$ to see if their values are equal:

$$\sqrt{a}\sqrt{b} = \sqrt{-4}\sqrt{-1} \quad \text{and} \quad \sqrt{ab} = \sqrt{(-4)(-1)}$$
$$= 2i \cdot i \qquad\qquad\qquad = \sqrt{4}$$
$$= 2i^2 \qquad\qquad\qquad\quad = 2$$
$$= -2$$

Since $-2 \neq 2$, the formula $\sqrt{ab} = \sqrt{a}\sqrt{b}$ is false when both a and b are negative. ∎

ORALS *Do each operation.*

1. $(3 + 4i) + (2 + 3i)$ 2. $(2 + 3i) - (1 - 2i)$
3. $(1 + i)(2 + i)$ 4. $(2 - i)(1 - i)$
5. $(2 + i)(1 - i)$ 6. $(1 + i)(2 - i)$

E X E R C I S E   9 . 5

In Exercises 1–10, solve each quadratic equation. Write all roots in bi or a + bi form.

1. $x^2 + 9 = 0$ 2. $x^2 + 16 = 0$ 3. $3x^2 = -16$ 4. $2x^2 = -25$

5. $x^2 + 2x + 2 = 0$ 6. $x^2 + 3x + 3 = 0$

7. $2x^2 + x + 1 = 0$ 8. $3x^2 + 2x + 1 = 0$

9. $3x^2 - 4x + 2 = 0$ 10. $2x^2 - 3x + 2 = 0$

In Exercises 11–18, simplify each expression.

11. i^{21} 12. i^{19} 13. i^{27} 14. i^{22}
15. i^{100} 16. i^{42} 17. i^{97} 18. i^{200}

In Exercises 19–60, express numbers in a + bi form, if necessary, and do the indicated operations. Give all answers in a + bi form.

19. $(3 + 4i) + (5 - 6i)$

20. $(5 + 3i) - (6 - 9i)$

21. $(7 - 3i) - (4 + 2i)$

22. $(8 + 3i) + (-7 - 2i)$

23. $(8 + \sqrt{-25}) + (7 + \sqrt{-4})$

24. $(-7 + \sqrt{-81}) - (-2 - \sqrt{-64})$

25. $(-8 - \sqrt{-3}) - (7 - \sqrt{-27})$

26. $(2 + \sqrt{-8}) + (-3 - \sqrt{-2})$

27. $3i(2 - i)$

28. $-4i(3 + 4i)$

29. $(2 + 3i)(3 - i)$

30. $(4 - i)(2 + i)$

31. $(2 - 4i)(3 + 2i)$

32. $(3 - 2i)(4 - 3i)$

33. $(2 + \sqrt{-2})(3 - \sqrt{-2})$

34. $(5 + \sqrt{-3})(2 - \sqrt{-3})$

35. $(-2 - \sqrt{-16})(1 + \sqrt{-4})$

36. $(-3 - \sqrt{-81})(-2 + \sqrt{-9})$

37. $(2 + \sqrt{-3})(3 - \sqrt{-2})$

38. $(1 + \sqrt{-5})(2 - \sqrt{-3})$

39. $(8 - \sqrt{-5})(-2 - \sqrt{-7})$

40. $(-1 + \sqrt{-6})(2 - \sqrt{-3})$

41. $\dfrac{1}{i}$

42. $\dfrac{1}{i^3}$

43. $\dfrac{4}{5i^3}$

44. $\dfrac{3}{2i}$

45. $\dfrac{3i}{8\sqrt{-9}}$

46. $\dfrac{5i^3}{2\sqrt{-4}}$

47. $\dfrac{-3}{5i^5}$

48. $\dfrac{-4}{6i^7}$

49. $\dfrac{-6}{\sqrt{-32}}$

50. $\dfrac{5}{\sqrt{-125}}$

51. $\dfrac{3}{5 + i}$

52. $\dfrac{-2}{2 - i}$

53. $\dfrac{-12}{7 - \sqrt{-1}}$

54. $\dfrac{4}{3 + \sqrt{-1}}$

55. $\dfrac{5i}{6 + 2i}$

56. $\dfrac{-4i}{2 - 6i}$

57. $\dfrac{3 - 2i}{3 + 2i}$

58. $\dfrac{2 + 3i}{2 - 3i}$

59. $\dfrac{3 + \sqrt{-2}}{2 + \sqrt{-5}}$

60. $\dfrac{2 - \sqrt{-5}}{3 + \sqrt{-7}}$

In Exercises 61–70, find each absolute value.

61. $|6 + 8i|$

62. $|12 + 5i|$

63. $|12 - 5i|$

64. $|3 - 4i|$

65. $|5 + 7i|$

66. $|6 - 5i|$

67. $|4 + \sqrt{-2}|$

68. $|3 + \sqrt{-3}|$

69. $|8 + \sqrt{-5}|$

70. $|7 - \sqrt{-6}|$

71. $|5 - 0i|$

72. $|0 - 5i|$

Writing Exercises   ■   *Write a paragraph using your own words.*

1. Explain how to add or subtract two complex numbers.

2. Explain how to find the absolute value of $3 + 2i$.

Something to Think About ■

1. The absolute value of a number x was defined as

$$|x| = x \quad \text{when } x \geq 0$$
$$|x| = -x \quad \text{when } x < 0$$

Show that this definition is consistent with the definition of the absolute value of a complex number.

2. Find the reciprocal of $3 + 5i$ and write it in $a + bi$ form.

Review Exercises ■   *Factor each polynomial.*

1. $3x^2 - 27$

2. $2x^2 - 8$

3. $2x^2 + x - 1$

4. $-x^2 - 4x + 21$

9.6 Graphing Quadratic Functions

■ Solving Quadratic Equations with a Graphing Calculator

GETTING READY   *If $y = f(x) = 2x^2 - x + 2$, find each value.*

1. $f(0)$ **2.** $f(1)$ **3.** $f(-1)$ **4.** $f(-2)$

If $x = -\dfrac{b}{2a}$, find x when a and b have the following values.

5. $a = 2, b = 8$

6. $a = 5, b = -20$

In Chapter 6, we graphed many parabolas. We now discuss in detail the parabolic graphs of equations of the form $y = ax^2 + bx + c$, where $a \neq 0$.

EXAMPLE 1   Graph $y = x^2 - 3$.

Solution   To find ordered pairs (x, y) that satisfy the equation, we pick several numbers x and find the corresponding values of y. If we let $x = 3$, we have

$$y = x^2 - 3$$
$$y = 3^2 - 3 \qquad \text{Substitute 3 for } x.$$
$$y = 6$$

The ordered pair $(3, 6)$ and others satisfying the equation appear in the table shown in Figure 9-4. To graph the equation, we plot each point and draw a smooth curve passing through them. The resulting parabola is the graph of $y = x^2 - 3$. The lowest point on the graph, called the **vertex of the parabola**, is the point $(0, -3)$.

$y = x^2 - 3$

x	y
3	6
2	1
1	-2
0	-3
-1	-2
-2	1
-3	6

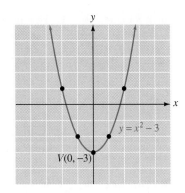

FIGURE 9-4 ■

EXAMPLE 2 Graph $y = x^2 - 4x + 4$ and find its vertex.

Solution To construct a table like the one shown in Figure 9-5, we pick several numbers x and find the corresponding values of y. To graph the equation, we plot the points and join them with a smooth curve.

$y = x^2 - 4x + 4$

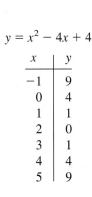

x	y
-1	9
0	4
1	1
2	0
3	1
4	4
5	9

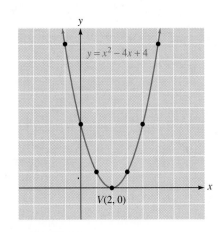

FIGURE 9-5

Since the graph is a parabola that opens upward, the vertex is the lowest point on the graph, the point (2, 0). ■

EXAMPLE 3 Graph $y = -x^2 + 2x - 1$ and find its vertex.

Solution We construct the table shown in Figure 9-6, plot the points, and draw the graph.

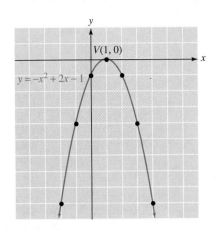

$$y = -x^2 + 2x - 1$$

x	y
-2	-9
-1	-4
0	-1
1	0
2	-1
3	-4
4	-9

FIGURE 9-6

Since the parabola opens downward, its vertex is its highest point, the point $(1, 0)$. ∎

The results of these first three examples suggest the following fact.

Theorem   The graph of the equation $y = ax^2 + bx + c$ is a parabola. It opens upward when $a > 0$, and it opens downward when $a < 0$.

It is usually easier to graph a parabola if we know the coordinates of its vertex. We can find the coordinates of the vertex of the graph of

1. $y = x^2 - 6x + 8$

if we complete the square in the following way.

$y = x^2 - 6x + 8$

$y = x^2 - 6x + 9 - 9 + 8$ Add 9 to complete the square on $x^2 - 6x$ and subtract 9.

$y = (x - 3)^2 - 1$ Factor $x^2 - 6x + 9$ and combine like terms.

Since $a > 0$ in Equation 1, the graph will be a parabola that opens upward. The vertex will be the lowest point on the parabola, and the y-coordinate of the vertex will be the smallest possible value of y. Because $(x - 3)^2 \geq 0$, the smallest value of y occurs when $(x - 3)^2 = 0$ or when $x = 3$. To find the corresponding value of y, we substitute 3 for x in the equation $y = (x - 3)^2 - 1$ and simplify.

$y = (x - 3)^2 - 1$

$y = (3 - 3)^2 - 1$ Substitute 3 for x.

$y = 0^2 - 1$

$y = -1$

The vertex of the parabola is the point $(3, -1)$.

A generalization of this discussion leads to the following fact.

Theorem   The graph of an equation of the form

$$y = a(x - h)^2 + k$$

is a parabola with its vertex at the point with coordinates (h, k). The parabola opens upward if $a > 0$, and it opens downward if $a < 0$.

EXAMPLE 4   Find the vertex of the parabola determined by $y = -4(x - 3)^2 - 2$. Does the parabola open upward or downward?

Solution   In the equation $y = a(x - h)^2 + k$, the coordinates of the vertex are given by the ordered pair (h, k). In the equation $y = -4(x - 3)^2 - 2$, $a = -4$, $h = 3$, and $k = -2$. Thus, the vertex is the point $(h, k) = (3, -2)$. Since $a = -4$ and $-4 < 0$, the parabola opens downward.   ∎

EXAMPLE 5   Find the vertex of the parabola determined by $y = 5(x + 1)^2 + 4$. Does the parabola open upward or downward?

Solution   The equation $y = 5(x + 1)^2 + 4$ is equivalent to the equation

$$y = 5[x - (-1)]^2 + 4$$

Since $h = -1$, $k = 4$, and $a = 5$, the vertex is the point $(h, k) = (-1, 4)$. Since $a = 5$ and $5 > 0$, the parabola opens upward.   ∎

EXAMPLE 6   Find the vertex of the parabola determined by $y = 2x^2 + 8x + 2$ and graph the parabola.

Solution   To make the coefficient of x^2 equal to 1, we factor 2 out of the binomial $2x^2 + 8x$. We then proceed as follows:

1. $\begin{aligned} y &= 2x^2 + 8x + 2 \\ &= 2(x^2 + 4x) + 2 \\ &= 2(x^2 + 4x + 4 - 4) + 2 \\ &= 2[(x + 2)^2 - 4] + 2 \\ &= 2(x + 2)^2 - 2 \cdot 4 + 2 \\ &= 2(x + 2)^2 - 6 \end{aligned}$
   Factor 2 out of $2x^2 + 8x$.
   Complete the square on $x^2 + 4x$.
   Factor $x^2 + 4x + 4$.
   Distribute the multiplication by 2.
   Simplify and combine like terms.

   or

$$y = 2[x - (-2)]^2 + (-6)$$

Since $h = -2$ and $k = -6$, the vertex of the parabola is the point $(-2, -6)$. Since $a = 2$, the parabola opens upward.

We can pick numbers x on either side of $x = -2$ to construct the table shown in Figure 9-7. To find the y-intercept, we substitute 0 for x in Equation 1 and solve for y: when $x = 0$, $y = 2$. Thus, the y-intercept is $(0, 2)$. We find some more ordered pairs, plot the points, and draw the parabola.

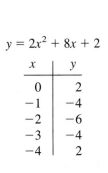

$$y = 2x^2 + 8x + 2$$

x	y
0	2
-1	-4
-2	-6
-3	-4
-4	2

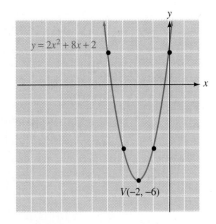

FIGURE 9-7 ∎

Much can be determined about the graph of $y = ax^2 + bx + c$ from the coefficients a, b, and c:

- The y-intercept is determined by the value of y attained when $x = 0$: the y-intercept is $(0, c)$.
- The x-intercepts (if any) are determined by the numbers x that make $y = 0$. To find them, we solve the quadratic equation $ax^2 + bx + c = 0$.
- By using the methods of Example 6, we can complete the square on x in the equation $y = ax^2 + bx + c$ and find the coordinates of the vertex of the parabola. We will see that the x-coordinate of the vertex is $-\frac{b}{2a}$.

We summarize these results as follows:

Graphing the Parabola
$$y = ax^2 + bx + c$$

The y-intercept is $(0, c)$.

The x-intercepts (if any) are determined by the solutions of $ax^2 + bx + c = 0$.

The x-coordinate of the vertex of the parabola $y = ax^2 + bx + c$ is $x = -\dfrac{b}{2a}$.

To find the y-coordinate of the vertex, substitute $-\dfrac{b}{2a}$ for x into the equation $y = ax^2 + bx + c$ and solve for y.

EXAMPLE 7 Graph the equation $y = x^2 - 2x - 3$.

Solution The equation is in the form $y = ax^2 + bx + c$, with $a = 1$, $b = -2$, and $c = -3$. Since $a > 0$, the parabola opens upward. To find the x-coordinate of the vertex, we substitute the values for a and b into the formula $x = -\frac{b}{2a}$.

$$x = -\frac{b}{2a}$$

$$x = -\frac{-2}{2(1)}$$

$$= 1$$

The x-coordinate of the vertex is $x = 1$. To find the y-coordinate, we substitute 1 for x into the equation and solve for y.

$$y = x^2 - 2x - 3$$
$$y = 1^2 - 2 \cdot 1^2 - 3$$
$$= 1 - 2 - 3$$
$$= -4$$

The vertex of the parabola is the point $(1, -4)$.

To graph the parabola, we find several other points with coordinates that satisfy the equation. One easy point to find is the y-intercept. It is the value of y when $x = 0$. Thus, the parabola passes through the point $(0, -3)$.

To find the x-intercepts of the graph, we set y equal to 0 and solve the resulting quadratic equation:

$$y = x^2 - 2x - 3$$
$$0 = x^2 - 2x - 3$$
$$0 = (x - 3)(x + 1) \qquad \text{Factor.}$$

$$x - 3 = 0 \quad \text{or} \quad x + 1 = 0 \qquad \text{Set each factor equal to 0.}$$
$$x = 3 \qquad | \qquad x = -1$$

Since the x-intercepts of the graph are $(3, 0)$ and $(-1, 0)$, the graph passes through these points. The graph appears in Figure 9-8. ∎

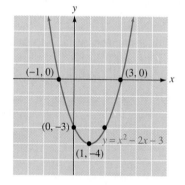

FIGURE 9-8

EXAMPLE 8 **Maximizing revenue** An electronics firm manufactures a type of radio with limited demand. Over the past 10 years, the firm has learned that it can sell x radios at a price of $\left(200 - \frac{1}{5}x\right)$ dollars. How many radios should the firm manufacture and sell to maximize its revenue? Find the maximum revenue.

Solution The revenue obtained is the product of the number of radios that the firms sells (x) and the price of each radio $\left(200 - \frac{1}{5}x\right)$. Thus, the revenue, R, is given by the formula

$$R = x\left(200 - \frac{1}{5}x\right) \qquad \text{or} \qquad R = -\frac{1}{5}x^2 + 200x$$

Since the graph of this equation is a parabola that opens downward, the maximum value of R will be the value of R determined by the vertex of the parabola. Because the x-coordinate of the vertex is at $x = \frac{-b}{2a}$, we have

$$x = \frac{-b}{2a}$$

$$= \frac{-200}{2\left(-\dfrac{1}{5}\right)}$$

$$= \frac{-200}{-\dfrac{2}{5}}$$

$$= (-200)\left(-\frac{5}{2}\right)$$

$$= 500$$

If the firm manufactures 500 radios, the maximum revenue will be

$$R = -\frac{1}{5}x^2 + 200x$$

$$= -\frac{1}{5}(500)^2 + 200(500)$$

$$= 50,000$$

The firm should manufacture 500 radios to get a maximum revenue of $50,000.

■

■ Solving Quadratic Equations with a Graphing Calculator

EXAMPLE 9 Solve the equation $x^2 - x - 3 = 0$.

Solution The solutions of $x^2 - x - 3 = 0$ are the numbers x that will make $y = 0$ in the quadratic function $y = x^2 - x - 3$. To find these numbers, we can graph the quadratic function and read the x-intercepts of the graph.

With a graphing calculator set for the standard RANGE settings, we graph the function $y = x^2 - x - 3$, as in Figure 9-9(a). We press the TRACE key and move the cursor close to the positive x-intercept until we read an x-coordinate of 2.4210526.

To obtain better results, we zoom to obtain the graph in Figure 9-9(b). We press the TRACE key again and move the cursor close to the intercept until we read an x-coordinate of 2.2894737.

To obtain even better results, we zoom again to obtain the graph in Figure 9-9(c), press the ⎡TRACE⎤ key, and move the cursor close to the intercept until we read an x-coordinate of 2.3092105.

Here $y = .02324273$, which is nearly 0. Thus, the positive x-intercept is approximately (2.3, 0), and a solution to the equation is approximately 2.3. For better results, we would do additional zooms.

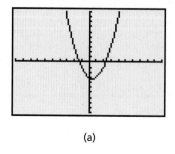
(a)

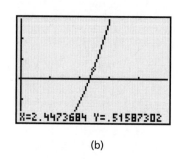
X=2.4473684 Y=.51587302
(b)

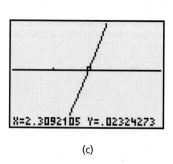
X=2.3092105 Y=.02324273
(c)

FIGURE 9-9

Similar steps will show that the negative x-intercept is approximately $(-1.3, 0)$, and that the second solution of the equation is $x = -1.3$. ∎

ORALS *The graph of each equation is a parabola. Does it open up or down?*

1. $y = 3x^2 - 2x + 4$ **2.** $y = 2x^2 + x - 5$

3. $y = -x^2 - 3x - 5$ **4.** $y = -3x^2 + 4x - 1$

5. $y - 4 = -\dfrac{3}{2}x^2 + x$ **6.** $y + 5 = \dfrac{3}{7}x^2 - x$

E X E R C I S E 9.6

In Exercises 1–12, graph each equation.

1. $y = x^2 + 1$

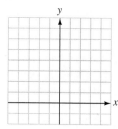

2. $y = x^2 - 4$

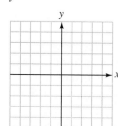

3. $y = -x^2$

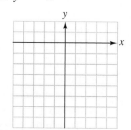

4. $y = -(x - 1)^2$

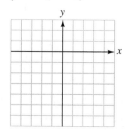

5. $y = x^2 + x$

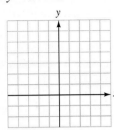

6. $y = x^2 - 2x$

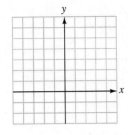

7. $y = -x^2 - 4x$

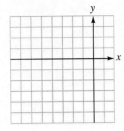

8. $y = -x^2 + 2x$

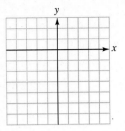

9. $y = x^2 + 4x + 4$

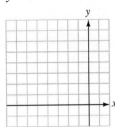

10. $y = x^2 - 6x + 9$

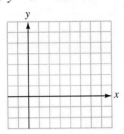

11. $y = x^2 - 4x + 6$

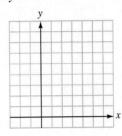

12. $y = x^2 + 2x - 3$

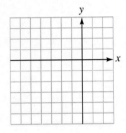

In Exercises 13–24, find the vertex of the graph of each equation. **Do not draw the graph.**

13. $y = -3(x - 2)^2 + 4$

14. $y = 4(x - 3)^2 + 2$

15. $y = 5(x + 1)^2 - 5$

16. $y = 4(x + 3)^2 + 1$

17. $y = (x - 1)^2$

18. $y = -(x - 2)^2$

19. $y = -7x^2 + 4$

20. $y = 5x^2 - 2$

21. $y = x^2 + 2x + 5$

22. $y = x^2 + 4x + 1$

23. $y = x^2 - 6x - 12$

24. $y = x^2 - 8x - 20$

In Exercises 25–36, complete the square, if necessary, to find the vertex of the graph of each equation. Then graph the equation.

25. $y = x^2 - 4x + 4$

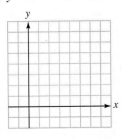

26. $y = x^2 + 6x + 9$

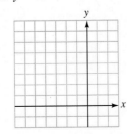

27. $y = -x^2 - 2x - 1$

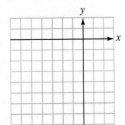

28. $y = -x^2 + 2x - 1$

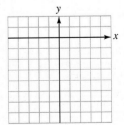

29. $y = x^2 + 2x - 3$

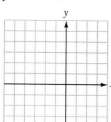

30. $y = x^2 + 6x + 5$

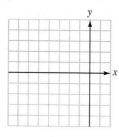

31. $y = -x^2 - 6x - 7$

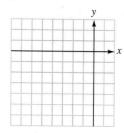

32. $y = -x^2 + 8x - 14$

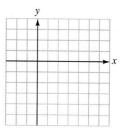

33. $y = 2x^2 + 8x + 6$

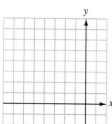

34. $y = 3x^2 - 12x + 9$

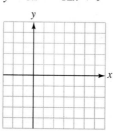

35. $y = -3x^2 + 6x - 2$

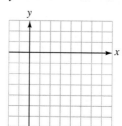

36. $y = -2x^2 - 4x + 2$

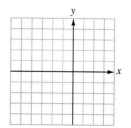

In Exercises 37–42, find the vertex and the x- and y-intercepts of the graph of each equation. Then graph the equation.

37. $y = x^2 - x - 2$

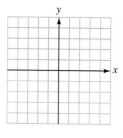

38. $y = x^2 - 6x + 8$

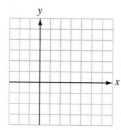

39. $y = -x^2 + 2x + 3$

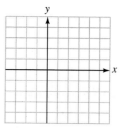

40. $y = -x^2 + 5x - 4$

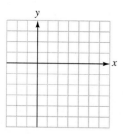

41. $y = 2x^2 + 3x - 2$

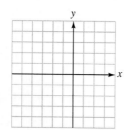

42. $y = 3x^2 - 7x + 2$

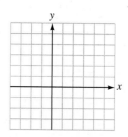

43. **Selling TVs**   A company has found that it can sell x TVs at a price of $\left(450 - \frac{1}{6}x\right)$. How many TVs must the company sell to maximize its revenue?

44. **Finding maximum revenue**   In Exercise 43, find the maximum revenue.

45. **Selling CD players**   A wholesaler sells CD players for $150 each. However, he gives volume discounts on purchases of between 500 and 1000 units according to the formula $\left(150 - \frac{1}{10}n\right)$, where n represents the number of units purchased. How many units would a retailer have to buy for the wholesaler to obtain maximum revenue?

46. **Finding maximum revenue**   In Exercise 45, find the maximum revenue.

 In Exercises 47–50, use a graphing calculator to solve each equation.

47. $x^2 - 5x + 6 = 0$ 48. $6x^2 + 5x - 6 = 0$ 49. $x^2 + 5x + 2 = 0$ 50. $2x^2 - 5x + 1 = 0$

Writing Exercises ■ *Write a paragraph using your own words.*

1. Explain why the y-intercept of the graph of $y = ax^2 + bx + c$ ($a \neq 0$) is $(0, c)$.

2. Define the vertex of a parabola and explain how to find its coordinates.

Something to Think About ■

1. The graph of $x = y^2$ is a parabola, but the equation does not define a function. Explain.

2. The graph of $x = -y^2$ is a parabola, but the equation does not define a function. Explain.

Review Exercises ■ *In Review Exercises 1–4, simplify each expression. Assume variables represent positive numbers.*

1. $\sqrt{12} + \sqrt{27}$

2. $3\sqrt{6y}\left(-4\sqrt{3y}\right)$

3. $\left(\sqrt{3} + 1\right)\left(\sqrt{3} - 1\right)$

4. $\dfrac{\sqrt{x} + 2}{\sqrt{x} - 2}$

MATHEMATICS IN SALES

Because American Appliance (first discussed at the beginning of the chapter) can sell x refrigerators for a price of $\$\left(1800 - \frac{3}{2}x\right)$, it will sell

0 refrigerators at a price of $1800, for a total revenue of $0;
50 refrigerators at a price of $1725, for a total revenue of $86,250;
100 refrigerators at a price of $1650, for a total revenue of $165,000;

and so on. In general, the total revenue R is the product of the price and the number of refrigerators sold.

$$R = x\left(1800 - \frac{3}{2}x\right)$$

1. $R = -\dfrac{3}{2}x^2 + 1800x$ Use the distributive property to remove parentheses.

(continued on next page)

MATHEMATICS IN SALES

Since the graph of Equation 1 is a parabola that opens downward, the maximum value of R will be the R-value that corresponds to the x-coordinate of the vertex:

$$x = \frac{-b}{2a} = \frac{-1800}{2\left(-\dfrac{3}{2}\right)} = \frac{-1800}{-3} = 600$$

The x-coordinate of the vertex is $x = \frac{-b}{2a}$.

The store will maximize revenue by selling 600 refrigerators at a price of

$$1800 - \frac{3}{2}x = 1800 - \frac{3}{2}(600) = \$900$$

The company should order 600 refrigerators from the distributor each month to keep up with demand and maximize revenue at 600($900) or $540,000 per month.

P R O J E C T ■ Better and Better and Better

One of the world's greatest scientific geniuses, Sir Isaac Newton (1642–1727) contributed to every major area of science and mathematics known in his time. Because of his early interest in science and mathematics, Newton enrolled at Trinity College, Cambridge, where he studied the mathematics of René Descartes and the astronomy of Galileo but did not show any great talent. In 1665, the plague closed the school, and Newton went home. There, his genius appeared. Within a few years, he had made major discoveries in mathematics, physics, and astronomy.

One of Newton's many contributions to mathematics is known as **Newton's method**, a way of finding better and better estimates of the solutions of certain equations. The process begins with a guess of the solution, and transforms that guess into a better estimate of the solution. When Newton's method is applied to this better answer, a third solution results, one that is more accurate than any of the previous ones. The method is applied again and again, producing better and better approximations of the solution.

To solve the quadratic equation $x^2 + x - 3 = 0$, for example, Newton's method uses the fraction $\dfrac{x^2 + 3}{2x + 1}$ to generate better solutions. We begin with a guess: 3, for example. We substitute 3 for x in the fraction.

$$\frac{x^2 + 3}{2x + 1} = \frac{3^2 + 3}{2(3) + 1} \approx 1.714286$$

The number 1.714286 is a better estimate of the solution than the original guess, 3. We apply Newton's method again:

$$\frac{x^2 + 3}{2x + 1} = \frac{1.714286^2 + 3}{2(1.714286) + 1} \approx 1.341014$$

The number 1.341014 is the best estimate yet of the solution. However, more passes through Newton's method produce even better estimates: 1.303173, and then 1.302776. This final answer is accurate to six decimal places.

- The quadratic equation $x^2 + x - 3 = 0$ has two solutions. Find the other solution by using Newton's method and another initial guess. (Try a negative number.)

- To solve the quadratic equation $ax^2 + bx + c = 0$, Newton's method uses the fraction $\dfrac{ax^2 - c}{2ax + b}$ to generate solutions. Solve $2x^2 - 2x - 1 = 0$ using Newton's method. Find two solutions accurate to six decimal places.

Chapter Summary

KEY IDEAS

(9.1) The two solutions of the equation $x^2 = c$ are $x = \sqrt{c}$ and $x = -\sqrt{c}$.

(9.2) To make a binomial $x^2 + 2bx$ into a trinomial square, add the square of one-half of the coefficient of x:

$$x^2 + 2bx + b^2 = (x + b)^2$$

To solve a quadratic equation by completing the square, follow these steps:

1. If necessary, divide both sides of the equation by the coefficient of x^2 to make its coefficient 1.

2. If necessary, get the constant on the right-hand side of the equation.

3. Complete the square.

4. Solve the resulting equation.

5. Check each solution.

(9.3) **The quadratic formula:** The solutions of $ax^2 + bx + c = 0$, $a \neq 0$, can be found by using the formula

$$x = \frac{-b \pm \sqrt{b^2 - 4ac}}{2a}$$

(9.5) Properties of complex numbers: If a, b, c, and d are real numbers and given that $i^2 = -1$, then

$$a + bi = c + di \quad \text{if and only if} \quad a = c \text{ and } b = d$$
$$(a + bi) + (c + di) = (a + c) + (b + d)i$$
$$(a + bi)(c + di) = (ac - bd) + (ad + bc)i$$
$$|a + bi| = \sqrt{a^2 + b^2}$$

(9.6) The graph of the equation $y = ax^2 + bc + c$ is a parabola. It opens upward when $a > 0$ and opens downward when $a < 0$.

The graph of an equation of the form $y = a(x - h)^2 + k$ is a parabola with vertex at the point (h, k). The parabola opens upward when $a > 0$ and opens downward when $a < 0$.

The x-coordinate of the vertex of the parabola $y = ax^2 + bx + c$ is $x = -\dfrac{b}{2a}$.

To find the y-coordinate of the vertex, substitute $-\dfrac{b}{2a}$ for x in the equation of the parabola and find y.

To use a graphing calculator to solve a quadratic equation, graph the equation and use the ZOOM and TRACE keys to find the x-coordinates of the x-intercepts.

■ Chapter 9 Review Exercises

(9.1) *In Review Exercises 1–6, use the square root method to solve each quadratic equation.*

1. $x^2 = 25$ **2.** $x^2 = 36$ **3.** $2x^2 = 18$ **4.** $4x^2 = 9$

5. $x^2 = 8$ **6.** $x^2 = 75$

In Review Exercises 7–12, use the square root method to solve each equation.

7. $(x - 1)^2 = 25$ **8.** $(x + 3)^2 = 36$ **9.** $2(x + 1)^2 = 18$ **10.** $4(x - 2)^2 = 9$

11. $(x - 8)^2 = 8$ **12.** $(x + 5)^2 = 75$

In Review Exercises 13–16, solve each equation. Note that each trinomial is a perfect square.

13. $x^2 + 2x + 1 = 9$ **14.** $x^2 - 6x + 9 = 4$ **15.** $x^2 - 8x + 16 = 20$ **16.** $x^2 - 2x + 1 = 18$

(9.2) *In Review Exercises 17–24, solve each quadratic equation by completing the square.*

17. $x^2 + 5x - 14 = 0$ **18.** $x^2 - 8x + 15 = 0$ **19.** $x^2 + 4x - 77 = 0$ **20.** $x^2 - 2x - 1 = 0$

21. $x^2 + 4x - 3 = 0$ **22.** $x^2 - 6x + 4 = 0$ **23.** $2x^2 + 5x - 3 = 0$ **24.** $2x^2 - 2x - 1 = 0$

(9.3) *In Review Exercises 25–32, use the quadratic formula to solve each quadratic equation.*

25. $x^2 - 2x - 15 = 0$ **26.** $x^2 - 6x - 7 = 0$ **27.** $x^2 - 15x + 26 = 0$ **28.** $2x^2 - 7x + 3 = 0$

29. $6x^2 - 7x - 3 = 0$ **30.** $x^2 + 4x + 1 = 0$ **31.** $x^2 - 6x + 7 = 0$ **32.** $x^2 + 3x = 0$

(9.4) *In Review Exercises 33–34, solve each problem.*

33. Integer problem   The product of two consecutive odd positive integers is 143. Find the integers.

34. Perimeter   The length of the rectangle in Illustration 1 is 14 cm greater than the width, and the diagonal is 26 cm. Find the perimeter.

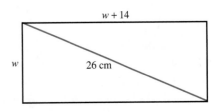

ILLUSTRATION 1

(9.5) *In Review Exercises 35–50, do the operations. Give all answers in a + bi form.*

35. $(5 + 4i) + (7 - 2i)$ **36.** $(6 - 40i) - (8 + 28i)$ **37.** $7i(-3 + 4i)$ **38.** $6i(2 + i)$

39. $(5 - 3i)(-6 + 2i)$ **40.** $\left(2 + \sqrt{128}i\right)\left(3 - \sqrt{98}i\right)$

41. $\dfrac{3}{4i}$ **42.** $\dfrac{-2}{5i^3}$ **43.** $\dfrac{6}{2 + i}$ **44.** $\dfrac{7}{3 - i}$

45. $\dfrac{4 + i}{4 - i}$ **46.** $\dfrac{3 - i}{3 + i}$ **47.** $\dfrac{3}{5 + \sqrt{-4}}$ **48.** $\dfrac{2}{3 - \sqrt{-9}}$

49. $|9 + 12i|$ **50.** $|24 - 10i|$

(9.6) *In Review Exercises 51–54, find the vertex of the graph of each equation.* **Do not draw the graph.**

51. $y = 5(x - 6)^2 + 7$ **52.** $y = 3(x + 3)^2 - 5$ **53.** $y = 2x^2 - 4x + 7$ **54.** $y = -3x^2 + 18x - 11$

In Review Exercises 55–56, graph each equation.

55. $y = x^2 + 8x + 10$

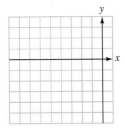

56. $y = -2x^2 - 4x - 6$

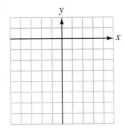

In Review Exercises 57–58, use a graphing calculator to solve each equation.

57. $x^2 - 4.5x + 4.5 = 0$ **58.** $2x^2 - 4x + 1.5 = 0$

Chapter 9 Test

1. Solve by factoring: $6x^2 + x - 1 = 0$.

2. Use the square root method to solve $x^2 = 16$.

3. Solve $(x - 2)^2 = 3$.

4. Solve $2(x^2 - 6) = x^2 + x - 6$.

In Problems 5–6, find the number required to complete the square.

5. $x^2 + 14x$

6. $x^2 - 7x$

7. Use the method of completing the square to solve $3a^2 + 6a - 12 = 0$.

8. Write the quadratic formula.

In Problems 9–10, use the quadratic formula to solve each equation.

9. $x^2 + 3x - 10 = 0$

10. $2x^2 - 5x = 12$

11. Solve $2x^2 + 5x + 1 = 2$.

12. Write the number $\sqrt{-49}$ in bi form.

13. Add: $(3 + 4i) + (-2 + 5i)$.

14. Subtract: $(4 - 3i) - (-5 + 2i)$.

15. Multiply: $(-2 - 5i)(-3 + 2i)$.

16. Rationalize the denominator: $\dfrac{-2}{3 + i}$.

17. Rationalize the denominator: $\dfrac{2 + i}{2 - i}$.

18. Evaluate: $|3 - 4i|$.

19. Find the vertex of the parabola determined by the equation $y = -4(x + 5)^2 - 4$.

20. Graph the equation $y = x^2 + 4x + 2$.

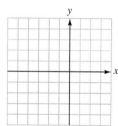

Cumulative Review Exercises

In Exercises 1–4, solve each equation.

1. $\dfrac{4}{5}x + 6 = 18$

2. $5 - \dfrac{x + 2}{3} = 7 - x$

3. $6x^2 - x - 2 = 0$

4. $5x^2 = 10x$

In Exercises 5–8, solve each inequality and graph the solution set.

5. $5x - 3 > 7$

6. $7x - 9 < 5$

7. $-2 < -x + 3 < 5$

8. $0 \le \dfrac{4 - x}{3} \le 2$

In Exercises 9–10, simplify each fraction.

9. $\dfrac{x^2 + 2x + 1}{x^2 + 4x + 3}$

10. $\dfrac{x^2 - x - 20}{x^2 + x - 30}$

In Exercises 11–14, do the operations.

11. $\dfrac{x^2 - x - 6}{x^2 - 4} \cdot \dfrac{x^2 + x - 6}{x^2 - 9}$

12. $\dfrac{x^2 - 4}{x - 1} \div \dfrac{x - 2}{x - 1}$

13. $\dfrac{x + y}{3} + \dfrac{x - y}{7}$

14. $\dfrac{x + 2}{x + 5} - \dfrac{x - 3}{x + 7}$

15. Graph the equation $4x - 3y = 12$.

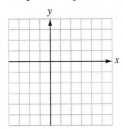

16. Graph the equation $3x + 4y = 4y + 12$.

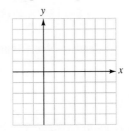

17. Find the slope of the line passing through points $(-2, 5)$ and $(4, 8)$.

18. Write the equation of the line passing through $(-2, 5)$ and $(4, 8)$.

19. Find the equation of a line passing through the origin and parallel to the graph of $y = 2x + 3$.

20. Graph the inequality $5x - 2y < 10$.

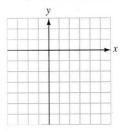

In Exercises 21–22, solve each system of equations.

21. $\begin{cases} x + y = 1 \\ x - y = 7 \end{cases}$

22. $\begin{cases} 4x + 9y = 8 \\ 2x - 6y = -3 \end{cases}$

In Exercises 23–24, solve each system by graphing.

23. $\begin{cases} x - y = 4 \\ 2x + y = 5 \end{cases}$

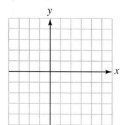

24. $\begin{cases} 3x + 2y \geq 6 \\ x + 3y \leq 6 \end{cases}$

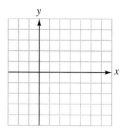

In Exercises 25–28, simplify each radical.

25. $\sqrt{\dfrac{1}{4}}$ **26.** $\sqrt{27x^3}$ **27.** $\sqrt{175x^2y^3}$ **28.** $\sqrt[3]{-80}$

In Exercises 29–34, do the operations and simplify when possible. (Assume $x \geq 0$.)

29. $\sqrt{27} - \sqrt{12}$ **30.** $\sqrt{25x^2} - \sqrt{16x^2}$ **31.** $\left(3\sqrt{6x}\right)\left(2\sqrt{3x}\right)$

32. $\sqrt{2}\left(\sqrt{2} + 2\right)$ **33.** $\left(2\sqrt{3} + 1\right)\left(\sqrt{3} - 1\right)$ **34.** $\left(\sqrt{x} + 2\right)\left(\sqrt{x} - 2\right)$

In Exercises 35–36, rationalize each denominator. (Assume $x > 0$.)

35. $\dfrac{3}{\sqrt{3x}}$ **36.** $\dfrac{x - 16}{\sqrt{x} - 4}$

In Exercises 37–38, solve each equation. (Assume $x \geq 0$.)

37. $\sqrt{6x + 1} + 3 = 8$ **38.** $\sqrt{3x + 1} = x - 1$

In Exercises 39–42, simplify each expression.

39. $4^{1/2}$ **40.** $\left(\dfrac{8}{27}\right)^{2/3}$ **41.** $x^{1/2}x^{1/2}$ **42.** $\left(x^{1/2}\right)^4$

In Exercises 43–44, find the distance between each pair of points.

43. $(-2, 3)$ and $(4, -5)$ **44.** $(-2, -8)$ and $(3, 4)$

In Exercises 45–48, find the number that must be added to each binomial to complete the square.

45. $a^2 + 6a$ **46.** $y^2 - 5y$ **47.** $t^2 - \dfrac{1}{2}t$ **48.** $x^2 + \dfrac{2}{3}x$

In Exercises 49–50, use the quadratic formula to solve each equation.

49. $x^2 + 2x - 8 = 0$ **50.** $-6x = x^2 + 4$

In Exercises 51–56, express each answer in a + bi form.

51. $\sqrt{-25}$

52. $3 - \sqrt{-16}$

53. $(2 + 3i) + (3 - 2i)$

54. $(3 + 5i) - (2 - 4i)$

55. $(2 + 3i)(3 - 2i)$

56. $\dfrac{1}{3 + i}$

In Exercises 57–58, graph each equation.

57. $y = x^2 - 2$

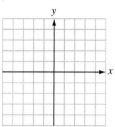

58. $y = x^2 + x - 6$

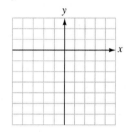

1. How many prime numbers are there between 20 and 30?
   a. 1   **b.** 2   **c.** 3   **d.** 4   **e.** none of the above

2. If $x = 3$, $y = -2$, and $z = -1$, find the value of $\dfrac{x + z}{y}$.
   a. 1   **b.** -1   **c.** -2   **d.** 2   **e.** none of the above

3. If $x = -2$, $y = -3$, and $z = -4$, find the value of $\dfrac{|x - z|}{|y|}$.
   a. -2   **b.** 2   **c.** $\dfrac{2}{3}$ **d.** $-\dfrac{2}{3}$ **e.** none of the above

4. The distributive property is written symbolically as
   a. $a + b = b + a$ **b.** $ab = ba$
   c. $a(b + c) = ab + ac$
   d. $(a + b) + c = a + (b + c)$ **e.** none of the above

5. Solve for x: $7x - 4 = 24$
   a. $x = 4$ **b.** $x = 5$ **c.** $x = -4$ **d.** $x = -5$
   e. none of the above

6. Solve for z: $6z - (9 - 3z) = -3(z + 2)$
   a. $\dfrac{4}{3}$ **b.** $\dfrac{3}{4}$ **c.** $-\dfrac{4}{3}$ **d.** $-\dfrac{1}{4}$ **e.** none of the above

7. Solve for r: $\dfrac{r}{5} - \dfrac{r - 3}{10} = 0$
   a. -3   **b.** -2   **c.** 3   **d.** 1   **e.** none of the above

8. Solve for x: $\dfrac{ax}{b} + c = 4$
   a. $\dfrac{b(4 - x)}{c}$ **b.** $\dfrac{4b - 4c}{b}$ **c.** $\dfrac{b(4 - c)}{a}$ **d.** none of the above

9. A man bought 25 pencils, some at 10 cents and some at 15 cents. The 25 pencils cost $3. How many 10-cent pencils did he buy?
   a. 5   **b.** 10   **c.** 15   **d.** 20   **e.** none of the above

10. Solve the inequality: $-3(x - 2) + 3 \geq 6$
    a. $x \geq -1$ **b.** $x \geq 1$ **c.** $x \leq -1$ **d.** $x \leq 1$
    e. none of the above

11. Simplify: $x^2 x^3 x^7$
    a. $12x$ **b.** x^{12} **c.** x^{42} **d.** x^{35} **e.** none of the above

12. Simplify: $\dfrac{(x^2)^7}{x^3 x^4}$
    a. 0   **b.** 1   **c.** x^7 **d.** x^2 **e.** none of the above

13. Simplify: $\dfrac{x^{-2} y^3}{xy^{-1}}$
    a. $x^3 y^4$ **b.** $\dfrac{x^3}{y^4}$ **c.** xy **d.** $\dfrac{y^4}{x^3}$ **e.** none of the above

14. Write 73,000,000 in scientific notation.
    a. 7.3×10^7 **b.** 7.3×10^{-7} **c.** 73×10^6
    d. 0.73×10^9 **e.** none of the above

15. If $P(x) = 2x^2 + 3x - 4$, find $P(-2)$.
    a. 2   **b.** -2   **c.** -18   **d.** 10   **e.** none of the above

16. Simplify: $2(y + 3) - 3(y - 2)$
    a. $-y$ **b.** $5y$ **c.** $5y + 12$ **d.** $-y + 12$ **e.** none of the above

17. Multiply: $-3x^2 y^2 (2xy^3)$
    a. $6x^3 y^5$ **b.** $5x^2 y^5$ **c.** $-6x^3 y^5$ **d.** $-6xy^5$
    e. none of the above

18. Multiply: $2a^3 b^2 (3a^2 b - 2ab^2)$
    a. $6a^6 b^3 - 4a^4 b^4$ **b.** $6a^5 b^3 - 4a^4 b^4$
    c. $6a^5 b^3 - 4a^4 b^2$ **d.** $6a^5 b^4 - 4a^4 b^4$ **e.** none of the above

19. Multiply: $(x + 7)(2x - 3)$
    a. $2x^2 - 11x - 21$ **b.** $2x^2 + 11x + 21$
    c. $2x^2 + 11x - 21$ **d.** $-2x^2 + 11x - 21$ **e.** none of the above

20. Divide: $(x^2 + 5x - 14) \div (x + 7)$
 a. $x + 2$ **b.** $x - 2$ **c.** $x + 1$ **d.** $x - 1$ **e.** none of the above

21. Factor completely: $r^2h + r^2a$
 a. $r(rh + ra)$ **b.** $r^2h(1 + a)$ **c.** $r^2a(h + 1)$
 d. $r^2(h + a)$ **e.** none of the above

22. Factor completely: $m^2n^4 - 49$
 a. $(mn^2 + 7)(mn^2 + 7)$ **b.** $(mn^2 - 7)(mn^2 - 7)$
 c. $(mn + 7)(mn - 7)$ **d.** $(mn^2 + 7)(mn^2 - 7)$
 e. none of the above

23. One of the factors of $x^2 - 5x + 6$ is
 a. $x + 3$ **b.** $x + 2$ **c.** $x - 6$ **d.** $x - 2$ **e.** none of the above

24. One of the factors of $36x^2 + 12x + 1$ is
 a. $6x - 1$ **b.** $6x + 1$ **c.** $x + 6$ **d.** $x - 6$
 e. none of the above

25. One of the factors of $2x^2 + 7xy + 6y^2$ is
 a. $2x + y$ **b.** $x - 3y$ **c.** $2x - y$ **d.** $x + 6y$
 e. none of the above

26. One of the factors of $8x^3 - 27$ is
 a. $2x + 3$ **b.** $4x^2 - 6x + 9$ **c.** $4x^2 + 12x + 9$
 d. $4x^2 + 6x + 9$ **e.** none of the above

27. One of the factors of $2x^2 + 2xy - 3x - 3y$ is
 a. $x - y$ **b.** $2x + 3$ **c.** $x - 3$ **d.** $2x - 3$ **e.** none of the above

28. Solve for x: $x^2 + x - 6 = 0$
 a. $x = -2, x = 3$ **b.** $x = 2, x = -3$ **c.** $x = 2,$
 $x = 3$ **d.** $x = -2, x = -3$ **e.** none of the above

29. Solve for x: $6x^2 - 7x - 3 = 0$
 a. $x = \dfrac{3}{2}, x = \dfrac{1}{3}$ **b.** $x = -\dfrac{3}{2}, x = \dfrac{1}{3}$
 c. $x = \dfrac{3}{2}, x = -\dfrac{1}{3}$ **d.** $x = -\dfrac{3}{2}, x = -\dfrac{1}{3}$ **e.** none of the above

30. Simplify the fraction: $\dfrac{x^2 - 16}{x^2 - 8x + 16}$
 a. $\dfrac{x + 4}{x - 4}$ **b.** 1 **c.** $\dfrac{x - 4}{x + 4}$ **d.** $-\dfrac{1}{8x}$ **e.** none of the above

31. Multiply: $\dfrac{x^2 + 11x - 12}{x - 5} \cdot \dfrac{x^2 - 5x}{x - 1}$
 a. $-x(x + 12)$ **b.** $x(x + 12)$ **c.** $\dfrac{x^2 + 11x - 12}{(x - 5)(x - 1)}$
 d. 1 **e.** none of the above

32. Divide: $\dfrac{t^2 + 7t}{t^2 + 5t} \div \dfrac{t^2 + 4t - 21}{t - 3}$
 a. $\dfrac{1}{t + 5}$ **b.** $t + 5$ **c.** 1 **d.** $\dfrac{(t + 7)(t + 7)}{t + 5}$
 e. none of the above

33. Simplify: $\dfrac{3x}{2} - \dfrac{x}{4}$
 a. $-x$ **b.** $\dfrac{5}{4}$ **c.** $2x$ **d.** $\dfrac{5x}{4}$ **e.** none of the above

34. Simplify: $\dfrac{a + 3}{2a - 6} - \dfrac{2a + 3}{a^2 - 3a} + \dfrac{3}{4}$
 a. $\dfrac{5a + 4}{4a}$ **b.** $\dfrac{4a + 4}{5}$ **c.** 0 **d.** $\dfrac{9 - a}{5a - a^2 - 2}$
 e. none of the above

35. Simplify: $\dfrac{x + \dfrac{1}{y}}{\dfrac{1}{x} + y}$
 a. $\dfrac{x^2y + x}{y + xy^2}$ **b.** 1 **c.** $\dfrac{xy + x}{y + xy}$ **d.** $\dfrac{x}{y}$ **e.** none of the above

36. Solve for x: $\dfrac{1}{x} + \dfrac{1}{2x} = \dfrac{1}{4}$
 a. 4 **b.** 5 **c.** 6 **d.** 7 **e.** none of the above

37. Solve for s: $\dfrac{2}{s + 1} + \dfrac{1 - s}{s} = \dfrac{1}{s^2 + s}$
 a. 1 **b.** 2 **c.** 3 **d.** 4 **e.** none of the above

38. Find the x-intercept of the graph of $3x - 4y = 12$.
 a. $(3, 0)$ **b.** $(-3, 0)$ **c.** $(4, 0)$ **d.** $(-4, 0)$
 e. none of the above

39. The graph of $y = -3x + 12$ does not pass through
 a. quadrant I   **b.** quadrant II   **c.** quadrant III
 d. quadrant IV   **e.** none of the above

40. Find the slope of the line passing through $P(-2, 4)$ and $Q(8, -6)$.
 a. -1 **b.** 1 **c.** $-\dfrac{1}{3}$ **d.** 3 **e.** none of the above

41. The equation of the line passing through $P(-2, 4)$ and $Q(8, -6)$ is
 a. $y = -x + 2$ **b.** $y = -x + 6$ **c.** $y = -x - 6$
 d. $y = -x - 2$ **e.** none of the above

42. If $f(x) = x^2 - 2x$, find $f(a + 2)$.
 a. $a^2 - 2a$ **b.** $a^2 + 2a$ **c.** $a + 2$ **d.** $a + 2a^2$
 e. none of the above

43. Solve the system
$$\begin{cases} 2x - 5y = 5 \\ 3x - 2y = -16 \end{cases}$$

for x.
a. $x = 8$ **b.** $x = -8$ **c.** $x = 4$ **d.** $x = -4$
e. none of the above

44. Solve the system
$$\begin{cases} 8x - y = 29 \\ 2x + y = 11 \end{cases}$$

for y.
a. $y = -4$ **b.** $y = 4$ **c.** $y = 3$ **d.** $y = -3$
e. none of the above

45. Simplify: $\sqrt{12}$
a. $2\sqrt{3}$ **b.** $4\sqrt{3}$ **c.** $6\sqrt{2}$ **d.** $4\sqrt{2}$ **e.** none of
the above

46. Simplify: $\sqrt{\dfrac{3}{4}}$

a. $\dfrac{\sqrt{3}}{4}$ **b.** $\dfrac{3}{2}$ **c.** $\dfrac{\sqrt{3}}{2}$ **d.** $\dfrac{9}{16}$ **e.** none of the
above

47. Simplify: $\sqrt{75x^3}$
a. $5\sqrt{x^3}$ **b.** $5x\sqrt{x}$ **c.** $x\sqrt{75x}$ **d.** $25x\sqrt{3x}$
e. none of the above

48. Simplify: $3\sqrt{5} - \sqrt{20}$
a. $3\sqrt{-15}$ **b.** $2\sqrt{5}$ **c.** $-\sqrt{5}$ **d.** $\sqrt{5}$ **e.** none
of the above

49. Rationalize the denominator: $\dfrac{11}{\sqrt{11}}$

a. $\dfrac{1}{11}$ **b.** $\sqrt{11}$ **c.** $\dfrac{\sqrt{11}}{11}$ **d.** 1 **e.** none of the
above

50. Rationalize the denominator: $\dfrac{7}{3 - \sqrt{2}}$

a. $3 - \sqrt{2}$ **b.** $7(3 - \sqrt{2})$ **c.** $3 + \sqrt{2}$
d. $7(3 + \sqrt{2})$ **e.** none of the above

51. Solve for x: $\sqrt{\dfrac{3x - 1}{5}} = 2$

a. 7 **b.** 4 **c.** -7 **d.** -4 **e.** none of the above

52. Solve for n: $3\sqrt{n} - 1 = 1$
a. $\dfrac{2}{3}$ **b.** $\sqrt{\dfrac{2}{3}}$ **c.** $\dfrac{4}{9}$ **d.** $n\sqrt{n + 1}$ **e.** none of
the above

53. Simplify: $(a^6 b^4)^{1/2}$
a. $(ab)^5$ **b.** $a^3 b^2$ **c.** $\dfrac{1}{a^3 b^2}$ **d.** $\dfrac{1}{a^6 b^4}$ **e.** none of
the above

54. Simplify: $\left(\dfrac{8}{125}\right)^{2/3}$

a. $\dfrac{4}{25}$ **b.** $\dfrac{25}{4}$ **c.** $\dfrac{2}{5}$ **d.** $\dfrac{5}{2}$ **e.** none of the above

55. What number must be added to $x^2 + 12x$ to make it a
perfect trinomial square?
a. 6 **b.** 12 **c.** 144 **d.** 36 **e.** none of the above

56. Write the quadratic formula.
a. $x = \dfrac{b \pm \sqrt{b^2 - 4ac}}{2a}$ **b.** $x = \dfrac{-b \pm \sqrt{b^2 - 4ac}}{2a}$

c. $x = \dfrac{-b \pm \sqrt{b^2 + 4ac}}{2a}$

d. $x = \dfrac{-b \pm \sqrt{b^2 - 4ac}}{2b}$ **e.** none of the above

57. One solution of the equation $x^2 - 2x - 2 = 0$ is
a. $2\sqrt{3}$ **b.** $2 + 2\sqrt{3}$ **c.** $1 - \sqrt{3}$ **d.** $2 - 2\sqrt{3}$
e. none of the above

58. The vertex of the graph of the equation
$y = x^2 - 2x + 1$ is
a. $(1, 0)$ **b.** $(0, 1)$ **c.** $(-1, 0)$ **d.** $(0, -1)$
e. none of the above

59. The graph of $y = -x^2 + 2x - 1$
a. does not intersect the y-axis   **b.** does not intersect
the x-axis   **c.** passes through the origin   **d.** passes
through $(2, 2)$ **e.** none of the above

60. The number $\sqrt{-36}$ written in bi form is
a. $36i$ **b.** -6 **c.** $6i$ **d.** $-6i$ **e.** none of the
above

SET NOTATION

In mathematics, any collection of objects is called a **set.** For example, the collection of natural numbers is referred to as the *set* of natural numbers. The objects contained within a set are called the **members** or the **elements** of the set. The symbol

$$\in$$

read as "is an element of," is used to indicate membership in a set. Elements of a set are often listed between **braces.**

EXAMPLE 1 The set of even numbers between 2 and 10 is designated by the notation

$$\{4, 6, 8\}$$

To indicate that 6 is an element of this set, write

$$6 \in \{4, 6, 8\}$$

To indicate that 7 is *not* an element of the set, write

$$7 \notin \{4, 6, 8\}$$ Read $\notin$ as "is not an element of."

To economize on symbols, we can designate the set $\{4, 6, 8\}$ by a capital letter. Thus, if $A = \{4, 6, 8\}$, we can write

$$6 \in A \quad \text{and} \quad 7 \notin A$$ ■

It is often difficult or even impossible to describe a set by listing its elements. For example, it is impossible to describe the set of composite numbers by listing them all, because there is no end to the list of composite numbers. To designate such sets, a notation called **set-builder notation** is used, which provides a rule that determines membership in the set. For example,

$$C = \{x \mid x \text{ is a composite number}\}$$

is read as "C is the set of all numbers x such that x is a composite number." The vertical bar in this notation means "such that." Set C is the set of composite numbers.

EXAMPLE 2   **a.** The set

$$P = \{2, 3, 5, 7\}$$

can also be described as

$$P = \{x \mid x \text{ is a prime number less than } 10\}$$

b. The set

$$V = \{a, e, i, o, u\}$$

can also be described as

$$V = \{x \mid x \text{ is a vowel}\}$$ ∎

A set can have several elements, or it can have no elements at all. A set that has no elements is called the **empty set,** and it is designated by either of the symbols

$$\{ \ \ \} \qquad \text{or} \qquad \emptyset$$

The set of all odd numbers that are even and the set of all unicorns are examples of the empty set.

Subset of a Set   If every element of set A is also an element of set B, then we say that **A is a subset of B.** The notation

$$A \subset B$$

indicates that A is a subset of B.

EXAMPLE 3   **a.** $\{a, b, c\}$ is a subset of $\{a, b, c, d\}$, because every element of the first set is also an element of the second set.

b. $\{3, 5, 7\}$ is *not* a subset of $\{1, 2, 3, 4, 5\}$, because $7 \notin \{1, 2, 3, 4, 5\}$.

c. If J is the set of integers and Q is the set of rational numbers, then

$$J \subset Q \qquad \text{Read } \subset \text{ as "is a subset of."}$$

because every integer is also a rational number.

d. Suppose that

$$O = \{x \mid x \text{ is an odd natural number}\}$$
$$P = \{x \mid x \text{ is a prime number}\}$$

Then

$$O \not\subset P \qquad \text{Read } \not\subset \text{ as "is not a subset of."}$$

because some odd numbers (such as the number 9) are not prime numbers. It is also true that

$$P \not\subset O$$

because one prime number (the number 2) is not an odd number.

e. Let A be any set. Because the empty set contains no elements that are not found in A, the empty set is a subset of A. If A is any set, then

$$\emptyset \subset A$$

In other words, the empty set is a subset of *any* set.

f. Note the difference between "is an element of" and "is a subset of":

$$a \in \{a, b, c\} \qquad \text{but} \qquad \{a\} \notin \{a, b, c\}$$
$$\{a\} \subset \{a, b, c\} \qquad \text{but} \qquad a \not\subset \{a, b, c\} \qquad \blacksquare$$

There are two operations that are used to combine sets. They are the operations of **union** and **intersection.**

Union of Two Sets   The **union of sets A and B** is the set that contains all of the elements that belong to either set A or set B or both. The notation $A \cup B$ designates the union of A and B.

If A and B are two sets, the notation $A \cup B$ is read as "the union of A and B," or just "A union B."

EXAMPLE 4   **a.** If $A = \{1, 2, 3\}$, and $B = \{3, 4\}$, then

$$A \cup B = \{1, 2, 3, 4\}$$

b. If $E = \{x \mid x \text{ is an even integer}\}$,
$O = \{x \mid x \text{ is an odd integer}\}$, and
$J = \{x \mid x \text{ is an integer}\}$, then

$$E \cup O = J$$

c. If A is *any* set, then

$$\emptyset \cup A = A$$

In other words, the only elements in $\emptyset \cup A$ are those elements in set A. $\blacksquare$

Intersection of Two Sets   The **intersection of sets A and B** is the set that contains only those elements that are in *both* sets A and B. The notation $A \cap B$ designates the intersection of sets A and B.

If A and B are two sets, the expression $A \cap B$ is read as "the intersection of A and B," or just "A intersect B."

EXAMPLE 5   **a.** If $A = \{1, 2, 3\}$ and $B = \{3, 4\}$, then

$$A \cap B = \{3\}$$

b. If $C = \{1, 2, 3, 4, 5\}$, and $D = \{2, 4, 6, 8\}$, then

$$C \cap D = \{2, 4\}$$

c. If $E = \{x \mid x$ is an even integer$\}$, and
$O = \{x \mid x$ is an odd integer$\}$, then

$$E \cap O = \emptyset$$

d. If A is any set, then

$$\emptyset \cap A = \emptyset$$

In other words, there are no elements common to the empty set and set A.   ∎

E X E R C I S E   I I . 1

In Exercises 1–8, find each set.

1. The set of even numbers between 1 and 9.

2. The set of odd numbers between 1 and 9.

3. The set of names of days of the week that begin with S.

4. The set of states bordering the Mississippi river.

5. The set of integers between 5 and 6.

6. The set of all eight-legged giraffes.

7. The set of prime numbers less than 23.

8. The set of all even prime numbers.

In Exercises 9–14, find each set.

9. $\{x \mid x$ is the largest state of the United States$\}$

10. $\{x \mid x$ is a letter in the word algebra$\}$

11. $\{x \mid x$ is a numeral on the face of a clock$\}$

12. $\{x \mid x$ is a natural number less than 10$\}$

13. $\{x \mid x$ is an even prime number$\}$

14. $\{x \mid x$ is a perfect square less than 20$\}$

In Exercises 15–20, describe each set by using set-builder notation.

15. $\{1, 2, 3, 4, 5, 6, 7, 8, 9\}$

16. $\{0\}$

17. $\{1, 4, 9, 16, 25, 36, 49, 64, 81\}$

18. $\{$I, II, III, IV, V, VI, VII, VIII, IX, X$\}$

19. $\{$January, March, May, July, August, October, December$\}$

20. $\{2, 3, 5, 7\}$

In Exercises 21–32, let A = {1, 2, 3} and let B = {2, 4, 6}. Make each a true statement by inserting one of the symbols ∈, ∉, ⊂, *or* ⊄ *between the expressions.*

21. 1 ▢ *A*

22. 4 ▢ *B*

23. 3 ▢ *B*

24. 4 ▢ *A*

25. {2, 3} ▢ *B*

26. {1, 4} ▢ *A*

27. {1} ▢ *A*

28. {3} ▢ *B*

29. ∅ ▢ *A*

30. { } ▢ *B*

31. *A* ▢ *A*

32. *A* ▢ *B*

In Exercises 33–40, let A = {a, b, c}, let B = {a, d, e, f}, and let C = {e, f, g}. Find each set.

33. $A \cup B$

34. $A \cup C$

35. $A \cap C$

36. $A \cap B$

37. $B \cup C$

38. $B \cap C$

39. $A \cup (B \cup C)$

40. $A \cap (B \cap C)$

POWERS AND ROOTS

n	n^2	$\sqrt{n}$	n^3	$\sqrt[3]{n}$	n	n^2	$\sqrt{n}$	n^3	$\sqrt[3]{n}$
1	1	1.000	1	1.000	51	2,601	7.141	132,651	3.708
2	4	1.414	8	1.260	52	2,704	7.211	140,608	3.733
3	9	1.732	27	1.442	53	2,809	7.280	148,877	3.756
4	16	2.000	64	1.587	54	2,916	7.348	157,464	3.780
5	25	2.236	125	1.710	55	3,025	7.416	166,375	3.803
6	36	2.449	216	1.817	56	3,136	7.483	175,616	3.826
7	49	2.646	343	1.913	57	3,249	7.550	185,193	3.849
8	64	2.828	512	2.000	58	3,364	7.616	195,112	3.871
9	81	3.000	729	2.080	59	3,481	7.681	205,379	3.893
10	100	3.162	1,000	2.154	60	3,600	7.746	216,000	3.915
11	121	3.317	1,331	2.224	61	3,721	7.810	226,981	3.936
12	144	3.464	1,728	2.289	62	3,844	7.874	238,328	3.958
13	169	3.606	2,197	2.351	63	3,969	7.937	250,047	3.979
14	196	3.742	2,744	2.410	64	4,096	8.000	262,144	4.000
15	225	3.873	3,375	2.466	65	4,225	8.062	274,625	4.021
16	256	4.000	4,096	2.520	66	4,356	8.124	287,496	4.041
17	289	4.123	4,913	2.571	67	4,489	8.185	300,763	4.062
18	324	4.243	5,832	2.621	68	4,624	8.246	314,432	4.082
19	361	4.359	6,859	2.668	69	4,761	8.307	328,509	4.102
20	400	4.472	8,000	2.714	70	4,900	8.367	343,000	4.121
21	441	4.583	9,261	2.759	71	5,041	8.426	357,911	4.141
22	484	4.690	10,648	2.802	72	5,184	8.485	373,248	4.160
23	529	4.796	12,167	2.844	73	5,329	8.544	389,017	4.179
24	576	4.899	13,824	2.884	74	5,476	8.602	405,224	4.198
25	625	5.000	15,625	2.924	75	5,625	8.660	421,875	4.217
26	676	5.099	17,576	2.962	76	5,776	8.718	438,976	4.236
27	729	5.196	19,683	3.000	77	5,929	8.775	456,533	4.254
28	784	5.292	21,952	3.037	78	6,084	8.832	474,552	4.273
29	841	5.385	24,389	3.072	79	6,241	8.888	493,039	4.291
30	900	5.477	27,000	3.107	80	6,400	8.944	512,000	4.309
31	961	5.568	29,791	3.141	81	6,561	9.000	531,441	4.327
32	1,024	5.657	32,768	3.175	82	6,724	9.055	551,368	4.344
33	1,089	5.745	35,937	3.208	83	6,889	9.110	571,787	4.362
34	1,156	5.831	39,304	3.240	84	7,056	9.165	592,704	4.380
35	1,225	5.916	42,875	3.271	85	7,225	9.220	614,125	4.397
36	1,296	6.000	46,656	3.302	86	7,396	9.274	636,056	4.414
37	1,369	6.083	50,653	3.332	87	7,569	9.327	658,503	4.431
38	1,444	6.164	54,872	3.362	88	7,744	9.381	681,472	4.448
39	1,521	6.245	59,319	3.391	89	7,921	9.434	704,969	4.465
40	1,600	6.325	64,000	3.420	90	8,100	9.487	729,000	4.481
41	1,681	6.403	68,921	3.448	91	8,281	9.539	753,571	4.498
42	1,764	6.481	74,088	3.476	92	8,464	9.592	778,688	4.514
43	1,849	6.557	79,507	3.503	93	8,649	9.644	804,357	4.531
44	1,936	6.633	85,184	3.530	94	8,836	9.695	830,584	4.547
45	2,025	6.708	91,125	3.557	95	9,025	9.747	857,375	4.563
46	2,116	6.782	97,336	3.583	96	9,216	9.798	884,736	4.579
47	2,209	6.856	103,823	3.609	97	9,409	9.849	912,673	4.595
48	2,304	6.928	110,592	3.634	98	9,604	9.899	941,192	4.610
49	2,401	7.000	117,649	3.659	99	9,801	9.950	970,299	4.626
50	2,500	7.071	125,000	3.684	100	10,000	10.000	1,000,000	4.642

ANSWERS TO SELECTED EXERCISES

1.1 GETTING READY (page 2)

1. 1, 2, 3, etc. **2.** $\frac{1}{2}, \frac{3}{4}$, etc. **3.** $-3°, -21°$, etc. **4.** $-\$12.95, -\$52.07$, etc.

ORALS (page 10)

11. -15 **12.** 25

EXERCISE 1.1 (page 10)

1. 1, 2, 6, 9 **3.** 1, 2, 6, 9 **5.** $-3, -1, 0, 1, 2, 6, 9$ **7.** $-3, -\frac{1}{2}, -1, 0, 1, 2, \frac{5}{3}, \sqrt{7}, 3.25, 6, 9$ **9.** $-3, -1, 1, 9$ **11.** 6, 9
13. 9; natural, odd, composite, and whole number **15.** 0; even integer, whole number **17.** 24; natural, even, composite, and whole number **19.** 3; natural, odd, prime, and whole number **21.** $=$ **23.** $<$ **25.** $>$ **27.** $=$ **29.** $=$ **31.** $<$ **33.** $=$
35. $7 > 3$ **37.** $8 \le 8$ **39.** $3 + 4 = 7$ **41.** $7 \ge 3$ **43.** $0 < 6$ **45.** $8 < 3 + 8$ **47.** $10 - 4 > 6 - 2$ **49.** $3 \cdot 4 > 2 \cdot 3$
51. $\frac{24}{6} > \frac{12}{4}$ **53.** $\rule{1.5cm}{0pt}$; 6, 6 **55.** $\rule{1.5cm}{0pt}$; 11, 11 **57.** $\rule{1.5cm}{0pt}$; 2, 2
59. $\rule{1.5cm}{0pt}$; 8, 8 **61.** $\rule{1.5cm}{0pt}$ **63.** $\rule{1.5cm}{0pt}$
65. $\rule{1cm}{0pt}$ **67.** $\rule{1.5cm}{0pt}$ **69.** $\rule{1cm}{0pt}$ **71.** $\rule{1cm}{0pt}$
73. 36 **75.** 0 **77.** 230 **79.** 8

REVIEW EXERCISES (page 13)

1. true **2.** false **3.** false **4.** false **5.** true **6.** true **7.** true **8.** true

1.2 GETTING READY (page 13)

1. 250 **2.** 148 **3.** 16,606 **4.** 105

ORALS (page 26)

1. $\frac{1}{2}$ **2.** $\frac{1}{2}$ **3.** $\frac{1}{2}$ **4.** $\frac{1}{3}$ **5.** $\frac{5}{12}$ **6.** $\frac{9}{20}$ **7.** $\frac{4}{9}$ **8.** $\frac{6}{25}$ **9.** $\frac{11}{9}$ **10.** $\frac{3}{7}$ **11.** $\frac{1}{6}$ **12.** $\frac{5}{4}$ **13.** 2.86 **14.** 1.24 **15.** 0.5 **16.** 3.9 **17.** 3.24
18. 3.25

EXERCISE 1.2 (page 26)

1. $\frac{1}{2}$ **3.** $\frac{3}{4}$ **5.** $\frac{4}{3}$ **7.** $\frac{9}{8}$ **9.** $\frac{3}{10}$ **11.** $\frac{8}{5}$ **13.** $\frac{3}{2}$ **15.** $\frac{1}{4}$ **17.** 10 **19.** $\frac{20}{3}$ **21.** $\frac{9}{10}$ **23.** $\frac{5}{8}$ **25.** $\frac{1}{4}$ **27.** $\frac{14}{5}$ **29.** 28 **31.** $\frac{1}{5}$ **33.** $\frac{6}{5}$
35. $\frac{1}{13}$ **37.** $\frac{5}{24}$ **39.** $\frac{19}{15}$ **41.** $\frac{17}{12}$ **43.** $\frac{22}{35}$ **45.** $\frac{1}{6}$ **47.** $\frac{9}{4}$ **49.** $\frac{29}{3}$ **51.** $\frac{4}{5}$ **53.** $5\frac{1}{5}$ **55.** $1\frac{2}{5}$ **57.** $1\frac{1}{4}$ **59.** $\frac{5}{9}$ **61.** $7\frac{2}{7}$ cm **63.** $3\frac{7}{10}$ km
65. $53\frac{1}{6}$ ft **67.** 2,514,820 **69.** 270 lb **71.** 158.65 **73.** 44.785 **75.** 44.88 **77.** 4.55 **79.** 350.49 **81.** 55.21 **83.** 3337.52
85. 10.02 **87.** 3865.11 ft^2 **89.** \$780.69 **91.** \$1170 **93.** \$2525.25 **95.** the Guernsey **97.** the second contractor **99.** the
regular furnace

REVIEW EXERCISES (page 29)

1. $=$ **2.** $-$ **3.** $=$ **4.** $>$

1.3 GETTING READY   (page 29)

1. 4   **2.** 9   **3.** 27   **4.** 8   **5.** $\frac{1}{4}$ **6.** $\frac{1}{27}$ **7.** $\frac{8}{125}$ **8.** $\frac{27}{1000}$

ORALS   (page 37)

1. 32   **2.** 81   **3.** 64   **4.** 125   **5.** 24   **6.** 36   **7.** 11   **8.** 1   **9.** 16   **10.** 24

EXERCISE 1.3   (page 37)

1. 16   **3.** 36   **5.** $\frac{1}{10,000}$ **7.** $x \cdot x$ **9.** $3 \cdot z \cdot z \cdot z \cdot z$ **11.** $5t \cdot 5t$ **13.** $5 \cdot 2x \cdot 2x \cdot 2x$ **15.** 36   **17.** 1000   **19.** 18   **21.** 216
23. 11   **25.** 3   **27.** 28   **29.** 64   **31.** 13   **33.** 16   **35.** 2   **37.** 16   **39.** 21   **41.** 56   **43.** 1   **45.** 1   **47.** 17   **49.** 9   **51.** 8
53. 8   **55.** $\frac{1}{144}$ **57.** 11   **59.** 1   **61.** $\frac{8}{9}$ **63.** 1   **65.** 4   **67.** 4   **69.** 12   **71.** 4   **73.** 11   **75.** 24   **77.** 12   **79.** 25   **81.** 1
83. 28   **85.** 35   **87.** 1   **89.** 2   **91.** 6   **93.** $(3 \cdot 8) + (5 \cdot 3)$ **95.** $(3 \cdot 8 + 5) \cdot 3$ **97.** $(4 + 3) \cdot (5 - 3)$ **99.** $(4 + 3) \cdot 5 - 3$
101. 16 in.   **103.** 15 m   **105.** 25 m^2 **107.** 60 ft^2 **109.** 88 m   **111.** 1386 ft^2 **113.** 6 cm^3 **115.** 905 m^3 **117.** 1056 cm^3
119. 40,764.51 ft^3 **121.** 480 ft^3 **123.** 8

SOMETHING TO THINK ABOUT   (page 40)

1. bigger   **2.** increasing powers get smaller

REVIEW EXERCISES   (page 41)

1.   **2.** $12 \geq 7$ **3.** prime number   **4.** $\frac{1}{10}$

1.4 GETTING READY   (page 41)

1. 17.52   **2.** 2.94   **3.** 3.4   **4.** 1   **5.** 96   **6.** 382

ORALS   (page 46)

1. 5   **2.** -3 **3.** 3   **4.** -11 **5.** 4   **6.** -12 **7.** 2   **8.** 16   **9.** -6 **10.** -6

EXERCISE 1.4   (page 47)

1. 12   **3.** -10 **5.** 2   **7.** -2 **9.** -12 **11.** 0.5   **13.** $\frac{12}{35}$ **15.** $-\frac{1}{12}$ **17.** 1   **19.** 2.2   **21.** 7   **23.** -1 **25.** -7 **27.** -8
29. -18 **31.** $\frac{1}{5}$ **33.** 1.3   **35.** -1 **37.** 3   **39.** 10   **41.** -3 **43.** -1 **45.** 9   **47.** 1   **49.** 7   **51.** 4   **53.** 12   **55.** -17 **57.** 5
59. $\frac{1}{2}$ **61.** $-\frac{22}{5}$ **63.** $-8\frac{3}{4}$ **65.** -4.2 **67.** 4   **69.** -7 **71.** 10   **73.** 0   **75.** 8   **77.** 64   **79.** 3   **81.** 2.45   **83.** 1   **85.** 9
87. -3 **89.** -15 **91.** 10   **93.** 1   **95.** $-\frac{3}{5}$ **97.** 3   **99.** -1 **101.** -1 **103.** $\frac{7}{6}$ **105.** -15.8 **107.** -7.1 **109.** \$175
111. $+9$ **113.** $-4°$ **115.** 2000   **117.** 1325 m   **119.** 4000 ft   **121.** 5°   **123.** 3187   **125.** 700   **127.** \$422.66   **129.** \$83,425.57

REVIEW EXERCISES   (page 50)

1. 20   **2.** 40   **3.** 24   **4.** 54

1.5 GETTING READY   (page 50)

1. 56   **2.** 54   **3.** 72   **4.** 63   **5.** 9   **6.** 6   **7.** 8   **8.** 8

ORALS   (page 55)

1. -3 **2.** 10   **3.** 18   **4.** -24 **5.** 24   **6.** -24 **7.** -2 **8.** 2   **9.** -9 **10.** -1

EXERCISE 1.5 (page 56)

1. 48 **3.** 56 **5.** 63 **7.** -49 **9.** -144 **11.** -16 **13.** 2 **15.** 1 **17.** 72 **19.** -24 **21.** -420 **23.** -96 **25.** 4 **27.** -9 **29.** -2 **31.** 5 **33.** -3 **35.** -8 **37.** -8 **39.** 6 **41.** 36 **43.** 18 **45.** 5 **47.** 7 **49.** -4 **51.** 2 **53.** -4 **55.** -20 **57.** 2 **59.** 1 **61.** -6 **63.** -30 **65.** 7 **67.** -10 **69.** -66 **71.** 6 **73.** -10 **75.** 14 **77.** -81 **79.** 88 **81.** -30 **83.** -21 **85.** $-\frac{1}{6}$ **87.** $-\frac{11}{12}$ **89.** $-\frac{7}{36}$ **91.** $-\frac{11}{48}$ **93.** $(+2)(+3) = +6$ **95.** $(-30)(15) = -450$ **97.** $(+23)(-120) = -2760$ **99.** $\frac{-18}{-3} = +6$ **101.** 2-point loss per day **103.** yes

REVIEW EXERCISES (page 58)

1. 1125 lb **2.** 650 lb **3.** 97 **4.** $>$

1.6 GETTING READY (page 58)

1. -2 **2.** 35 **3.** $-\frac{1}{2}$ **4.** 24 **5.** -4 **6.** -8 **7.** -15 **8.** -15

ORALS (page 62)

1. 1 **2.** -14 **3.** -11 **4.** 7 **5.** 16 **6.** 64 **7.** -12 **8.** 36

EXERCISE 1.6 (page 62)

1. $x + y$ **3.** $x(2y)$ **5.** $y - x$ **7.** $\frac{y}{x}$ **9.** $z + \frac{x}{y}$ **11.** $z - xy$ **13.** $3xy$ **15.** $\frac{x+y}{y+z}$ **17.** $xy + \frac{y}{z}$ **19.** $c + 4$ **21.** $\$9987t$ **23.** $\frac{x}{5}$ ft **25.** $\$(3d + 5)$ **27.** the sum of x and 3 **29.** the quotient obtained when x is divided by y **31.** the product of 2, x, and y **33.** the quotient obtained when 5 is divided by the sum of x and y **35.** the quotient obtained when the sum of 3 and x is divided by y **37.** the product of x, y, and the sum of x and y **39.** $x + z$; 10 **41.** $y - z$; 2 **43.** $yz - 3$; 5 **45.** $\frac{xy}{z}$; 16 **47.** 1; 6 **49.** 3; -1 **51.** 4; 3 **53.** 3; -4 **55.** 4; 3 **57.** 19 and x **59.** 29, x, y, and z **61.** 3, x, y, and z **63.** 17, x, and z **65.** 5, 1, and 8 **67.** x and y **69.** 75 **71.** x and y

REVIEW EXERCISES (page 65)

1. 532 **2.** 2859 **3.** $\frac{1}{2}$ **4.** $\frac{1}{2}$

1.7 GETTING READY (page 65)

1. 17 **2.** 17 **3.** 38.6 **4.** 38.6 **5.** 56 **6.** 56 **7.** 0 **8.** 1 **9.** 777 **10.** 777

EXERCISE 1.7 (page 70)

1. 10 **3.** -24 **5.** 144 **7.** 3 **9.** both are 12 **11.** both are 29 **13.** both are 60 **15.** both are 0 **17.** both are -6 **19.** both are -12 **21.** $3x + 3y$ **23.** $x^2 + 3x$ **25.** $-xa - xb$ **27.** $4x^2 + 4x$ **29.** $-5t - 10$ **31.** $-2ax - 2a^2$ **33.** $-2, \frac{1}{2}$ **35.** $-\frac{1}{3}, 3$ **37.** 0, none **39.** $\frac{5}{2}, -\frac{2}{5}$ **41.** 0.2, -5 **43.** $-\frac{4}{3}, \frac{3}{4}$ **45.** commut. prop. of $+$ **47.** commut. prop. of $\times$ **49.** distrib. prop. **51.** commut. prop. of $+$ **53.** identity for $\times$ **55.** additive inverse **57.** $3x + 3 \cdot 2$ **59.** xy^2 **61.** $(y + x)z$ **63.** $x(yz)$ **65.** x

REVIEW EXERCISES (page 72)

1. $x + y^2 \geq z$ **2.** the product of 3 and the sum of x and z **3.** 0 **4.** $-y$ **5.** positive **6.** negative

CHAPTER 1 REVIEW EXERCISES (page 75)

1. 1, 2, 3, 4, 5 **2.** 2, 3, 5 **3.** 1, 3, 5 **4.** 4 **5.** $<$ **6.** $<$ **7.** $=$ **8.** $>$ **9.**

10. **11.** **12.** **13.** 11 **14.** 31 **15.** $\frac{5}{3}$ **16.** 11 **17.** $\frac{1}{3}$ **18.** 1 **19.** $\frac{10}{21}$ **20.** $\frac{11}{21}$ **21.** 48.61 **22.** 12.99 **23.** 18.55 **24.** 3.7 **25.** 4.70 **26.** 26.36 **27.** 3.57 **28.** 3.75 **29.** 6.85 hr **30.** 57 **31.** 81 **32.** $\frac{4}{9}$ **33.** -25 **34.** 25 **35.** 32 **36.** 7 **37.** 6 **38.** 3 **39.** 98 **40.** 38 **41.** 3

42. 15 **43.** 58 **44.** 4 **45.** 7 **46.** 3 **47.** 81 **48.** 8 **49.** 22 **50.** 1 **51.** 40.6 ft **52.** 15,133.6 ft^3 **53.** −5 **54.** −7 **55.** $\frac{3}{2}$
56. 1 **57.** 1 **58.** $-\frac{1}{7}$ **59.** 1.2 **60.** −3.54 **61.** 7 **62.** $\frac{7}{2}$ **63.** $\frac{1}{4}$ **64.** $-\frac{2}{3}$ **65.** 6 **66.** $\frac{3}{2}$ **67.** 26 **68.** 7 **69.** −4 **70.** −1
71. −2 **72.** 5 **73.** 4 **74.** 6 **75.** −6 **76.** 3 **77.** 2 **78.** 6 **79.** −7 **80.** 39 **81.** 6 **82.** −2 **83.** −8 **84.** 1 **85.** xz
86. $x + 2y$ **87.** $2(x + y)$ **88.** $x - yz$ **89.** the product of 3, x, and y **90.** 5 decreased by the product of y and z **91.** 5 less than
the product of y and z **92.** the sum of x, y, and z, divided by twice their product **93.** 3 **94.** 7 **95.** 1 **96.** 9 **97.** clos. prop.
98. commut. prop. of × **99.** assoc. prop. of + **100.** distrib. prop. **101.** commut. prop. of + **102.** assoc. prop. of ×
103. commut. prop. of + **104.** identity for × **105.** additive inverses **106.** identity for +

CHAPTER 1 TEST (page 78)

1. 31, 37, 41, 43, 47 **2.** 2 **3.** **4.** **5.** −23 **6.** 0 **7.** <
8. < **9.** = **10.** = **11.** $\frac{13}{20}$ **12.** 1 **13.** $\frac{4}{5}$ **14.** $\frac{9}{2} = 4\frac{1}{2}$ **15.** −1 **16.** $-\frac{1}{13}$ **17.** 77.7 **18.** 301.57 ft^2 **19.** 64 cm^2
20. 1539 in.^{3} **21.** −2 **22.** −14 **23.** −4 **24.** 12 **25.** 5 **26.** −23 **27.** $\frac{xy}{x+y}$ **28.** $5y - (x + y)$ **29.** $24x + 14y$
30. \$$(12a + 8b)$ **31.** 3 **32.** 4 **33.** 0 **34.** 5 **35.** commut. prop. of × **36.** distrib. prop. **37.** commut. prop. of +
38. multiplicative inverse

2.1 GETTING READY (page 81)

1. −3 **2.** 7 **3.** 4 **4.** 7 **5.** 17 **6.** −23 **7.** 3 **8.** −5

ORALS (page 88)

1. −2 **2.** 16 **3.** 0 **4.** −25 **5.** 0 **6.** 690 **7.** 444 **8.** 111 **9.** 100 **10.** $33\frac{1}{3}\%$

EXERCISE 2.1 (page 88)

1. yes **3.** no **5.** yes **7.** yes **9.** yes **11.** no **13.** yes **15.** no **17.** yes **19.** yes **21.** yes **23.** yes **25.** 10 **27.** 5
29. 74 **31.** −28 **33.** 2 **35.** $\frac{5}{6}$ **37.** $-\frac{1}{5}$ **39.** $\frac{1}{2}$ **41.** 1 **43.** 25 **45.** −2 **47.** $-\frac{1}{2}$ **49.** 98 **51.** 5 **53.** $\frac{5}{2}$ **55.** 4912 **57.** 15
59. −1 **61.** 27 **63.** −11 **65.** −5 **67.** 0 **69.** 15 **71.** −33 **73.** 1 **75.** $\frac{1}{2}$ **77.** $-\frac{3}{2}$ **79.** $-\frac{1}{2}$ **81.** 80 **83.** 19 **85.** 320
87. 380 **89.** 150 **91.** 20% **93.** 8% **95.** 200% **97.** \$45,149 **99.** 117 **101.** 1519 **103.** 55% **105.** \$270 **107.** 2800

REVIEW EXERCISES (page 91)

1. 15; integer, composite **2.** 16; integer, composite **3.** −1; integer **4.** 1; integer **5.** closure prop. of + **6.** distrib. prop.
7. commut. prop. of + **8.** assoc. prop. of +

2.2 GETTING READY (page 91)

1. 22 **2.** 36 **3.** 5 **4.** $\frac{13}{2}$ **5.** −1 **6.** −1 **7.** $\frac{7}{9}$ **8.** $-\frac{19}{3}$

ORALS (page 97)

1. add 7 **2.** subtract 3 **3.** add 3 **4.** multiply by 7 **5.** add 5 **6.** subtract 5 **7.** multiply by 3 **8.** subtract 2 **9.** 3 **10.** 13

EXERCISE 2.2 (page 97)

1. 1 **3.** −1 **5.** 3 **7.** −2 **9.** 2 **11.** −5 **13.** $\frac{3}{2}$ **15.** 2 **17.** 3 **19.** −54 **21.** −9 **23.** −33 **25.** 10 **27.** −4 **29.** 28
31. 5 **33.** 7 **35.** −8 **37.** 10 **39.** 4 **41.** 10 **43.** $\frac{17}{5}$ **45.** $-\frac{2}{3}$ **47.** 0 **49.** 6 **51.** $\frac{3}{5}$ **53.** 5 **55.** \$250 **57.** 7 days
59. 29 min **61.** \$7400 **63.** no chance; he needs 112 **65.** \$50 **67.** 15% to 6%

SOMETHING TO THINK ABOUT (page 99)

1. $\frac{7x + 25}{22} = \frac{1}{2}$ **2.** 9%

REVIEW EXERCISES   (page 99)

1. 50 cm   **2.** 8.51 in.^{2} **3.** 80.325 in.^{2} **4.** 635.664 cm^3

2.3 GETTING READY   (page 100)

1. $3x + 4x$ **2.** $7x + 2x$ **3.** $8w - 3w$ **4.** $10y - 4y$ **5.** $7x$ **6.** $9x$ **7.** $5w$ **8.** $6y$

ORALS   (page 107)

1. x **2.** 0   **3.** 0   **4.** unlike terms   **5.** 12   **6.** $6x$ **7.** 3   **8.** impossible   **9.** 2   **10.** $\frac{1}{3}$

EXERCISE 2.3   (page 107)

1. $20x$ **3.** $3x^2$ **5.** $9x + 3y$ **7.** $7x + 6$ **9.** $7z - 15$ **11.** $12x + 121$ **13.** $6y + 62$ **15.** $-2x + 7y$ **17.** $2 + y$ **19.** $5x + 7$
21. $5x^2 + 24x$ **23.** -2 **25.** 3   **27.** 1   **29.** 1   **31.** $\frac{1}{3}$ **33.** 2   **35.** 6   **37.** 35   **39.** -9 **41.** 0   **43.** -20 **45.** -41 **47.** 9
49. -1 **51.** 8   **53.** 5   **55.** 4   **57.** -3 **59.** 1   **61.** -5 **63.** $\frac{8}{3}$ **65.** 0   **67.** -3 **69.** 2   **71.** identity   **73.** impossible equation
75. 16   **77.** impossible equation   **79.** identity   **81.** identity   **83.** 4 ft and 8 ft   **85.** $1211.50   **87.** $900   **89.** $5.45   **91.** 250
93. 125 lb   **95.** 7.3 ft and 10.7 ft   **97.** $1.50   **99.** $20,000

SOMETHING TO THINK ABOUT   (page 110)

1. 0   **2.** 0

REVIEW EXERCISES   (page 110)

1. 0   **2.** 125   **3.** 2   **4.** 6

2.4 GETTING READY   (page 110)

1. 3   **2.** -5 **3.** r **4.** $-a$ **5.** 7   **6.** 12   **7.** d **8.** s

ORALS   (page 115)

1. $a = \dfrac{d - c}{b}$ **2.** $b = \dfrac{d - c}{a}$ **3.** $c = d - ab$ **4.** $d = ab + c$ **5.** $a = \dfrac{c}{d} - b$ **6.** $b = \dfrac{c}{d} - a$ **7.** $c = d(a + b)$ **8.** $d = \dfrac{c}{a + b}$

EXERCISE 2.4   (page 115)

1. $I = \dfrac{E}{R}$ **3.** $w = \dfrac{V}{lh}$ **5.** $b = P - a - c$ **7.** $w = \dfrac{P - 2l}{2}$ **9.** $t = \dfrac{A - P}{Pr}$ **11.** $r = \dfrac{C}{2\pi}$ **13.** $w = \dfrac{2gK}{v^2}$ **15.** $R = \dfrac{P}{I^2}$
17. $g = \dfrac{wv^2}{2K}$ **19.** $M = \dfrac{Fd^2}{Gm}$ **21.** $d^2 = \dfrac{GMm}{F}$ **23.** $r = \dfrac{G}{2b} + 1$ or $r = \dfrac{G + 2b}{2b}$ **25.** $t = \dfrac{d}{r}$; $t = 3$ **27.** $t = \dfrac{i}{pr}$; $t = 2$
29. $c = P - a - b$; $c = 3$ **31.** $h = \dfrac{2K}{a + b}$; $h = 8$ **33.** $I = \dfrac{E}{R}$; $I = 4$ amp   **35.** $r = \dfrac{C}{2\pi}$; $r = 2.28$ ft   **37.** $R = \dfrac{P}{I^2}$; $R = 13.78$
ohms   **39.** $m = \dfrac{Fd^2}{GM}$ **41.** $D = \dfrac{L - 3.25r - 3.25R}{2}$; $D = 6$ ft

SOMETHING TO THINK ABOUT   (page 117)

1. 90,000,000,000 joules   **2.** Although only twice as fast, the energy is four times greater.

REVIEW EXERCISES   (page 117)

1. $5x - 5y$ **2.** unlike terms   **3.** $-x - 13$ **4.** $4x + \frac{7}{11}y^2$

2.5 GETTING READY   (page 118)
1. 7, 9   **2.** 6, 10   **3.** $3k$ **4.** $72s$

ORALS   (page 122)
1. $x + 2$ and $x + 4$ **2.** $x + (x + 2) + (x + 4) = 42$ **3.** $x = 12$ **4.** 12, 14, and 16

EXERCISE 2.5   (page 122)
1. 26 and 28   **3.** 39, 40, and 41   **5.** 7   **7.** 13   **9.** 19 ft   **11.** 29 m by 18 m   **13.** 17 in. by 39 in.   **15.** 60°   **17.** 20   **19.** 1100
21. 12   **23.** 90   **25.** 15 days   **27.** 960 pairs   **29.** 300 plates   **31.** 7500 gal   **33.** A

REVIEW EXERCISES   (page 125)
1. 200 cm³   **2.** $\frac{1584}{7}$ cm³   **3.** $7x - 6$ **4.** $x - 11$ **5.** $-\frac{3}{2}$ **6.** $2x + 5$

2.6 GETTING READY   (page 126)
1. $840   **2.** 385 mi   **3.** 3.2 hr   **4.** 9.5 lb

ORALS   (page 130)
1. $7d$ **2.** $18,000r$ **3.** $\dfrac{A}{6}$ ft   **4.** $\left(\dfrac{P}{2} - 9\right)$ ft

EXERCISE 2.6   (page 130)
1. $4500 at 9% and $19,500 at 8%   **3.** $3750 in each account   **5.** $5000   **7.** 6% and 7%   **9.** 3   **11.** 6.5 hr   **13.** 7.5
15. 500 mph   **17.** 20   **19.** 50   **21.** 7.5 oz   **23.** 40 lb lemon drops and 60 lb jelly beans   **25.** $1.20   **27.** 80

REVIEW EXERCISES   (page 132)
1. $1488   **2.** 82

2.7 GETTING READY   (page 133)
1.   **2.**   **3.**   **4.**

ORALS   (page 139)
1. $x < 2$ **2.** $x > 1$ **3.** $x > 2$ **4.** $x < -2$ **5.** $x < 6$ **6.** $x > -1$

EXERCISE 2.7   (page 139)
1.   **3.**   **5.**   **7.**   **9.**   **11.**
13.   **15.**   **17.**   **19.**   **21.**   **23.**
25.   **27.**   **29.**   **31.**   **33.**   **35.**
37.   **39.**   **41.**   **43.**   **45.**
47.   **49.**   **51.**   **53.**
55.   **57.**   **59.**   **61.** $s \geq 98\%$ **63.** $r \geq 27$ mpg
65. $0 \text{ ft} < s \leq 19 \text{ ft}$ **67.** $0.1 \text{ mi} \leq x \leq 2.5 \text{ mi}$ **69.** $3.3 \text{ mi} < x < 4.1 \text{ mi}$ **71.** $66.2° < F < 71.6°$ **73.** $37.052 \text{ in.} < C < 38.308 \text{ in.}$
75. $68.18 \text{ kg} < w < 86.36 \text{ kg}$ **77.** $5 \text{ ft} < w < 9 \text{ ft}$

SOMETHING TO THINK ABOUT (page 142)

2. $0 < x < 1$

REVIEW EXERCISES (page 142)

1. $5x^2 - 2y^2$ **2.** $2xy + 2$ **3.** $-x + 14$ **4.** $-xy + \frac{4}{5}x + \frac{9}{5}y$

CHAPTER 2 REVIEW EXERCISES (page 145)

1. solution **2.** not a solution **3.** not a solution **4.** solution **5.** solution **6.** not a solution **7.** 1 **8.** 0 **9.** 4 **10.** -2 **11.** 5 **12.** -2 **13.** $\frac{1}{2}$ **14.** $\frac{3}{2}$ **15.** 18 **16.** -35 **17.** $-\frac{1}{2}$ **18.** 6 **19.** 245 **20.** 1300 **21.** 37% **22.** 12.5% **23.** 3 **24.** 2 **25.** -1 **26.** 1 **27.** 1 **28.** -2 **29.** 2 **30.** 7 **31.** -2 **32.** -1 **33.** 5 **34.** 3 **35.** 13 **36.** -12 **37.** 5 **38.** 7 **39.** 15 **40.** 30 **41.** $\frac{15}{2}$ **42.** 44 **43.** \$320 **44.** 6.5% **45.** 96.4% **46.** 53.8% **47.** $14x$ **48.** $19a$ **49.** $5b$ **50.** $-2x$ **51.** $-2y$ **52.** not like terms **53.** $4y^2 - 6$ **54.** 4 **55.** $9x$ **56.** $6 - 7x$ **57.** 7 **58.** 13 **59.** -3 **60.** -41 **61.** 9 **62.** -7 **63.** 1 **64.** -2 **65.** 7 **66.** 4 **67.** -8 **68.** -18 **69.** 12 **70.** $\frac{4}{3}$ **71.** 3 **72.** 0 **73.** $\frac{9}{2}$ **74.** $-\frac{7}{2}$ **75.** $-\frac{1}{3}$ **76.** $\frac{7}{9}$ **77.** 6 **78.** -6 **79.** 35 **80.** -16 **81.** $R = \dfrac{E}{I}$ **82.** $t = \dfrac{i}{pr}$ **83.** $R = \dfrac{P}{I^2}$ **84.** $r = \dfrac{d}{t}$ **85.** $h = \dfrac{V}{lw}$ **86.** $m = \dfrac{y - b}{x}$ **87.** $h = \dfrac{V}{\pi r^2}$ **88.** $r = \dfrac{a}{2\pi h}$ **89.** $G = \dfrac{Fd^2}{Mm}$ **90.** $m = \dfrac{RT}{PV}$ **91.** $V = \dfrac{T}{n} + 3$ or $V = \dfrac{T + 3n}{n}$ **92.** $n = \dfrac{T}{V - 3}$ **93.** 8 **94.** 147 **95.** 85 **96.** 8 ft **97.** 13 in. **98.** 382 units **99.** 40 units on each machine; 80 units total **100.** \$16,000 at 7%, \$11,000 at 9% **101.** 20 **102.** 10 lb of each **103.** 1 **104.** -3 **105.** 4 **106.** 6 **107.** 3 **108.** 3 **109.** 6 11 **110.** -2 1 **111.** 0 9 **112.** 3 4

CHAPTER 2 TEST (page 148)

1. solution **2.** solution **3.** not a solution **4.** solution **5.** -36 **6.** -12 **7.** -7 **8.** -2 **9.** 1 **10.** -3 **11.** -4 **12.** -2 **13.** -2 **14.** 0 **15.** $6x - 15$ **16.** $8x - 10$ **17.** $3x^2 - 6x$ **18.** $3x^2 - 36x - 36$ **19.** $-18x$ **20.** $-36x^2 + 13x$ **21.** $t = \dfrac{d}{r}$ **22.** $l = \dfrac{P - 2w}{2}$ **23.** $h = \dfrac{A}{2\pi r}$ **24.** $r = \dfrac{A - P}{Pt}$ **25.** $v = \dfrac{RT}{P}$ **26.** $h = \dfrac{3A}{\pi r^2}$ **27.** 17 and 19 **28.** \$5250 **29.** $\frac{3}{5}$ hr **30.** $7\frac{1}{2}$ liters **31.** 3 **32.** -2 **33.** -3 4 **34.** 3/5 3

3.1 GETTING READY (page 151)

1. 8 **2.** 9 **3.** 6 **4.** 6 **5.** 12 **6.** 32 **7.** 18 **8.** 3

ORALS (page 157)

1. base x, exponent 3 **2.** base 3, exponent x **3.** base b, exponent c **4.** base ab, exponent c **5.** 36 **6.** 36 **7.** 9 **8.** 27

EXERCISE 3.1 (page 157)

1. base 4, exponent 3 **3.** base x, exponent 5 **5.** base $2y$, exponent 3 **7.** base x, exponent 4 **9.** base x, exponent 1 **11.** base x, exponent 3 **13.** $5 \cdot 5 \cdot 5$ **15.** $x \cdot x \cdot x \cdot x \cdot x \cdot x \cdot x$ **17.** $-4 \cdot x \cdot x \cdot x \cdot x \cdot x$ **19.** $(3t)(3t)(3t)(3t)(3t)$ **21.** 2^3 **23.** x^4 **25.** $(2x)^3$ **27.** $-4t^4$ **29.** 625 **31.** 13 **33.** 561 **35.** -725 **37.** x^7 **39.** x^{10} **41.** t^3 **43.** a^{12} **45.** y^9 **47.** $12x^7$ **49.** $-4y^5$ **51.** $6x^9$ **53.** 3^8 **55.** y^{15} **57.** a^{21} **59.** x^{25} **61.** $243z^{30}$ **63.** x^{31} **65.** r^{36} **67.** s^{33} **69.** x^3y^3 **71.** r^6s^4 **73.** $16a^2b^4$ **75.** $-8r^6s^9t^3$ **77.** $\dfrac{a^3}{b^3}$ **79.** $\dfrac{x^{10}}{y^{15}}$ **81.** $\dfrac{-32a^5}{b^5}$ **83.** $\dfrac{b^6}{27a^3}$ **85.** x^2 **87.** y^4 **89.** $3a$ **91.** ab^4 **93.** $\dfrac{10r^{13}s^3}{3}$ **95.** $\dfrac{x^{12}y^{16}}{2}$ **97.** $\dfrac{y^3}{8}$ **99.** $-\dfrac{8r^3}{27}$

REVIEW EXERCISES   (page 158)

1. $-3/2$

2.   **3.** the product of 3 and the sum of x and y **4.** the sum of y and the product of 3 and x **5.** $|2x| + 3$ **6.** $y + z - (y^2 + z^2)$

3.2 GETTING READY   (page 159)

1. $\frac{1}{3}$ **2.** $\frac{1}{y}$ **3.** 1   **4.** $\frac{1}{xy}$

ORALS   (page 162)

1. $\frac{1}{2}$ **2.** $\frac{1}{4}$ **3.** 2   **4.** 1   **5.** x **6.** $\frac{1}{y^7}$ **7.** 1   **8.** $\frac{y}{x}$

EXERCISE 3.2   (page 162)

1. 8   **3.** 1   **5.** 1   **7.** 512   **9.** 2   **11.** 1   **13.** 1   **15.** -2 **17.** $\frac{1}{x^2}$ **19.** $\frac{1}{b^5}$ **21.** $\frac{1}{16y^4}$ **23.** $\frac{1}{a^3b^6}$ **25.** $\frac{1}{y}$

27. $\frac{1}{r^6}$ **29.** y^5 **31.** 1   **33.** $\frac{1}{a^2b^4}$ **35.** $\frac{1}{x^6y^3}$ **37.** $\frac{1}{x^3}$ **39.** $\frac{1}{y^2}$ **41.** a^8b^{12} **43.** $-\frac{y^{10}}{32x^{15}}$ **45.** a^{14} **47.** $\frac{1}{b^{14}}$ **49.** $\frac{256x^{28}}{81}$

51. $\frac{16y^{14}}{z^{10}}$ **53.** $\frac{x^{14}}{128y^{28}}$ **55.** $\frac{16u^4v^8}{81}$ **57.** $\frac{1}{9a^2b^2}$ **59.** $\frac{c^{15}}{216a^9b^3}$ **61.** $\frac{1}{512}$ **63.** $\frac{17y^{27}z^5}{x^{35}}$ **65.** x^{3m} **67.** u^{5m} **69.** y^{2m+2}

71. y^m **73.** $\frac{1}{x^{3n}}$ **75.** x^{2m+2} **77.** x^{8n-12} **79.** y^{4n-8}

REVIEW EXERCISES   (page 163)

1. 2   **2.** 7   **3.** $\frac{6}{5}$ **4.** -1 **5.** $s = \frac{f(P-L)}{i}$ or $s = \frac{fP-fL}{i}$ **6.** $i = \frac{f(P-L)}{s}$ or $i = \frac{fP-fL}{s}$

3.3 GETTING READY   (page 164)

1. 100   **2.** 1000   **3.** 10   **4.** $\frac{1}{100}$ **5.** 500   **6.** 8000   **7.** 30   **8.** $\frac{7}{100}$

ORALS   (page 167)

1. 3.72×10^2 is larger   **2.** 37.2 is larger   **3.** 4.72×10^3 is larger   **4.** 3.72×10^3 is larger   **5.** 3.72×10^{-1} is larger   **6.** 2.72×10^{-2} is larger

EXERCISE 3.3   (page 168)

1. 2.3×10^4 **3.** 1.7×10^6 **5.** 6.2×10^{-2} **7.** 5.1×10^{-6} **9.** 4.25×10^3 **11.** 2.5×10^{-3} **13.** 230   **15.** 812,000
17. 0.00115   **19.** 0.000976   **21.** 25,000,000   **23.** 0.00051   **25.** 2.57×10^{13} mi.   **27.** 114,000,000 mi   **29.** 6.22×10^{-3} mi
31. 714,000   **33.** 30,000   **35.** 200,000   **37.** 1.44×10^{11} **39.** 1.9008×10^{11} ft   **41.** 3.3×10^{-1} km/sec

REVIEW EXERCISES   (page 169)

1. 5   **2.** 3   **3.** commut. prop. of $+$ **4.** distrib. prop.   **5.** 6   **6.** 4

3.4 GETTING READY   (page 169)

1. $2x^2y^3$ **2.** $3xy^3$ **3.** $2x^2 + 3y^2$ **4.** $x^3 + y^3$ **5.** $6x^3y^3$ **6.** $5x^2y^2z^4$ **7.** $5x^2y^2$ **8.** $x^3y^3z^3$

EXERCISE 3.4 (page 173)

1. binomial   **3.** trinomial   **5.** monomial   **7.** binomial   **9.** trinomial   **11.** none of these   **13.** 4th   **15.** 3rd   **17.** 8th   **19.** 6th
21. 12th   **23.** 0th   **25.** 7   **27.** -8 **29.** $5w - 3$ **31.** $-5y - 3$ **33.** -4 **35.** -5 **37.** $-r^2 - 4$ **39.** $-9s^2 - 4$ **41.** 3
43. 11   **45.** $b^2 + 2b + 3$ **47.** $\frac{1}{16}w^2 + \frac{1}{2}w + 3$ **49.** -1 **51.** $5u^2 - 2$ **53.** $-20z^6 - 2$ **55.** $5x^2y^2 - 2$ **57.** $5x + 5h - 2$
59. $5x + 5h - 4$ **61.** $10y + 5z - 2$ **63.** $10y + 5z - 4$ **65.** 16 ft   **67.** 160 ft

REVIEW EXERCISES (page 174)

1. 8   **2.** 4   **3.**   **4.**   **5.** x^{18} **6.** a^{12} **7.** y^9 **8.** $\dfrac{1}{16t^8}$

3.5 GETTING READY (page 175)

1. $5x$ **2.** $2y$ **3.** $25x$ **4.** $5z$ **5.** $12r$ **6.** impossible   **7.** 0   **8.** impossible

ORALS (page 178)

1. $4x^3$ **2.** $4xy$ **3.** $2y$ **4.** $2 - 2x$ **5.** $-2y^2$ **6.** $6x^2 + 4y$ **7.** $4x^2 + y$ **8.** $2y$

EXERCISE 3.5 (page 178)

1. like terms, $7y$ **3.** unlike terms   **5.** like terms, $13x^3$ **7.** like terms, $8x^3y^2$ **9.** like terms, $65t^6$ **11.** unlike terms   **13.** $9y$
15. $-12t^2$ **17.** $16u^3$ **19.** $7x^5y^2$ **21.** $14rst$ **23.** $-6a^2bc$ **25.** $15x^2$ **27.** $4x^2y^2$ **29.** $95x^8y^4$ **31.** $7x + 4$ **33.** $2a + 7$
35. $7x - 7y$ **37.** $-19x - 4y$ **39.** $6x^2 + x - 5$ **41.** $7b + 4$ **43.** $3x + 1$ **45.** $5x + 15$ **47.** $3x - 3y$ **49.** $5x^2 - 25x - 20$
51. $5x^2 + x + 11$ **53.** $-7x^3 - 7x^2 - x - 1$ **55.** $2x^2y + xy + 13y^2$ **57.** $5x^2 + 6x - 8$ **59.** $-x^3 + 6x^2 + x + 14$
61. $-12x^2y^2 - 13xy + 36y^2$ **63.** $6x^2 - 2x - 1$ **65.** $t^3 + 3t^2 + 6t - 5$ **67.** $-3x^2 + 5x - 7$ **69.** $6x - 2$ **71.** $-5x^2 - 8x - 19$
73. $4y^3 - 12y^2 + 8y + 8$ **75.** $3a^2b^2 - 6ab + b^2 - 6ab^2$ **77.** $-6x^2y^2 + 4xy^2z - 20xy^3 + 2y$ **79.** $6x + 3h - 10$

REVIEW EXERCISES (page 180)

1. -8 **2.** 8   **3.** -9 **4.** 2   **5.**   **6.** $m = \dfrac{2K}{v^2}$

3.6 GETTING READY (page 181)

1. $6x$ **2.** $3x^4$ **3.** $5x^3$ **4.** $8x^5$ **5.** $3x + 15$ **6.** $x^2 + 5x$ **7.** $4y - 12$ **8.** $2y^2 - 6y$

ORALS (page 186)

1. $6x^3 - 2x^2$ **2.** $10y^3 - 15y$ **3.** $7x^2y + 7xy^2$ **4.** $-4xy + 6y^2$ **5.** $x^2 + 5x + 6$ **6.** $x^2 - x - 6$ **7.** $2x^2 + 7x + 6$ **8.** $9x^2 - 1$
9. $x^2 + 6x + 9$ **10.** $x^2 - 10x + 25$

EXERCISE 3.6 (page 187)

1. $12x^5$ **3.** $-24b^6$ **5.** $6x^5y^5$ **7.** $-3x^4y^7z^8$ **9.** $x^{10}y^{15}$ **11.** $a^5b^4c^7$ **13.** $3x + 12$ **15.** $-4t - 28$ **17.** $3x^2 - 6x$
19. $-6x^4 + 2x^3$ **21.** $3x^2y + 3xy^2$ **23.** $6x^4 + 8x^3 - 14x^2$ **25.** $2x^7 - x^2$ **27.** $-6r^3t^2 + 2r^2t^3$ **29.** $-6x^4y^4 - 6x^3y^5$
31. $a^2 + 9a + 20$ **33.** $3x^2 + 10x - 8$ **35.** $6a^2 + 2a - 20$ **37.** $6x^2 - 7x - 5$ **39.** $2x^2 + 3x - 9$ **41.** $6t^2 + 7st - 3s^2$
43. $x^2 + xz + xy + yz$ **45.** $u^2 + 2tu + uv + 2tv$ **47.** $-4r^2 - 20rs - 21s^2$ **49.** $-12t^2 + 7tu - u^2$ **51.** $4x^2 + 11x + 6$
53. $12x^2 + 14xy - 10y^2$ **55.** $x^3 - 1$ **57.** $x^2 + 8x + 16$ **59.** $t^2 - 6t + 9$ **61.** $r^2 - 16$ **63.** $x^2 + 10x + 25$ **65.** $4s^2 + 4s + 1$
67. $16x^2 - 25$ **69.** $x^2 - 4xy + 4y^2$ **71.** $4a^2 - 12ab + 9b^2$ **73.** $16x^2 - 25y^2$ **75.** $2x^2 - 6x - 8$ **77.** $3a^3 - 3ab^2$
79. $4t^3 + 11t^2 + 18t + 9$ **81.** $-3x^3 + 25x^2y - 56xy^2 + 16y^3$ **83.** $x^3 - 8y^3$ **85.** $5t^2 - 11t$ **87.** $x^2y + 3xy^2 + 2x^2$
89. $2x^2 + xy - y^2$ **91.** $8x$ **93.** $5s^2 - 7s - 9$ **95.** -3 **97.** -8 **99.** -1 **101.** 0   **103.** 1   **105.** 5 and 6   **107.** 90 ft
109. $\frac{3}{2}$ in.

REVIEW EXERCISES   (page 189)

1. distrib. prop.   **2.** assoc. prop. of $+$ **3.** commut. prop. of $\times$ **4.** additive identity prop.   **5.** 0   **6.** $m = \dfrac{Fd^2}{GM}$

3.7 GETTING READY   (page 190)

1. $2xy^2$ **2.** y **3.** $\dfrac{3xy}{2}$ **4.** $\dfrac{x}{y}$ **5.** xy **6.** 3

ORALS   (page 192)

1. $2x^2$ **2.** $2y$ **3.** $5bc^2$ **4.** $-2pq$ **5.** 1   **6.** $3x$

EXERCISE 3.7   (page 192)

1. $\dfrac{1}{3}$ **3.** $-\dfrac{5}{3}$ **5.** $\dfrac{3}{4}$ **7.** 1   **9.** $-\dfrac{1}{4}$ **11.** $\dfrac{42}{19}$ **13.** $\dfrac{x}{z}$ **15.** $\dfrac{r^2}{s}$ **17.** $\dfrac{2x^2}{y}$ **19.** $-\dfrac{3u^3}{v^2}$ **21.** $\dfrac{4r}{y^2}$ **23.** $-\dfrac{13}{3rs}$ **25.** $\dfrac{x^4}{y^6}$ **27.** $a^8 b^8$

29. $-\dfrac{3r}{s^9}$ **31.** $-\dfrac{x^3}{4y^3}$ **33.** $\dfrac{125}{8b^3}$ **35.** $\dfrac{xy^2}{3}$ **37.** a^8 **39.** z^3 **41.** $\dfrac{2}{y} + \dfrac{3}{x}$ **43.** $\dfrac{1}{5y} - \dfrac{2}{5x}$ **45.** $\dfrac{1}{y^2} + \dfrac{2y}{x^2}$ **47.** $3a - 2b$

49. $\dfrac{1}{y} - \dfrac{1}{2x} + \dfrac{2z}{xy}$ **51.** $3x^2 y - 2x - \dfrac{1}{y}$ **53.** $5x - 6y + 1$ **55.** $\dfrac{10x^2}{y} - 5x$ **57.** $-\dfrac{4x}{3} + \dfrac{3x^2}{2}$ **59.** $xy - 1$ **61.** $\dfrac{x}{y} - \dfrac{11}{6} + \dfrac{y}{2x}$

63. 2

REVIEW EXERCISES   (page 194)

1. 52   **2.** 10   **3.** 2.65×10^{-4} **4.** 5670   **5.** 1   **6.** $a^{18} b^{24}$

3.8 GETTING READY   (page 194)

1. 13   **2.** 21   **3.** 19   **4.** 13

ORALS   (page 198)

1. $2 + \dfrac{3}{x}$ **2.** $3 - \dfrac{5}{x}$ **3.** $2 + \dfrac{1}{x+1}$ **4.** $3 + \dfrac{2}{x+1}$ **5.** x **6.** x

EXERCISE 3.8   (page 198)

1. $x + 2$ **3.** $y + 12$ **5.** $a + b$ **7.** $3a - 2$ **9.** $b + 3$ **11.** $x - 3y$ **13.** $2x + 1$ **15.** $x - 7$ **17.** $3x + 2y$ **19.** $2x - y$

21. $x + 5y$ **23.** $x - 5y$ **25.** $x^2 + 2x - 1$ **27.** $2x^2 + 2x + 1$ **29.** $x^2 + xy + y^2$ **31.** $x + 1 + \dfrac{-1}{2x + 3}$ **33.** $2x + 2 + \dfrac{-3}{2x + 1}$

35. $x^2 + 2x + 1$ **37.** $x^2 + 2x - 1 + \dfrac{6}{2x + 3}$ **39.** $2x^2 + 8x + 14 + \dfrac{31}{x - 2}$ **41.** $x + 1$ **43.** $2x - 3$ **45.** $x^2 - x + 1$

47. $a^2 - 3a + 10 + \dfrac{-30}{a + 3}$ **49.** $5x^2 - x + 4 + \dfrac{16}{3x - 4}$

REVIEW EXERCISES   (page 200)

1. 20, 21, 22, 24, 25, 26, 27, 28, 30   **2.**   **3.** 5   **4.** 1   **5.** -5 **6.** -5

7. $8x^2 - 6x + 1$ **8.** $-11y^3 - 4y^2 - y$

CHAPTER 3 REVIEW EXERCISES (page 202)

1. 125 **2.** 243 **3.** 64 **4.** -64 **5.** 13 **6.** 25 **7.** 162 **8.** 18 **9.** x^5 **10.** x^3y **11.** y^{10} **12.** y^5 **13.** $2b^{12}$ **14.** $-y^2z^5$
15. $256s^3$ **16.** $-3y^6$ **17.** x^{15} **18.** $4x^4y^2$ **19.** 9 **20.** 1 **21.** x^4 **22.** $\dfrac{x^2}{y^2}$ **23.** $\dfrac{y^2}{x^2}$ **24.** $5yz^4$ **25.** x **26.** y **27.** x^{10} **28.** x^{14}
29. $\dfrac{1}{x^4}$ **30.** $\dfrac{1}{x^5}$ **31.** $\dfrac{1}{8s^3}$ **32.** $\dfrac{1}{9z^2}$ **33.** 7.28×10^2 **34.** 9.37×10^3 **35.** 1.36×10^{-2} **36.** 9.42×10^{-3} **37.** 7.73×10^0
38. 7.53×10^5 **39.** 1.8×10^{-4} **40.** 6×10^4 **41.** 726,000 **42.** 0.000391 **43.** 2.68 **44.** 57.6 **45.** 7.39 **46.** 0.000437
47. 0.03 **48.** 160 **49.** 7th, monomial **50.** 2nd, binomial **51.** 5th, trinomial **52.** 3rd, binomial **53.** 11 **54.** 2 **55.** -4
56. $6t + 2$ **57.** 402 **58.** 0 **59.** 82 **60.** $80t^4 - 2t$ **61.** $7x$ **62.** simplest terms **63.** $4x^2y^2$ **64.** x^2yz **65.** $5x^2$ **66.** $5x + 35$
67. $8x^2 - 6x$ **68.** $4a^2 + 4a - 6$ **69.** $9x^2 + 3x + 9$ **70.** $6x^3 + 8x^2 + 3x - 72$ **71.** $10x^3y^5$ **72.** x^7yz^5 **73.** $5x + 15$
74. $6x + 12$ **75.** $3x^4 - 5x^2$ **76.** $2y^4 + 10y^3$ **77.** $-x^2y^3 + x^3y^2$ **78.** $-3x^2y^2 + 3x^2y$ **79.** $x^2 + 5x + 6$ **80.** $2x^2 - x - 1$
81. $6a^2 - 6$ **82.** $6a^2 - 6$ **83.** $2a^2 - ab - b^2$ **84.** $6x^2 + xy - y^2$ **85.** $-9a^2 + b^2$ **86.** $x^2 - 25$ **87.** $y^2 - 4$
88. $x^2 + 8x + 16$ **89.** $x^2 - 6x + 9$ **90.** $y^3 + 2y^2 + y$ **91.** $4y^2 + 4y + 1$ **92.** $y^4 - 1$ **93.** $3x^3 + 7x^2 + 5x + 1$ **94.** $8a^3 - 27$
95. 1 **96.** -1 **97.** 7 **98.** 5 **99.** 1 **100.** 0 **101.** $\frac{3}{2y} + \frac{3}{x}$ **102.** $2 - \frac{3}{y}$ **103.** $-3a - 4b + 5c$ **104.** $-\frac{x}{y} - \frac{y}{x}$
105. $x + 1 + \frac{3}{x+2}$ **106.** $x - 5$ **107.** $2x + 1$ **108.** $x + 5 + \frac{3}{3x-1}$ **109.** $3x^2 + 2x + 1 + \frac{2}{2x-1}$ **110.** $3x^2 - x - 4$

CHAPTER 3 TEST (page 204)

1. $2x^3y^4$ **2.** 134 **3.** y^6 **4.** $6b^7$ **5.** $32x^{21}$ **6.** $8r^{18}$ **7.** 3 **8.** $\dfrac{2}{y^3}$ **9.** y^3 **10.** $\dfrac{64a^3}{b^3}$ **11.** 2.8×10^4 **12.** 2.5×10^{-3}
13. 7400 **14.** 0.000093 **15.** binomial **16.** 10th degree **17.** 0 **18.** $-3x^2y^2$ **19.** $-7x + 2y$ **20.** $-3x + 6$
21. $5x^3 + 2x^2 + 2x - 5$ **22.** $-x^2 - 5x + 4$ **23.** $-4x^5y$ **24.** $3y^4 - 6y^3 + 9y^2$ **25.** $6x^2 - 7x - 20$ **26.** $2x^3 - 7x^2 + 14x - 12$
27. $\frac{1}{2}$ **28.** $\frac{y}{2x}$ **29.** $\frac{a}{4b} - \frac{b}{2a}$ **30.** $x - 2$

CUMULATIVE REVIEW EXERCISES (page 205)

1. integer, rational, real, positive **2.** rational, real, negative **3.** **4.** **5.** 0
6. $\frac{10}{3}$ **7.** $8\frac{1}{10}$ **8.** 35.65 **9.** 0 **10.** -2 **11.** 16 **12.** 0 **13.** 24.75 **14.** 5275 **15.** 5 **16.** 37, y **17.** $-2x + 2y$ **18.** $x - 5$
19. x^2y^3 **20.** $4x^2$ **21.** 13 **22.** 41 **23.** $\frac{7}{4}$ **24.** 2 **25.** $h = \frac{2A}{b+B}$ **26.** $x = \frac{y-b}{m}$ **27.** -9 **28.** 1 **29.** 280 **30.** 2
31. **32.** **33.** y^{14} **34.** xy **35.** $\dfrac{a^7}{b^6}$ **36.** x^2y^2 **37.** $x^2 + 4x - 14$ **38.** $6x^2 + 10x - 56$
39. $x^3 - 8$ **40.** $2x + 1$ **41.** 4.8×10^{18} m **42.** 4 in. **43.** 879.6 in.^{2} **44.** \$512

4.1 GETTING READY (page 210)

1. $5x + 15$ **2.** $7y - 56$ **3.** $3x^2 - 2x$ **4.** $5y^2 + 9y$ **5.** $ab + 9a$ **6.** $3x + x^2 + xy$ **7.** $x^2y - 4xy$ **8.** $2x^2y^2 - 5xy^3$

ORALS (page 214)

1. 2^23^2 **2.** 3^3 **3.** 3^4 **4.** 3^25 **5.** 3 **6.** $3ab$ **5.** $5(3xy + 2)$ **6.** $5xy(3 + 2y)$

EXERCISE 4.1 (page 215)

1. $2^2 \cdot 3$ **3.** $3 \cdot 5$ **5.** $2^3 \cdot 5$ **7.** $2 \cdot 7^2$ **9.** $3^2 \cdot 5^2$ **11.** $2^5 \cdot 3^2$ **13.** $3(x + 2)$ **15.** $x(y - z)$ **17.** $t^2(t + 2)$ **19.** $r^2(r^2 - 1)$
21. $a^2b^3z^2(az - 1)$ **23.** $8xy^2z^3(3xyz + 1)$ **25.** $6uvw^2(2w - 3v)$ **27.** $3(x + y - 2z)$ **29.** $a(b + c - d)$ **31.** $2y(2y + 4 - x)$
33. $3r(4r - s + 3rs^2)$ **35.** $abx(1 - b + x)$ **37.** $2xyz^2(2xy - 3y + 6)$ **39.** $7a^2b^2c^2(10a + 7bc - 3)$ **41.** $-(a + b)$
43. $-(2x - 5y)$ **45.** $-(2a - 3b)$ **47.** $-(3m + 4n - 1)$ **49.** $-(3xy - 2z - 5w)$ **51.** $-(3ab + 5ac - 9bc)$ **53.** $-3xy(x + 2y)$
55. $-4a^2b^2(b - 3a)$ **57.** $-2ab^2c(2ac - 7a + 5c)$ **59.** $-7ab(2a^5b^5 - 7ab^2 + 3)$ **61.** $-5a^2b^3c(1 - 3abc + 5a^2)$ **63.** 2, -3
65. 4, -1 **67.** $\frac{5}{2}$, -2 **69.** 1, -2, 3 **71.** 0, 3 **73.** 0, $\frac{5}{2}$ **75.** 0, 7 **77.** 0, $-\frac{8}{3}$ **79.** 0, 2 **81.** 0, $\frac{1}{5}$

REVIEW EXERCISES   (page 216)

1. 7   **2.** 1   **3.** 11   **4.** 0

4.2 GETTING READY   (page 217)

1. $3x + 3y + ax + ay$ **2.** $xy + x + 5y + 5$ **3.** $5x + 5 - yx - y$ **4.** $x^2 + 2x - yx - 2y$ **5.** $3x^2 + 2xy - y^2$
6. $5y - 35 - y^2 + 7y$

ORALS   (page 219)

1. $x + 3$ **2.** $a - 1$ **3.** $x - 2$ **4.** $y + 5$ **5.** $x - 7$ **6.** $2y + 9$

EXERCISE 4.2   (page 219)

1. $(x + y)(2 + b)$ **3.** $(x + y)(3 - a)$ **5.** $(r - 2s)(3 - x)$ **7.** $(x - 3)(x - 2)$ **9.** $2(a^2 + b)(x + y)$ **11.** $3(r + 3s)(x^2 - 2y^2)$
13. $(a + b + c)(3x - 2y)$ **15.** $7xy(r + 2s - t)(2x - 3)$ **17.** $(x + 1)(x + 3 - y)$ **19.** $(x^2 - 2)(3x - y + 1)$ **21.** $(x + y)(2 + a)$
23. $(r + s)(7 - k)$ **25.** $(r + s)(x + y)$ **27.** $(2x + 3)(a + b)$ **29.** $(b + c)(2a + 3)$ **31.** $(x + y)(2x - 3)$ **33.** $(v - 3w)(3t + u)$
35. $(3p + q)(3m - n)$ **37.** $(m - n)(p - 1)$ **39.** $(a - b)(x - y)$ **41.** $x^2(a + b)(x + 2y)$ **43.** $4a(b + 3)(a - 2)$
45. $(x^2 + 1)(x + 2)$ **47.** $y(x^2 - y)(x - 1)$ **49.** $(x + 2)(x + y + 1)$ **51.** $(m - n)(a + b + c)$ **53.** $(d + 3)(a - b - c)$
55. $(a + b + c)(x^2 - y)$ **57.** $(r - s)(2 + b)$ **59.** $(x + y)(a + b)$ **61.** $(a - b)(c - d)$ **63.** $r(r + s)(a - b)$ **65.** $(b + 1)(a + 3)$
67. $(r - s)(p - q)$

REVIEW EXERCISES   (page 221)

1. u^9 **2.** $\dfrac{1}{y^2}$ **3.** $\dfrac{a}{b}$ **4.** 1

4.3 GETTING READY   (page 221)

1. $a^2 - b^2$ **2.** $4r^2 - s^2$ **3.** $9x^2 - 4y^2$ **4.** $16x^4 - 9$

ORALS   (page 225)

1. $(x + 3)(x - 3)$ **2.** $(y + 6)(y - 6)$ **3.** $(z + 2)(z - 2)$ **4.** $(p + q)(p - q)$ **5.** $(5 + t)(5 - t)$ **6.** $(6 + r)(6 - r)$
7. $(10 + y)(10 - y)$ **8.** $(10 + y^2)(10 - y^2)$

EXERCISE 4.3   (page 225)

1. $(x + 4)(x - 4)$ **3.** $(y + 7)(y - 7)$ **5.** $(2y + 7)(2y - 7)$ **7.** $(3x + y)(3x - y)$ **9.** $(5t + 6u)(5t - 6u)$ **11.** $(4a + 5b)(4a - 5b)$
13. prime   **15.** $(a^2 + 2b)(a^2 - 2b)$ **17.** $(7y + 15z^2)(7y - 15z^2)$ **19.** $(14x^2 + 13y)(14x^2 - 13y)$ **21.** $8(x + 2y)(x - 2y)$
23. $2(a + 2y)(a - 2y)$ **25.** $3(r + 2s)(r - 2s)$ **27.** $x(x + y)(x - y)$ **29.** $x(2a + 3b)(2a - 3b)$ **31.** $3m(m + n)(m - n)$
33. $x^2(2x + y)(2x - y)$ **35.** $2ab(a + 11b)(a - 11b)$ **37.** $(x^2 + 9)(x + 3)(x - 3)$ **39.** $(a^2 + 4)(a + 2)(a - 2)$
41. $(a^2 + b^2)(a + b)(a - b)$ **43.** $(9r^2 + 16s^2)(3r + 4s)(3r - 4s)$ **45.** $(a^2 + b^4)(a + b^2)(a - b^2)$
47. $(x^4 + y^4)(x^2 + y^2)(x + y)(x - y)$ **49.** $2(x^2 + y^2)(x + y)(x - y)$ **51.** $b(a^2 + b^2)(a + b)(a - b)$
53. $3n(4m^2 + 9n^2)(2m + 3n)(2m - 3n)$ **55.** $3ay(a^4 + 2y^4)$ **57.** $3a^2(a^4 + b^2)(a^2 + b)(a^2 - b)$
59. $2y^2(x^4 + 4y^2)(x^2 + 2y)(x^2 - 2y)$ **61.** $a^2b^2(a^2 + b^2c^2)(a + bc)(a - bc)$ **63.** $a^2b^3(b^2 + 25)(b + 5)(b - 5)$
65. $3rs(9r^2 + 4s^2)(3r + 2s)(3r - 2s)$ **67.** $(4x - 4y + 3)(4x - 4y - 3)$ **69.** $(a + 3)(a + 3)(a - 3)$ **71.** $(y + 4)(y - 4)(y - 3)$
73. $3(x + 2)(x - 2)(x + 1)$ **75.** $3(m + n)(m - n)(m + a)$ **77.** $2(m + 4)(m - 4)(mn^2 + 4)$ **79.** $5, -5$
81. $7, -7$ **83.** $\frac{1}{2}, -\frac{1}{2}$ **85.** $\frac{2}{3}, -\frac{2}{3}$ **87.** $7, -7$ **89.** $\frac{9}{2}, -\frac{9}{2}$

REVIEW EXERCISES   (page 226)

1. $p = w\left(k - h - \dfrac{v^2}{2g}\right)$ **2.** $h = k - \dfrac{p}{w} - \dfrac{v^2}{2g}$

4.4 GETTING READY   (page 227)

1. $x^2 + 12x + 36$ **2.** $y^2 - 14y + 49$ **3.** $a^2 - 6a + 9$ **4.** $x^2 + 9x + 20$ **5.** $r^2 - 7r + 10$ **6.** $m^2 - 4m - 21$
7. $a^2 + ab - 12b^2$ **8.** $u^2 - 8uv + 15v^2$ **9.** $x^2 - 2xy - 24y^2$ **10.** $2a^2 - ab - b^2$

ORALS   (page 233)

1. 4 **2.** $-, -$ **3.** $-, 3$ **4.** $-, 2$ **5.** 6, 1 **6.** 6, 1 or 1, 6

EXERCISE 4.4   (page 233)

1. $(x + 3)(x + 3)$ **3.** $(y - 4)(y - 4)$ **5.** $(t + 10)(t + 10)$ **7.** $(u - 9)(u - 9)$ **9.** $(x + 2y)(x + 2y)$ **11.** $(r - 5s)(r - 5s)$
13. $(x + 2)(x + 1)$ **15.** $(a - 5)(a + 1)$ **17.** $(z + 11)(z + 1)$ **19.** $(t - 7)(t - 2)$ **21.** prime **23.** $(y - 6)(y + 5)$
25. $(a + 8)(a - 2)$ **27.** $(t - 10)(t + 5)$ **29.** prime **31.** $(y + z)(y + z)$ **33.** $(x + 2y)(x + 2y)$ **35.** $(m + 5n)(m - 2n)$
37. $(a - 6b)(a + 2b)$ **39.** $(u + 5v)(u - 3v)$ **41.** $-(x + 5)(x + 2)$ **43.** $-(y + 5)(y - 3)$ **45.** $-(t + 17)(t - 2)$
47. $-(r - 10)(r - 4)$ **49.** $-(a + 3b)(a + b)$ **51.** $-(x - 7y)(x + y)$ **53.** $(x - 4)(x - 1)$ **55.** $(y + 9)(y + 1)$
57. $(c + 5)(c - 1)$ **59.** $-(r - 2s)(r + s)$ **61.** $(r + 3x)(r + x)$ **63.** $(a - 2b)(a - b)$ **65.** $2(x + 3)(x + 2)$ **67.** $3y(y + 1)(y + 1)$
69. $-5(a - 3)(a - 2)$ **71.** $3(z - 4t)(z - t)$ **73.** $4y(x + 6)(x - 3)$ **75.** $-4x(x + 3y)(x - 2y)$ **77.** $(x + 2)(ax + 2a + b)$
79. $(a + 5)(a + 3 + b)$ **81.** $(a + b + 2)(a + b - 2)$ **83.** $(b + y + 2)(b - y - 2)$ **85.** 12, 1 **87.** 5, -3 **89.** $-3, 7$ **91.** 8, 1
93. $-3, -5$ **95.** $-4, 2$ **97.** 0, $-1, -2$ **99.** 0, 9, -3 **101.** 1, $-2, -3$

REVIEW EXERCISES   (page 235)

1.
8

2.
-1

3.
-3

4.
5

5.
17

6.
$-33/2$

7.
-2 4

8.
-6 -1

4.5 GETTING READY   (page 235)

1. $6x^2 + 7x + 2$ **2.** $6y^2 - 19y + 10$ **3.** $8t^2 + 6t - 9$ **4.** $4r^2 + 4r - 15$ **5.** $6m^2 - 13m + 6$ **6.** $16a^2 + 16a + 3$

ORALS   (page 242)

1. 2, 3 **2.** 2, 3 or 3, 2 **3.** $+, -$ **4.** $+, -$ **5.** 3, 1 **6.** 3, 1

EXERCISE 4.5   (page 243)

1. $(2x - 1)(x - 1)$ **3.** $(3a + 1)(a + 4)$ **5.** $(z + 3)(4z + 1)$ **7.** $(3y + 2)(2y + 1)$ **9.** $(3x - 2)(2x - 1)$ **11.** $(3a + 2)(a - 2)$
13. $(2x + 1)(x - 2)$ **15.** $(2m - 3)(m + 4)$ **17.** $(5y + 1)(2y - 1)$ **19.** $(3y - 2)(4y + 1)$ **21.** $(5t + 3)(t + 2)$
23. $(8m - 3)(2m - 1)$ **25.** $(3x - y)(x - y)$ **27.** $(2u + 3v)(u - v)$ **29.** $(2a - b)(2a - b)$ **31.** $(3r + 2s)(2r - s)$
33. $(2x + 3y)(2x + y)$ **35.** $(4a - 3b)(a - 3b)$ **37.** $(3x + 2)(x - 5)$ **39.** $(2a - 5)(4a - 3)$ **41.** $(4y - 3)(3y - 4)$ **43.** prime
45. $(2a + 3b)(a + b)$ **47.** $(3p - q)(2p + q)$ **49.** prime **51.** $(4x - 5y)(3x - 2y)$ **53.** $2(2x - 1)(x + 3)$ **55.** $y(y + 12)(y + 1)$
57. $3x(2x + 1)(x - 3)$ **59.** $3r^3(5r - 2)(2r + 5)$ **61.** $4(a - 2b)(a + b)$ **63.** $4(2x + y)(x - 2y)$ **65.** $-2mn(4m + 3n)(2m + n)$
67. $-2uv^3(7u - 3v)(2u - v)$ **69.** $(2x + y + 4)(2x + y - 4)$ **71.** $(3 + a + 2b)(3 - a - 2b)$
73. $(2x + y + a + b)(2x + y - a - b)$ **75.** $2z(x - y + 3z)(x - y - 3z)$ **77.** $(x + 5)(x + 4)$ **79.** $(2r + 5)(r + 2)$
81. $(3x - 5)(2x + 1)$ **83.** $(3t + 4)(4t - 1)$ **85.** $\frac{1}{2}, 2$ **87.** $\frac{1}{5}, 1$ **89.** $-\frac{1}{3}, 3$ **91.** $\frac{2}{3}, -\frac{1}{5}$ **93.** $-\frac{3}{2}, \frac{2}{3}$ **95.** $\frac{1}{8}, 1$ **97.** 0, $-3, -\frac{1}{3}$
99. 0, $-3, -3$

REVIEW EXERCISES   (page 245)

1. $n = \frac{l - f + d}{d}$ **2.** $f = \frac{2S}{n} - l$

4.6 GETTING READY   (page 245)

1. $x^3 - 27$ **2.** $x^3 + 8$ **3.** $y^3 + 64$ **4.** $r^3 - 125$ **5.** $a^3 - b^3$ **6.** $a^3 + b^3$

ORALS (page 247)

1. $(x - y)(x^2 + xy + y^2)$ **2.** $(x + y)(x^2 - xy + y^2)$ **3.** $(a + 2)(a^2 - 2a + 4)$ **4.** $(b - 3)(b^2 + 3b + 9)$
5. $(1 + 2x)(1 - 2x + 4x^2)$ **6.** $(2 - r)(4 + 2r + r^2)$ **7.** $(xy + 1)(x^2y^2 - xy + 1)$ **8.** $(5 - 2t)(25 + 10t + 4t^2)$

EXERCISE 4.6 (page 248)

1. $(y + 1)(y^2 - y + 1)$ **3.** $(a - 3)(a^2 + 3a + 9)$ **5.** $(2 + x)(4 - 2x + x^2)$ **7.** $(s - t)(s^2 + st + t^2)$ **9.** $(3x + y)(9x^2 - 3xy + y^2)$
11. $(a + 2b)(a^2 - 2ab + 4b^2)$ **13.** $(4x - 3)(16x^2 + 12x + 9)$ **15.** $(3x - 5y)(9x^2 + 15xy + 25y^2)$ **17.** $(a^2 - b)(a^4 + a^2b + b^2)$
19. $(x^3 + y^2)(x^6 - x^3y^2 + y^4)$ **21.** $2(x + 3)(x^2 - 3x + 9)$ **23.** $-(x - 6)(x^2 + 6x + 36)$ **25.** $8x(2m - n)(4m^2 + 2mn + n^2)$
27. $xy(x + 6y)(x^2 - 6xy + 36y^2)$ **29.** $3rs^2(3r - 2s)(9r^2 + 6rs + 4s^2)$ **31.** $a^3b^2(5a + 4b)(25a^2 - 20ab + 16b^2)$
33. $yz(y^2 - z)(y^4 + y^2z + z^2)$ **35.** $2mp(p + 2q)(p^2 - 2pq + 4q^2)$ **37.** $(x + 1)(x^2 - x + 1)(x - 1)(x^2 + x + 1)$
39. $(x^2 + y)(x^4 - x^2y + y^2)(x^2 - y)(x^4 + x^2y + y^2)$ **41.** $(x + y)(x^2 - xy + y^2)(3 - z)$ **43.** $(m + 2n)(m^2 - 2mn + 4n^2)(1 + x)$
45. $(a + 3)(a^2 - 3a + 9)(a - b)$ **47.** $(x + 2)(x - 2)(y + z)$ **49.** $(r + s)(r - s)(x - a)$ **51.** $(x - 1)(x + 1)$
53. $(y + 1)(y - 1)(y - 3)(y^2 + 3y + 9)$

REVIEW EXERCISES (page 249)

1. 0.0000000000001 cm   **2.** 1×10^8

4.7 GETTING READY (page 249)

1. $3ax(x + a)$ **2.** $(x + 3y)(x - 3y)$ **3.** $(x - 2)(x^2 + 2x + 4)$ **4.** $2(x + 2)(x - 2)$ **5.** $(x - 5)(x + 2)$ **6.** $(2x - 3)(3x - 2)$
7. $2(3x - 1)(x - 2)$ **8.** $(a + b)(x + y)(x - y)$

ORALS (page 251)

1. common factor   **2.** difference of squares   **3.** sum of cubes   **4.** grouping   **5.** prime   **6.** common factor   **7.** difference of
squares   **8.** difference of cubes

EXERCISE 4.7 (page 251)

1. $3(2x + 1)$ **3.** $(x - 7)(x + 1)$ **5.** $(3t - 1)(2t + 3)$ **7.** $(2x + 5)(2x - 5)$ **9.** $(t - 1)(t - 1)$ **11.** $(a - 2)(a^2 + 2a + 4)$
13. $(y^2 - 2)(x + 1)(x - 1)$ **15.** $7p^4q^2(10q - 5 + 7p)$ **17.** $2a(b + 6)(b - 2)$ **19.** $-4p^2q^3(2pq^4 + 1)$
21. $(2a - b + 3)(2a - b - 3)$ **23.** prime   **25.** $-2x^2(x - 4)(x^2 + 4x + 16)$ **27.** $2t^2(3t - 5)(t + 4)$ **29.** $(x - a)(a + b)(a - b)$
31. $(2p^2 - 3q^2)(4p^4 + 6p^2q^2 + 9q^4)$ **33.** $(5p - 4y)(25p^2 + 20py + 16y^2)$ **35.** $-x^2y^2z(16x^2 - 24x^3yz^3 + 15yz^6)$
37. $(9p^2 + 4q^2)(3p + 2q)(3p - 2q)$ **39.** prime   **41.** $2(3x + 5y^2)(9x^2 - 15xy^2 + 25y^4)$ **43.** prime   **45.** $t(7t - 1)(3t - 1)$
47. $(x + y)(x - y)(x + y)(x^2 - xy + y^2)$ **49.** $2(a + b)(a - b)(c + 2d)$

REVIEW EXERCISES (page 253)

1. $\frac{8}{3}$ **2.** impossible equation   **3.** 0, 7   **4.** $\frac{5}{2}, -\frac{7}{3}$

4.8 GETTING READY (page 253)

1. s^2 **2.** $2w + 4$ **3.** $x(x + 1)$ **4.** $w(w + 3)$

ORALS (page 256)

1. $A = lw$ **2.** $A = \frac{1}{2}bh$ **3.** $A = s^2$ **4.** $A = lwh$ **5.** $P = 2l + 2w$ **6.** $P = 4s$

EXERCISE 4.8 (page 257)

1. 5, 7   **3.** 9   **5.** 9 sec   **7.** $\frac{15}{4}$ sec and 10 sec   **9.** 2 sec   **11.** 4 m by 9 m   **13.** 48 ft   **15.** $h = 5$ ft, $b = 12$ ft   **17.** 9 sq units
19. 20 cm   **21.** 4 m   **23.** 6 cm by 8 cm   **25.** 9 cm

REVIEW EXERCISES   (page 259)

1. -10 **2.** 3   **3.** 675 cm^2 **4.** \$10,000

CHAPTER 4 REVIEW EXERCISES   (page 262)

1. $5 \cdot 7$ **2.** $3^2 \cdot 5$ **3.** $2^5 \cdot 3$ **4.** $2 \cdot 3 \cdot 17$ **5.** $3 \cdot 29$ **6.** $3^2 \cdot 11$ **7.** $2 \cdot 5^2 \cdot 41$ **8.** 2^{12} **9.** $3(x + 3y)$ **10.** $5a(x^2 + 3)$
11. $7x(x + 2)$ **12.** $3x(x - 1)$ **13.** $2x(x^2 + 2x - 4)$ **14.** $a(x + y - z)$ **15.** $a(x + y - 1)$ **16.** $xyz(x + y)$
17. $5a(a + b^2 + 2cd - 3)$ **18.** $7xy(a + 3x - 5x^2 + y)$ **19.** $(x + y)(a + b)$ **20.** $(x + y)(x + y + 1)$ **21.** $2x(x + 2)(x + 3)$
22. $3x(y + z)(1 - 3y - 3z)$ **23.** $(p + 3q)(3 + a)$ **24.** $(r - 2s)(a + 7)$ **25.** $(x + a)(x + b)$ **26.** $(y + 2)(x - 2)$
27. $y(3x - y)(x - 2)$ **28.** $5(x + 2)(x - 3y)$ **29.** $(x + 3)(x - 3)$ **30.** $(xy + 4)(xy - 4)$ **31.** $(x + 2 + y)(x + 2 - y)$
32. $(z + x + y)(z - x - y)$ **33.** $6y(x + 2y)(x - 2y)$ **34.** $(x + y + z)(x + y - z)$ **35.** $(x + 3)(x + 7)$ **36.** $(x - 3)(x + 7)$
37. $(x + 6)(x - 4)$ **38.** $(x - 6)(x + 2)$ **39.** $(2x + 1)(x - 3)$ **40.** $(3x + 1)(x - 5)$ **41.** $(2x + 3)(3x - 1)$ **42.** $3(2x - 1)(x + 1)$
43. $x(x + 3)(6x - 1)$ **44.** $x(4x + 3)(x - 2)$ **45.** $(x + a + y)(x + a - y)$ **46.** $(x + 1)(ax + 3a - b)$ **47.** $(x + y)(a + b)$
48. $a(a + b)(2x + a)$ **49.** $(c - 3)(c^2 + 3c + 9)$ **50.** $(d + 2)(d^2 - 2d + 4)$ **51.** $2(x + 3)(x^2 - 3x + 9)$
52. $2ab(b - 1)(b^2 + b + 1)$ **53.** $0, -2$ **54.** $0, 3$ **55.** $3, -3$ **56.** $5, -5$ **57.** $3, 4$ **58.** $5, -3$ **59.** $-4, 6$ **60.** $2, 8$
61. $3, -\frac{1}{2}$ **62.** $1, -\frac{3}{2}$ **63.** $\frac{1}{2}, -\frac{1}{2}$ **64.** $\frac{2}{3}, -\frac{2}{3}$ **65.** $0, 3, 4$ **66.** $0, -2, -3$ **67.** $0, \frac{1}{2}, -3$ **68.** $0, -\frac{2}{3}, 1$ **69.** 5 and 7   **70.** $\frac{1}{3}$
71. 15 ft   **72.** 15 sec   **73.** 3 ft by 9 ft   **74.** 3 ft by 6 ft

CHAPTER 4 TEST   (page 264)

1. $2^2 \cdot 7^2$ **2.** $3 \cdot 37$ **3.** $5a(12b^2c^3 + 6a^2b^2c - 5)$ **4.** $3x(a + b)(x - 2y)$ **5.** $(x + y)(a + b)$ **6.** $(x + 5)(x - 5)$
7. $3(a + 3b)(a - 3b)$ **8.** $(4x^2 + 9y^2)(2x + 3y)(2x - 3y)$ **9.** $(x + 3)(x + 1)$ **10.** $(x - 11)(x + 2)$ **11.** $(x + 9y)(x + y)$
12. $6(x - 4y)(x - y)$ **13.** $(3x + 1)(x + 4)$ **14.** $(2a - 3)(a + 4)$ **15.** $(2x - y)(x + 2y)$ **16.** $(4x - 3)(3x - 4)$
17. $6(2a - 3b)(a \div 2b)$ **18.** $(x - 4)(x^2 + 4x + 16)$ **19.** $8(3 + a)(9 - 3a + a^2)$ **20.** $z^3(x^3 - yz)(x^6 + x^3yz + y^2z^2)$ **21.** $0, -3$
22. $-1, -\frac{3}{2}$ **23.** $3, -3$ **24.** $3, -6$ **25.** $\frac{9}{5}, -\frac{1}{2}$ **26.** $-\frac{9}{10}, 1$ **27.** $\frac{1}{5}, -\frac{9}{2}$ **28.** $-\frac{1}{10}, 9$ **29.** 12 sec   **30.** 10 m

5.1 GETTING READY   (page 267)

1. $\frac{3}{4}$ **2.** 2   **3.** $\frac{5}{11}$ **4.** $\frac{1}{2}$

ORALS   (page 273)

1. $\frac{2}{3}$ **2.** 2   **3.** $\frac{z}{w}$ **4.** $2x$ **5.** $\frac{x}{y}$ **6.** $\frac{1}{y}$ **7.** 1   **8.** -1

EXERCISE 5.1   (page 273)

1. $\frac{4}{5}$ **3.** $\frac{4}{5}$ **5.** $\frac{2}{13}$ **7.** $\frac{2}{9}$ **9.** $-\frac{1}{3}$ **11.** $2x$ **13.** $-\frac{x}{3}$ **15.** $\frac{5}{a}$ **17.** $\frac{2}{z}$ **19.** $\frac{a}{3}$ **21.** $\frac{2}{3}$ **23.** $\frac{3}{2}$ **25.** in lowest terms   **27.** $\frac{3x}{y}$ **29.** $\frac{7x}{8y}$
31. $\frac{1}{3}$ **33.** 5   **35.** $\frac{x}{2}$ **37.** $\frac{3x}{5y}$ **39.** $\frac{2}{3}$ **41.** -1 **43.** -1 **45.** -1 **47.** $\frac{x + 1}{x - 1}$ **49.** $\frac{x - 5}{x + 2}$ **51.** $\frac{2x}{x - 2}$ **53.** $\frac{x}{3}$ **55.** $\frac{x + 2}{x^2}$ **57.** $\frac{x - 4}{x + 4}$
59. $\frac{2(x + 2)}{x - 1}$ **61.** in lowest terms   **63.** $\frac{3 - x}{3 + x}$ or $-\frac{x - 3}{x + 3}$ **65.** $\frac{4}{3}$ **67.** $x + 3$ **69.** $x + 1$ **71.** $a - 2$ **73.** $\frac{b + 2}{b + 1}$ **75.** $\frac{y + 3}{x - 3}$

REVIEW EXERCISES   (page 274)

1. $(a + b) + c = a + (b + c)$ **2.** $a(b + c) = ab + ac$ **3.** 0   **4.** 1   **5.** $\frac{5}{3}$ **6.** $-\frac{3}{5}$

5.2 GETTING READY   (page 275)

1. $\frac{2}{3}$ **2.** $\frac{14}{3}$ **3.** 3   **4.** 6   **5.** $\frac{5}{2}$ **6.** 1   **7.** $\frac{3}{4}$ **8.** 2

ORALS   (page 277)

1. $\frac{10}{21}$ **2.** $\frac{3}{2}$ **3.** $\frac{7}{5}$ **4.** x **5.** 5   **6.** 2

EXERCISE 5.2   (page 277)

1. $\frac{8}{15}$ **3.** $\frac{45}{91}$ **5.** $\frac{2}{5}$ **7.** $-\frac{2}{3}$ **9.** $\frac{3}{11}$ **11.** $\frac{15}{4}$ **13.** $\frac{5}{7}$ **15.** $\frac{3x}{2}$ **17.** $\frac{xy}{z}$ **19.** $\frac{3y}{10}$ **21.** $\frac{14}{9}$ **23.** 26 **25.** x^2y^2 **27.** $2xy^2$ **29.** $-3y^2$

31. $\frac{b^3c}{a^4}$ **33.** $\frac{r^3t^4}{s}$ **35.** $\frac{(z+7)(z+2)}{7z}$ **37.** x **39.** $\frac{x}{5}$ **41.** $x+2$ **43.** $\frac{3}{2x}$ **45.** $x-2$ **47.** x **49.** $\frac{(x-2)^2}{x}$

51. $\frac{(m-2)(m-3)}{2(m+2)}$ **53.** 1 **55.** $\frac{1}{3}$ **57.** $\frac{c^2}{ab}$ **59.** $\frac{x+1}{2(x-2)}$ **61.** 1 **63.** $\frac{1}{x-4}$ **65.** $\frac{x^2-2x+4}{x-2}$ **67.** $\frac{-1}{(x+1)(x+3)}$

69. $\frac{x+y}{x(x-y)}$ **71.** $\frac{-(x-y)(x^2+xy+y^2)}{a+b}$

REVIEW EXERCISES   (page 279)

1. $-6x^5y^6z$ **2.** $-4xy^3$ **3.** $\frac{1}{81y^4}$ **4.** a^3 **5.** $\frac{1}{x^m}$ **6.** 1

5.3 GETTING READY   (page 280)

1. $\frac{1}{3}$ **2.** $\frac{9}{25}$ **3.** 4 **4.** 15 **5.** $\frac{7}{64}$ **6.** $\frac{1}{8}$ **7.** $\frac{9}{4}$ **8.** 1

ORALS   (page 283)

1. $\frac{9}{49}$ **2.** 1 **3.** $\frac{1}{4}$ **4.** $\frac{3}{16}$ **5.** 4 **6.** $\frac{16}{3}$

EXERCISE 5.3   (page 283)

1. $\frac{2}{3}$ **3.** $\frac{3}{10}$ **5.** $\frac{6}{5}$ **7.** $\frac{16}{35}$ **9.** $\frac{3}{5}$ **11.** $\frac{7}{5}$ **13.** $\frac{7}{3}$ **15.** $\frac{3x}{2}$ **17.** $\frac{3}{2y}$ **19.** 3 **21.** $\frac{6}{y}$ **23.** 6 **25.** $\frac{2x}{3}$ **27.** $\frac{2y^2}{15z}$ **29.** $\frac{2}{y}$ **31.** $\frac{2}{3x}$ **33.** $\frac{2(z-2)}{z}$

35. $\frac{5z(z-7)}{z+2}$ **37.** $\frac{x+2}{3}$ **39.** 1 **41.** $\frac{x-2}{x-3}$ **43.** $x+5$ **45.** 1 **47.** $\frac{3}{7}$ **49.** $\frac{9}{2x}$ **51.** $\frac{x}{36}$ **53.** $\frac{(x+1)(x-1)}{5(x-3)}$ **55.** 2 **57.** $\frac{2x(1-x)}{5(x-2)}$

59. $\frac{y^2}{3}$ **61.** $\frac{x+2}{x-2}$ **63.** 1 **65.** $\frac{1}{(x+1)^2}$ **67.** $-x-y$ **69.** $a+2$ **71.** $\frac{-p}{m+n}$

REVIEW EXERCISES   (page 285)

1. $4y^3+4y^2-8y+32$ **2.** $8a^3+9a^2+6a+5$ **3.** $-6m^3+5m^2+3m-2$ **4.** $2p^3+3p^2q-q^3$ **5.** $5y^2+22y+114+\frac{569}{y-5}$

6. $6x^2-24x+92-\frac{363}{x+4}$

5.4 GETTING READY   (page 285)

1. $\frac{4}{5}$ **2.** 1 **3.** $\frac{7}{8}$ **4.** 2 **5.** $\frac{1}{9}$ **6.** $\frac{1}{2}$ **7.** $-\frac{2}{13}$ **8.** $\frac{13}{10}$

ORALS   (page 288)

1. 2 **2.** 1 **3.** $\frac{13}{x}$ **4.** $\frac{y+1}{5}$ **5.** 1 **6.** 2 **7.** $\frac{6}{z}$ **8.** $\frac{r-1}{t}$

EXERCISE 5.4   (page 288)

1. $\frac{2}{3}$ **3.** $\frac{3}{5}$ **5.** $\frac{1}{3}$ **7.** 2 **9.** 2 **11.** $\frac{12}{7}$ **13.** $\frac{25}{4}$ **15.** $4x$ **17.** $-\frac{2y}{3}$ **19.** 0 **21.** $\frac{4x}{y}$ **23.** $\frac{2y}{x}$ **25.** $\frac{x^2}{2y}$ **27.** $\frac{2y+6}{5z}$ **29.** 9 **31.** $\frac{1}{7}$

33. $-\frac{4}{3}$ **35.** $\frac{2}{13}$ **37.** $-\frac{24}{23}$ **39.** 1 **41.** $-\frac{17}{41}$ **43.** $\frac{10}{7}$ **45.** $\frac{x}{y}$ **47.** $\frac{y}{x}$ **49.** $\frac{x}{y}$ **51.** $-\frac{2}{5z}$ **53.** $\frac{1}{y}$ **55.** $-\frac{1}{z}$ **57.** $\frac{x+3}{xy}$ **59.** 1 **61.** 0

63. $\frac{8}{5}$ **65.** $\frac{4x}{3}$ **67.** $\frac{2x}{3y}$ **69.** $\frac{4x-2y}{y+2}$ **71.** $\frac{2x+10}{x-2}$ **73.** $\frac{xy}{x-y}$ **75.** $-\frac{1}{a-b}$

REVIEW EXERCISES (page 290)

1. 7^2 **2.** 2^6 **3.** $2^3 \cdot 17$ **4.** $2 \cdot 11^2$ **5.** $2 \cdot 3 \cdot 17$ **6.** $3^2 \cdot 5 \cdot 7$ **7.** $2^4 \cdot 3^2$ **8.** $5 \cdot 29$

5.5 GETTING READY (page 290)

1. equal **2.** equal **3.** not equal **4.** equal **5.** equal **6.** not equal **7.** equal **8.** equal

ORALS (page 295)

1. equal **2.** equal **3.** not equal **4.** equal **5.** equal **6.** not equal **7.** equal **8.** equal

EXERCISE 5.5 (page 295)

1. $\dfrac{4}{6}$ **3.** $\dfrac{125}{20}$ **5.** $\dfrac{2x}{x^2}$ **7.** $\dfrac{5x}{xy}$ **9.** $\dfrac{8xy}{x^2y}$ **11.** $\dfrac{3x(x+1)}{(x+1)^2}$ **13.** $\dfrac{2y(x+1)}{x^2+x}$ **15.** $\dfrac{z(z+1)}{z^2-1}$ **17.** $\dfrac{(x+2)^2}{x^2-4}$ **19.** $\dfrac{2(x+2)}{x^2+3x+2}$ **21.** 60

23. 42 **25.** $6x$ **27.** x^2y^2 **29.** $18xy$ **31.** x^2-1 **33.** x^2+6x **35.** $(x+1)(x-2)^2$ **37.** $(x+1)(x+5)(x-5)$ **39.** $\dfrac{7}{6}$

41. $-\dfrac{1}{6}$ **43.** $\dfrac{5y}{9}$ **45.** $\dfrac{7a}{60}$ **47.** $\dfrac{53x}{42}$ **49.** $\dfrac{4xy+6x}{3y}$ **51.** $\dfrac{2-3x^2}{x}$ **53.** $\dfrac{3x^2+2xy}{6y^2}$ **55.** $\dfrac{4y+10}{15y}$ **57.** $\dfrac{x^2+4x+1}{x^2y}$

59. $\dfrac{2x^2-1}{x(x+1)}$ **61.** $\dfrac{-x^2+3x+1}{x-2}$ **63.** $\dfrac{2xy+x-y}{xy}$ **65.** $\dfrac{x+2}{x-2}$ **67.** $\dfrac{2x^2+2}{(x-1)(x+1)}$ **69.** $\dfrac{10x+4}{(x-2)(x+2)}$

71. $\dfrac{2x^2-4x+8}{(x-2)(x-2)(x+2)}$ **73.** $\dfrac{x}{x-2}$ **75.** $\dfrac{5x+3}{x+1}$ **77.** $\dfrac{-10y+18}{y(y-3)}$ **79.** $-\dfrac{1}{2(x-2)}$ **81.** $\dfrac{3}{x-3}$

5.6 GETTING READY (page 298)

1. 4 **2.** -18 **3.** 7 **4.** -8 **5.** $3+3x$ **6.** $2-y$ **7.** $12x-2$ **8.** $3y+2x$

ORALS (page 302)

1. $\dfrac{4}{3}$ **2.** 4 **3.** $\dfrac{1}{4}$ **4.** 3

EXERCISE 5.6 (page 302)

1. $\dfrac{8}{9}$ **3.** $\dfrac{3}{8}$ **5.** $\dfrac{5}{4}$ **7.** $\dfrac{5}{7}$ **9.** $\dfrac{x^2}{y}$ **11.** $\dfrac{5t^2}{27}$ **13.** $\dfrac{1-3x}{5+2x}$ **15.** $\dfrac{1+x}{2+x}$ **17.** $\dfrac{3-x}{x-1}$ **19.** $\dfrac{1}{x+2}$ **21.** $\dfrac{1}{x+3}$ **23.** $\dfrac{xy}{y+x}$

25. $\dfrac{y}{x-2y}$ **27.** $\dfrac{x^2}{x^2-2x+1}$ **29.** $\dfrac{7x+3}{-x-3}$ **31.** $\dfrac{x-2}{x+3}$ **33.** -1 **35.** $\dfrac{y}{x^2}$ **37.** $\dfrac{x+1}{1-x}$ **39.** $\dfrac{a^2-a+1}{a^2}$ **41.** 2 **43.** $\dfrac{y-5}{y+5}$

SOMETHING TO THINK ABOUT (page 304)

1. $\dfrac{1}{2}, \dfrac{2}{3}, \dfrac{3}{5}, \dfrac{5}{8}$ **2.** Fibonacci sequence: $1, 2, 3, 5, \ldots$; $\dfrac{8}{13}$

REVIEW EXERCISES (page 304)

1. t^9 **2.** a^6 **3.** $-2r^7$ **4.** s^6 **5.** $\dfrac{81}{256r^8}$ **6.** $\dfrac{y^{10}}{16}$ **7.** $\dfrac{r^{10}}{9}$ **8.** $\dfrac{25}{16x^{12}}$

5.7 GETTING READY (page 304)

1. $3x+1$ **2.** $8x-1$ **3.** $3+2x$ **4.** $y-6$ **5.** 19 **6.** $7x+6$ **7.** y **8.** $3x+5$

ORALS   (page 309)

1. multiply by 10   **2.** multiply by $x(x - 1)$ **3.** multiply by 9   **4.** multiply by 15

EXERCISE 5.7   (page 309)

1. 4   **3.** -20 **5.** 6   **7.** 60   **9.** -12 **11.** 0   **13.** -7 **15.** -1 **17.** 12   **19.** 0   **21.** -3 **23.** 3   **25.** no solution, 0 is extraneous   **27.** 1   **29.** 5   **31.** no solution; -2 is extraneous   **33.** no solution; 5 is extraneous   **35.** -1 **37.** 6   **39.** 2   **41.** -3 **43.** 1   **45.** no solution; -2 is extraneous   **47.** 1, 2   **49.** 3; -3 is extraneous   **51.** 3, -4 **53.** 1   **55.** 0   **57.** $-2, 1$ **59.** $a = \dfrac{b}{b - 1}$ **61.** $f = \dfrac{d_1 d_2}{d_1 + d_2}$

SOMETHING TO THINK ABOUT   (page 310)

1. 1   **2.** -1

REVIEW EXERCISES   (page 311)

1. $(b + 3)(a + 2)$ **2.** $(y + z)(z + 1)$ **3.** $(r + s)(m + n)$ **4.** $(a + b)(c - d)$ **5.** $(2 - a)(a + b)$ **6.** $(a - 1)(x + y)$

5.8 GETTING READY   (page 311)

1. $\frac{1}{5}$ **2.** $\$(.05x)$ **3.** $\$(\frac{y}{.05})$ **4.** $\frac{y}{52}$ hr

ORALS   (page 314)

1. $I = Pr$ **2.** $d = rt$ **3.** $C = qd$

EXERCISE 5.8   (page 314)

1. 2   **3.** 5   **5.** $\frac{2}{3}, \frac{3}{2}$ **7.** $2\frac{2}{9}$ hr   **9.** $2\frac{6}{11}$ days   **11.** $7\frac{1}{2}$ hr   **13.** 4 mph   **15.** 7% and 8%   **17.** 5   **19.** 30   **21.** 25 mph

REVIEW EXERCISES   (page 316)

1. $-1, 6$ **2.** $5, -5$ **3.** $-2, -3, -4$ **4.** $2, -4$ **5.** $0, 0, 1$ **6.** $0, 5, -5$ **7.** $1, -1, 2, -2$ **8.** $0, -6, \frac{1}{6}$

5.9 GETTING READY   (page 316)

1. 10   **2.** $\frac{7}{3}$ **3.** $\frac{20}{7}$ **4.** 5   **5.** $\frac{14}{3}$ **6.** 5   **7.** 21   **8.** $\frac{3}{2}$

ORALS   (page 321)

1. $\frac{5}{7}$ **2.** $\frac{50}{1}$ **3.** $\frac{1}{3}$ **4.** $\frac{7}{10}$ **5.** proportion   **6.** not a proportion   **7.** not a proportion   **8.** proportion

EXERCISE 5.9   (page 321)

1. $\frac{5}{7}$ **3.** $\frac{1}{2}$ **5.** $\frac{2}{3}$ **7.** $\frac{1}{3}$ **9.** $\frac{1}{5}$ **11.** $\frac{3}{7}$ **13.** $\frac{1}{18}$ **15.** $\frac{3}{4}$ **17.** $\frac{6}{5}$ **19.** $\frac{25}{1}$ **21.** not a proportion   **23.** a proportion   **25.** not a proportion   **27.** not a proportion   **29.** a proportion   **31.** a proportion   **33.** a proportion   **35.** 4   **37.** 6   **39.** -3 **41.** 9   **43.** 0   **45.** -17 **47.** $-\frac{3}{2}$ **49.** $-\frac{3}{5}$ **51.** 65   **53.** 39   **55.** $-\frac{5}{2}, -1$ **57.** $-5, -1$ **59.** 4; 0 is extraneous   **61.** \$17   **63.** \$6.50   **65.** $7\frac{1}{2}$ gal   **67.** \$309   **69.** $49\frac{7}{24}$ ft   **71.** 162   **73.** Not exact but close enough   **75.** $21\frac{3}{7}$ ft   **77.** 528 ft   **79.** 880 ft   **81.** 180 mg

REVIEW EXERCISES   (page 324)

1. 24 in.   **2.** 18 cm   **3.** 22 cm   **4.** 8π ft ≈ 25.1 ft   **5.** 36 in.^{2} **6.** 14 cm^2 **7.** $\frac{65}{2}$ cm^2 **8.** 16π ft$^2 \approx 50.3$ ft^2

CHAPTER 5 REVIEW EXERCISES   (page 327)

1. $\frac{2}{5}$ **2.** $-\frac{2}{3}$ **3.** $-\frac{1}{3}$ **4.** $\frac{7}{3}$ **5.** $\frac{1}{2x}$ **6.** $\frac{5}{2x}$ **7.** $\frac{x}{x+1}$ **8.** $\frac{1}{x}$ **9.** 2 **10.** 1 **11.** $\frac{x+3}{x-5}$ **12.** $\frac{x+7}{x+3}$ **13.** $\frac{x}{x-1}$ **14.** in lowest terms **15.** $\frac{3x}{y}$
16. $\frac{6}{x^2}$ **17.** 1 **18.** $\frac{x}{x+y}$ **19.** $\frac{3y}{2}$ **20.** $\frac{1}{x}$ **21.** $x+2$ **22.** $x^2+10x+25$ **23.** 8 **24.** 70 **25.** $3x^2y^2$ **26.** $x(2x+1)$

27. $(x+2)(x-3)$ **28.** $x(x+1)(x+3)$ **29.** 1 **30.** $\frac{2(x+1)}{x-7}$ **31.** $\frac{x^2+x-1}{x(x-1)}$ **32.** $\frac{x-7}{7x}$ **33.** $\frac{x-2}{x(x+1)}$ **34.** $\frac{x^2+4x-4}{2x^2}$
35. $\frac{x+1}{x}$ **36.** 0 **37.** $\frac{9}{4}$ **38.** $\frac{3}{2}$ **39.** $\frac{1+x}{1-x}$ **40.** $\frac{x(x+3)}{2x^2-1}$ **41.** x^2+3 **42.** $\frac{a(a+bc)}{b(b+ac)}$ **43.** 3 **44.** 1 **45.** 3 **46.** 4, $-\frac{3}{2}$
47. -2 **48.** 0 **49.** $T_1=\frac{T_2}{1-E}$ **50.** $R=\frac{HB}{B-H}$ **51.** $9\frac{9}{19}$ hr **52.** $5\frac{5}{6}$ days **53.** 5 mph **54.** 40 mph **55.** $\frac{1}{2}$ **56.** $\frac{4}{5}$ **57.** $\frac{2}{3}$
58. $\frac{5}{6}$ **59.** $\frac{9}{2}$ **60.** 0 **61.** 7 **62.** $-4, 3$ **63.** 30 tons **64.** 210 **65.** 0.05 g **66.** 4700 ft **67.** 132.5 ft **68.** 3564 ft

CHAPTER 5 TEST   (page 330)

1. $\frac{8x}{9y}$ **2.** $\frac{x+1}{2x+3}$ **3.** 3 **4.** $\frac{5y^2}{4t}$ **5.** $\frac{x+1}{3(x-2)}$ **6.** $\frac{3t^2}{5y}$ **7.** $\frac{x^2}{3}$ **8.** $x+2$ **9.** $\frac{10x-1}{x-1}$ **10.** $\frac{13}{2y+3}$ **11.** $\frac{2x^2+x+1}{x(x+1)}$
12. $\frac{2x+6}{x-2}$ **13.** $\frac{2x^3}{y^3}$ **14.** $\frac{x+y}{y-x}$ **15.** -5 **16.** 6 **17.** 4 **18.** $B=\frac{HR}{R-H}$ **19.** $\frac{2}{3}$ **20.** yes **21.** $\frac{2}{3}$ **22.** $3\frac{15}{16}$ hr **23.** 5 mph
24. 8050 ft

6.1 GETTING READY   (page 333)

1. 1 **2.** 5 **3.** -3 **4.** 2

EXERCISE 6.1   (page 343)

1. QI
3. QII
5. QIII
7. QIV

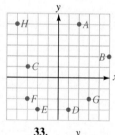

9. QII
11. QIV
13. QIII
15. QI

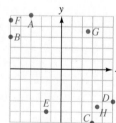

17. y-axis
19. x-axis
21. x-axis
23. both axes

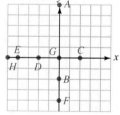

25. $(2, 3)$
27. $(-2, -3)$
29. $(0, 0)$
31. $(-5, -5)$

33.

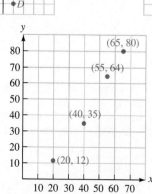

35. $12, $24, $36, $60
37. $50, $90, $130, $230
39. 200, 190, 180, 160

41.

x	y
3	5
1	3
-2	0

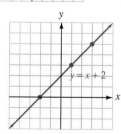

43.

x	y
2	-4
1	-2
-3	6

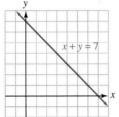

$y = -2x$

45.

x	y
3	5
-1	-3
-2	-5

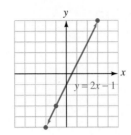

$y = 2x - 1$

47.

x	y
8	2
0	-2
-2	-3

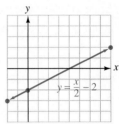

$y = \frac{x}{2} - 2$

49.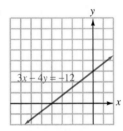

$x + y = 7$

51.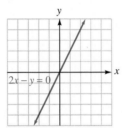

$x - y = 7$

53.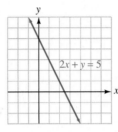

$2x + y = 5$

55.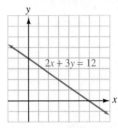

$2x + 3y = 12$

57.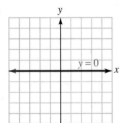

$3x - 4y = -12$

59.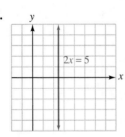

$2x - y = 0$

61.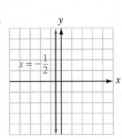

$y = -5$

63.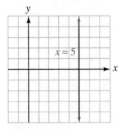

$x = 5$

65.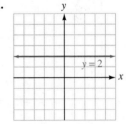

$y = 0$

67.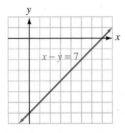

$2x = 5$

69.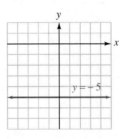

$x = -\frac{1}{2}$

71.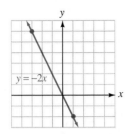

$y = 2$

73. $y = -2$ **75.** $y = 0$

SOMETHING TO THINK ABOUT   (page 347)

1. $(6, 6)$ **2.** $(6, 8)$ **3.** $\left(-\frac{1}{2}, \frac{5}{2}\right)$ **4.** $\left(-\frac{5}{2}, \frac{3}{2}\right)$ **5.** $(7, 6)$ **6.** $(4, -3)$

REVIEW EXERCISES   (page 348)

1. -1 **2.** 6   **3.** $\frac{4y}{3} + 2x$ **4.** $3x - 2$

6.2 GETTING READY (page 348)

1. 1 **2.** 3 **3.** −3 **4.** −$\frac{1}{2}$

ORALS (page 357)

1. $\frac{3}{2}$ **2.** −$\frac{5}{6}$ **3.** 4, (0, 9) **4.** −$\frac{2}{3}$, (0, −1) **5.** 3, (0, 1) **6.** −4, (0, 2)

EXERCISE 6.2 (page 357)

1. 1
3. −1
5. 1
7. 2
9. $\frac{1}{5}$
11. 0
13. −1
15. undefined

17.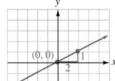

19.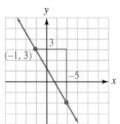

21.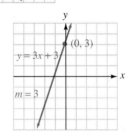

23.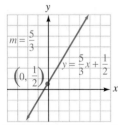

25.

27.

29.

31.

33.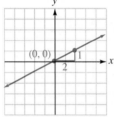

35.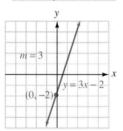

37.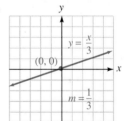

39.

41. $y = -3x + 7$, −3, (0, 7)
43. $y = \frac{5}{2}x - 13$, $\frac{5}{2}$, (0, −13)
45. $y = 2$, 0, (0, 2)
47. $y = x - \frac{7}{2}$, 1, $\left(0, -\frac{7}{2}\right)$
49. $y = \frac{2}{3}x + 7$, $\frac{2}{3}$, (0, 7)
51. $y = \frac{1}{3}x - \frac{5}{3}$, $\frac{1}{3}$, $\left(0, -\frac{5}{3}\right)$
53. $\frac{1}{220}$
55. $\frac{5}{12}$

57. $1200 **59.** $26,750 **61.** $6000 **63.** $150

REVIEW EXERCISES (page 360)

1. $3x(x - 2)$ **2.** $(y + 5)(y - 5)$ **3.** $(2z + 1)(z - 3)$ **4.** $3(3t - 2)(t - 1)$ **5.** 0, −2 **6.** 6, −6 **7.** 1, 6 **8.** $\frac{1}{4}$, −$\frac{4}{3}$

6.3 GETTING READY   (page 360)

1. $3x - 2y = 4$ **2.** $2x + y = 7$ **3.** $2x - y = -14$ **4.** $3x + y = 12$ **5.** $y = 2x - 12$ **6.** $y = -\frac{3}{4}x + \frac{1}{4}$ **7.** $y = 4x$
8. $y = -2x + 8$

ORALS   (page 367)

1. $y = 2x + 4$ **2.** $y = -\frac{1}{2}x + 2$ **3.** parallel   **4.** perpendicular

EXERCISE 6.3   (page 367)

1. $4x - y = -5$ **3.** $7x + y = -2$ **5.** $x + 2y = -1$ **7.** $3x + 5y = -2$ **9.** $y = 3$ **11.** $y = 0$ **13.** $2x - y = 1$ **15.** $5x - y = 6$
17. $2x + y = 0$ **19.** $x + 2y = 6$ **21.** $7x - 5y = 19$ **23.** $y = 3$ **25.** neither   **27.** parallel   **29.** perpendicular   **31.** parallel
33. parallel   **35.** neither   **37.** perpendicular   **39.** perpendicular   **41.** parallel   **43.** $2x - 15y = -120$ **45.** $4x + 9y = -2$
47. $5x - y = 0$ **49.** $5x - y = 4$ **51.** $5x - 2y = 8$ **53.** $2x - y = -1$ **55.** $y = 2$ **57.** $5x + y = 15$ **59.** $x + 5y = 0$
61. $5x + y = 3$ **63.** \$74 million   **65.** \$800   **67.** \$8000

REVIEW EXERCISES   (page 369)

1. $\dfrac{x + 1}{x}$ **2.** $\dfrac{1}{x}$ **3.** $\dfrac{6}{x + 1}$ **4.** $\dfrac{4x}{(x + 1)(x - 1)}$

6.4 GETTING READY   (page 370)

1. -1 **2.** 0   **3.** 0   **4.** 3   **5.** 0   **6.** 2   **7.** 2   **8.** 3

ORALS   (page 375)

1. 2   **2.** 3   **3.** 6   **4.** 27   **5.** 0   **6.** 2   **7.** 4   **8.** 6

EXERCISE 6.4   (page 375)

1. 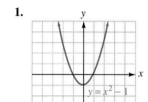   **3.** 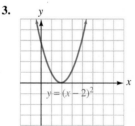   **5.** 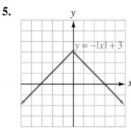   **7.**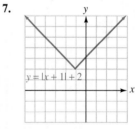

17.   **19.**   **21.** 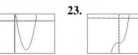   **23.**   **25.**   **27.**

29.   **31.**   **33.** $(-2, -4)$ **35.** $(3, 9)$ **37.** $(-2, -9)$ **39.** $(-2, 6)$

REVIEW EXERCISES   (page 377)

1. 41, 43, 47   **2.** $(a + b) + c = a + (b + c)$ **3.** $a \cdot b = b \cdot a$ **4.** 0   **5.** 1   **6.** $\frac{3}{5}$

6.5 GETTING READY   (page 377)

1. below   **2.** above   **3.** below   **4.** on   **5.** on   **6.** above   **7.** below   **8.** above

ORALS   (page 382)

1. no   **2.** no   **3.** yes   **4.** yes   **5.** no   **6.** yes   **7.** no   **8.** yes

EXERCISE 6.5   (page 382)

1.
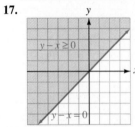

3.
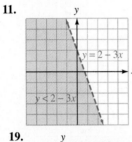

5.
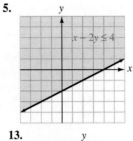

7.
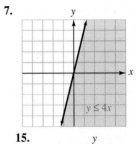

9.
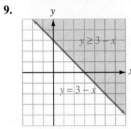

11.
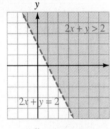

13.
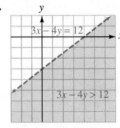

15.
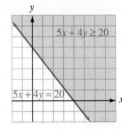

17.
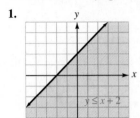

19.
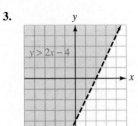

21.
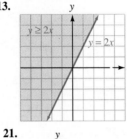

23.
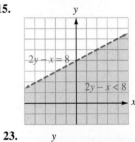

25.
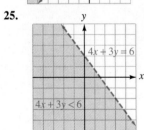

27.
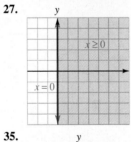

29.
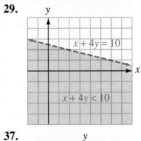

31.
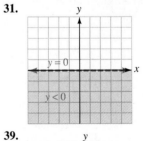

33.
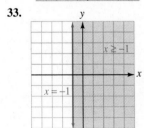

35.
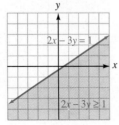

37.
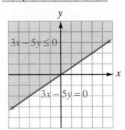

39.
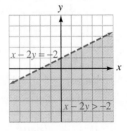

41.

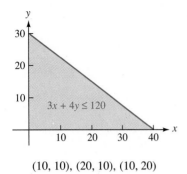

(10, 10), (20, 10), (10, 20)

43.

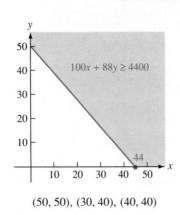

(50, 50), (30, 40), (40, 40)

45.

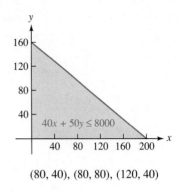

(80, 40), (80, 80), (120, 40)

REVIEW EXERCISES (page 387)

1. 7 **2.** 42 **3.** $3x^2 + 4x + 3$ **4.** $3x^2 + 2x + 2$

6.6 GETTING READY (page 387)

1. -2 **2.** 2 **3.** -1 **4.** 2

ORALS (page 391)

1. yes **2.** no **3.** yes **4.** no **5.** 1 **6.** 3 **7.** -1 **8.** -3

EXERCISE 6.6 (page 391)

1. yes **3.** yes **5.** yes **7.** no **9.** yes **11.** no **13.** domain = all reals; range = all reals ≤ 2 **15.** domain = all reals; range = all reals ≥ -2 **17.** 9, 0, -3 **19.** 3, -3, -5 **21.** 22, 7, 2 **23.** 3, 9, 11 **25.** 1, 4, 9 **27.** 0, -9, 26 **29.** 4, 1, 16 **31.** 1, 10, 15 **33.** 4, 3, 4 **35.** 2, -1, 2 **37.** $\frac{1}{5}, \frac{1}{4}, 1$ **39.** $-2, -\frac{1}{2}, \frac{2}{5}$ **41.** $2w, 2w + 2$ **43.** $3w - 5, 3w - 2$ **45.** $-2w + 3, -2w + 1$ **47.** $|w|, |w + 1|$ **49.** 12 **51.** $2b - 2a$ **53.** $2b$ **55.** 1 **57.** 192 ft

REVIEW EXERCISES (page 393)

1. -2 **2.** -3 **3.** 6 **4.** no solution

6.7 GETTING READY (page 393)

1. 4 **2.** 16 **3.** $\frac{A}{bh}$ **4.** $\frac{PV}{T}$

ORALS (page 398)

1. inverse **2.** direct **3.** combined **4.** joint **5.** direct **6.** combined **7.** joint **8.** direct

EXERCISE 6.7 (page 398)

1. $d = kn$ **3.** $l = \frac{k}{w}$ **5.** $A = kr^2$ **7.** $d = kst$ **9.** $I = \frac{kV}{R}$ **11.** 35 **13.** 42 **15.** 675 **17.** 1 **19.** $\frac{80}{3}$ **21.** 24 **23.** 4 **25.** 80 **27.** $\frac{5}{2}$ **29.** 576 ft **31.** $2450 **33.** $2\frac{1}{2}$ hr **35.** $53\frac{1}{3}$ m^3 **37.** $270 **39.** $3\frac{9}{11}$ amp

REVIEW EXERCISES (page 400)

1. $8x + 5$ **2.** $-13x - 23$ **3.** $-3x^3 + 6x^2 - 6x$ **4.** $-8a^3 - 17a^2 + 9a$

CHAPTER 6 REVIEW EXERCISES (page 403)

1–6.

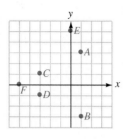

7. $(3, 1)$
8. $(-4, 5)$
9. $(-3, -4)$
10. $(2, -3)$
11. $(0, 0)$
12. $(0, 4)$
13. $(-5, 0)$
14. $(0, -3)$

15.

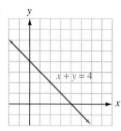

16.

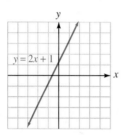

17.

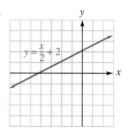

18.

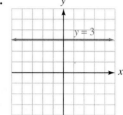

19.

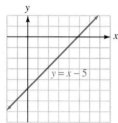

20.

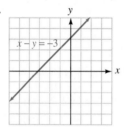

21.

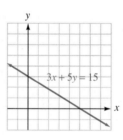

22.

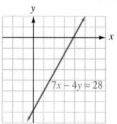

23. -1
24. $-\frac{5}{4}$
25. $-\frac{1}{2}$
26. 0

27.

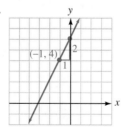

28.

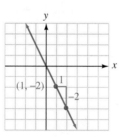

29.

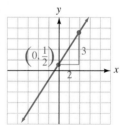

30.

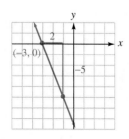

31. $5, (0, 2)$

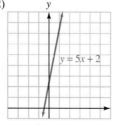

32. $-\frac{1}{2}, (0, 4)$

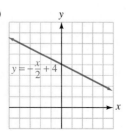

33. $0, (0, -3)$

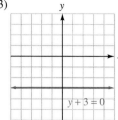

34. $-\frac{1}{3}, \left(0, \frac{1}{3}\right)$

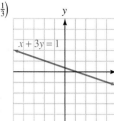

35. $3x + y = 2$ **36.** $y = -7$ **37.** $7x - y = 0$
38. $x - 2y = 3$ **39.** $3x - y = 0$ **40.** $x + 3y = 3$
41. $x - 9y = -9$ **42.** $3x + 5y = 2$ **43.** neither
44. parallel   **45.** perpendicular   **46.** parallel
47. neither   **48.** perpendicular   **49.** parallel
50. perpendicular   **51.** $7x - y = 9$
52. $3x + 2y = 1$ **53.** $5x + 2y = 0$
54. $3x + y = -4$

55.

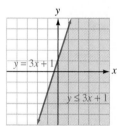

56.

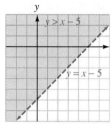

57.

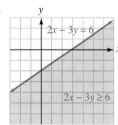

58.

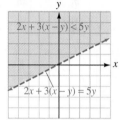

59.

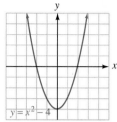

60.

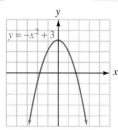

61.

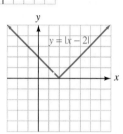

62.

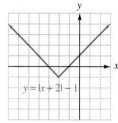

63. yes   **64.** yes   **65.** no   **66.** no   **67.** 1   **68.** 7   **69.** 7   **70.** $w^2 - w + 1$ **71.** -2 **72.** 10   **73.** $a^2 - a + 2$ **74.** $-a^2 + a$
75. 400   **76.** 600   **77.** 81   **78.** $\frac{5}{7}$

CHAPTER 6 TEST   (page 407)

1.

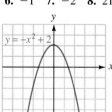

2.

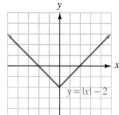

3.

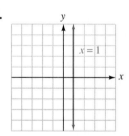

4.

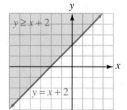

5. $\frac{4}{3}$ **6.** -1 **7.** -2 **8.** 21   **9.** 2   **10.** $-\frac{1}{2}$ **11.** $x - 2y = -6$ **12.** $7x - y = -19$ **13.** $x = -3$ **14.** $3x + y = 4$

15.

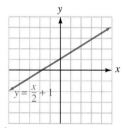

16.

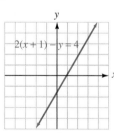

17.

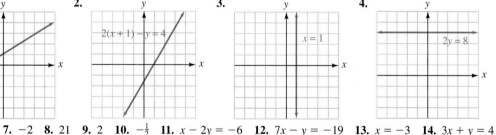

18.

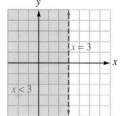

19. yes
20. no   **21.** 11   **22.** -2 **23.** $3a - 3b$ **24.** $3t^2 + 2$ **25.** 1   **26.** 400

CUMULATIVE REVIEW EXERCISES (page 408)

1. $6x^3 - 2x - 1$ **2.** $2x^3 + 2x^2 + x - 1$ **3.** $13x^2 - 8x + 1$ **4.** $16x^2 - 24x + 2$ **5.** $-12x^5y^5$ **6.** $-35x^5 + 10x^4 + 10x^2$
7. $6x^2 + 14x + 4$ **8.** $15x^2 - 2xy - 8y^2$ **9.** $x + 4$ **10.** $x^2 + x + 1$ **11.** $3xy(x - 2y)$ **12.** $(a + b)(3 + x)$ **13.** $(a + b)(2 + b)$
14. $(5p^2 + 4q)(5p^2 - 4q)$ **15.** $(x - 12)(x + 1)$ **16.** $(x - 3y)(x + 2y)$ **17.** $(3a + 4)(2a - 5)$ **18.** $(4m + n)(2m - 3n)$
19. $(p - 3q)(p^2 + 3pq + 9q^2)$ **20.** $8(r + 2s)(r^2 - 2rs + 4s^2)$ **21.** $-1, -2$ **22.** $\frac{3}{2}, -4$ **23.** $\dfrac{x + 1}{x - 1}$ **24.** $\dfrac{x - 3}{x - 2}$ **25.** $\dfrac{(x - 2)^2}{x - 1}$

26. $\dfrac{(p + 2)(p - 3)}{3(p + 3)}$ **27.** 1 **28.** $\dfrac{2(x^2 + 1)}{(x + 1)(x - 1)}$ **29.** $\dfrac{-1}{2(a - 2)}$ **30.** $\dfrac{y + x}{y - x}$ **31.** 2 **32.** 2 **33.** $-\frac{3}{2}$ **34.** $\frac{8}{11}$

35.

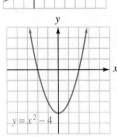

36.

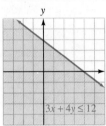

37. $\frac{1}{2}$ **38.** $-\frac{1}{2}$ **39.** $y = \frac{2}{3}x + 5$ **40.** $3x - 4y = -22$
41. perpendicular **42.** parallel

43.

44.

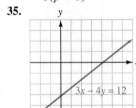

45. yes **46.** no **47.** -3 **48.** 15
49. 5 **50.** $8x^2 - 3$ **51.** 12 **52.** 2

7.1 GETTING READY (page 413)

1. -3 **2.** -2 **3.** 1 **4.** 6 **5.** $\frac{3}{2}$ **6.** $-\frac{4}{3}$

ORALS (page 420)

1. yes **2.** yes **3.** no **4.** no

EXERCISE 7.1 (page 420)

1. yes **13.**
3. yes
5. no
7. yes
9. no
11. no

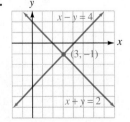

15.

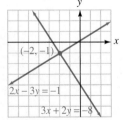

17.

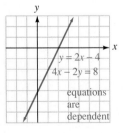

19.

21.
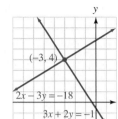

23.
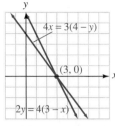

25.
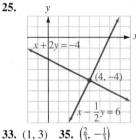

27.
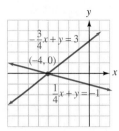

29.
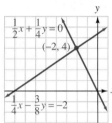

31.
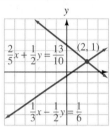

33. $(1, 3)$ **35.** $\left(\frac{2}{3}, -\frac{1}{3}\right)$

REVIEW EXERCISES   (page 423)

1. x^{12} **2.** y^2 **3.** 1   **4.** t^3

7.2 GETTING READY   (page 423)

1. $6x + 4$ **2.** $-25 - 10x$ **3.** $2x - 4$ **4.** $3x - 12$

ORALS   (page 428)

1. $2z + 1$ **2.** $z + 2$ **3.** $3t + 3$ **4.** $\frac{t}{3} + 4$

EXERCISE 7.2   (page 428)

1. $(2, 4)$ **3.** $(3, 0)$ **5.** $(-3, -1)$ **7.** inconsistent   **9.** $(-2, 3)$ **11.** $(3, 2)$ **13.** $(3, -2)$ **15.** $(-1, 2)$ **17.** $(-1, -1)$
19. dependent   **21.** $(1, 1)$ **23.** $(4, -2)$ **25.** $(-3, -1)$ **27.** $(-1, -3)$ **29.** $\left(\frac{1}{2}, \frac{1}{3}\right)$ **31.** $(1, 4)$ **33.** $(4, 2)$ **35.** $(-5, -5)$
37. $(-6, 4)$ **39.** $\left(\frac{1}{5}, 4\right)$ **41.** $(5, 5)$

REVIEW EXERCISES   (page 429)

1. $8xy(xy - 4yz + 2z^2)$ **2.** $(x - y)(a - b)$ **3.** $(a^3 + 5)(a^3 - 5)$ **4.** $(b^2 + 25)(b + 5)(b - 5)$ **5.** $(r + 5s)(r - 3s)$
6. $(4m - 3n)(m - 3n)$

7.3 GETTING READY   (page 429)

1. $5x = 10$ **2.** $y = 6$ **3.** $2x = 33$ **4.** $18y = 28$

ORALS   (page 434)

1. 1   **2.** 2   **3.** 3   **4.** 5

EXERCISE 7.3   (page 434)

1. $(1, 4)$ **3.** $(-2, 3)$ **5.** $(-1, 1)$ **7.** $(2, 5)$ **9.** $(-3, 4)$ **11.** $(0, 8)$ **13.** $(2, 3)$ **15.** $(3, -2)$ **17.** $(2, 7)$ **19.** inconsistent
21. dependent   **23.** $(4, 0)$ **25.** $\left(\frac{10}{3}, \frac{10}{3}\right)$ **27.** $(5, -6)$ **29.** $(-5, 0)$ **31.** $\left(-1, 2\right)$ **33.** $\left(1, -\frac{5}{2}\right)$ **35.** $(-1, 2)$ **37.** $(0, 1)$
39. $(-2, 3)$ **41.** $(2, 2)$

REVIEW EXERCISES   (page 435)

1. 4 **2.** 6 **3.** $\frac{2}{3}, 1$ **4.** $\frac{2}{3}, 2$ **5.** $-\frac{5}{2}, \frac{2}{5}$ **6.** $2, \frac{2}{3}$

7.4 GETTING READY   (page 436)

1. $x + y$ **2.** $x - y$ **3.** $A = lw$ **4.** $P = 2l + 2w$ **5.** $\$(.06a + .07b)$ **6.** $(.05x + .12y)$ liters

ORALS   (page 443)

1. $2x$ **2.** $y + 1$ **3.** $2x + 3y$ **4.** $\$(3x + 2y)$ **5.** $\$(4x + 5y)$

EXERCISE 7.4   (page 443)

1. 32, 64   **3.** 5, 8   **5.** 140   **7.** \$15, \$5   **9.** \$5.40, \$6.20   **11.** 10 ft, 15 ft   **13.** \$150,000   **15.** 25 ft by 30 ft   **17.** 60 ft^2
19. 9.9 years   **21.** 80+   **23.** \$2000   **25.** 250   **27.** 10 mph   **29.** 50 mph   **31.** 5 liters 40% solution, 10 liters 55% solution
33. 32 lb peanuts, 16 lb cashews   **35.** \$640   **37.** 15   **39.** 9%

SOMETHING TO THINK ABOUT   (page 446)

2. 2 nails

REVIEW EXERCISES   (page 447)

1. 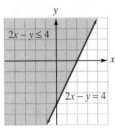 **2.** 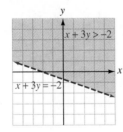 **3.** 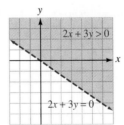 **4.**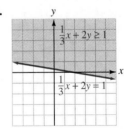

7.5 GETTING READY   (page 447)

1. $x < 6$ **2.** $x < -3$ **3.** $x \geq 3$ **4.** $x \geq -5$

ORALS   (page 452)

1. yes   **2.** no   **3.** yes   **4.** yes   **5.** no   **6.** yes

EXERCISE 7.5 (page 453)

1.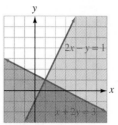

3.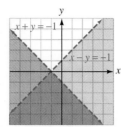

5.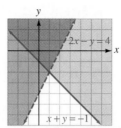

7.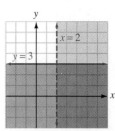

9.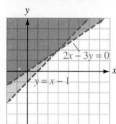

11.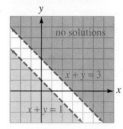

13.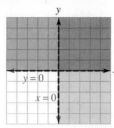

15.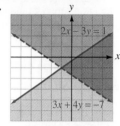

17.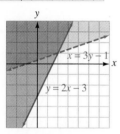

19.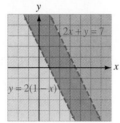

21.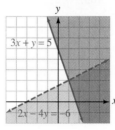

23.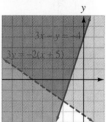

25.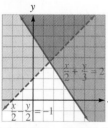

27.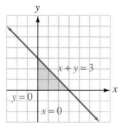

29.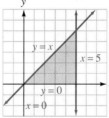

31. 1 $10 CD and 2 $15 CDs; 4 $10 CDs and 1 $15 CD

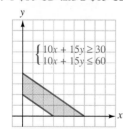

33. 2 desk chairs and 4 side chairs; 1 desk chair and 5 side chairs

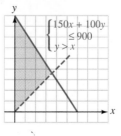

REVIEW EXERCISES (page 456)

1. $\dfrac{1}{z^3}$ **2.** $\dfrac{1}{t^2}$ **3.** $\dfrac{81m^8}{n^{12}}$ **4.** y^{18} **5.** $\dfrac{xz}{y^2}$ **6.** $\dfrac{x^2 + y^2}{y^2 - x^2}$

7.6 GETTING READY   (page 456)

1. no   **2.** yes   **3.** no   **4.** no   **5.** $5x - z = 12$ **6.** $5x + 4y = 22$

ORALS   (page 460)

1. 2   **2.** 8   **3.** 0   **4.** -9

EXERCISE 7.6   (page 461)

1. $(1, 1, 2)$ **3.** $(0, 2, 2)$ **5.** $(3, 2, 1)$ **7.** inconsistent system   **9.** dependent equations   **11.** $(2, 6, 9)$ **13.** $-2, 4, 16$ **15.** 9, 10, 11   **17.** 1000 \$2 chips, 2000 \$3 chips, 3000 \$4 chips   **19.** 50 expensive footballs, 75 middle-priced footballs, 1000 cheap footballs   **21.** 250 \$5 tickets, 375 \$3 tickets, 125 \$2 tickets

REVIEW EXERCISES   (page 462)

1. 2   **2.** 3   **3.** $-\frac{5}{3}$ **4.** $(0, 0)$

CHAPTER 7 REVIEW EXERCISES   (page 465)

1. yes   **5.**   **6.**   **7.**   **8.**
2. no
3. yes
4. yes

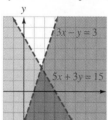
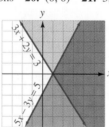
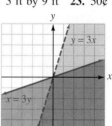
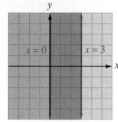

9. $(-1, -2)$ **10.** $(-2, 5)$ **11.** $(1, -1)$ **12.** $(-2, 1)$ **13.** $(3, -5)$ **14.** $\left(3, \frac{1}{2}\right)$ **15.** $(-1, 7)$ **16.** $\left(-\frac{1}{2}, \frac{7}{2}\right)$ **17.** $(0, 9)$
18. inconsistent system   **19.** dependent equations   **20.** $(0, 0)$ **21.** 3, 15   **22.** 3 ft by 9 ft   **23.** 50¢   **24.** \$66   **25.** \$1.69
26. \$750   **27.**   **28.**   **29.**   **30.**

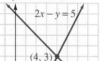
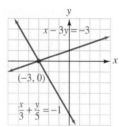
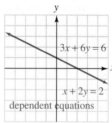
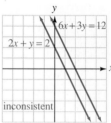

31. $(1, 0, -3)$ **32.** $(2, -1, 1)$ **33.** dependent equations   **34.** $(0, 0, 1)$

CHAPTER 7 TEST   (page 467)

1. yes   **2.** no   **3.**   **4.**   **5.** $(-2, -3)$ **6.** $(12, 10)$ **7.** $(2, 4)$ **8.** $(-3, 3)$
9. inconsistent system   **10.** consistent system   **11.** 65
12. \$4000

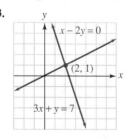
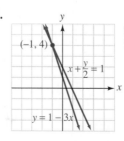

13.

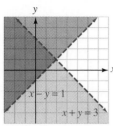

14.

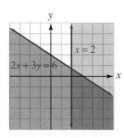

15. $(1, 3, 0)$ **16.** $(-1, 0, -2)$

8.1 GETTING READY (page 470)

1. 9 **2.** 16 **3.** 8 **4.** 125 **5.** 81 **6.** 256 **7.** -27 **8.** -32

ORALS (page 474)

1. 5 **2.** 2 **3.** $2x$ **4.** $6x^2$ **5.** 4 **6.** 3 **7.** 2 **8.** 5 **9.** 1 **10.** 4 **11.** 2 **12.** 1

EXERCISE 8.1 (page 474)

1. 3 **3.** 7 **5.** 6 **7.** 9 **9.** -5 **11.** -9 **13.** 14 **15.** 16 **17.** -17 **19.** 100 **21.** 18 **23.** -60 **25.** 1.414 **27.** 2.236
29. 2.449 **31.** 3.317 **33.** 4.796 **35.** 9.747 **37.** 80.175 **39.** -99.378 **41.** 4.621 **43.** 0.599 **45.** 0.996 **47.** -0.915
49. rational **51.** rational **53.** irrational **55.** imaginary **57.** 1 **59.** 3 **61.** -2 **63.** -4 **65.** 5 **67.** 1 **69.** -4 **71.** 9
73. 2 **75.** -2 **77.** 1 **79.** -2 **81.** xy **83.** x^2z^2 **85.** $-x^2y$ **87.** $2x$ **89.** $-3x^2y$ **91.** xyz **93.** $-xyz^2$ **95.** $-5x^2z^6$
97. $6z^{18}$ **99.** $-4z$ **101.** $3yz^2$ **103.** $-2p^2q$ **105.** $x + 9$ **107.** $x + 11$ **109.** $x + 3$ **111.** $x + 10$ **113.** $x + 7y$

REVIEW EXERCISES (page 475)

1. -1 **2.** 5 **3.** 9 **4.** $2t^2 + t - 1$

8.2 GETTING READY (page 475)

1. 10 **2.** 2 **3.** 5 **4.** 12 **5.** 3 **6.** 4 **7.** $3xy^2$ **8.** $-2x^2y^3$

ORALS (page 480)

1. $2\sqrt{2}$ **2.** $2\sqrt{3}$ **3.** $\dfrac{\sqrt{5}}{3}$ **4.** $\dfrac{\sqrt{7}}{5}$ **5.** $2x$ **6.** $3a^2$

EXERCISE 8.2 (page 480)

1. $2\sqrt{5}$ **3.** $5\sqrt{2}$ **5.** $3\sqrt{5}$ **7.** $7\sqrt{2}$ **9.** $4\sqrt{3}$ **11.** $10\sqrt{2}$ **13.** $8\sqrt{3}$ **15.** $2\sqrt{22}$ **17.** 18 **19.** $7\sqrt{3}$ **21.** $6\sqrt{5}$ **23.** $12\sqrt{3}$
25. $48\sqrt{2}$ **27.** $-70\sqrt{10}$ **29.** $14\sqrt{5}$ **31.** $-45\sqrt{2}$ **33.** $5\sqrt{x}$ **35.** $a\sqrt{b}$ **37.** $3x\sqrt{y}$ **39.** $x^3y^2\sqrt{2}$ **41.** $48x^2y\sqrt{y}$
43. $-9x^2y^3z\sqrt{2xy}$ **45.** $6ab^2\sqrt{3ab}$ **47.** $-\dfrac{8n\sqrt{5m}}{5}$ **49.** $\dfrac{5}{3}$ **51.** $\dfrac{9}{8}$ **53.** $\dfrac{\sqrt{26}}{5}$ **55.** $\dfrac{2\sqrt{5}}{7}$ **57.** $\dfrac{4\sqrt{3}}{9}$ **59.** $\dfrac{4\sqrt{2}}{5}$ **61.** $\dfrac{5\sqrt{5}}{11}$
63. $\dfrac{7\sqrt{5}}{6}$ **65.** $\dfrac{6x\sqrt{2x}}{y}$ **67.** $\dfrac{5mn^2\sqrt{5}}{8}$ **69.** $\dfrac{4m\sqrt{2}}{3n}$ **71.** $\dfrac{2rs^2\sqrt{3}}{9}$ **73.** $2x$ **75.** $-4x\sqrt[3]{x^2}$ **77.** $3xyz^2\sqrt[3]{2y}$ **79.** $-3yz\sqrt[3]{3x^2z}$
81. $\dfrac{3m}{2n^2}$ **83.** $\dfrac{rs\sqrt[3]{2rs^2}}{5t}$ **85.** $\dfrac{5ab}{2}$

REVIEW EXERCISES (page 481)

1. $\dfrac{1}{2xz}$ **2.** $\dfrac{5a}{9bc}$ **3.** $\dfrac{a+1}{a+3}$ **4.** $\dfrac{y+6}{y+3}$ **5.** 1 **6.** $\dfrac{x-3}{x-2}$ **7.** 2 **8.** $-\dfrac{16}{u+1}$

8.3 GETTING READY (page 482)

1. $7x$ **2.** $3y$ **3.** $6xy$ **4.** $9t^2$ **5.** $2\sqrt{5}$ **6.** $3\sqrt{5}$ **7.** $2x\sqrt{2y}$ **8.** $3x\sqrt{2y}$

ORALS (page 485)

1. $5\sqrt{5}$ **2.** $5\sqrt{7}$ **3.** $2\sqrt{xy}$ **4.** $4\sqrt{5yz}$ **5.** $3\sqrt{2}$ **8.** $-\sqrt{3}$

EXERCISE 8.3 (page 485)

1. $5\sqrt{3}$ **3.** $9\sqrt{3}$ **5.** $7\sqrt{5}$ **7.** $12\sqrt{5}$ **9.** $8\sqrt{5}$ **11.** $10\sqrt{10}$ **13.** $37\sqrt{5}$ **15.** $12\sqrt{7}$ **17.** $38\sqrt{2}$ **19.** $85\sqrt{2}$ **21.** $9\sqrt{5}$
23. $7\sqrt{6} + 4\sqrt{15}$ **25.** $\sqrt{2}$ **27.** $3 - 5\sqrt{2}$ **29.** $2\sqrt{2}$ **31.** $-2\sqrt{3}$ **33.** $\sqrt{3}$ **35.** $4\sqrt{10}$ **37.** $-7\sqrt{5}$ **39.** $22\sqrt{6}$
41. $-18\sqrt{2}$ **43.** $13\sqrt{10}$ **45.** $3\sqrt{2} - \sqrt{3}$ **47.** $10\sqrt{2} - \sqrt{3}$ **49.** $-6\sqrt{6}$ **51.** $10\sqrt{2} + 5\sqrt{3}$ **53.** $7\sqrt{3} - 6\sqrt{2}$
55. $6\sqrt{6} - 2\sqrt{3}$ **57.** $3x\sqrt{2}$ **59.** $3x\sqrt{2x}$ **61.** $3x\sqrt{2y} - 3x\sqrt{3y}$ **63.** $x^2\sqrt{2x}$ **65.** $19x\sqrt{6}$ **67.** $-5y\sqrt{10y}$
69. $2x\sqrt{5xy} + 3x^2y\sqrt{5xy} - 4x^3y^2\sqrt{5xy}$ **71.** $5\sqrt[3]{2}$ **73.** $\sqrt[3]{3}$ **75.** $2\sqrt[3]{5} + 5$ **77.** $x\sqrt[3]{x} - x^2\sqrt[3]{x}$ **79.** $2xy\sqrt[3]{3xy^2}$ **81.** $x^2y\sqrt[3]{5xy}$

REVIEW EXERCISES (page 487)

1. 8 **2.** 36 **3.** -9 **4.** 7

8.4 GETTING READY (page 487)

1. x^5 **2.** y^7 **3.** x^3 **4.** y^3 **5.** $x^2 + 2x$ **6.** $6y^5 - 8y^4$ **7.** $x^2 - x - 6$ **8.** $6x^2 + 13xy + 6y^2$

ORALS (page 493)

1. 5 **2.** 10 **3.** $x + 2\sqrt{x}$ **4.** $4\sqrt{y} - y$ **5.** $x - 1$ **6.** $1 - z$ **7.** $\dfrac{\sqrt{5}}{5}$ **8.** $\sqrt{x}$

EXERCISE 8.4 (page 493)

1. 3 **3.** 4 **5.** 8 **7.** 4 **9.** x^3 **11.** b^7 **13.** $4\sqrt{15}$ **15.** $-60\sqrt{2}$ **17.** 24 **19.** $-8x$ **21.** $700x\sqrt{10}$ **23.** $4x^2\sqrt{y}$
25. $2 + \sqrt{2}$ **27.** $9 - \sqrt{3}$ **29.** $7 - 3\sqrt{7}$ **31.** $3\sqrt{5} - 5$ **33.** $3\sqrt{2} + \sqrt{3}$ **35.** $7 - 2\sqrt[3]{7}$ **37.** $x\sqrt{3} - 2\sqrt{x}$ **39.** $6x + 6\sqrt{x}$
41. $6\sqrt{x} + 3x$ **43.** $3x\sqrt{7} + x\sqrt{42}$ **45.** 1 **47.** 1 **49.** $\sqrt[3]{4} + 2\sqrt[3]{2} + 1$ **51.** $7 - x^2$ **53.** $2 - 2\sqrt{2x} + x$ **55.** $6x - 7$
57. $4x - 18$ **59.** $8xy + 4\sqrt{2xy} + 1$ **61.** $16x - 8$ **63.** $\frac{2x}{3}$ **65.** $\dfrac{3y\sqrt{2}}{5}$ **67.** $\frac{2y}{x}$ **69.** $\frac{2x}{3}$ **71.** $\dfrac{y\sqrt{xy}}{3}$ **73.** $\dfrac{5\sqrt{2y}}{x}$ **75.** $\dfrac{\sqrt{3}}{3}$
77. $\dfrac{2\sqrt{7}}{7}$ **79.** $\sqrt[3]{25}$ **81.** $\sqrt[3]{3}$ **83.** $\dfrac{3\sqrt{2}}{8}$ **85.** $2\sqrt[3]{2}$ **87.** $\dfrac{\sqrt{15}}{3}$ **89.** $\dfrac{10\sqrt{x}}{x}$ **91.** $\dfrac{3\sqrt{2x}}{2x}$ **93.** $\dfrac{\sqrt{2xy}}{3y}$ **95.** $\dfrac{\sqrt{6}}{2x}$ **97.** $\sqrt{3x}$
99. $\dfrac{\sqrt[3]{20}}{2}$ **101.** $\sqrt[3]{x}$ **103.** $\dfrac{\sqrt[3]{2xy^2}}{xy}$ **105.** $\dfrac{-\sqrt[3]{5ab}}{ab}$ **107.** $\dfrac{3(\sqrt{3} + 1)}{2}$ **109.** $\sqrt{7} - 2$ **111.** $6 + 2\sqrt{3}$ **113.** $2 - \sqrt{2}$
115. $\dfrac{\sqrt{3} - 3}{2}$ **117.** $5\sqrt{3} - 5\sqrt{2}$ **119.** $\sqrt{10} + \sqrt{6}$ **121.** $5 - 2\sqrt{6}$ **123.** $\dfrac{3x - 2\sqrt{3x} + 1}{3x - 1}$ **125.** $\dfrac{2x + 2\sqrt{2x} - 15}{2x - 9}$
127. $\dfrac{2\sqrt{2z} - 1 - 2z}{2z - 1}$ **129.** $\dfrac{y^2 - 2y\sqrt{15} + 15}{y^2 - 15}$

REVIEW EXERCISES (page 496)

1. $(x - 7)(x + 3)$ **2.** $(y + 9)(y - 3)$ **3.** $3xy(2x - 3)$ **4.** $(x + 2)(x^2 - 2x + 4)$

8.5 GETTING READY (page 496)

1. x **2.** $3y$ **3.** $x - 2$ **4.** $y + 3$ **5.** x **6.** $x - 2$ **7.** $x^2 + 2x + 1$ **8.** $4y^2 - 12y + 9$

ORALS (page 500)

1. 16 **2.** 5 **3.** 3 **4.** $\frac{1}{2}$ **5.** 2 **6.** 1

EXERCISE 8.5   (page 500)

1. 9   **3.** 49   **5.** no solution   **7.** 1   **9.** 30   **11.** no solution   **13.** 5   **15.** 6   **17.** 3   **19.** -1 **21.** 46   **23.** -3 **25.** 0
27. no solution   **29.** 2   **31.** -1 **33.** 10   **35.** 3   **37.** 0, -1 **39.** 2   **41.** 2, 1   **43.** 3   **45.** 65   **47.** 22   **49.** 161   **51.** 256 ft
53. about 1.8 ft   **55.** 980 watts   **57.** 56 mph   **59.** 2×10^6 m   **61.** $v^2 = c^2 - f^2 c^2$

REVIEW EXERCISES   (page 502)

1. $(2, 3)$ **2.** $(-1, 2)$ **3.** $(3, -2)$ **4.** $(-3, -5)$

8.6 GETTING READY   (page 503)

1. x^6 **2.** x^{12} **3.** x^4 **4.** 1   **5.** $\dfrac{1}{x^3}$ **6.** $x^3 y^6$ **7.** $\dfrac{x^8}{y^6}$ **8.** x^{15}

ORALS   (page 506)

1. 4   **2.** 5   **3.** 3   **4.** 3   **5.** 4   **6.** 8   **7.** $\frac{1}{3}$ **8.** $\frac{1}{4}$

EXERCISE 8.6   (page 506)

1. 9   **3.** -12 **5.** $\frac{1}{2}$ **7.** $\frac{2}{7}$ **9.** 3   **11.** -5 **13.** -2 **15.** $\frac{1}{4}$ **17.** $\frac{3}{4}$ **19.** 2   **21.** 2   **23.** -3 **25.** 729   **27.** 125   **29.** 25
31. 100   **33.** 4   **35.** 8   **37.** 27   **39.** -8 **41.** $\frac{4}{9}$ **43.** $\frac{8}{125}$ **45.** 6   **47.** 25   **49.** 7   **51.** 25   **53.** 8   **55.** 125   **57.** 6   **59.** 192
61. $\frac{1}{2}$ **63.** $\frac{1}{9}$ **65.** $\frac{1}{64}$ **67.** $\frac{1}{81}$ **69.** x **71.** x^2 **73.** x^4 **75.** x^2 **77.** y^2 **79.** $x^{2/5}$ **81.** $x^{2/7}$ **83.** x **85.** y^7 **87.** y^2 **89.** $x^{17/12}$
91. $b^{3/10}$ **93.** $t^{4/15}$ **95.** x **97.** $a^{14/15}$ **99.** $\dfrac{3}{b^{101/60}}$

SOMETHING TO THINK ABOUT   (page 508)
1. 2^x **2.** $\left(\frac{1}{2}\right)^y$

REVIEW EXERCISES   (page 508)

1. $3(z - 4t)(z - t)$ **2.** $(a^2 + b^2)(a + b)(a - b)$ **3.** 5   **4.** 4

ORALS   (page 513)

1. 5   **2.** 13   **3.** 10   **4.** 17

EXERCISE 8.7   (page 513)

1. 5   **3.** 13   **5.** 20   **7.** 28   **9.** $2\sqrt{14}$ **11.** 5   **13.** 10   **15.** 13   **17.** 10   **19.** 5   **21.** 12 ft   **23.** 30 ft   **25.** $90\sqrt{2}$ ft or 127.3 ft
27. 5.8 mi   **29.** 2.45 in.   **31.** no   **33.** $3\sqrt{2}$ ft   **35.** 24 in.

REVIEW EXERCISES   (page 516)

1.
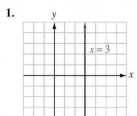

2.
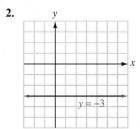

3.
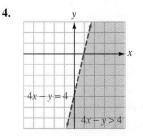

4.
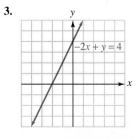

CHAPTER 8 REVIEW EXERCISES (page 519)

1. 5 **2.** 8 **3.** -12 **4.** -17 **5.** 16 **6.** -8 **7.** 13 **8.** -15 **9.** 3 **10.** -5 **11.** 3 **12.** 2 **13.** 4.583 **14.** -3.873
15. -7.570 **16.** 27.421 **17.** $4\sqrt{2}$ **18.** $5\sqrt{2}$ **19.** $10\sqrt{5}$ **20.** $4\sqrt{7}$ **21.** $4x\sqrt{5}$ **22.** $3y\sqrt{7}$ **23.** $-5t\sqrt{10t}$ **24.** $-10z^2\sqrt{7z}$
25. $10x\sqrt{2y}$ **26.** $5y\sqrt{3z}$ **27.** $2y\sqrt[3]{x^2}$ **28.** $5xy\sqrt[3]{2x}$ **29.** $\frac{4}{5}$ **30.** $\frac{10}{7}$ **31.** $\frac{10}{3}$ **32.** $\frac{\sqrt[3]{2}}{2}$ **33.** $\frac{2\sqrt{15}}{7}$ **34.** $\frac{4\sqrt{5}}{15}$ **35.** $\frac{11x\sqrt{2}}{13}$
36. $\frac{15a^2\sqrt{2}}{14}$ **37.** 0 **38.** $2\sqrt{3}$ **39.** $18\sqrt{5}$ **40.** $\sqrt{7}$ **41.** $5x\sqrt{2y}$ **42.** $y^2\sqrt{5xy}$ **43.** $5\sqrt[3]{2}$ **44.** $6x\sqrt[3]{2}$ **45.** $-6\sqrt{6}$ **46.** $10x$
47. $36x\sqrt{2}$ **48.** $-6y^3\sqrt{6}$ **49.** $4\sqrt[3]{2}$ **50.** $-24x\sqrt[3]{x}$ **51.** -2 **52.** $2y\sqrt{3}+15\sqrt{2y}$ **53.** -2 **54.** $15+6x\sqrt{15}+9x^2$
55. $\sqrt[3]{9}+\sqrt[3]{3}-2$ **56.** $\sqrt[3]{25}-1$ **57.** $49+12\sqrt{5}$ **58.** $13-4\sqrt{3}$ **59.** $x-2\sqrt{2x}+2$ **60.** $7+2\sqrt{7a}+a$ **61.** $\frac{\sqrt{7}}{7}$
62. $\frac{\sqrt{2}}{2}$ **63.** $2\sqrt[3]{4}$ **64.** $\frac{5\sqrt[3]{2}}{2}$ **65.** $\frac{\sqrt{5}}{5}$ **66.** $\frac{3\sqrt{3}+3}{2}$ **67.** $5-\sqrt{15}$ **68.** $\frac{4+\sqrt{7}}{3}$ **69.** 6 **70.** -3 **71.** no solution
72. 4 **73.** 4; -1 is extraneous **74.** 8; 0 is extraneous **75.** 2; -2 is extraneous **76.** 9; -1 is extraneous **77.** 7 **78.** -10
79. 216 **80.** $\frac{32}{243}$ **81.** 64 **82.** 25 **83.** x **84.** $\frac{1}{r}$ **85.** 6 **86.** 5 **87.** x **88.** $\frac{1}{a^2b^4}$ **89.** $x^{11/15}$ **90.** $t^{1/12}$ **91.** $\frac{1}{x^{4/5}}$ **92.** $\frac{1}{r^{1/4}}$
93. 35 **94.** 60 **95.** 1 **96.** $2\sqrt{6}$ **97.** 5 **98.** 13 **99.** 2 **100.** 29 **101.** 68 in. **102.** 45 ft

CHAPTER 8 TEST (page 521)

1. 10 **2.** -20 **3.** -3 **4.** $9x$ **5.** $2x\sqrt{2}$ **6.** $3x\sqrt{6xy}$ **7.** $4\sqrt{2}$ **8.** $3y\sqrt{x}$ **9.** x^2y^2 **10.** $\frac{2x^2}{y}$ **11.** $5\sqrt{3}$ **12.** $-x\sqrt{2x}$
13. $-24x\sqrt{6}$ **14.** $2\sqrt{6}+3\sqrt{2}$ **15.** -1 **16.** $2x-4\sqrt{x}-6$ **17.** $\sqrt{2}$ **18.** $\frac{y\sqrt{xy}}{4x}$ **19.** $2\sqrt{5}+4$ **20.** $\frac{x\sqrt{3}-2\sqrt{3x}}{x-4}$
21. 36 **22.** 66 **23.** 5 **24.** 5 **25.** 5; 0 is extraneous **26.** 29 **27.** 11 **28.** $\frac{1}{81}$ **29.** y^6 **30.** a^{10} **31.** $p^{17/12}$ **32.** $\frac{1}{x^{4/15}}$
33. 13 in. **34.** 10 units **35.** 5 units **36.** 10 ft

9.1 GETTING READY (page 524)

1. $2x(2x-1)$ **2.** $(x+3)(x-3)$ **3.** $(x+3)(x-2)$ **4.** $(2x-3)(x+3)$ **5.** 5 **6.** -6 **7.** $2\sqrt{5}$ **8.** $-5\sqrt{2}$

ORALS (page 529)

1. $-5, 5$ **2.** 2, 3 **3.** $10, -10$ **4.** $7, -7$

EXERCISE 9.1 (page 529)

1. ±3 **3.** $0, -3$ **5.** 2, 3 **7.** $-1, \frac{2}{3}$ **9.** $-\frac{1}{3}, -\frac{3}{2}$ **11.** $\frac{2}{5}, -\frac{1}{2}$ **13.** ±1 **15.** ±3 **17.** $\pm2\sqrt{5}$ (±4.5) **19.** ±3 **21.** ±2
23. $\pm\sqrt{a}$ **25.** $-6, 4$ **27.** 7, -11 **29.** $2\pm2\sqrt{2}$ (4.8 and -0.8) **31.** $3a, -a$ **33.** $-b\pm4c$ **35.** 0, -4 **37.** 2, $-\frac{2}{3}$
39. $\frac{-1\pm2\sqrt{5}}{2}$ (1.7 and -2.7) **41.** ±3 **43.** ±5 **45.** $-1\pm2\sqrt{2}$ (1.8 and -3.8)

REVIEW EXERCISES (page 530)

1. y^2-2y+1 **2.** z^2+4z+4 **3.** $x^2+2xy+y^2$ **4.** $a^2-2ab+b^2$ **5.** $4r^2-4rs+s^2$ **6.** $m^2+6mn+9n^2$

9.2 GETTING READY (page 530)

1. 9 **2.** 25 **3.** 1 **4.** $\frac{25}{4}$ **5.** 16 **6.** 36 **7.** $\frac{1}{16}$ **8.** $\frac{1}{9}$

ORALS (page 534)

1. 4 **2.** 9 **3.** 16 **4.** 25 **5.** $\frac{25}{4}$ **6.** $\frac{9}{4}$

EXERCISE 9.2   (page 535)

1. $x^2 + 2x + 1$ **3.** $x^2 - 4x + 4$ **5.** $x^2 + 7x + \frac{49}{4}$ **7.** $a^2 - 3a + \frac{9}{4}$ **9.** $b^2 + \frac{2}{3}b + \frac{1}{9}$ **11.** $c^2 - \frac{5}{2}c + \frac{25}{16}$ **13.** $-2, -4$ **15.** $2, 6$
17. $5, -3$ **19.** $3, 4$ **21.** $1, -6$ **23.** $1, -2$ **25.** $-4, -4$ **27.** $2, -\frac{1}{2}$ **29.** $-2, \frac{1}{4}$ **31.** $-2 \pm \sqrt{3}$ $(-0.3 \text{ and } -3.7)$
33. $1 \pm \sqrt{5}$ $(3.2 \text{ and } -1.2)$ **35.** $2 \pm \sqrt{7}$ $(4.6 \text{ and } -0.6)$ **37.** $-1 \pm \sqrt{2}$ $(0.4 \text{ and } -2.4)$ **39.** $1, -4$ **41.** $\frac{3}{2}, -\frac{2}{3}$
43. $\dfrac{-3 \pm \sqrt{3}}{2}$ $(-0.6 \text{ and } -2.4)$

REVIEW EXERCISES   (page 536)

1. -4 **2.** $\frac{3}{4}, -\frac{1}{5}$ **3.** 3 **4.** 6

9.3 GETTING READY   (page 536)

1. -8 **2.** -7 **3.** 8 **4.** 0

ORALS   (page 540)

1. $a = 3, b = 4, c = -12$ **2.** $a = -2, b = -4, c = 10$ **3.** $a = 5, b = -1, c = -1$ **4.** $a = 1, b = 3, c = -9$

EXERCISE 9.3   (page 541)

1. $a = 1, b = 4, c = 3$ **3.** $a = 3, b = -2, c = 7$ **5.** $a = 4, b = -2, c = 1$ **7.** $a = 3, b = -5, c = -2$ **9.** $a = 7, b = 13,$
$c = 21$ **11.** $a = 1, b = -1, c = -5$ **13.** $2, 3$ **15.** $-3, -4$ **17.** $1, -\frac{1}{2}$ **19.** $-1, -\frac{2}{3}$ **21.** $\frac{1}{2}, -\frac{3}{2}$ **23.** $2, -\frac{2}{5}$ **25.** $\dfrac{-3 \pm \sqrt{5}}{2}$
$(-0.4 \text{ and } -2.6)$ **27.** $\dfrac{-5 \pm \sqrt{37}}{2}$ $(0.5 \text{ and } -5.5)$ **29.** no real solutions   **31.** $\dfrac{-1 \pm \sqrt{41}}{4}$ $(1.4 \text{ and } -1.9)$ **33.** $-2 \pm \sqrt{3}$
$(-0.3 \text{ and } -3.7)$ **35.** no real solutions   **37.** $-1 \pm \sqrt{2}$ $(0.4 \text{ and } -2.4)$ **39.** $\dfrac{3 \pm \sqrt{15}}{3}$ $(2.3 \text{ and } -0.3)$

REVIEW EXERCISES   (page 542)

1. $r = \dfrac{A - p}{pt}$ **2.** $M = \dfrac{Fd^2}{Gm}$ **3.** $3x - 5y = -60$ **4.** $4x - 3y = 0$

9.4 GETTING READY   (page 542)

1. 72 cm^2 **2.** $16t(9 - t)$ **3.** 30 ft **4.** $\$300$

ORALS   (page 546)

1. $x, x + 1, x + 2$ **2.** $x, x + 2, x + 4$ **3.** $A = lw$ **4.** $P = 2l + 2w$ **5.** $a^2 + b^2 = c^2$

EXERCISE 9.4   (page 547)

1. $6, 8$ **3.** 9 **5.** no   **7.** 4 ft by 8 ft   **9.** 6 in.   **11.** 10 ft by 18 ft   **13.** about 9.5 sec   **15.** 288 ft   **17.** 324 ft   **19.** 68 sec
21. 218 sec   **23.** 6 in. by 8 in.   **25.** 30 and 40 nautical miles   **27.** 7%   **29.** $\$210,000$ **31.** 2 in.   **33.** 4 hr

REVIEW EXERCISES   (page 550)

1. $4\sqrt{5}$ **2.** $12xy\sqrt{x}$ **3.** $\dfrac{\sqrt{7x}}{7}$ **4.** $\dfrac{x + 4\sqrt{x} + 4}{x - 4}$

9.5 GETTING READY   (page 550)

1. $5x - 1$ **2.** $x - 7$ **3.** $x^2 - x - 20$ **4.** $6x^2 - 5x + 1$

ORALS (page 557)

1. $5 + 7i$ **2.** $1 + 5i$ **3.** $1 + 3i$ **4.** $1 - 3i$ **5.** $3 - i$ **6.** $3 + i$

EXERCISE 9.5 (page 557)

1. $\pm 3i$ **3.** $\pm \dfrac{4\sqrt{3}}{3}i$ **5.** $-1 \pm i$ **7.** $-\dfrac{1}{4} \pm \dfrac{\sqrt{7}}{4}i$ **9.** $\dfrac{2}{3} \pm \dfrac{\sqrt{2}}{3}i$ **11.** i **13.** $-i$ **15.** 1 **17.** i **19.** $8 - 2i$ **21.** $3 - 5i$
23. $15 + 7i$ **25.** $-15 + 2\sqrt{3}i$ **27.** $3 + 6i$ **29.** $9 + 7i$ **31.** $14 - 8i$ **33.** $8 + \sqrt{2}i$ **35.** $6 - 8i$
37. $6 + \sqrt{6} + (3\sqrt{3} - 2\sqrt{2})i$ **39.** $-16 - \sqrt{35} + (2\sqrt{5} - 8\sqrt{7})i$ **41.** $0 - i$ **43.** $0 + \frac{4}{5}i$ **45.** $\frac{1}{8} + 0i$ **47.** $0 + \frac{3}{5}i$
49. $0 + \dfrac{3\sqrt{2}}{4}i$ **51.** $\frac{15}{26} - \frac{3}{26}i$ **53.** $-\frac{42}{25} - \frac{6}{25}i$ **55.** $\frac{1}{4} + \frac{3}{4}i$ **57.** $\frac{5}{13} - \frac{12}{13}i$ **59.** $\dfrac{6 + \sqrt{10}}{9} + \dfrac{2\sqrt{2} - 3\sqrt{5}}{9}i$ **61.** 10 **63.** 13
65. $\sqrt{74}$ **67.** $3\sqrt{2}$ **69.** $\sqrt{69}$ **71.** 5

SOMETHING TO THINK ABOUT (page 559)

2. $\frac{3}{34} - \frac{5}{34}i$

REVIEW EXERCISES (page 559)

1. $3(x + 3)(x - 3)$ **2.** $2(x + 2)(x - 2)$ **3.** $(2x - 1)(x + 1)$ **4.** $-(x - 3)(x + 7)$

9.6 GETTING READY (page 559)

1. 2 **2.** 3 **3.** 5 **4.** 12 **5.** -2 **6.** 2

ORALS (page 566)

1. up **2.** up **3.** down **4.** down **5.** down **6.** up

EXERCISE 9.6 (page 566)

1.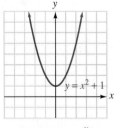
3.
5.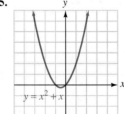
7.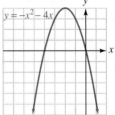

9.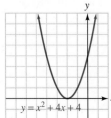
11.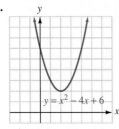
13. $(2, 4)$ **15.** $(-1, -5)$ **17.** $(1, 0)$ **19.** $(0, 4)$ **21.** $(-1, 4)$

23. $(3, -21)$ **25.**

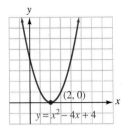

27.

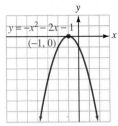

29.

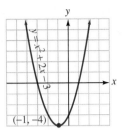

31.

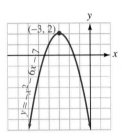

33.

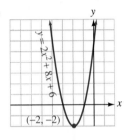

35.

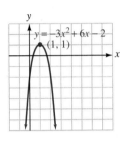

37. $\left(\frac{1}{2}, -\frac{9}{4}\right), (2, 0), (-1, 0), (0, -2)$

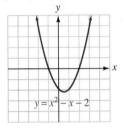

39. $(1, 4), (3, 0), (-1, 0), (0, 3)$ **41.** $\left(-\frac{3}{4}, -\frac{25}{8}\right), (-2, 0), \left(\frac{1}{2}, 0\right), (0, -2)$ **43.** 1350   **45.** 750   **47.** 2, 3   **49.** $-0.44, -4.56$

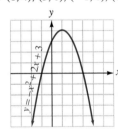

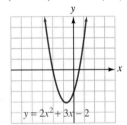

REVIEW EXERCISES   (page 569)

1. $5\sqrt{3}$ **2.** $-36y\sqrt{2}$ **3.** 2   **4.** $\dfrac{x + 4\sqrt{x} + 4}{x - 4}$

CHAPTER 9 REVIEW EXERCISES   (page 572)

1. ± 5 **2.** ± 6 **3.** ± 3 **4.** $\pm \frac{3}{2}$ **5.** $\pm 2\sqrt{2}$ (± 2.8)   **6.** $\pm 5\sqrt{3}$ (± 8.7)   **7.** $-4, 6$ **8.** $3, -9$ **9.** $2, -4$ **10.** $\frac{7}{2}, \frac{1}{2}$ **11.** $8 \pm 2\sqrt{2}$ (10.8 and 5.2)   **12.** $-5 \pm 5\sqrt{3}$ (3.7 and -13.7)   **13.** $2, -4$ **14.** $5, 1$ **15.** $4 \pm 2\sqrt{5}$ (8.5 and -0.5)   **16.** $1 \pm 3\sqrt{2}$ (5.2 and -3.2)   **17.** $2, -7$ **18.** $3, 5$ **19.** $7, -11$ **20.** $1 \pm \sqrt{2}$ (2.4 and -0.4)   **21.** $-2 \pm \sqrt{7}$ (0.6 and -4.6)   **22.** $3 \pm \sqrt{5}$ (5.2 and 0.8)   **23.** $\frac{1}{2}, -3$ **24.** $\dfrac{1 \pm \sqrt{3}}{2}$ (1.4 and -0.4)   **25.** $5, -3$ **26.** $7, -1$ **27.** $13, 2$ **28.** $\frac{1}{2}, 3$ **29.** $\frac{3}{2}, -\frac{1}{3}$ **30.** $-2 \pm \sqrt{3}$ (-0.3 and -3.7)   **31.** $3 \pm \sqrt{2}$ (4.4 and 1.6)   **32.** $0, -3$ **33.** 11 and 13   **34.** 68 cm   **35.** $12 + 2i$ **36.** $-2 - 68i$ **37.** $-28 - 21i$ **38.** $-6 + 12i$ **39.** $-24 + 28i$ **40.** $118 + 10\sqrt{2}i$ **41.** $0 - \frac{3}{4}i$ **42.** $0 - \frac{2}{5}i$ **43.** $\frac{12}{5} - \frac{6}{5}i$ **44.** $\frac{21}{10} + \frac{7}{10}i$ **45.** $\frac{15}{17} + \frac{8}{17}i$ **46.** $\frac{4}{5} - \frac{3}{5}i$ **47.** $\frac{15}{29} - \frac{6}{29}i$ **48.** $\frac{1}{3} + \frac{1}{3}i$ **49.** 15   **50.** 26   **51.** $(6, 7)$ **52.** $(-3, -5)$ **53.** $(1, 5)$ **54.** $(3, 16)$

55. 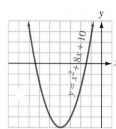 **56.** 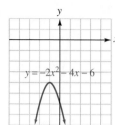 **57.** 3, 1.5   **58.** 0.5, 1.5

CHAPTER 9 TEST   (page 574)

1. $\frac{1}{3}, -\frac{1}{2}$ **2.** 4, -4 **3.** $2 \pm \sqrt{3}$ (3.7 and 0.3)   **4.** 3, -2 **5.** 49   **6.** $\frac{49}{4}$ **7.** $-1 \pm \sqrt{5}$ (1.2 and -3.2)

8. $x = \dfrac{-b \pm \sqrt{b^2 - 4ac}}{2a}$ **9.** 2, -5 **10.** $-\frac{3}{2}$, 4   **11.** $\dfrac{-5 \pm \sqrt{33}}{4}$ (0.2 and -2.7)   **12.** $7i$ **13.** $1 + 9i$ **14.** $9 - 5i$

15. $16 + 11i$ **16.** $-\frac{3}{5} + \frac{1}{5}i$ **17.** $\frac{3}{5} + \frac{4}{5}i$ **18.** 5   **19.** $(-5, -4)$ **20.**

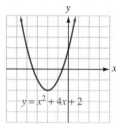

CUMULATIVE REVIEW EXERCISES   (page 574)

1. 15   **2.** 4   **3.** $\frac{2}{3}, -\frac{1}{2}$ **4.** 0, 2   **5.**    **6.**   **7.**    **8.**   **9.** $\frac{x+1}{x+3}$

10. $\frac{x+4}{x+6}$ **11.** 1   **12.** $x + 2$ **13.** $\frac{10x + 4y}{21}$ **14.** $\frac{7x + 29}{(x+5)(x+7)}$ **15.** 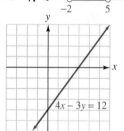   **16.**   **17.** $\frac{1}{2}$

18. $x - 2y = -12$ **19.** $y = 2x$ **20.** 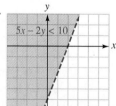   **21.** $(4, -3)$ **22.** $\left(\frac{1}{2}, \frac{2}{3}\right)$ **23.**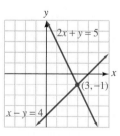

24.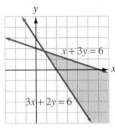

25. $\frac{1}{2}$ **26.** $3x\sqrt{3x}$ **27.** $5xy\sqrt{7y}$ **28.** $-2\sqrt[3]{10}$ **29.** $\sqrt{3}$ **30.** x **31.** $18x\sqrt{2}$ **32.** $2 + 2\sqrt{2}$

33. $5 - \sqrt{3}$ **34.** $x - 4$ **35.** $\dfrac{\sqrt{3x}}{x}$ **36.** $\sqrt{x} + 4$ **37.** 4   **38.** 5; 0 is extraneous   **39.** 2   **40.** $\frac{4}{9}$ **41.** x **42.** x^2 **43.** 10

44. 13   **45.** 9   **46.** $\frac{25}{4}$ **47.** $\frac{1}{16}$ **48.** $\frac{1}{9}$ **49.** $2, -4$ **50.** $-3 \pm \sqrt{5}$ (-0.8 and -5.2)   **51.** $0 + 5i$ **52.** $3 - 4i$ **53.** $5 + i$

54. $1 + 9i$ **55.** $12 + 5i$ **56.** $\frac{3}{10} - \frac{1}{10}i$ **57.** 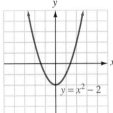   **58.**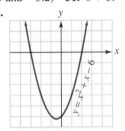

1. b   **2.** b   **3.** c   **4.** c   **5.** a   **6.** e   **7.** a   **8.** c   **9.** c   **10.** d   **11.** b   **12.** c   **13.** d   **14.** a   **15.** b   **16.** d   **17.** c   **18.** b
19. c   **20.** b   **21.** d   **22.** d   **23.** d   **24.** b   **25.** e   **26.** d   **27.** d   **28.** b   **29.** c   **30.** a   **31.** b   **32.** a   **33.** d   **34.** a   **35.** d
36. c   **37.** b   **38.** c   **39.** c   **40.** a   **41.** a   **42.** b   **43.** e   **44.** c   **45.** a   **46.** c   **47.** e   **48.** d   **49.** b   **50.** c   **51.** a   **52.** c
53. b   **54.** a   **55.** d   **56.** b   **57.** c   **58.** a   **59.** e   **60.** c

1. $\{2, 4, 6, 8\}$ **3.** {Sunday, Saturday}   **5.** $\emptyset$ **7.** $\{2, 3, 5, 7, 11, 13, 17, 19\}$ **9.** {Alaska}   **11.** $\{1, 2, 3, 4, 5, 6, 7, 8, 9, 10, 11, 12\}$ **13.** $\{2\}$ **15.** $\{x | x \text{ is a natural number less than } 10\}$ **17.** $\{x | x \text{ is the square of a natural number less than } 10\}$ **19.** $\{x | x$ is a month with 31 days}   **21.** $\in$ **23.** $\notin$ **25.** $\not\subset$ **27.** $\subset$ **29.** $\subset$ **31.** $\subset$ **33.** $\{a, b, c, d, e, f\}$ **35.** $\emptyset$ **37.** $\{a, d, e, f, g\}$
39. $\{a, b, c, d, e, f, g\}$

■ ■ ■ ■ ■ ■ ■ INDEX

■ ■ ■ ■ ■ ■ ■ INDEX OF APPLICATIONS

3.6 MULTIPLYING POLYNOMIALS

$(x + y)^2 = x^2 + 2xy + y^2$
$(x - y)^2 = x^2 - 2xy + y^2$
$(x + y)(x - y) = x^2 - y^2$

3.7 DIVIDING POLYNOMIALS BY MONOMIALS

$\dfrac{a}{b} = \dfrac{1}{b} \cdot a \quad (b \neq 0)$

4.1–4.6 FACTORING POLYNOMIALS

$ax + bx = x(a + b)$

Zero-factor property: Let a and b be real numbers. Then:

If $ab = 0$, then $a = 0$ or $b = 0$.

$(a + b)x + (a + b)y = (a + b)(x + y)$
$ax + ay + cx + cy = a(x + y) + c(x + y)$
$\qquad\qquad\qquad = (x + y)(a + c)$
$x^2 - y^2 = (x + y)(x - y)$
$x^2 + 2xy + y^2 = (x + y)(x + y)$
$x^2 - 2xy + y^2 = (x - y)(x - y)$
$x^3 + y^3 = (x + y)(x^2 - xy + y^2)$
$x^3 - y^3 = (x - y)(x^2 + xy + y^2)$

5.1–5.5 PROPERTIES OF FRACTIONS

If no denominators are 0, then

$\dfrac{a}{b} = \dfrac{-a}{-b} = -\dfrac{a}{-b} = \dfrac{-a}{b}$

$-\dfrac{a}{b} = \dfrac{-a}{b} = \dfrac{a}{-b} = -\dfrac{-a}{-b}$

$\dfrac{ax}{bx} = \dfrac{a}{b} \qquad\qquad \dfrac{a}{1} = a$

$\dfrac{a}{b} \cdot \dfrac{c}{d} = \dfrac{ac}{bd} \qquad \dfrac{a}{b} \div \dfrac{c}{d} = \dfrac{ad}{bc}$

$\dfrac{a}{d} + \dfrac{b}{d} = \dfrac{a + b}{d} \qquad \dfrac{a}{d} - \dfrac{b}{d} = \dfrac{a - b}{d}$

5.9 RATIO AND PROPORTION

If $\dfrac{a}{b} = \dfrac{c}{d}$, then $ad = bc$.

6.1 GRAPHING LINEAR EQUATIONS

General form of the equation of a line:

$Ax + By = C$

Equation of a vertical line: $x = a$

Equation of a horizontal line: $y = b$

6.2 THE SLOPE OF A LINE

If $P(x_1, y_1)$ and $Q(x_2, y_2)$ are two points on a line, the **slope** of the line is

$m = \dfrac{y_2 - y_1}{x_2 - x_1} \quad (x_2 \neq x_1)$

Slope–intercept form of the equation of a line:

$y = mx + b$

Horizontal lines have a slope of 0.

Vertical lines have undefined slope.

6.3 WRITING EQUATIONS OF LINES

Point–slope form of the equation of a line:

$y - y_1 = m(x - x_1)$

Two lines with the same slope are parallel.

The product of the slopes of perpendicular lines is -1.

6.6 RELATIONS AND FUNCTIONS

$y = f(a)$ is the value of $y = f(x)$ when $x = a$.

6.7 VARIATION

Direct variation: $y = kx$

Inverse variation: $y = \dfrac{k}{x}$

Joint variation: $y = kxz$

Combined variation: $y = \dfrac{kx}{z}$